Rydberg states of atoms and molecules

Rydberg states of atoms and molecules

Editors

R. F. Stebbings and F. B. Dunning

Department of Space Physics and Astronomy
Rice University

Cambridge University Press

Cambridge
London New York New Rochelle
Melbourne Sydney

CAMBRIDGE UNIVERSITY PRESS
Cambridge, New York, Melbourne, Madrid, Cape Town, Singapore,
São Paulo, Delhi, Dubai, Tokyo, Mexico City

Cambridge University Press
The Edinburgh Building, Cambridge CB2 8RU, UK

Published in the United States of America by Cambridge University Press, New York

www.cambridge.org
Information on this title: www.cambridge.org/9780521189736

First published 1983
First paperback edition 2011

A catalogue record for this publication is available from the British Library

Library of Congress Cataloguing in Publication data
Main entry under title:
Rydberg states of atoms and molecules.
Includes index.
1. Rydberg states. 2. Atomic spectra.
3. Molecular spectra. I. *Stebbings, R. F.*
II. Dunning, F. B.
QC454.A8R898 539.7 82–1181

ISBN 978-0-521-24823-5 Hardback
ISBN 978-0-521-18973-6 Paperback

Contents

Contributors

J. Berlande
Institut de Recherche Fondamentale
Service de Physique des Atomes
et des Surfaces
CEN/Saclay
91191 Gif-sur-Yvette Cedex, France

A. Dalgarno
Center for Astrophysics
60 Garden Street
Cambridge, MA 02138, USA

R. J. Damburg
Physics Institute
Latvian SSR Academy of Sciences
Riga, Salaspils, LSSR, USSR

J.-F. Delpech
Institut d'Electronique Fondamentale
Bat. 220, Université Paris XI
F-91405 Orsay, France

F. B. Dunning
Department of Space Physics
and Astronomy
Rice University
P.O. Box 1892
Houston, TX 77251, USA

C. Fabre
Laboratoire de Spectroscopie
Hertzienne de L'Ecole
Normale Supérieure
Associé au CNRS No. 18
24 Rue Lhomond
75231 Paris Cedex 05, France

M. R. Flannery
School of Physics
Georgia Institute of Technology
Atlanta, GA 30332, USA

Robert S. Freund
Bell Laboratories
600 Mountain Avenue
Murray Hill, NJ 07974, USA

T. F. Gallagher
Molecular Physics Center
SRI International
333 Ravenswood Avenue
Menlo Park, CA 94025, USA

F. Gounand
Institut de Recherche Fondamentale
Service de Physique des Atomes
et des Surfaces
CEN/Saclay
91191 Gif-sur-Yvette Cedex, France

S. Haroche
Laboratoire de Spectroscopie
Hertzienne de L'Ecole
Normale Supérieure
Associé au CNRS No. 18
24 Rue Lhomond
75231 Paris Cedex 05, France

A. P. Hickman
Molecular Physics Laboratory
SRI International
333 Ravenswood Avenue
Menlo Park, CA 94025, USA

D. Kleppner
Research Laboratory of Electronics and Department of Physics
Massachusetts Institute of Technology
Cambridge, MA 02139, USA

Peter M. Koch
Physics Department
Yale University
P.O. Box 6666
217 Prospect Street
New Haven, CT 06511, USA
Present address: Physics Department
State University of New York at Stony Brook
Stony Brook, NY 11794, USA

V. V. Kolosov
Physics Institute
Latvian SSR Academy of Sciences
Riga, Salaspils, LSSR, USSR

Michael G. Littman
Department of Mechanical and Aerospace Engineering
Princeton University
Princeton, NJ 08544, USA

Michio Matsuzawa
Department of Engineering Physics and Institute for Laser Science
University of Electro-Communications
1-5-1 Chofugoaka, Chofu-shi
Tokyo 182, Japan

R. E. Olson
Department of Physics
University of Missouri
Rolla, MO 65401, USA

J. Pascale
Service de Physique des Atomes et des Surfaces
CEN/Saclay
91191 Gif-sur-Yvette Cedex, France

R. F. Stebbings
Department of Space Physics and Astronomy
Rice University
P.O. Box 1892
Houston, TX 77251, USA

Myron L. Zimmerman
Venturcom
139 Main Street
Cambridge, MA 02142, USA

Preface

In this book we have attempted to provide a complete and comprehensive overview of the current state of knowledge, both experimental and theoretical, regarding Rydberg states of atoms and molecules. This undertaking is feasible and timely because the field has advanced to the stage where many of the more fundamental questions have been resolved, but it has not yet reached the point where the sheer volume of the literature precludes its treatment in a single book.

The field has recently experienced a very rapid, even explosive, growth of interest. Although some of the references date back several decades, the vast body of the literature is of very recent origin. Indeed, more than half the papers published in this field have appeared since 1975. Much of the early work was theoretical and was directed toward problems of astrophysical interest. From the standpoint of the experimentalist, however, the field lay essentially dormant until the arrival of the tunable laser, which provided the resolution essential for many detailed Rydberg studies. This stimulated much experimental activity, which in turn encouraged further theory.

We are fortunate to have enlisted the help of many of the acknowledged leaders in this field in preparing this volume, and to them we express our appreciation.

R. F. Stebbings
F. B. Dunning

1

Rydberg atoms in astrophysics

A. DALGARNO

1.1. Introduction

The interstellar medium is a hostile environment for compound systems. The medium is subjected to ultraviolet radiation from stars; to x rays from extragalactic objects, the galactic corona, and cooling supernova remnants; to energetic cosmic rays from supernova explosions; and to shock waves, driven by supernova explosions, expanding H II regions, stellar winds, collisions of interstellar clouds, and supersonic gas flows associated with star formation. The space between the stars, though, is nearly empty and there is room in the interstellar medium for atoms in high-Rydberg states to survive until they radiate. In interstellar space, the Rydberg levels are formed by the process of radiative recombination, which produces a significant population of atoms in levels with a wide range of values of the principal quantum number n. Because broadened lines are difficult to distinguish against the continuum emission from the ionized gas, pressure broadening restricts the possibility of detecting levels of high n to extended low-density regions. Nevertheless, emission from Rydberg states of atomic hydrogen has been observed from levels with n as large as 390.

Emissions from Rydberg atoms have been detected in the spectral range spanning the Lyman series in the ultraviolet near 1000 Å, or 100 nm, and the long-wave high-n series at meter wavelengths. The emissions constitute a powerful diagnostic probe of the nature of the physical environments in which the levels are excited and in which they radiate. Because the emission frequencies depend on the mass of the emitter, its identity can be established, and hydrogen, helium, carbon, and a group of heavier elements have been detected by the observation of line emissions in the radio-frequency spectrum.

Emissions from Rydberg atoms have been observed from a wide variety of objects including galactic H II regions, planetary nebulae, the diffuse interstellar gas, molecular clouds, and other galaxies, and these observations relate directly to a broad range of astronomical questions arising in theories of star formation, the structure and evolution of the galaxy, extragalactic phenomena, and cosmology.

In discussing Rydberg atoms in astrophysics, it is convenient to make a distinction between radio recombination lines corresponding to levels with $n \geqslant 30$ and optical recombination lines with $n < 30$.

1.2. Radio recombination lines

1.2a. Transition frequencies

The frequency of a transition ν from the level with principal quantum number $n+\Delta n$ of atomic hydrogen to the level with principal quantum number n, when n is sufficiently large that fine structure is negligible, is given by

$$\nu = R_{\mathrm{H}} c\left(\frac{1}{n^2} - \frac{1}{(n+\Delta n)^2}\right) \tag{1}$$

where R_{H} is the Rydberg constant for the hydrogen atom,

$$R_{\mathrm{H}} = 109\,677.58\ \mathrm{cm}^{-1} \tag{2}$$

In gigahertz,

$$\nu = 3.288\,051 \times 10^6\left(\frac{1}{n^2} - \frac{1}{(n+\Delta n)^2}\right) \tag{3}$$

For large n, the outer electron of a multielectron neutral atom moves in an effective field characterized by a charge of unity, and deviations caused by core polarization and other effects are negligible. The transition frequencies are given by Eq. (1) with R_{H} replaced by

$$R_{\mathrm{M}} = 109\,737.31/[1+(m/M)]\ \mathrm{cm}^{-1} \tag{4}$$

where m is the electron mass and M the mass of the nucleus. The corresponding frequencies of the lines of hydrogen, helium, carbon, sulfur, and an atom of infinite mass are in the ratios 0.999 45 : 0.999 86 : 0.999 95 : 0.999 99 : 1.0. The frequency separations are sufficient to permit the identification of narrow lines from the different elements. Recombination lines of ionized helium have also been detected. For ^{4}He II, the frequencies are given by

$$\nu = 438\,889.5\, c\left(\frac{1}{n^2} - \frac{1}{(n+\Delta n)^2}\right) \tag{5}$$

A simple approximate formula for the frequencies of transitions between high-lying Rydberg levels of neutral atoms, useful when high accuracy is not required, is

$$\nu = 6.58\,\Delta n\,(100/n)^3\ \mathrm{GHz} \tag{6}$$

Small n transitions ($n \simeq 40$) produce high-frequency lines at millimeter wavelengths, and high n transitions ($n \simeq 300$) produce low-frequency lines at meter wavelengths.

The spectroscopic notation used in radio-recombination-line studies is the conventional one: the $n+\Delta n \rightarrow n$ transition of atom A is described as A$n\alpha$ for $\Delta n=1$, as A$n\beta$ for $\Delta n=2$, A$n\gamma$ for $\Delta n=3$, and so on.

1.2b. Radiative and collision processes

In astrophysical environments, the Rydberg levels are populated by radiative recombination. For capture into the level with principal and azimuthal quantum numbers n and l, respectively, of an atom M, the process may be represented by

$$M^+ + e \rightarrow M(nl) + h\nu \tag{7}$$

If ϵ is the kinetic energy of the electron before capture, conservation of energy requires that

$$h\nu = \epsilon + I_n \tag{8}$$

where

$$I_n = hcR_M/n^2 \tag{9}$$

is the ionization threshold of the nl levels.

The recombination cross sections $\sigma_r(nl\,|\,\epsilon)$ may be determined by detailed balancing arguments from the cross sections $\sigma_i(nl\,|\,\nu)$ for the reverse photoionization process

$$M(nl) + h\nu \rightarrow M^+ + e(\epsilon) \tag{10}$$

Detailed calculations of $\sigma_i(nl\,|\,\gamma)$ for atomic hydrogen have been carried out by Karzas and Latter[1] and Burgess.[2]

The free electrons in astrophysical plasmas exchange energy efficiently in elastic electron–electron encounters and rapidly assume a Maxwellian velocity distribution $f(\epsilon)$ characterized by an electron temperature T. A recombination coefficient $\alpha(nl\,|\,T)$ may be defined as the Maxwellian average of the product of the electron velocity v and $\sigma_r(nl\,|\,\epsilon)$:

$$\alpha(nl\,|\,T) = \int f(\epsilon)v\sigma_r(nl\,|\,\epsilon)\,dv \tag{11}$$

which is a function of T and which is such that the rate of entry into level nl by radiative recombination is given by $N_eN(M^+)\alpha(nl\,|\,T)$ cm^{-3} s^{-1}, where N_e is the electron density, $N(M^+)$ the positive ion density, and $\alpha(nl\,|\,T)$ has units of cubic centimeters per second.

Values of $\alpha(nl\,|\,T)$ for atomic hydrogen have been tabulated by Burgess,[2] and a computer program for their calculation has been constructed by Flower and Seaton.[3] When the thermal energy kT is small compared with I_n, the recombination coefficient for capture into all the angular momentum levels l associated with a given n,

$$\alpha(n\,|\,T) = \sum_{l=0}^{n-1} \alpha(nl\,|\,T) \tag{12}$$

varies as $n^{-1}T^{-1/2}$. When kT is large compared with I_n,[4]

$$\alpha(n\,|\,T) \simeq T^{-3/2}[\ln(n^2T/157\,809) - 0.5772 + 8.56 \times 10^{-3}T^{1/3} - 2.3 \times 10^{-5}T^{2/3}] \tag{13}$$

The excited Rydberg levels may radiate by spontaneous allowed electric dipole transitions to lower levels

$$M(nl) \rightarrow M(n'l') + h\nu_{nn'} \tag{14}$$

giving rise to emission lines at frequencies $\nu_{nn'}$. The transition probabilities $A(nl, n'l')$ have been worked out for atomic hydrogen by several authors.[1,2,5] The total transition probability out of level nl is given by

$$\mathrm{A}(nl) = \sum_{n'}\sum_{l'} \mathrm{A}(nl, n'l') \tag{15}$$

and its inverse is the radiative lifetime of level nl,

$$\tau_{\mathrm{r}}(nl) = 1/\mathrm{A}(nl) \tag{16}$$

In an ionized plasma of vanishingly low density, collisions are negligible and the level populations are determined by a balance between capture and radiative cascade. In actual plasmas, collisions play a significant role in modifying the populations. Collisions with protons (and other ionized atoms) are particularly efficient in changing the angular momentum of a high-Rydberg atom without changing its energy. The collision process

$$\mathrm{H}^+ + \mathrm{H}(nl) \rightarrow \mathrm{H}^+ + \mathrm{H}(nl') \tag{17}$$

has been investigated by Pengelly and Seaton[6] using a semiclassical perturbation theory. The theory requires the introduction of a cutoff procedure at large impact parameters that depends on the electron density. The rate coefficient $q(n, l \rightarrow l'\,|\,T)$ for process (17) increases approximately as n^4 with increasing n, and depends weakly on N_{e}.

The mean collision time for a change in angular momentum is given by

$$\tau_{\mathrm{c}}(nl) = 1/N(\mathrm{H}^+)q(n, l \rightarrow l'\,|\,T) \tag{18}$$

Figure 1.1 shows the product $\tau_{\mathrm{r}}(nl)q(n, l \rightarrow l'\,|\,T)$ for a temperature of 10^4 K and an electron density of 10^4 cm^{-3}. The product increases rapidly with increasing n and the proton density

$$N_{\mathrm{c}}(n) = \left(\sum_{l}\sum_{l'} \frac{2l+1}{2n^2}\,\tau_{\mathrm{r}}(nl)q(n, l \rightarrow l'\,|\,T)\right)^{-1} \tag{19}$$

at which the mean collision and radiative lifetimes of level n are equal, defines a critical density below which the angular-momentum-level populations are

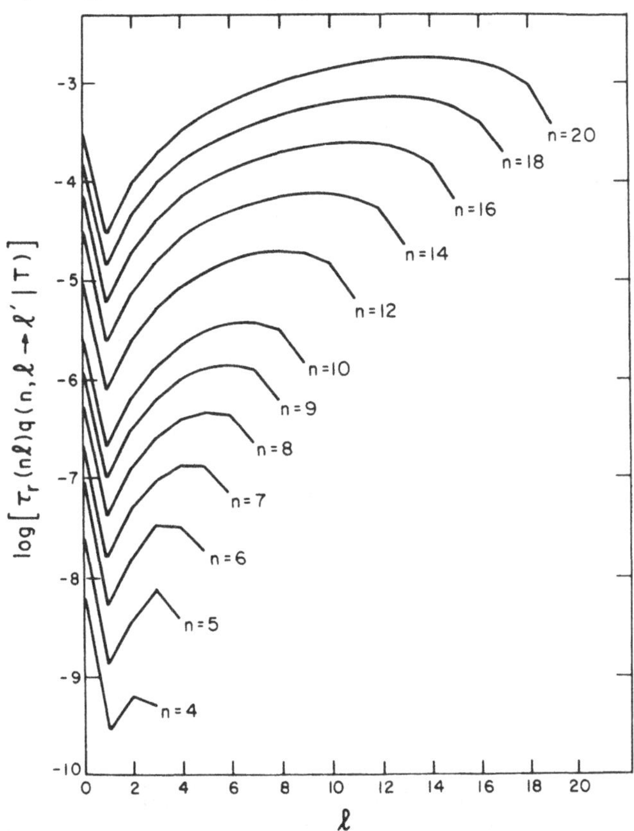

Fig. 1.1. Values of the product of the radiative lifetime and the rate coefficient for changes in azimuthal quantum number l as a function of l and n. As n increases, the product increases and collisions are more effective in bringing about a statistical population of l levels. (From Pengelly and Seaton[6])

determined by radiation and above which they are determined by collisions. Figure 1.2 is a graph of $N_c(n)$.[6] The critical density varies markedly with n. For radio recombination lines with n on the order of 100, $N_c \simeq 10^{-3}$ cm^{-3}, whereas for optical recombination lines with n on the order of 10, $N_c \simeq 10^6$ cm^{-3}.

Collisions bring about a statistical equilibrium of the l level populations and the populations $N(nl)$ are related by

$$N(nl) = \frac{2l+1}{n^2} N(n) \tag{20}$$

where

$$N(n) = \sum_l N(nl) \tag{21}$$

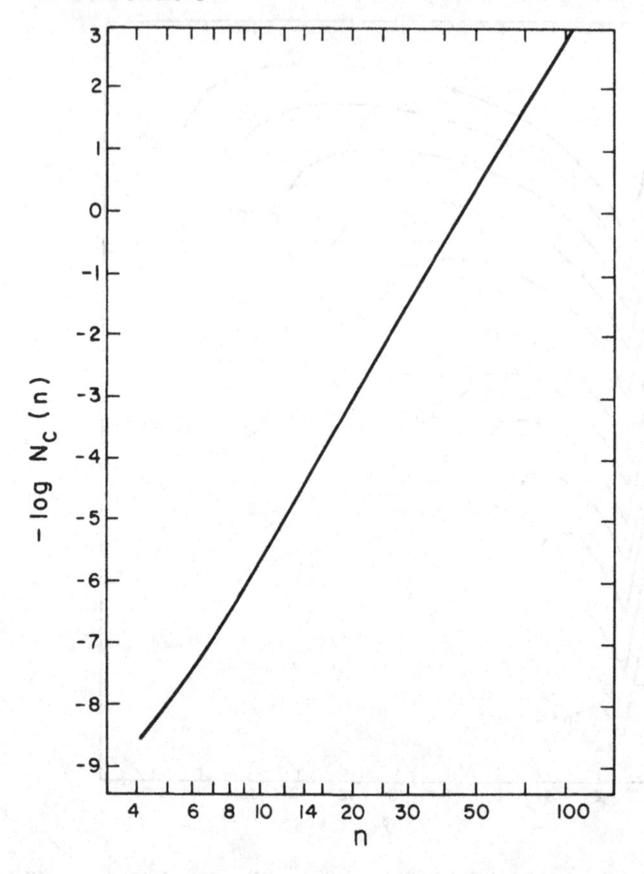

Fig. 1.2. Values of the critical density of H^+ at which the collision and radiative lifetimes of level n are equal as a function of n. (From Pengelly and Seaton[6])

The analysis of radio recombination lines can usually be simplified by introducing the average spontaneous radiative transition probability for the transitions from level n to level n',

$$A(n,n') = \sum_{l} \sum_{l'} \frac{2l+1}{n^2} A(nl,n'l') \tag{22}$$

The transition probability can be written in terms of the absorption oscillator strength $f_{n'n}$ according to

$$A(n,n') = \frac{8\pi e^2 \nu_{nn'}^2}{mc^3} \frac{\omega_{n'}}{\omega_n} f_{nn'} \tag{23}$$

where ω_n and $\omega_{n'}$ are the statistical weights of the upper and lower states, respectively. Menzel[7] has developed a useful formula for $f_{n'n}$ valid when $n \gg 1$ and $\Delta n' = n - n' \ll n$. It may be written as

Table 1.1. The functions $M(\Delta n')$ and $a(\Delta n')$ of Eq. (24)

$\Delta n'$	$M(\Delta n')$	$a(\Delta n')$
1	1.907 75 (−1)	0.187 73
2	2.633 21 (−2)	0.373 92
3	8.105 62 (−3)	0.521 87
4	3.491 68 (−3)	0.882 71
5	1.811 85 (−3)	1.072 77
6	1.058 47 (−3)	1.498 75

$$f_{n'n} = n'M(\Delta n')\left(1 + \frac{3\Delta n'}{2n'} - \frac{a(\Delta n')}{n'^2} + \cdots\right) \tag{24}$$

where $M(\Delta n')$ and $a(\Delta n')$ take the values listed in Table 1.1. Convenient representations of $A(n,n')$ in other limiting circumstances have been given by Burgess and Summers.[8]

For a given initial level n, the intensities diminish with increasing Δn, the emissivities $h\nu_{nn'}A(n,n')$ in the $n-1\alpha$, $n-2\beta$, and $n-3\gamma$ lines having the approximate ratios 0.919:0.053:0.024.

For large n, the radiative lifetime of level n, $\tau(n)=[\sum_{n'}A(n,n')]^{-1}$, increases approximately as $n^{4.5}$ and collision processes must become important in determining the populations of the high-n levels. Estimates of the rate coefficients for transitions owing to electron impacts

$$e + H(n) \rightleftarrows e + H(n+\Delta n) \tag{25}$$

have been made by Gee et al.[9] for $n \geqslant 5$ using methods of classical mechanics and the correspondence principle,[10] which are considered to be reliable for temperatures in the range $10^6/n^2 < T < 3\times 10^9$. Transitions in which Δn is large are important, particularly at low temperatures.

Proton collisions, which effectively determine the l population, also modify the population of the high-n levels. Rate coefficients for

$$H^+ + H(n) \rightleftarrows H^+ + H(n+\Delta n) \tag{26}$$

have been derived by Burgess and Summers,[8] based upon an impact parameter description of the scattering.

Ionizing collisions

$$e + H(n) \rightarrow H^+ + e + e \tag{27}$$

are important in the determination of the high-n populations. Cross sections for ionization have been computed using the binary-encounter model of the collision process.[11,12] The inverse of (27), the three-body recombination

$$e + e + H^+ \rightarrow H(n) + e \tag{28}$$

must then be included. Its rate coefficient q_{in} can be obtained from considerations of detailed balancing in thermodynamic equilibrium. Thus

$$q_{in}(T) = (h^2/2\pi mkT)^{3/2} n^2 \exp(I_n/kT) q_{ni}(T) \tag{29}$$

where $q_{ni}(T)$ is the rate coefficient for (27). Recombination then occurs through a combination of radiative and collision-induced transitions and is described as collisional–radiative recombination.[13] Detailed studies of its influence in dense plasmas have been carried out by Bates, Kingston and McWhirter,[14] Burgess and Summers,[8] and Jacobs and Davis.[15]

Because the collision cross sections increase with n and the radiative transition probabilities decrease, the populations of the high-n levels tend to a thermal equilibrium characterized by the electron temperature. It is convenient to introduce factors b_n, which are the ratios of the actual populations to those that occur where local thermodynamic equilibrium (LTE) prevails. Thus we write, for the density of level n,

$$N(n) = N_e N(\mathrm{H}^+)(h^2/2\pi mkT)^{3/2} n^2 \exp(I_n/kT) b_n \tag{30}$$

In LTE, $b_n = 1$.

1.2c. Emission intensities from Rydberg levels

The volume emissivity of a recombination line is given by

$$j_\nu^{\mathrm{L}} = \frac{A(n',n)}{4\pi} N(n') h\nu\phi_\nu = b_n j_\nu^{\mathrm{L}^*} \tag{31}$$

where ϕ_ν is the line profile normalized so that

$$\int \phi_\nu \, d\nu = 1 \tag{32}$$

and $j_\nu^{\mathrm{L}^*}$ is the emissivity in conditions of local thermodynamic equilibrium. The absorption coefficient per unit length is given by the formula

$$\kappa_\nu^{\mathrm{L}^*} = \frac{c^2}{2\nu^2}\left(\frac{N(n)}{\omega_n} - \frac{N(n')}{\omega_{n'}}\right)\omega_{n'} \frac{A(n',n)}{4\pi} \nu\phi_\nu \tag{33}$$

in which stimulated emission appears as a negative absorption. If $\kappa_\nu^{\mathrm{L}^*}$ is the absorption coefficient in LTE,

$$j_\nu^{\mathrm{L}^*} = \kappa_\nu^{\mathrm{L}^*} B_\nu(T) \tag{34}$$

where $B_\nu(T)$ is the Planck function

$$B_\nu(T) = \frac{2h\nu^3}{c^2} \frac{1}{\exp(h\nu/kT) - 1} \tag{35}$$

Then (33) can be written as

$$\kappa_\nu^{\mathrm{L}} = b_n \beta_{nn'} \kappa_\nu^{\mathrm{L}^*} \tag{36}$$

where

$$\beta_{nn'} = \frac{1 - (b_{n'}/b_n)\exp(-h\nu/kT)}{1 - \exp(-h\nu/kT)} \tag{37}$$

Ordinarily, in recombination line studies, $h\nu \ll kT$ and (37) may be simplified to

$$\beta_{nn'} = 1 - \frac{kT}{h\nu}\,\frac{b_{n'} - b_n}{b_n} \tag{38}$$

which, if $n' - n \ll n$, may be approximated by

$$\beta_n = 1 - kT\,d\ln b_n/dE_n \tag{39}$$

where $E_n = h\nu$ is the transition energy.

Departures from LTE often lead to negative values of β_n, and in the presence of a source of continuum radiation, the line intensities can be enhanced by stimulated emission and maser action can occur.[16]

1.2d. Continuum sources of radiation

In the presence of continuum radiation, the radio-recombination-line radiation from a plasma is modified also by changes in level populations caused by absorption in discrete and in photoionizing transitions.

Continuum radiation is produced internally in an ionized gas by the elastic scattering of electrons in collisions with positive ions. The accelerated electrons emit photons in free–free transitions and give rise to a continuous emission. In a plasma of variable density, the continuum emission from the denser regions may, by stimulated emission, strongly enhance the line emission from the diffuse regions.[17] The continuum emissivity j_ν^c and the continuum opacity κ_ν^c depend on the photon frequency, the electron and ion densities, and the electron temperature.[18] The ratio of the line and continuum intensities is often used as a measure of the electron temperatures.[19]

In addition to the effects of density and temperature gradients, the level populations and the line intensities for high-n levels are modified by absorption of the universal blackbody radiation field, which is characterized by a temperature of 2.8 K. The level populations and the line intensities may be modified also by external radiation sources. The effects will depend on the location of the ionized region and its geometrical relationship to the sources. For H II regions surrounding stars, the population of neutral hydrogen atoms is usually large enough to affect the transmission of radiation from transitions that terminate in the ground state. It is conventional in astronomy to distinguish two limiting cases.[20] In case A, all radiation escapes freely, and in case B, the radiation emitted in transitions directly into the ground state is promptly reabsorbed. Intermediate cases occur and in quasars absorptions by the metastable 2s state of hydrogen may be important. Case B usually applies to radio-recombination-line emissions. The differences between cases A and B in level populations diminish rapidly with increasing n.

For other ionized regions the geometry may be less certain. A weakly ionized H I or C II region is often adjacent to a highly ionized H II region. The observational consequences depend on whether the H I region lies behind or in front of the H II region as seen by the observer.

1.2e. Line profiles

To complete the theoretical formulation, we need a description of the line profile ϕ_ν. The line profile, in practice the linewidth, is a powerful additional diagnostic probe of the ionized regions.

For Rydberg levels, natural broadening is negligible and broadening occurs through Doppler and pressure broadening. Doppler broadening arises from thermal and nonthermal motions and produces a Gaussian profile

$$\phi_\nu^{D} = (\alpha/\pi^{1/2}\nu_0)\exp\{-[\alpha(\nu-\nu_0)/\nu_0]^2\} \tag{40}$$

where

$$\alpha = (Mc^2/2kT_D)^{1/2} \tag{41}$$

depends on the effective Doppler temperature T_D and ν_0 is the frequency at line center. Pressure broadening arises from the Stark effect because of quasi-static electric fields and from collisions with electrons and protons. Electron impacts provide the major contribution to the broadening[21] and produce a Lorentz profile

$$\phi_\nu^{l} = (\delta/\pi)[(\nu-\nu_0)^2+\delta^2]^{-1} \tag{42}$$

where the parameter δ is related to the Maxwellian average of the rate coefficient for total inelastic scattering $\langle vQ\rangle$ and the electron density N_e by

$$\delta = (1/2\pi)\langle vQ\rangle N_e \tag{43}$$

For temperatures near 10^4 K, the calculations of Brocklehurst and Leeman[22] may be represented by[17]

$$\delta = 4.7(n/100)^{4.4}(10^4/T)^{0.1}N_e\nu_0 \text{ Hz} \tag{44}$$

A more elaborate theory[23] suggests that the errors in Eq. (44) do not exceed 10%. The convolution of the Doppler and Lorentz profiles is the Voigt profile

$$\phi_\nu = \alpha H(a,x)/\nu_0\pi^{1/2} \tag{45}$$

where

$$a = \alpha\delta/\nu_0 \tag{46}$$

$$x = \alpha(\nu-\nu_0)/\gamma_0 \tag{47}$$

and

$$H(a,x) = \frac{a}{\pi}\int_{-\infty}^{+\infty}\frac{\exp(-t^2)\,dt}{a^2+(t-x)^2} \tag{48}$$

Useful representations of $H(a,x)$ have been given by Kielkopf.[24]

1.2f. Level populations

Many detailed calculations of the level population factors b_n have been carried out with various assumptions about the radiative and collision rates and with various assumptions about the plasma density and temperature and the external radiation field.[25] In statistical equilibrium, the number of transitions into level n per unit volume per unit time from recombination and from radiative and collisional processes from other levels is equal to the number of transitions per unit volume per unit time out of level n from radiative, collisional, and ionization processes. A system of equations with the form

$$\mathbf{KN} = \mathbf{R} \tag{49}$$

is obtained where $\mathbf{N}$ is a column vector with components N_n, $\mathbf{R}$ the capture rate into level n, and $\mathbf{K}$ a matrix containing the transition rates into and out of all the levels of the system. Several procedures have been used to solve (49) and a general computer program has been published by Brocklehurst and Salem.[26]

Figure 1.3 illustrates the behavior of b_n and $1-\beta_n$ with n conditions appropriate to a galactic H II region. Results are given for case B for a range of electron densities in a plasma at a temperature of 10^4 K in the absence of radiation sources.[27] With increasing n, b_n tends to unity but falls substantially below unity for smaller values of n as collision processes become too slow to compete with radiative processes and downward radiative transitions are not compensated by upward absorptions. For low values of n, $1-\beta_n$ exceeds unity, β_n is negative, and amplification by stimulated emission of radiation could occur. As the electron density increases, the range of n over which b_n is less than unity and the range of n over which β_n is negative diminish. Collisions drive the populations toward local thermodynamic equilibrium and the possibility of maser action is suppressed.

Cold, partly ionized regions are also detectable by observations of recombination lines, and Fig. 1.4 illustrates b_n and $1-\beta_n$ for case A in a plasma with various electron densities at 30 K embedded within a blackbody radiation field corresponding to a temperature of 100 K and subjected to radiation from an adjacent H II region at 10^4 K.[27] The intensity of the radiation from an H II region depends on the emission measure

$$Em = \int N_e^2\, ds \tag{50}$$

integrated through the region. Figure 1.4 corresponds to an emission measure of 10^6 cm^{-6} pc^{-1} (1 pc $= 3.0856 \times 10^{18}$ cm). Collisions in which Δn is large are important in low-temperature regions. The behavior of b_n and β_n is similar to that in Fig. 1.3 though the departures from LTE are more severe and stimulated emission will substantially enhance the line-emission intensities.

The influence of the 2.8-K blackbody background radiation field on b_n has been explored by Burgess and Summers[8] for H II regions and by Ungerechts

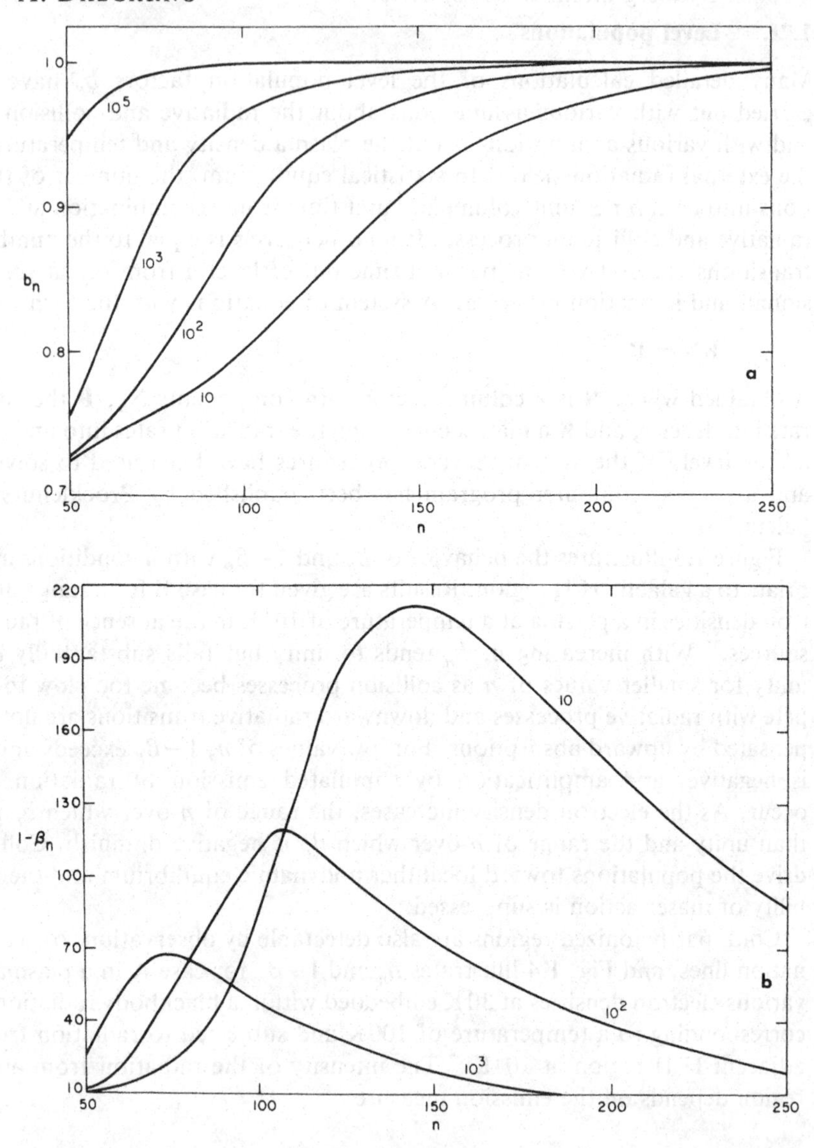

Fig. 1.3. Values of the level-population parameters b_n (*a*) and $1-\beta_n$ (*b*) for a plasma with $T=10^4$ K in case B, for several electron densities. (From Salem and Brocklehurst[27])

and Walmsley[28] for C II regions. The 2.8-K radiation affects only the high-n levels. The collision rate coefficients are large for the high-n levels and the radiation has significant effects only for very low electron densities below 1 cm^{-3}. The effects of far-infrared radiation from an adjacent H II region are apparently of comparable magnitude.[28]

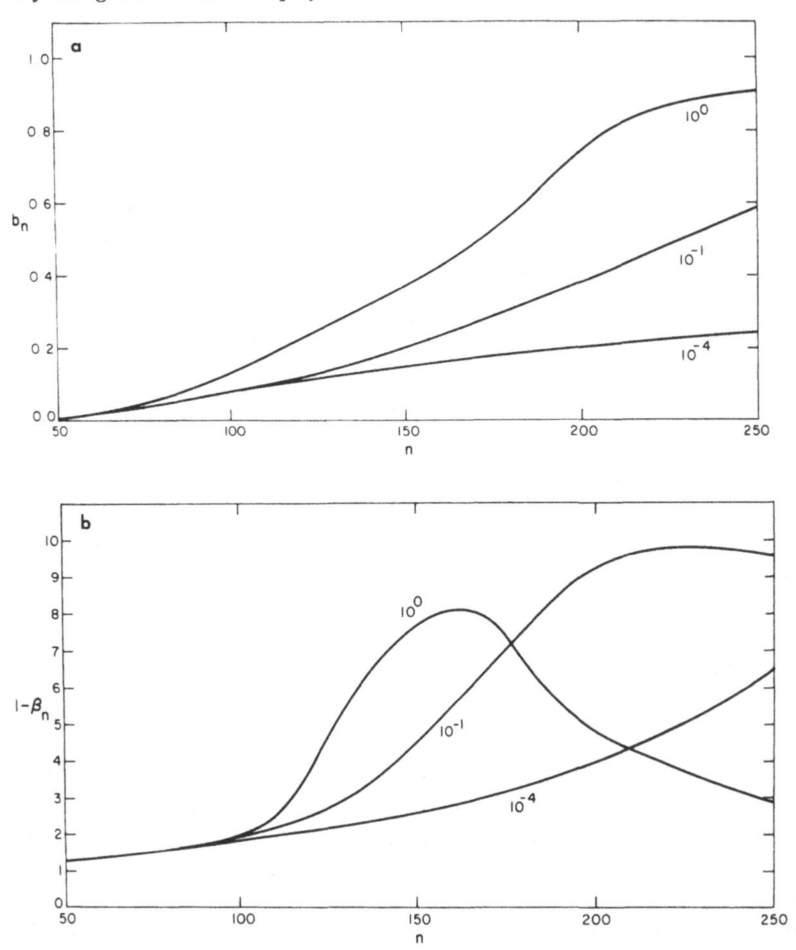

Fig. 1.4. Values of b_n (a) and $1-\beta_n$ (b) for a plasma with $T=30$ K in case A embedded in a 100-K blackbody radiation field and subject to radiation from an adjacent H II region at 10^4 with an emission measure of 10^6 cm^{-6} pc^{-1}. (From Salem and Brocklehurst[27])

1.2g. Helium and heavy elements

In addition to neutral hydrogen, radio recombination lines have been detected from neutral and ionized helium, neutral carbon, and a group of neutral heavier elements including sulfur. Ionized helium is strictly hydrogenlike and the electron moves in a Coulomb field corresponding to a charge $Z=2$. Apart from the mass factor, the frequencies of He II recombination lines are four times those of H I lines. The radiative rate coefficients can be obtained from those of hydrogen by simple scaling procedures, the oscillator strengths being unaltered.

Electron-impact collision processes

$$e + He^+(nl) \rightarrow e + He^+(n'l') \quad (51)$$

are modified by the Coulomb-attraction and proton-impact collision processes

$$H^+ + He^+(nl) \rightarrow H^+ + He^+(n'l') \quad (52)$$

by the Coulomb repulsion. The He II recombination lines that have been detected come from ionized regions with temperatures of the order of 10^4 K. At such temperatures, the Coulomb effects are small for high-Rydberg levels.[6] In contrast, the spontaneous radiative transition probabilities increase by a factor of Z^6. Weisheit and Walmsley[29] showed that, in the absence of strong external radiation sources, the departure coefficients $b_n(N_e, Z)$ for a hydrogenlike system of nuclear charge Z in a plasma of density N_e cm^{-3} are related to the hydrogen coefficients $b_n(N_e, 1)$ by

$$b_n(N_e, Z) = b_n(N_e/Z^6, 1) \quad (53)$$

Because of the more rapid radiative decays, the He II level populations experience greater departures from LTE.

The physics of the high-Rydberg levels of the neutral many-electron systems is similar to that of hydrogen, except for the possible effects of an additional recombination process, dielectronic recombination. Dielectronic recombination,

$$M^+ + e \rightarrow M^* \rightarrow M' + h\nu \quad (54)$$

occurs by capture into a doubly excited resonance state of M and stabilization is achieved by a radiative transition of the inner electron.[13] Doubly excited levels usually lie high in energy, and dielectronic recombination proceeds more rapidly than radiative recombination only at high temperatures. Where dielectronic recombination is significant, it can populate high-Rydberg levels efficiently, large overpopulations can result, and recombination lines may appear in absorption.[30] The solar corona is a possible environment,[31] though radio recombination lines from the sun have not yet been detected.

In H II regions, dielectronic recombination may affect the populations of the Rydberg levels of carbon by capture into the configurations ($2s2p^3$).[32] In practice, the regions in which the carbon recombination lines are detected appear to contain cold, weakly ionized gas, residing in the mostly neutral shells surrounding H II regions or in the C II regions produced by early B-type stars located in molecular clouds,[19,33] and dielectronic recombination is not important.

An analogous process may be as follows. Singly ionized carbon has an excited fine-structure $^2P_{3/2}$ level, which lies at an energy of 0.0079 eV, equivalent to 92 K, above the ground $^2P_{1/2}$ level. Electrons may lose energy in exciting the $^2P_{1/2}$–$^2P_{3/2}$ transition and be captured temporarily into a high-Rydberg level of the $^2P_{3/2}$ core. The lifetime of the resonance state can be lengthened by collisions that alter the angular momentum of the captured electron

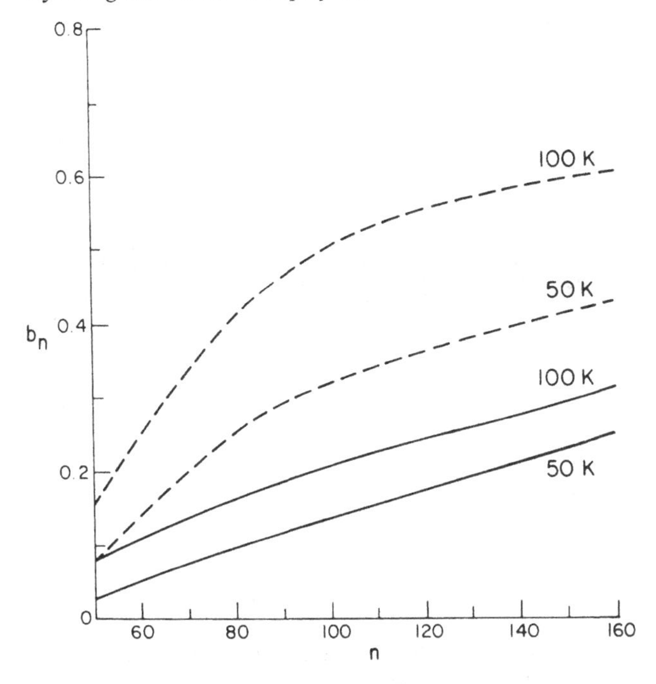

Fig. 1.5. Level population factors b_n for Rydberg levels of carbon in a plasma with $N_e = 0.1\ \text{cm}^{-3}$ and $T = 50$ and 100 K. The dashed lines include the dielectric capture process; the solid lines exclude it. The $^2P_{1/2}$ and $^2P_{3/2}$ levels are assumed to be populated in LTE. (From Watson, Western, and Christensen[34])

$$H^+ + C(^2P_{3/2}, nl) \rightarrow H^+ + C(^2P_{3/2}, nl') \qquad (55)$$

Figure 1.5 is a comparison for temperatures of 50 and 100 K of calculated level populations in a plasma of density $N_e = 0.1\ \text{cm}^{-3}$, excluding and including the angular-momentum-changing collisions Eq. (55).[34] For Fig. 1.5, the relative populations of the $^2P_{1/2}$ and $^2P_{3/2}$ levels of C II were assumed to be those obtaining in LTE. If the population of the $^2P_{3/2}$ falls below the LTE value, a range of n exists for which db_n/dn is negative and the recombination lines should appear in absorption.[34] A line at 26.131 MHz has been observed in absorption toward the young supernova remnant Cassiopeia A by Konovalenko and Sodin,[35] which may be due to the 631α line of carbon produced in H I clouds lying along the line of sight.[36]

1.2h. Zeeman effect

The observation of radio recombination lines may provide a means of determining the magnetic field through the Zeeman effect. The Zeeman pattern for high $n\alpha$ lines of atomic hydrogen has been calculated by Greve[37] and by Troland and Heiles[38] and for high $n\alpha$ and $n\beta$ lines of H, He, and C by Greve and Pauls.[39]

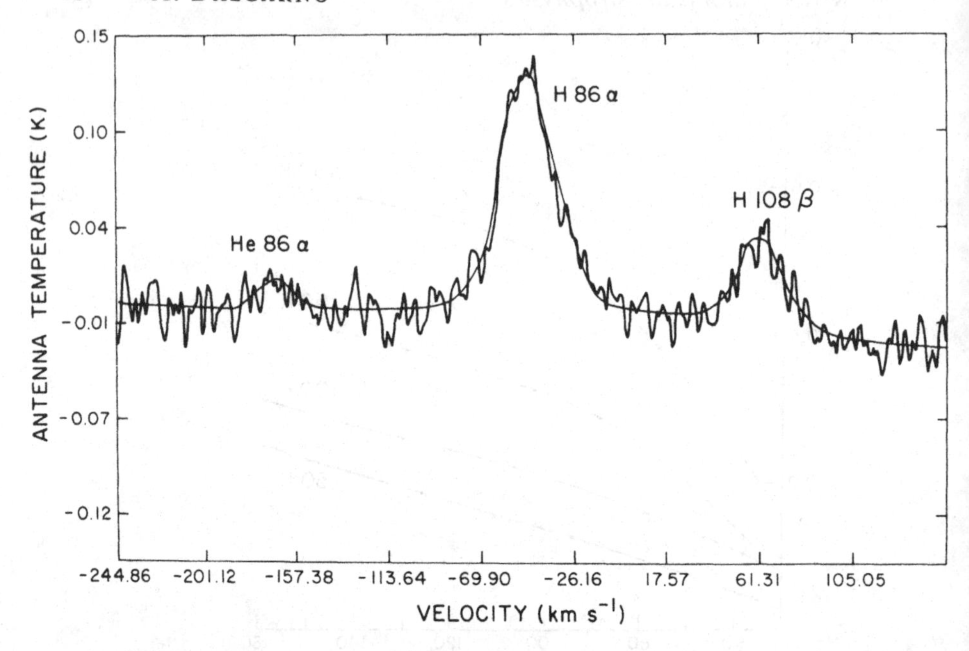

Fig. 1.6. The spectrum toward W3N, showing recombination lines H 86α, H 108β, and He 86α. (Courtesy of E. J. Chaisson)

Attempts to measure the Zeeman splitting of radio recombination lines in the interstellar medium have been unsuccessful, but upper limits to the strength of the magnetic field have been obtained.[38]

1.3. Observations of radio recombination lines

1.3a. Galactic H II regions

Galactic H II regions are regions of ionized hydrogen in the interstellar gas containing one or more O or early B-type stars with effective temperatures exceeding 25 000 K.[40] The stars emit ultraviolet radiation, which ionizes the gas around them. The brightest H II regions are observed in the visible through emission of optical recombination lines of hydrogen and forbidden lines of O II, N II, and other constituents. That detectable radio recombination lines of hydrogen and helium should be emitted by galactic H II regions was predicted by Kardashev[41] in 1959 and successful observations were reported in 1964 by Dravskikh and Dravskikh[42] and Höglund and Mezger.[43] Extensive measurements have been made in subsequent years spanning the frequency range from 0.1 to 100 GHz, using both single-dish and aperture-synthesis telescopes, and considerable information has been gathered on the physical conditions in H II regions and in the galaxy.[19, 32, 44] A typical spectrum is illustrated in Fig. 1.6. It shows the 86α and 108β recombination lines of hydrogen and the 86α line of helium observed toward the H II region W3N.

There has been much debate over the applicability of the theory of radio recombination lines, most of it concerning the importance of stimulated emission in the solution of the equations of radiative transfer and in the interpretation of the observations.[19] Electron temperatures are often derived from the ratio of the line and continuum optical depths τ_ν^{L} and τ_ν^{c}, respectively, obtained by integrating the line and continuum opacities κ_ν^{L} and κ_ν^{c} along the line of sight through the nebula. In a homogeneous nebula not exposed to external radiation, the ratio of the actual temperature T to the temperature T^* derived from $\tau_\nu^{\mathrm{L}}/\tau_\nu^{\mathrm{c}}$ on the LTE assumption may be written as[19]

$$T^*/T = \{b_n[1-(\beta_n\tau_\nu^{\mathrm{c}}/2)]\}^{-0.87} \tag{56}$$

if τ_ν^{c} is small compared with unity. Expression (56) demonstrates that departures from LTE will show up as a dependence of τ on the level n. An apparent dependence of τ on n can arise from density inhomogeneities and the different spatial resolutions of the observations, which tend to decrease toward lower frequencies.

Observations of selected transitions with different values of Δn can be made at neighboring frequencies and the relative strengths of the lines may be a more certain test of departures from LTE. Stimulated emission usually enhances α lines more than β lines and departures from LTE will be manifested in an anomalously small β-to-α intensity ratio.[32] Stimulated emission occurs most strongly at low frequencies corresponding to transitions between high-n levels. Electron-impact broadening increases rapidly with n and mitigates the enhancement arising from stimulated emission. The broadening transfers power from the line center into the wings and the line merges into the background continuum, becoming undetectable. Thus in an inhomogeneous nebula, low-frequency line observations reflect the tenuous regions of low electron density.[17] In contrast, high-frequency line observations are dominated by emission from the denser regions.

It appears that for most galactic H II regions the observed lines are those for which neither stimulated emission nor impact broadening is significant and the assumption of LTE is appropriate for their analysis.[45] A similar conclusion applies to the bright extended H II regions seen in the nearby galaxies, the Magellanic clouds, through emission of recombination lines of hydrogen and helium.[46] The derived temperatures of galactic H II regions studied by radio-recombination-line observations range between about 13 000 and 4000 K.[47,48] For the densest part of the Orion nebula, which is an exceptionally bright nebula, $T = 9000$ K and $N_e \simeq 2\times10^4$ cm^{-3} and for NGC 6604, which is an extended object of low visibility, $T = 4000$ K and $N_e \simeq 10^2$ cm^{-3}.[45] The electron temperatures tend to increase with emission measure,[47] presumably because of the reduction in cooling efficiency caused by collisional deactivation of excited metastable species.

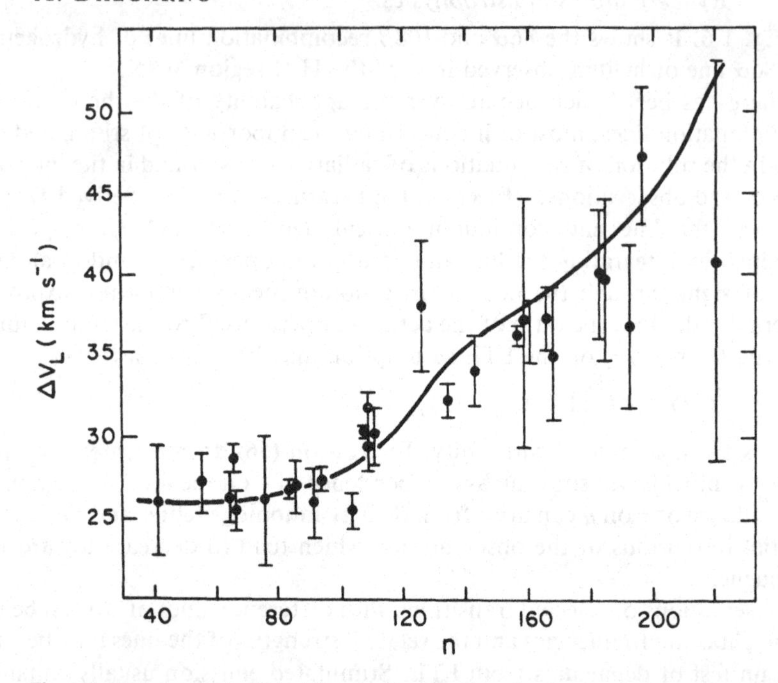

Fig. 1.7. The measured widths of recombination lines from Rydberg levels of hydrogen in the Orion nebula. The solid curve corresponds to a theoretical model of the region. (From Shaver[45])

Linewidths provide an independent assessment of the electron density. Figure 1.7 summarizes the observational data for the Orion nebula and presents the measured half-widths of H $n\alpha$ lines as a function of n.[48,49] A model with a temperature of 8200 K, a Doppler width of 26 km s^{-1}, and a central density N_e of 1.5×10^4 cm^{-2} decreasing outward reproduces satisfactorily the observed widths.[45]

The He/H abundance ratio. The helium content of the galaxy is believed to be primordial helium created during the early stages of the universe, enriched by the effects of chemical evolution. Recombination lines of helium have been measured for many H II regions.[50] The interpretation of the observations is complicated by the same effects of instrumental resolution, departures from LTE, density inhomogeneities, and line broadening that entered the analysis of the hydrogen line recombination data. It has been established nevertheless that a gradient exists in the ratio of He^+ to H^+ as a function of distance from the galactic center.[51] The ratio increases from 0.02 near the galactic center to 0.06 at 5 kpc and reaches a maximum of 0.09 at 9 kpc before decreasing at larger radii.

The conversion of the derived He^+/H^+ ratio to a helium-to-hydrogen ratio is not immediate because the region containing He^+ is often smaller than that

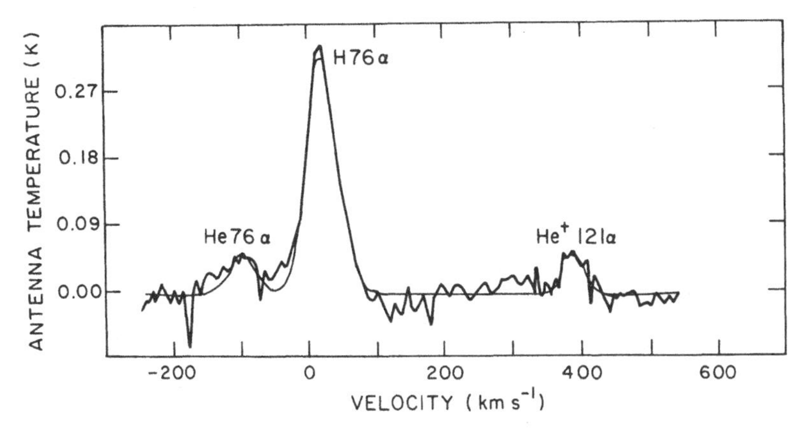

Fig. 1.8. The radio-frequency spectrum of the planetary nebula NGC 7027 showing the lines H 76α, He 76α, and He^+ 121α. (From Terzian[56])

containing H^+,[52] ionization of helium requiring photons of energy of at least 24.6 eV. Absorption by dust, which increases in its effectiveness toward shorter wavelengths, may accentuate the lack of He^+.[53] Another cause for the underabundance of He^+ in H II regions toward the galactic center is a decrease in the fraction of energetic photons, produced by the OB stars responsible for the creation of the ionized regions. The decrease is associated with the observed gradient in the metal abundances.[51] With appropriate corrections, the ratio of helium to hydrogen by number is estimated as 0.13 at the galactic center and 0.10 at the distance of the sun.

1.3b. Planetary nebulae

Planetary nebulae are isolated nebulae consisting of shells of gas composed of material ejected from the central star. The stars are old and their effective temperatures are high, some as high as 3×10^5 K. The central stars are evolving rapidly toward the white-dwarf stage. The surrounding ionized gas is expanding and planetary nebulae are observable over short-time scales on the order of 10^4 years. The high temperatures of the central stars produce He^{2+} in addition to H^+ and He^+. The first detection of radio recombination lines of He^+ in a planetary nebula, NGC 7027, was reported by Chaisson and Malkan[54] in 1976.

Figure 1.8 is a reproduction of the radio-frequency spectrum observed toward the planetary nebula NGC 7027 at frequencies near 14690 MHz.[55,56] Recombination lines 76α of H, 76α of He, and 121α of He^+ are present.

The emission lines from planetary nebulae are weak and detection is difficult. Lines of hydrogen and helium from levels with $n \leqslant 110$ have been detected with the large radio telescopes and distinct broadening has been seen. The widths increase with n as expected from electron impact broadening. For the

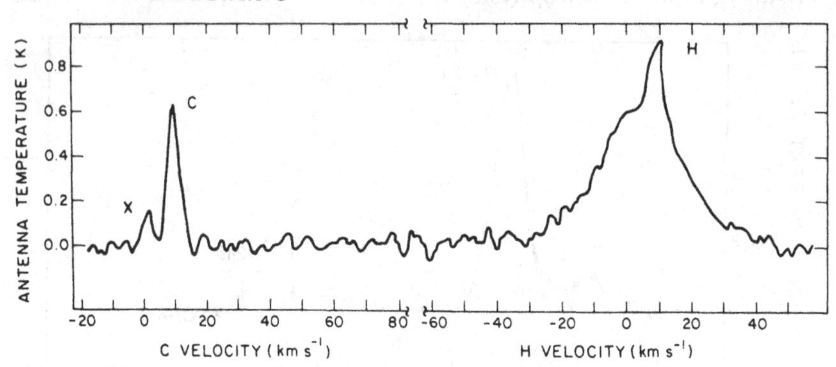

Fig. 1.9. The radio-frequency spectrum of NGC 2024. (From Pankonin, Walmsley, Wilson, and Thomasson[61])

planetary nebula NGC 7027, an electron density of 5×10^4 cm^{-3} has been derived from the measured linewidths,[57] a density that is similar to values obtained from analyses of the optical emission of forbidden lines of O III and N II.[57]

1.3c. Carbon II regions

Stellar photons with energies less than the ionization potential of hydrogen, 13.6 eV, escape from H II regions and ionize elements with lesser ionization potential, of which the most abundant is carbon with an ionization potential of 11.26 eV. The region containing the ionized carbon is called a C II region. The cosmic abundance of carbon relative to hydrogen is 3×10^{-4} and the ionizing photons are absorbed preferentially by interstellar grains, which are mixed with the gas. The size of the C II shell surrounding the H II region is sensitive to the scattering properties of the grains and to the density. The heat source in the C II region is much reduced from that in the H II region and the temperature is low.

Recombination lines of carbon were first reported in 1967.[58] Subsequently, C II regions have been seen toward about 15 galactic H II regions by observations of emission lines ranging in n from C 273α at 328 MHz to C 66α at 23.4 GHz.[59] From analyses of the level populations and particularly of the intensity ratio of the C 157α and C 197β lines, it appears that stimulated emission is often substantial. The derived temperatures do not exceed 200 K. The derived electron densities are sensitive to the calculations of the b_n factors, which are uncertain at low temperatures. The values are on the order of 10 cm^{-3}, which implies a neutral density of 10^5 cm^{-3}.

Occasionally associated with the carbon emission lines are narrow recombination lines of hydrogen superimposed upon the broader hydrogen lines originating in the H II region.[60,61] Figure 1.9 reproduces the spectrum measured toward NGC 2024, with frequency expressed in velocity units.[61] The velocities of the narrow hydrogen component and the carbon lines, though not identical, are distinctly different from the' velocity of the broad hydrogen component. The source of the hydrogen ionization has not been established. The cold, ionized hydrogen may be located in the ionization front of the expanding H II region[62] or it may be produced deeper in the surrounding interstellar gas by soft x rays originating in the hot cavity created by the stellar wind.[63]

C II regions are also produced by early B-type stars, which do not emit many photons energetic enough to ionize hydrogen. Such C II regions have been detected embedded in dense molecular clouds and their physical properties have been inferred from analyses of the carbon recombination lines. The exciting star often appears as a near-infrared point source with an extended source of far-infrared thermal emission from dust grains. Coincident with the C II region is a localized volume of enhanced molecular line emission. The derived densities of C^+ are approximately 1–10 cm^{-3}, implying substantial neutral densities.[64]

Carbon recombination lines have been detected also in two diffuse interstellar coulds, those lying toward the stars ζ Ophiuchi and o Persei, by Crutcher,[65] who measured the intensities of the C166α line. The measurements suggest that the ionization source is a combination of the average interstellar radiation field and radiation from the parent star.[65] The C631α line identified toward Cas A is probably formed from carbon, photoionized by the interstellar radiation field.[36]

Heavy-element recombination lines. The heavy elements magnesium, silicon, sulfur, and iron have ionization potentials less than 13.6 eV and should be present in singly ionized form in C II regions, though the lower cosmic abundances lead to weaker recombination lines. Recombination line emissions from elements heavier than carbon were detected in 1972 in the H II regions NGC 2024 and W3,[66] and they have been observed in Orion A and in several dark clouds.[59,67,68] The heavy element primarily responsible for the observed emission lines is probably sulfur. The S II regions are larger in size than the C II regions.[68]

1.3d. Supernova remnants

After a supernova explosion, a supernova remnant is left behind, consisting of an expanding, cooling, recombining shell of ionized gas that emits nonthermal continuum radiation in the radio-frequency spectrum.[69] Weak recombination lines have been seen toward a few of the remnants, but it seems that the lines mostly originate in very extended H II regions of low brightness and low density, which lie along the line of sight but are not directly related to the supernova remnant.[70]

1.3e. Galactic center

The center of our galaxy is a source of strong continuum emission that may amplify the recombination line intensities from ionized regions lying along the

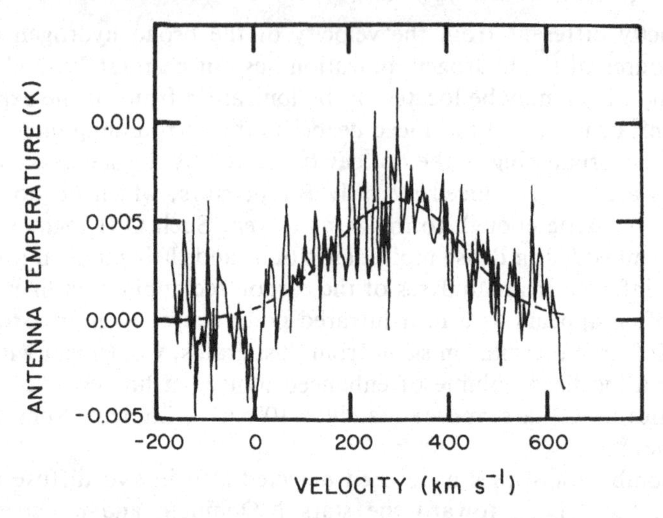

Fig. 1.10. The emission spectrum of M82, showing the line H 125α, expressed as a function of the velocity relative to the local standard of rest. The antenna temperature measures the intensity. (From Rodriguez and Chaisson[75])

extended line of sight to the galactic center. The galactic center is an active, crowded region and the interpretation of the observational data is not straightforward. From recombination-line studies, it appears that amplification by stimulated emission is occurring in directions toward the galactic center and that the ionized regions, sampled by the observations, are conventional H II regions with temperatures of about 5000 K and electron densities of the order $10\ cm^{-3}$.[71,72]

1.3f. Extragalactic radio recombination lines

Because of their great utility as diagnostic probes of the emitting gas, considerable effort has been made to extend the observations of radio recombination lines to external galaxies, beyond the Magellanic clouds.[73] The lines are usually weak and broad and detection is difficult. However, if a background source exists, then substantial amplification of the lines by stimulated emission may occur and distant galaxies may become observable. M82, an irregular II galaxy, located at a distance of 3.3 Mpc, has been detected in several α lines of hydrogen.[74,75] Figure 1.10 shows the H 125α emission line from M82.[75] A satisfactory representation of the observations can be achieved by a model in which the recombination line intensities of frequencies below 10 GHz are largely attributable to emission, stimulated by a background continuum source at the nucleus of the galaxy. The derived temperature of the emitting gas is about 5000 K and the electron density is of the order $10^2\ cm^{-3}$.

A similar model in which stimulated emission by a continuum source at the nucleus is largely responsible for the observed recombination line intensities is appropriate to the galaxy NGC 253.[76]

A more distant galaxy, the Markarian–Seyfert galaxy Mrk 668 (= *OQ* 208), has been observed recently by Bell and Seaquist[77] in the H 83α and H 99α recombination lines. The red shift $z = 0.0763$ derived from the lines is close to the value 0.0767 obtained optically.[78] Stimulated emission is again responsible for the intensities.[77]

The possibility of detecting radio recombination lines from quasars has been explored by Shaver,[79] who draws attention to the contribution from heavy elements such as magnesium for which the population of high-n levels is augmented by dielectronic recombination.[80] The observation of recombination lines from quasars could lead to more stringent limits on the variability with time of the fundamental constants.

1.4. Optical recombination lines

Radio recombination lines are a direct manifestation of the importance of Rydberg levels in astrophysics. However, optical recombination lines have been, for much longer, a source of information about the physics of astronomical phenomena.[40] With the development of detection techniques in the far-infrared and submillimeter regions of the spectrum, low-lying Rydberg levels assume a still larger importance. The theory for the population of low-n levels is essentially that of high-n levels but more complex, because the l levels are not statistically distributed, and more straightforward, because stimulated emission and impact broadening are usually unimportant. For high n, atomic systems tend to have similar properties, but for low n, greater attention must be given to the particular characteristics of the different elements.

The level populations are far from local thermodynamic equilibrium. Calculations of the departure coefficients b_{nl} have been carried out for H, He, and He^+ by Pengelly and Seaton,[6] Brocklehurst,[81] Giles,[82] and Seaton.[83] Because the effects of stimulated emission are small at optical wavelengths, it is useful to introduce an effective recombination coefficient $\alpha_M(n \rightarrow n' | T)$ such that

$$4\pi j_{nn'}/h\nu_{nn'} = N_e N(M^+)\alpha_M(n \rightarrow n' | T) \tag{57}$$

where $j_{nn'}$ is the emission coefficient in the line at frequency $\nu_{nn'}$. The effective recombination coefficient takes into account the population of level n by cascading. Tables of $\alpha(n \rightarrow n' | T)$ have been constructed[81,82] and they may be used in conjunction with measurements of line intensities to obtain the emission measure, Eq. (50), once the temperature has been derived. Optical emissions in the Balmer series $n \rightarrow 2$ with n up to about 50 have been measured for the Orion nebula and for NGC 7027.[84] At low resolution, the lines overlap and form a continuum. The derived values of N_e and T are consistent with more accurate determinations from studies of the radio spectrum.

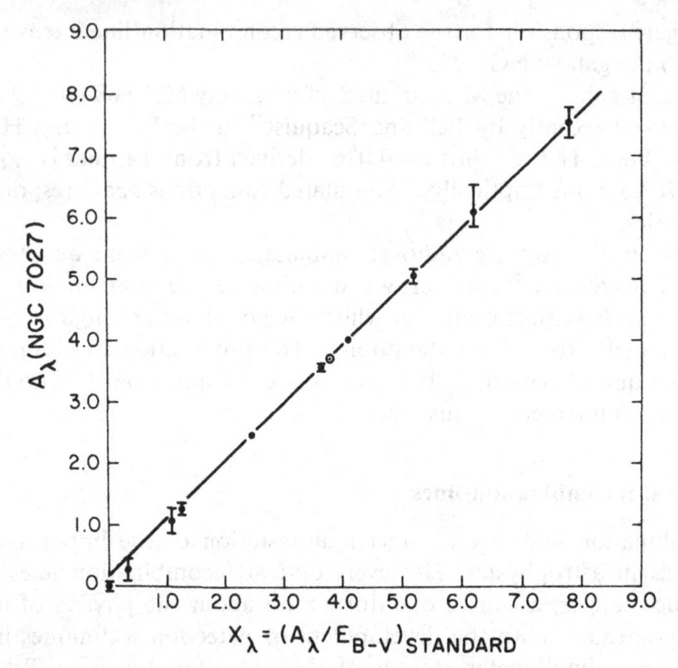

Fig. 1.11. A comparison of the extinction toward NGC 7027 with the standard interstellar extinction law. (From Seaton[85])

In the optical region of the spectrum, emissions from metastable levels of the ionized heavy elements provide a more accurate measure of the electron density and temperature. Observations of optical recombination lines of H, He, and He^+ give useful additional information on the physical models, but of more value, because the theoretical description of the recombination spectrum is so precise and detailed, are the departures, from the theoretical spectra, that can be used to extract the differential extinction with wavelength. The extinction arises from dust in the interstellar medium or in the object under observation.

A widely used measure is the Balmer decrement, which is the run with n of the emissivity ratio j_{n2}/j_{42}. The Balmer decrement is insensitive to temperature and density, and deviations from the predicted curve may be attributed to extinction or reddening.

Seaton[85] used recombination-line theory to derive a reddening correction for NGC 7027 from a comparison of the theory with the measured intensities of recombination lines of H and He^+. If I_λ is the intrinsic intensity and I_λ^0 the measured intensity at wavelength λ, the extinction A_λ is defined in magnitudes by

$$I_\lambda^0 = 10^{-0.4A_\lambda} I_\lambda \tag{58}$$

Figure 1.11 is a comparison of the derived values of A_λ for NGC 7027 with the

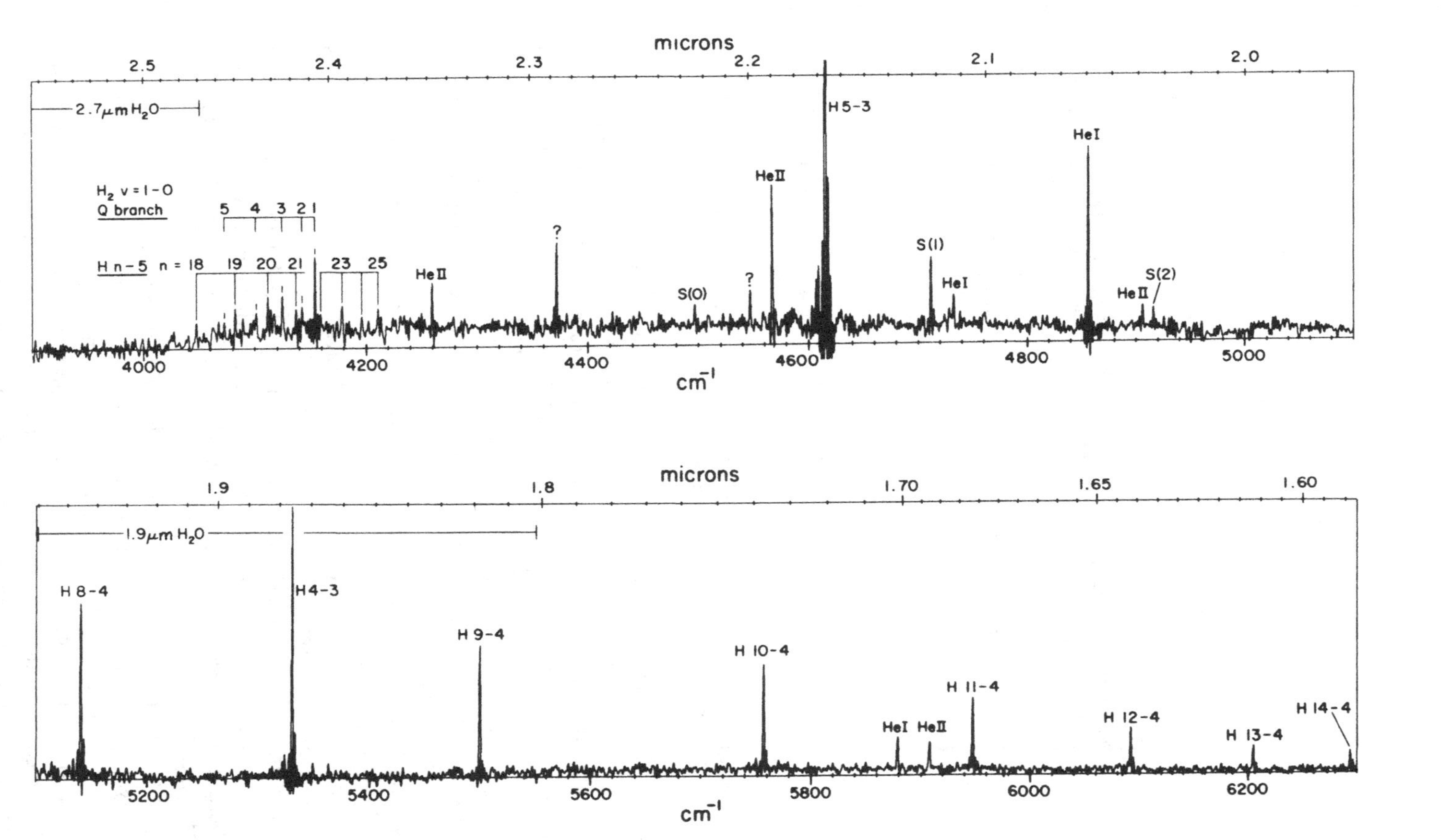

Fig. 1.12. The spectrum of NGC 7027 showing recombination lines of hydrogen and ionized helium in emission. (From Smith, Larson, and Fink[86])

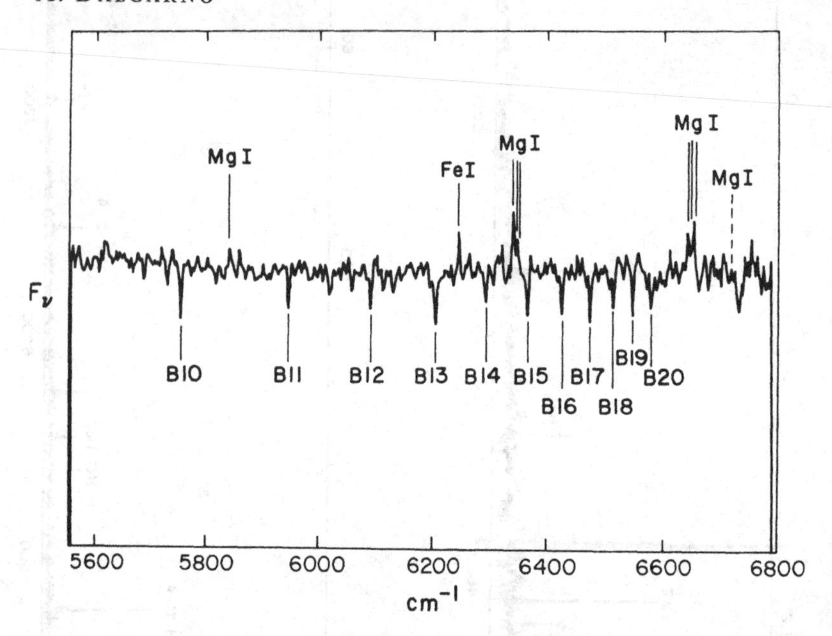

Fig. 1.13. The spectrum of IRC + 10420 showing absorption lines of the Brackett series of hydrogen. (From Thompson and Boroson[87])

standard interstellar extinction X_λ. Most of the data lie on a straight line but there are small departures at large λ, which may be due to local obscuration.

The observation of Rydberg atoms in the infrared is illustrated in Figure 1.12, which is a reproduction of the spectrum of the planetary nebula NGC 7027 between 1.6 and 2.5 μm.[86] Emissions from levels of hydrogen with n up to 25 and of ionized helium with n up to 15 are present. They coexist with lines of molecular hydrogen together with which they present an interesting problem of interpretation.

Rydberg levels have been detected in absorption in the infrared. The spectrum between 1.5 and 1.8 μm of the infrared object called IRC + 10420 is shown in Fig. 1.13.[87] Absorption lines in the Brackett series $4 \rightarrow n$ for values of n up to 20 are present.

Studies in the infrared have led to the discovery of highly ionized carbon in the atmospheres of Wolf–Rayet stars where, in addition to emission from Rydberg levels of He and He^+, the 9–8, 10–9, 12–10, and possibly 15–13 lines of C^{3+} have been observed.[88]

Figures 1.11 and 1.12 illustrate the considerable power of Rydberg atoms in astrophysics. Because of the numerous predictable coincidences, the identifications of the lines are quite secure despite the noisy spectra. We can look forward to an increased role for Rydberg atoms in astrophysics as measurements are carried out with greater spatial and spectral resolution and as the ultraviolet and submillimeter regions of the spectrum are opened up. The precision with which theory can reproduce the modes of population and depopulation and the resulting emission and absorption profiles and its ability to unify observations over the entire spectral range are critical elements in the utility of Rydberg atoms as diagnostics of the densities, temperatures, and radiation fields of the extraordinary variety of physical conditions in which Rydberg atoms emit and absorb radiation.

Acknowledgments

Drs. John Black and Eric Chaisson were very helpful to me in the preparation of this chapter. I am indebted to the Astronomy Section of the National Science Foundation for their support.

References and notes

1. W. J. Karzas and R. Latter, *Astrophys. J. Suppl. Ser. 6,* 167 (1961).
2. A. Burgess, *Mem. R. Astron. Soc. 69,* 1 (1964).
3. D. R. Flower and M. J. Seaton, *Comput. Phys. Commun. 1,* 31 (1969).
4. M. J. Seaton, *Mon. Not. R. Astron. Soc. 119,* 81 (1959).
5. L. C. Green, P. P. Rush, and C. D. Chandler, *Astrophys. Suppl. Ser. 3,* 37 (1957).
6. R. M. Pengelly and M. J. Seaton, *Mon. Not. R. Astron. Soc. 127,* 165 (1964).
7. D. H. Menzel, *Astrophys. J. Suppl. Ser. 18,* 221 (1968).
8. A. Burgess and H. P. Summers, *Mon. Not. R. Astron. Soc. 174,* 345 (1976).
9. C. S. Gee, I. C. Percival, J. G. Lodge, and D. Richards, *Mon. Not. R. Astron. Soc. 175,* 209 (1975).
10. I. C. Percival and D. Richards, *Adv. At. Mol. Phys. 11,* 1 (1975).
11. A. Burgess and I. C. Percival, *Adv. At. Mol. Phys. 4,* 109 (1968).
12. M. Brocklehurst, *Mon. Not. R. Astron. Soc. 148,* 417 (1970).
13. D. R. Bates and A. Dalgarno, in *Atomic and Molecular Processes,* ed. D. R. Bates (New York: Academic Press, 1962).
14. D. R. Bates, A. E. Kingston, and R. W. P. McWhirter, *Proc. R. Soc. London, Ser. A267,* 297 (1962).
15. V. L. Jacobs and J. Davis, *Phys. Rev. A 18,* 697 (1978).
16. L. Goldberg, *Astrophys. J. 144,* 1225 (1966).
17. M. Brocklehurst and M. J. Seaton, *Mon. Not. R. Astron. Soc. 157,* 179 (1972).
18. R. Gayet, *Astron. Astrophys. 9,* 312 (1970); L. Oster, *Astron. Astrophys. 9,* 318 (1970); M. Salem, *Mon. Not. R. Astron. Soc. 173,* 513 (1975).
19. R. L. Brown, F. J. Lockman, and G. R. Knapp, *Annu. Rev. Astron. Astrophys. 16,* 445 (1978).
20. J. G. Baker and D. H. Menzel, *Astrophys. J. 88,* 52 (1938).
21. H. Griem, *Astrophys. J. 148,* 547 (1967); M. J. Seaton, in *Radio Recombination Lines,* ed. P. A. Shaver (Holland: Reidel, 1980), p. 3.
22. M. Brocklehurst and S. Leeman, *Astrophys. Lett. 9,* 35 (1971).
23. G. Peach, *Astrophys. Lett. 10,* 129 (1972).
24. J. F. Kielkopf, *J. Opt. Soc. 63,* 987 (1973).
25. M. J. Seaton, *Mon. Not. R. Soc. 127,* 177 (1964); D. Hoang-Binh, *Astrophys. Lett. 2,* 231 (1968); J. E. Dyson, *Astrophys. J. 155,* 47 (1969); T. J. Sejnowski and R. M. Hjellming, *Astrophys. J. 156,* 915 (1969); R. M. Hjellming, M. H. Andrews, and

T. J. Sejnowski, *Astrophys. Lett. 3,* 111 (1969); M. H. Andrews and R. M. Hjellming, *Astrophys. Lett. 4,* 159 (1969); M. Brocklehurst, *Nature 225,* 618 (1970); M. Brocklehurst, *Mon. Not. R. Astron. Soc. 148,* 417 (1970); M. Brocklehurst, *Mon. Not. R. Astron. Soc. 153,* 471 (1971); M. Brocklehurst, *Astrophys. Lett. 14,* 81 (1973); A. Dupree, *Astrophys. J. 173,* 293 (1972); D. Hoang-Binh and C. M. Walmsley, *Astron. Ap. 35,* 49 (1974); A. I. Ahmed, *Astrophys. J. 194,* 503 (1974); P. A. Shaver, *Pramana 5,* 1 (1975); P. A. Shaver, *Astron. Astrophys. 46,* 1 (1976); A. Burgess and H. P. Summers, *Mon. Not. R. Astron. Soc. 174,* 345 (1976); M. Salem and M. Brocklehurst, *Astrophys. J. Suppl. Ser. 39,* 633 (1979); H. Ungerechts and C. M. Walmsley, *Astron. Astrophys. 80,* 325 (1980).
26. M. Brocklehurst and M. Salem, *Comput. Phys. Commun. 13,* 39 (1977).
27. M. Salem and M. Brocklehurst, *Astrophys. J. Suppl. Ser. 39,* 633 (1979).
28. H. Ungerechts and C. M. Walmsley, *Astron. Astrophys. 80,* 325 (1979).
29. J. C. Weisheit and C. M. Walmsley, *Astron. Astrophys. 61,* 141 (1977).
30. L. Goldberg and A. K. Dupree, *Nature 215,* 41 (1967).
31. A. K. Dupree, *Astrophys. J. Lett. 152,* L1251 (1968); V. K. Khersonskii and D. A. Varshalovich, *Sov. Astron. 24,* 359 (1980).
32. A. K. Dupree, *Astrophys. J. 158,* 491 (1969); A. K. Dupree and L. Goldberg, *Annu. Rev. Astron. Astrophys. 8,* 231 (1970).
33. V. Pankonin, R. L. Brown, E. Falgrone, M. R. Viner, and V. A. Hughes, in *Radio Recombination Lines,* ed. P. A. Shaver (Holland: Reidel, 1980).
34. W. D. Watson, L. R. Western, and R. B. Christensen, *Astrophys. J. 240,* 956 (1980).
35. A. A. Konovalenko and L. G. Sodin, *Nature 283,* 360 (1979).
36. D. H. Blake, R. M. Crutcher, and W. D. Watson, *Nature 287,* 707 (1980).
37. A. Greve, *Sol. Phys. 44,* 371 (1975).
38. T. H. Troland and C. Heiles, *Astrophys. J. 214,* 703 (1977).
39. A. Greve and T. Pauls, *Astron. Astrophys. 82,* 388 (1980).
40. D. Osterbrock, *Astrophysics of Gaseous Nebulae* (San Francisco: Freeman, 1974).
41. N. S. Kardashev, *Astron. Zh. 36,* 838 (1959) [*Sov. Astron. 3,* 813 (1959)].
42. Z. V. Dravskikh and A. F. Dravskikh, *Astron. Tsirk. 282,* 2 (1964).
43. B. Höglund and P. E. Mezger, *Science 150,* 359 (1965).
44. P. A. Shaver, ed., *Radio Recombination Lines* (Holland: Reidel, 1980).
45. P. A. Shaver, *Astron. Astrophys. 90,* 34 (1980); *91,* 279 (1980).
46. P. G. Mezger, T. L. Wilson, F. F. Gardner, and D. K. Milne, *Astrophys. Lett. 5,* 117 (1970); W. K. Huchtmeier and E. Churchwell, *Astron. Astrophys. 35,* 417 (1974); R. X. McGee, L. M. Newton, and R. A. Batchelor, *Aust. J. Phys. 28,* 185 (1975).
47. R. R. Silverglate and Y. Terzian, *Astrophys. J. Suppl. Ser. 39,* 157 (1979); Z. Abraham, J. R. D. Lepine, and M. A. Braz, *Mon. Not. R. Astron. Soc. 193,* 737 (1980); T. L. Wilson, in *Radio Recombination Lines,* ed. P. A. Shaver (Holland: Reidel, 1980), p. 205.
48. A. Pedlar and R. D. Davies, in *Radio Recombination Lines,* ed. P. A. Shaver (Holland: Reidel, 1980), p. 239.
49. F. J. Lockman and R. L. Brown, *Astrophys. J. 201,* 134 (1975); T. Pauls and T. L. Wilson, *Astron. Astrophys. 60,* L31 (1977).
50. E. Churchwell, P. G. Mezger, and W. Huchtmeier, *Astron. Astrophys. 32,* 283 (1974); E. Churchwell, L. F. Smith, J. S. Mathis, P. G. Mezger, and W. Huchtmeier, *Astron. Astrophys. 70,* 719 (1978).

51. N. Panagia, in *Radio Recombination Lines,* ed. P. A. Shaver (Holland: Reidel, 1980), p. 99.
52. J. S. Mathis, *Astrophys. J. 167,* 261 (1971).
53. P. G. Mezger, L. F. Smith, and E. Churchwell, *Astron. Astrophys. 32,* 269 (1974).
54. E. J. Chaisson and A. Malkan, *Astrophys. J. 210,* 108 (1976).
55. C. M. Walmsley, E. Churchwell, and Y. Terzian, *Astron. Astrophys. 96,* 278 (1981).
56. Y. Terzian, in *Radio Recombination Lines,* ed. P. A. Shaver (Holland: Reidel, 1980), p. 75.
57. J. S. Miller and W. G. Mathews, *Astrophys. J. 172,* 593 (1972); D. Péquignot, S. M. V. Aldrovandi, and G. Stasinska, *Astron. Astrophys. 63,* 313 (1978).
58. P. Palmer, B. Zuckerman, H. Penfield, A. E. Lilley, and P. G. Mezger, *Nature 215,* 40 (1967).
59. V. Pankonin, in *Radio Recombination Lines,* ed. P. A. Shaver (Holland: Reidel, 1980), p. 111.
60. J. A. Ball, D. Cesarsky, A. K. Dupree, L. Goldberg, and A. E. Lilley, *Astrophys. J. 162,* L25 (1970).
61. V. Pankonin, C. M. Walmsley, T. L. Wilson, and P. Thomasson, *Astron. Astrophys. 57,* 341 (1977).
62. J. K. Hill, *Astrophys. J. 212,* 692 (1977).
63. E. Krugel and G. Tenorio-Tagle, *Astron. Atrophys. 70,* 51 (1979).
64. R. L. Brown, in *Radio Recombination Lines,* ed. P. A. Shaver (Holland: Reidel, 1980), p. 127.
65. R. M. Crutcher, *Astrophys. J. Lett. 217,* L109 (1977).
66. E. J. Chaisson, J. H. Black, A. K. Dupree, and D. A. Cesarsky, *Astrophys. J. Lett. 173,* L131 (1972).
67. E. Falgarone, in *Radio Recombination Lines,* ed. P. A. Shaver (Holland: Reidel, 1980), p. 141.
68. E. Falgarone, D. Cesarsky, P. J. Encrenaz, and R. Lucas, *Astron. Astrophys. 65,* L13 (1978).
69. R. Chevalier, *Annu. Rev. Astron. Astrophys. 15,* 175 (1977).
70. D. A. Cesarsky and C. J. Cesarsky, *Astrophys. J. Lett. 183,* L143 (1973); R. C. Bignell, *Astrophys. J. 186,* 889 (1973); D. Downes and T. L. Wilson, *Astron. Astrophys. 34,* 133 (1974); A. Parrish and V. Pankonin, *Astrophys. J. 198,* 349 (1975); V. Pankonin, *Astron. Astrophys. 38,* 445 (1975); V. Pankonin and D. Downes, *Astron. Astrophys. 47,* 303 (1976); D. A. Cesarsky, *Astron. Astrophys. 50,* 259 (1976).
71. L. Hart and A. Pedlar, *Mon. Not. R. Astron. Soc. 193,* 781 (1980).
72. T. Pauls and P. G. Mezger, *Astron. Astrophys. 85,* 26 (1980).
73. M. B. Bell, in *Radio Recombination Lines,* ed. P. A. Shaver (Holland: Reidel, 1980), p. 259.
74. P. A. Shaver, E. Churchwell, and A. H. Roos, *Astron. Astrophys. 55,* 435 (1977); M. B. Bell and E. R. Seaquist, *Astron. Astrophys. 56,* 461 (1977); M. B. Bell and E. R. Seaquist, *Astrophys. J. 223,* 378 (1978); E. J. Chaisson and L. F. Rodriguez, *Astrophys. J. Lett. 214,* 411 (977).
75. L. F. Rodriguez and E. J. Chaisson, *Astrophys. J. 238,* 41 (1980).
76. E. R. Seaquist and M. B. Bell, *Astron. Astrophys. 60,* L1 (1977); U. Mebold, P. A. Shaver, M. B. Bell, and E. R. Seaquist, *Astron. Astrophys. 82,* 272 (1980).
77. M. B. Bell and E. R. Seaquist, *Astrophys. J. 238,* 818 (1980).

78. D. E. Osterbrock and R. Cohen, *Mon. Not. R. Astron. Soc. 187,* 61P (1979).
79. P. A. Shaver, in *Radio Recombination Lines,* ed. P. A. Shaver (Holland: Reidel, 1980), p. 247.
80. P. A. Shaver, *Astron. Astrophys. 68,* 97 (1978).
81. M. Brocklehurst, *Mon. Not. R. Astron. Soc. 153,* 471 (1971); *Mon. Not. R. Astron. Soc. 157,* 211 (1972).
82. K. Giles, *Mon. Not. R. Astron. Soc. 180,* 57P (1977).
83. M. J. Seaton, *Mon. Not. R. Astron. Soc. 185,* 5P (1978).
84. J. S. Miller, *Astrophys. J. Lett. 165,* 401 (1970); J. S. Miller and W. G. Mathews, *Astrophys. J. 172,* 591 (1972); L. E. Goad, L. Goldberg, and J. L. Greenstein, *Astrophys. J. 175* (1972).
85. M. J. Seaton, *Mon. Not. R. Astron. Soc. 187,* 785 (1979).
86. H. A. Smith, H. P. Larson, and U. Fink, *Astrophys. J. 244,* 835 (1981).
87. R. I. Thompson and T. A. Boroson, *Astrophys. J. Lett. 216,* L75 (1977).
88. M. Cohen and S. N. Vogel, *Mon. Not. R. Astron. Soc. 185,* 47 (1978); P. M. Williams and D. A. Allen, *Observatory 100,* 202 (1980).

2

Theoretical studies of hydrogen Rydberg atoms in electric fields

R. J. DAMBURG AND V. V. KOLOSOV

2.1 Introduction

A uniform electric field removes the degeneracy of energy levels of hydrogen atoms relative to orbital and magnetic quantum numbers. States that differ only in the sign of the magnetic quantum number remain degenerate. Thus the energy level of the hydrogen atom specified by the principal quantum number n is split in the uniform electric field into $n(n+1)/2$ sublevels. Each sublevel can be characterized by a set of parabolic quantum numbers n_1, n_2, m such that the principal quantum number $n = n_1 + n_2 + m + 1$.

In the presence of the electric field, all discrete levels shift. They evolve into narrow bands of a dense continuum that become wider and more diffuse as the electric field becomes more intense. This phenomenon is known as the Stark effect.[1,2] To characterize quantitatively the influence of the electric field with the given strength F on the atom in a certain quantum state $n_1 n_2 m$, it is necessary to determine the center of the level E_0 and its width Γ. The level width Γ defines the ionization probability of the quasi-stationary atomic state in the electric field: $\omega = \Gamma/\hbar$.[3] The value of Γ is inversely proportional to the lifetime of the atomic state for such fields where the ionization probability exceeds the probability of radiative transitions. In the opposite case, the level width of an excited state depends also on the lifetime of this state relative to light emission increasing with the growth of n: $\tau_{\text{em}} \simeq n^{9/2}$.[2] In our consideration, we shall neglect the relativistic corrections and the spin of the electron and present results that are valid for fields such that the Stark shifts considerably exceed the fine structure intervals.

The Schrödinger equation for the hydrogen atom in a uniform electric field F parallel to the z axis is of the form

$$\left(\Delta + \frac{2}{r} - 2Fz + 2E\right)\Psi = 0 \tag{1}$$

where (except in specially mentioned cases) atomic units will be used throughout.

Equation (1) is separable in parabolic coordinates,[1-3]

$$x = (\xi\eta)^{1/2}\cos\varphi, \qquad y = (\xi\eta)^{1/2}\sin\varphi, \qquad z = \tfrac{1}{2}(\xi - \eta)$$

Defining the wave function Ψ as the product

$$\Psi = (\xi\eta)^{-1/2} V(\xi) U(\eta) e^{\pm im\varphi} \tag{2}$$

and substituting (2) into (1), we obtain, for $V(\xi)$ and $U(\eta)$, the equations

$$\frac{d^2V}{d\xi^2} + \left(\frac{E}{2} + \frac{\beta_1}{\xi} + \frac{1-m^2}{4\xi^2} - \frac{F}{4}\xi\right)V = 0 \tag{3}$$

$$\frac{d^2U}{d\eta^2} + \left(\frac{E}{2} + \frac{\beta_2}{\eta} + \frac{1-m^2}{4\eta^2} - \frac{F}{4}\eta\right)U = 0 \tag{4}$$

where β_1 and β_2 are separation constants coupled by the requirement

$$\beta_1 + \beta_2 = 1 \tag{5}$$

The functions $V(\xi)$ and $U(\eta)$ should be finite at the origin

$$V(\xi) \underset{\xi \to 0}{\simeq} \xi^{(m+1)/2}, \qquad U(\eta) \underset{\eta \to 0}{\simeq} \eta^{(m+1)/2} \tag{6}$$

The asymptotic solution for $V(\xi)$ falls off exponentially as

$$V(\xi) \underset{\xi \to \infty}{\simeq} \frac{A}{\xi^{1/4}} \exp\left(-\frac{F^{1/2}}{3}\xi^{3/2} + \frac{E}{F^{1/2}}\xi^{1/2}\right) \tag{7}$$

The asymptotic behavior of $U(\eta)$ for $\eta \to \infty$ shows that the problem as a whole is of an unbound character

$$U(\eta) \underset{\eta \to \infty}{\simeq} \frac{B}{\eta^{1/4}} \sin\left(\frac{F^{1/2}}{3}\eta^{3/2} + \frac{E}{F^{1/2}}\eta^{1/2} + \Phi\right) \tag{8}$$

where A, B, Φ are constants for given F and E. Thus, the Stark problem for the hydrogen atom is reduced to the solution of Eqs. (3) and (4) coupled by the requirement (5) with boundary conditions (6)–(8).

We shall not dwell on early papers on the Stark effect. The significance of these first investigations for the development of quantum mechanics is well known.[1,2] However, some of the early papers will be referred to in connection with specific questions. Still, it should be mentioned that those investigations carried out by Traubenberg et al.[4-6] remained for a long time the only direct experimental studies on this problem.

The advent of tunable lasers made it possible to populate selected highly excited atomic Rydberg states in an efficient way. It gave a new stimulus to the experimental study of the classical problem of the Stark effect for hydrogen.[7,8] On the other hand, these mentioned circumstances to some extent also stimulated theoretical studies. Though the Stark effect for hydrogen was explained qualitatively in the early days of quantum mechanics, the results obtained then by using low-order perturbation theory and WKB approximations do not always agree with current experimental results. For this reason, numerous theoretical papers have recently appeared on the Stark effect. We do not intend to review all the papers on the subject; our intention is to elucidate some of the results obtained recently on this problem.

2.2. Exact numerical solution

2.2a. Methods

Modern computers make possible an exact numerical solution to the problem of the nonrelativistic hydrogen atom in a uniform electric field. The problem is essentially simplified because of the separation of variables.

Calculations for the ground state of the hydrogen atom based on numerical integration of Eqs. (3) and (4) coupled to condition (5) were given in Refs. 9 and 10. In the former,[9] the cases for strong and weak fields were considered separately. In the case of weak fields ($F \leqslant 0.02$), Eqs. (3) and (4) were treated as eigenvalue problems and the Stark shift was determined without regard for the ionization probabilities. In the case of strong fields ($0.03 \leqslant F \leqslant 0.12$), the wave function $U(\eta)$, obtained by numerical integration of Eq. (4) through expansion in the power series, was matched with its asymptotes at sufficiently large η. Thus we can obtain the dependence of the asymptotic phase shift Φ on E. Values of E_0 and Γ were determined by the Breit–Wigner parametrization of this dependence in the vicinity of resonance. In the latter reference,[10] the formalism of Weyl's theory was used, according to which, values of E_0 and Γ were determined by poles of Weyl's "m function" in a complex plane. When solving Eq. (3), the eigenvalue of the separation constant β_1 for the given values of E and F was calculated. In Eq. (4), the m function was determined by matching two linearly independent solutions, obtained in numerical integration, with the asymptotes of $U(\eta)$ expressed in terms of Airy functions. Numerical integration of Eqs. (3) and (4) was performed by the Runge–Kutta method and the results obtained for the fields $F \leqslant 0.12$ in Refs. 9 and 10 practically coincide.

A number of calculations for the ground state of hydrogen have been done using variational methods. In Ref. 11, to determine Stark energies for $0.01 \leqslant F \leqslant 0.25$, the minimum-variance principle was applied. In Ref. 12, the authors extended this method to the calculation of the level width for the ground state, and their results for the case of strong fields appear to be in satisfactory agreement with the results of the previously mentioned works. However, in the case of weak fields, the results obtained in Ref. 12 for the level width Γ essentially disagree with those of Refs. 9 and 10. Then, applying the method of complex-scale transformation with the use of the extended virial theorem,[13] the authors achieved excellent agreement with the results of Refs. 9 and 10 for three values of the field intensity. In Ref. 14–17, to perform variational calculations, the complex-scale transformation was also used. (This method is also known as the coordinate rotation method or the dilatation transformation.) The idea is that the transformation $r \rightarrow re^{i\theta}$ (a coordinate rotation) in the Hamiltonian does not affect its complex eigenvalues corresponding to quasi-stationary states but does transform the eigenfunctions into square integrable functions. The complex eigenvalue obtained gives both the Stark energy and the width without the use of continuum functions. The main shortcoming of this method is the slow convergence of results with increase of the basis set of the trial functions.[16] Also, convergence here is more delicate than for the discrete spectra of Hermitian Hamiltonians because the results are only stationary, with the bounding principles being lost.[17]

In Refs. 18 and 19, numerical calculations were done for the state $n=5$, $n_1=3$, $n_2=0$, and $m=1$. In both papers, Eq. (4) was solved in the same way: the solution that was obtained by integration of Eq. (4) using the Numerov method was matched with the asymptotes at a point sufficiently removed from the outer turning point. Equation (3) was solved differently. In Ref. 19 it was solved numerically by the Numerov method and the value of the separation parameter was determined by matching the numerical solution with the asymptotes [Eq. (7)]. In Ref. 18 Eq. (3) was solved by the perturbation theory. In both papers, the lifetime was defined as $\tau(E)=N_0(E)/I_{out}(E)$, where I_{out} is the flux corresponding to the outward component of the wave function and N_0 the number of nonionized atoms. The Stark energy E_0 is the value of E for which $\tau(E)$ has its maximum value $\tau(E_0)$ that determines the value of Γ: $\tau(E_0)=1/\Gamma$. The results obtained in Ref. 18 differ from the results of Refs. 19, 24, and 25 and are obviously erroneous.

We should also mention several papers that deal with the calculation of optical absorption by hydrogenic excitons in a uniform electric field.[20-23] In the effective-mass approximation, the Wannier exciton is described by an equation similar to Eq. (1). Calculations given in Refs. 20–23 are based on the numerical solution of Eqs. (3) and (4) with boundary conditions (7) and (8). In these papers, only low-excited states have been considered. The results obtained concern mainly the behavior of wave functions, with numerical values for E_0 and Γ not given.

In Ref. 24, a method for the exact numerical solution of the problem was presented, allowing reliable results to be obtained for any quantum state. Equations equivalent to (3) and (4) were solved by expanding their solutions in power series. To determine E_0 and Γ, the Breit–Wigner parametrization of the asymptotic amplitude $B(E)$ and phase shift $\Phi(E)$ were used. A more detailed description of the method is given in Sect. 2.2c. In a recently published paper,[25] the wave function in Eq. (2) was normalized to a Dirac function. Values of E_0 and Γ were obtained by the Breit–Wigner parametrization of the normalization constant $\mathcal{C}(E)$. Equations (3) and (4) were solved numerically by the Numerov method. By counting the number of nodes of the functions $V(\xi)$ and $U(\eta)$ in the internal regions, the authors checked the choice of initial intervals for $\Delta\beta_1$ and ΔE, where β_1 and E_0 were to be found. For a number of states, the agreement between the results of Refs. 24 and 25 was very good.

2.2b. Determination of resonance parameters E_0 and Γ

It has already been mentioned at the start of this chapter that in the presence of the uniform electric field the spectrum of eigenvalues of the hydrogen atom becomes continuous. However, when the ionization probability is small, it is possible to introduce the concept of quasi-stationary states. When defining the quasi-stationary state, the solution of the Schrödinger equation at infinity is usually required to be in the form of the outgoing wave, which corresponds to the particle leaving the decaying system.[3] Because this boundary condition is complex, the eigenvalues of the Schrödinger equation E_R are also complex:

$$E_R = E_0 - \tfrac{1}{2}i\Gamma \tag{9}$$

where E_0 and Γ are, respectively, the position and width of the resonance corresponding to the quasi-stationary state. The value of Γ is positive, and it determines the probability of the decay of the system. Eigenvalues of quasi-stationary states correspond to the poles of the function $S(E)=e^{2i\Phi}$ on the second sheet of the complex-energy plane. For $|E-E_0| \ll \Gamma$, the following expression for $S(E)$ can be obtained:[3]

$$S(E) = e^{2i\Phi_0}(E - E_0 - \tfrac{1}{2}i\Gamma)/(E - E_0 + \tfrac{1}{2}i\Gamma) \tag{10}$$

Formula (10) is equivalent to the Breit–Wigner parametrization of the asymptotic phaseshift $\Phi(E)$ and asymptotic amplitude $B(E)$ [see Eq. (8)] in the vicinity of the resonance energy E_0:

$$\Phi(E) = \Phi_0 + \tan^{-1}[\Gamma/2(E_0 - E)] \tag{11}$$

$$B^2(E) = B_0^2[(E_0 - E)^2 + \tfrac{1}{4}\Gamma^2] \tag{12}$$

where Φ_0 and B_0 are constants independent of energy. Using formulae (11) and (12), from $\Phi(E)$ or $B^2(E)$ obtained in the numerical solution of (3) and (4), the values of the resonance parameters E_0 and Γ may be determined. It should be noted, however, that with an increase in electric field strength the increasing asymmetry of $\Phi(E)$ and $B^2(E)$ in the vicinity of the resonance increases, in its turn, both the discrepancy between the values of E_0 and Γ, obtained on the basis of the Breit–Wigner parametrization, and the resonance parameters corresponding to the exact complex-energy value (9). This discrepancy does not exceed the value of Γ^2 and results in an essential difference only for the fields for which the very notion of the quasi-stationary state has lost its strict physical meaning. A number of works have recently appeared involving attempts to analyze mathematically Eq. (1) (see, e.g., Refs. 26–30). In particular, the problem of the behavior of the function $E_0(F) - \frac{1}{2}i\Gamma(F)$ as $F \to \infty$ has been studied.[27] It is clear that in these works the authors did not use the Breit–Wigner approximation [Eqs. (11) and (12)]; rather, they dealt with exact eigenvalues of the Schrödinger equation (1). Calculations of exact complex eigenvalues for the ground and first excited states were given in Ref. 30. The authors introduced uniform, strongly asymptotic approximants whose existence is related to the Borel summability of the perturbation series. Values of E_0 and Γ were calculated at different electric field strengths up to a value of $F \simeq 10^{20}$. Here E_0 and Γ reach the values of $\sim 10^{14}$. At small F, there is excellent agreement with the results obtained by using the Breit–Wigner approximation. Thus, at $F=0.03$, the results for E_0 coincide up to the ninth figure and the results for Γ coincide at least up to the sixth figure.

Though E_0 and Γ have no evident physical meaning at very large Γ, exact calculations at moderately large fields may be of use for certain physical applications. Wave functions obtained by this method may be used, for example, to calculate the photoionization cross section of the hydrogen atom in a uniform electric field.[31,32]

2.2c. Calculation procedure

In Ref. 24, to separate the variables of Eq. (1), "squared" parabolic coordinates (μ, ν, φ), which seem to be more convenient for the numerical solution, were used

$$x = \mu\nu \cos\varphi, \qquad y = \mu\nu \sin\varphi, \qquad z = \tfrac{1}{2}(\mu^2 - \nu^2) \tag{13}$$

Inserting the wave function $\Psi = (\mu\nu)^{-1/2} M(\mu) N(\nu) e^{\pm im\varphi}$ into Eq. (1) we obtain, after separation of the variables

$$\left(\frac{d^2}{d\mu^2} + \frac{1-4m^2}{4\mu^2} + 2E\mu^2 - F\mu^4 + Z_1\right) M(\mu) = 0 \tag{14}$$

$$\left(\frac{d^2}{d\nu^2} + \frac{1-4m^2}{4\nu^2} + 2E\nu^2 - F\nu^4 + Z_2\right) N(\nu) = 0 \tag{15}$$

where Z_1 and Z_2 are the separation constants coupled by the requirement

$$Z_1 + Z_2 = 4 \tag{16}$$

The behavior of $M(\mu)$ and $N(\nu)$ at the origin and at infinity in these coordinates is defined by the formulae:

$$\begin{aligned} M(\mu) &\underset{\mu \to 0}{\simeq} \mu^{m+1/2} \\ M(\mu) &\underset{\mu \to \infty}{\simeq} \frac{A}{\mu} \exp\left(-\frac{F^{1/2}}{3}\mu^3 + \frac{E}{F^{1/2}}\mu\right) \end{aligned} \tag{17}$$

$$\begin{aligned} N(\nu) &\underset{\nu \to 0}{\simeq} \nu^{m+1/2} \\ N(\nu) &\underset{\nu \to \infty}{\simeq} \frac{B}{\nu} \sin\left(\frac{F^{1/2}}{3}\nu^3 + \frac{E}{F^{1/2}}\nu + \Phi\right) \end{aligned} \tag{18}$$

where the asymptotic amplitudes A and B and the phase shift Φ are constant for fixed F and E.

By solving (14) and (15) and taking (16)–(18) into account, we obtain $B(E)$ and $\Phi(E)$ for a given value of F. For this purpose we first obtain the separation constant Z_1 by solving (14) for fixed E. Then, substituting $Z_2=4-Z_1$ into (15) and matching the obtained numerical solution with the asymptotic form (18), we obtain the values of B and Φ. Finally, using the Breit–Wigner parametrization of the obtained $B^2(E)$ and $\Phi(E)$, we determine values of E_0 and Γ.

Next, we shall consider in more detail the solutions of Eqs. (14) and (15). Equation (14) with the variable μ behaves like a one-dimensional Schrödinger equation with bound states. Because values of E and F are assigned, Z_1 serves as the eigenvalue. Equation (14) for the state defined by the set of quantum numbers n_1, n_2, and m is solved in the following way. Based upon the results obtained by the perturbation theory and the results of previous calculations, we choose the interval ΔZ_1, where Z_1 is expected to be found. In this interval, by the linear interpolation method, we search for the value of Z_1 that would make $M(\mu)$ equal to zero at some distant value of μ_k. This procedure is efficient for determining Z_1 with high accuracy.

The value of $M(\mu)$ is determined in numerical step-by-step integration of (14) by expansion in the power series. Expanding $M(\mu)$ in the vicinity of zero

$$M(\mu) = \mu^{m+1/2} \sum_{n=0}^{\infty} C_{0n}\mu^n \tag{19}$$

and in the vicinity of other points $\mu_i \neq 0$ corresponding to the following steps

$$M(\mu) = \sum_{n=0}^{\infty} C_{in}(\mu-\mu_i)^n \tag{20}$$

by further substitution of (19) and (20) into (14), we obtain recurrence formulae for the expansion coefficients C_{0n} and C_{in}:

$$C_{0n} = -\frac{Z_1 C_{0,n-2} + 2EC_{0,n-4} - FC_{0,n-6}}{n(n+2m)} \tag{21}$$

$$\begin{aligned} C_{in} = -\frac{1}{n(n-1)\mu_i^2}\{ & \mu_i(n-1)(2n-3)C_{i,n-1} \\ & + [(n-2)^2 - m^2 + Z_1\mu_i^2 + 2E\mu_i^4 - F\mu_i^6]C_{i,n-2} \\ & + (2Z_1\mu_i + 8E\mu_i^3 - 6F\mu_i^5)C_{i,n-3} + (Z_1 + 12E\mu_i - 15F\mu_i^4)C_{i,n-4} \\ & + (8E\mu_i - 20F\mu_i^3)C_{i,n-5} + (2E - 15F\mu_i^2)C_{i,n-6} \\ & - 6F\mu_i C_{i,n-7} - FC_{i,n-8}\} \end{aligned} \tag{22}$$

Imposing $C_{00}=1$ and using recurrence relation (21), we obtain all the coefficients of expansion (19) and, consequently, the values of $M(\mu)$ and its derivative at the point μ_1 that define, respectively, C_{10} and C_{11}. To determine $M(\mu)$ at some distant point μ_k, the expansion procedure has to be repeated several times. For the second and subsequent steps, expansion (20) and, consequently, recurrence relation (22) are used. Our computations were made mainly on a BESM–IV using 36 binary bits in the mantissa, which is equivalent to 10.8 significant figures in the decimal system. This is the reason terms smaller by a factor of 10^{11} or more than the largest term in the series were omitted. We also used this condition to terminate expansions (19) and (20). The value of the expansion step was constant in all cases $(\mu-\mu_i)=1$. The choice of a necessary number of steps and, consequently, of the point where the value of $M(\mu)$ is set equal to zero, was done automatically.

The procedure of searching for Z_1 is repeated to the next point μ_{k+1}, and the obtained value is compared with the value of Z_1 at the point μ_k. If both values of Z_1 coincide with a desired accuracy, the procedure is terminated; otherwise, it continues to the next point μ_{k+1}, etc. The number of steps depends on the state of the atom, the value of the field strength, and the desired degree of accuracy of Z_1 determination.

Equation (15) with boundary condition (18) has a continuous spectrum with respect to the energy. The function playing the role of the potential energy in (15) has the form of a barrier and tends to $-\infty$ as $\nu \to \infty$. The numerical solution of Eq. (15) is matched with its asymptote at a certain point beyond the outer turning point. The numerical integration of Eq. (15) is performed similarly to (14) by expanding $N(\nu)$ in a power series. Recurrence relations for the corresponding coefficients differ from (21) and (22) by the changes $Z_1 \to Z_2$ and $F \to -F$. Because of the rapid oscillations of $N(\nu)$ for large ν, the value of the steps is chosen equal to 0.1. To avoid the necessity of integrating Eq. (15) to very large values of ν, we use an asymptotic expansion for $N(\nu)$

$$N(\nu) = \sin\left(\frac{F^{1/2}}{3}\nu^3 + \frac{E}{F^{1/2}}\nu\right)\sum_{k=1}\frac{a_k}{\nu^k} + \cos\left(\frac{F^{1/2}}{3}\nu^3 + \frac{E}{F^{1/2}}\nu\right)\sum_{k=1}\frac{b_k}{\nu^k} \tag{23}$$

Substituting (23) into (15), we obtain the following recurrence relation for a_n and b_n:

$$a_n = \frac{1}{2F^{1/2}(n-1)}\left\{\left[(n-3)(n-2) - \frac{4m^2-1}{4}\right]b_{n-3} - \frac{2E}{F^{1/2}}(n-2)a_{n-2} - \left(\frac{E^2}{F} - Z_2\right)b_{n-1}\right\}$$

$$b_n = \frac{1}{2F^{1/2}(n-1)}\left\{\left[\frac{4m^2-1}{4} - (n-3)(n-2)\right]a_{n-3} - \frac{2E}{F^{1/2}}(n-2)b_{n-2} + \left(\frac{E^2}{F} - Z_2\right)a_{n-1}\right\} \tag{24}$$

All coefficients in (23) can be expressed in terms of a_1 and b_1 by using (24). The amplitude B and the phase shift Φ, in their turn, are also expressed by a_1 and b_1 as

$$B^2 = a_1^2 + b_1^2, \qquad \Phi = \tan^{-1}(b_1/a_1)$$

Table 2.1. Values of E_0 and Γ for $n=1$

$F\times 10^2$	E_0	Γ	$F\times 10^2$	E_0	Γ
2.0	−0.500 908 887 5	5.5×10^{-13}	5.0	−0.506 105 39	7.717×10^{-5}
2.5	−0.501 429 292	3.318×10^{-10}	5.5	−0.507 553 6	2.201×10^{-4}
3.0	−0.502 074 273	2.2376×10^{-8}	6.0	−0.509 202 3	5.146×10^{-4}
3.5	−0.502 851 420	4.3497×10^{-7}	6.5	−0.511 046	1.033×10^{-3}
4.0	−0.503 771 590	3.8926×10^{-6}	7.0	−0.513 064	1.842×10^{-3}
4.5	−0.504 850 148	2.0776×10^{-5}	7.5	−0.515 23	2.99×10^{-3}

The choice of the matching point ν_k depends both on the number of terms in the asymptotic expansion (23) and on the values of F and E. For each state the dependence of the matching point on the mentioned values was analyzed to provide the desired accuracy in determination of a_1 and b_1.

We start the calculations of E_0 and Γ for any of the atomic states with a low electric field, where perturbation theory provides a reliable indication of initial intervals ΔZ_1 and ΔE, where E_0 and Z_1 are to be found. Inside the interval ΔE, we choose three values $E_1<E_2<E_3$ for which we solve a system of equations (14)-(16) and obtain $B^2(E_1)$, $B^2(E_2)$, $B^2(E_3)$. Inserting these into Eq. (12), we obtain approximate values of E_0 and Γ. If $E_0-E_2=\delta>0$, we repeat the procedure for E_2, E_0, $E_0+\delta$; if $E_0-E_2=\delta<0$, then we repeat the procedure for $E_0+\delta$, E_0, E_2. This allows us to narrow the interval ΔE rapidly and to estimate Γ. The results obtained are practically independent of the choice of initial E_1, E_2, E_3. The only requirement is

$$B^2(E_2) < \max\{B^2(E_1), B^2(E_3)\}$$

i.e., E_1 and E_3 should not be in the vicinity of the neighboring resonance. To obtain a still more accurate value of E_0, we look directly for those values of E that would correspond to the minimum $B^2(E)$ in the narrow interval ΔE. Then we use formula (11) for a more precise determination of Γ.

Starting with low electric fields, we gradually move to higher ones. For each new value of F, we choose new initial intervals ΔE and ΔZ_1. These are chosen based upon the difference between the results obtained by the perturbation theory and the exact numerical results for lower fields. The choice of the intervals is done automatically. Some testing is required only when passing to the calculation of a new state. This procedure does not require counting the number of nodes in $M(\mu)$ and $N(\nu)$.

2.2d. Results

In Tables 2.1-2.11, the results of the numerical calculations of E_0 and Γ obtained by the method described in the preceding section are given. Atomic units are used everywhere, except for the values of electric field strength in Tables 2.7, 2.9, and 2.11. The bulk of the results are published for the first

Table 2.2. Values of E_0 and Γ for $n=2$

	$n_1=0, n_2=1, m=0$		$n_1=0, n_2=0, m=1$		$n_1=1, n_2=0, m=0$	
$F\times10^3$	E_0	Γ	E_0	Γ	E_0	Γ
2.5	−0.133 061 759	9.92×10^{-11}	−0.125 496 728	1.05×10^{-11}	−0.118 009 878	1.08×10^{-12}
3.0	−0.134 825 559	1.320×10^{-8}	−0.125 721 746	1.72×10^{-9}	−0.116 733 001	2.12×10^{-10}
3.5	−0.136 651 161	3.876×10^{-7}	−0.125 993 649	6.020×10^{-8}	−0.115 497 621	8.756×10^{-9}
4.0	−0.138 548 793	4.439×10^{-6}	−0.126 316 885	8.100×10^{-7}	−0.114 305 339	1.364×10^{-7}
4.5	−0.140 533 13	2.708×10^{-5}	−0.126 698 247	5.759×10^{-6}	−0.113 158 789	1.110×10^{-6}
5.0	−0.142 618 3	1.058×10^{-4}	−0.127 146 593	2.614×10^{-5}	−0.112 061 922	5.729×10^{-6}

Table 2.3. Values of E_0 and Γ for $n=5$, $n_1=2$, $n_2=2$, $m=0$

$F\times 10^4$	$E_0\times 10^2$	Γ	$F\times 10^4$	$E_0\times 10^2$	Γ
1.2	−2.025 931 09	2.05×10^{-11}	1.5	−2.041 553 93	2.828×10^{-8}
1.3	−2.030 658 33	3.526×10^{-10}	1.6	−2.047 800 50	1.553×10^{-7}
1.4	−2.035 854 87	3.802×10^{-9}	1.7	−2.054 655 81	6.626×10^{-7}

Table 2.4. Values of E_0 and Γ for $n=10$, $n_1=9$, $n_2=0$, $m=0$

$F\times 10^5$	$E_0\times 10^3$	Γ	$F\times 10^5$	$E_0\times 10^3$	Γ
1.3	−3.372 991 60	6.5×10^{-14}	1.7	−2.915 659 47	1.979×10^{-9}
1.4	−3.256 887 83	1.62×10^{-12}	1.8	−2.804 329 50	1.133×10^{-8}
1.5	−3.141 960 52	2.52×10^{-11}	1.9	−2.694 274 62	5.135×10^{-8}
1.6	−3.028 213 00	2.627×10^{-10}	2.0	−2.585 573 8	1.900×10^{-7}

Table 2.5. Values of E_0 and Γ for $n=10$, $n_1=0$, $n_2=9$, $m=0$

$F\times 10^5$	$E_0\times 10^3$	Γ	$F\times 10^5$	$E_0\times 10^3$	Γ
0.8	−6.153 514 04	3.77×10^{-13}	1.1	−6.645 975 28	7.173×10^{-7}
0.9	−6.311 892 43	2.220×10^{-10}	1.2	−6.826 384 6	7.145×10^{-6}
1.0	−6.475 414 45	2.4045×10^{-8}	1.3	−7.014 913	2.907×10^{-5}

Table 2.6. Values of E_0 and Γ for $n=15$, $n_1=7$, $n_2=7$, $m=0$

$F\times 10^6$	$E_0\times 10^3$	Γ	$F\times 10^6$	$E_0\times 10^3$	Γ
2.2	−2.285 672 23	3.67×10^{-13}	2.6	−2.314 423 57	4.713×10^{-9}
2.3	−2.292 165 42	6.04×10^{-12}	2.7	−2.322 945 2	2.665×10^{-8}
2.4	−2.299 094 38	7.26×10^{-11}	2.8	−2.332 163 1	1.214×10^{-7}
2.5	−2.306 496 95	6.627×10^{-10}	2.9	−2.342 207 8	4.480×10^{-7}

Table 2.7. Values of E_0 and Γ for $n=25$, $n_1=24$, $n_2=0$, $m=0$

$F\times 10^{-3}$ (V cm^{-1})	$E_0\times 10^{-4}$	Γ	$F\times 10^{-3}$ (V cm^{-1})	$E_0\times 10^{-4}$	Γ
2.8	−3.567 310 24	4.62×10^{-13}	3.4	−2.711 532 7	7.71×10^{-9}
3.0	−3.278 604 97	2.24×10^{-11}	3.6	−2.433 816 5	6.429×10^{-8}
3.2	−2.993 276 19	5.53×10^{-10}	3.8	−2.160 630 2	3.284×10^{-7}

Table 2.8. Values of E_0 and Γ for $n=25$, $n_1=12$, $n_2=11$, $m=1$

$F\times10^7$	$E_0\times10^4$	Γ	$F\times10^7$	$E_0\times10^4$	Γ
3.2	−8.169 285 54	1.31×10^{-14}	3.8	−8.288 351 70	6.856×10^{-9}
3.4	−8.204 083 91	2.424×10^{-12}	4.0	−8.340 794 2	1.1789×10^{-7}
3.6	−8.243 434 47	1.891×10^{-10}	4.2	−8.403 187	8.8967×10^{-7}

Table 2.9. Values of E_0 and Γ for $n=30$, $n_1=0$, $n_2=29$, $m=0$

$F\times10^{-2}$ (V cm^{-1})	$E_0\times10^4$	Γ	$F\times10^{-2}$ (V cm^{-1})	$E_0\times10^4$	Γ
6.4	−7.318 560 53	8×10^{-16}	7.2	−7.571 615 93	1.544×10^{-10}
6.6	−7.380 577 22	2.59×10^{-14}	7.4	−7.637 371 23	1.5748×10^{-9}
6.8	−7.443 361 69	6.35×10^{-13}	7.6	−7.704 531 46	1.2116×10^{-8}
7.0	−7.507 002 17	1.145×10^{-11}	7.8	−7.773 497 1	6.872×10^{-8}

Table 2.10. Values of E_0 and Γ for $n=30$, $n_1=29$, $n_2=0$, $m=0$

$F\times10^7$	$E_0\times10^4$	Γ	$F\times10^7$	$E_0\times10^4$	Γ
2.8	−2.262 372 51	3.8×10^{-13}	3.2	−1.836 896 95	7.746×10^{-10}
2.9	−2.155 043 63	3.5×10^{-12}	3.3	−1.732 202 75	3.323×10^{-9}
3.0	−2.048 344 99	2.53×10^{-11}	3.4	−1.628 257 9	1.216×10^{-8}
3.1	−1.942 289 06	1.529×10^{-10}	3.5	−1.525 128 6	3.811×10^{-8}

Table 2.11. Values of E_0 and Γ for $n=40$, $n_1=39$, $n_2=0$, $m=0$

$F\times10^{-2}$ (V cm^{-1})	$E_0\times10^4$	Γ	$F\times10^{-2}$ (V cm^{-1})	$E_0\times10^4$	Γ
5.0	−1.085 934 66	3.81×10^{-13}	5.5	−0.901 497 304	2.106×10^{-10}
5.1	−1.048 791 02	1.58×10^{-12}	5.6	−0.865 008 694	5.935×10^{-10}
5.2	−1.011 773 42	6.05×10^{-12}	5.7	−0.828 663 990	1.558×10^{-9}
5.3	−0.974 883 549	2.13×10^{-11}	5.8	−0.792 471 07	3.816×10^{-9}
5.4	−0.938 123 754	6.956×10^{-11}	5.9	−0.756 439 38	8.715×10^{-9}

time. Those results for the numerical calculation for a number of other states and electric field strengths obtained by the same method have been previously published.[24, 33-35]

2.3. Perturbation theory

The Stark effect was the first application of the perturbation theory (PT) in quantum mechanics.[1] By using the first terms of PT, many qualitative features of the Stark effect may be successfully explained. At the same time, to understand what quantitative results can be obtained by PT, a large number of PT terms are required. Though the Stark problem can be solved exactly, it does not, for different reasons, deaden the interest in PT. First, the use of sufficiently accurate input data obtained by PT considerably simplifies exact numerical calculations [see Sect. 2.2c]. On the other hand, as will be shown in this section, perturbation series (or the sequences of diagonal Padé approximants) allow the Stark energy to be determined with an accuracy comparable to the level width. For some practical purposes, such accuracy appears to be sufficient. Moreover, it is either difficult or impossible to obtain exact numerical results for the majority of problems in atomic and molecular physics, and, therefore, to solve these problems, PT and, more recently, Padé approximations have been used.[36] The analysis given in this section may appear to be instructive also for these cases, especially with respect to the application of high-order PT.

2.3a. Analytical derivation of perturbation terms

To derive terms of perturbation series for the Stark energy of hydrogen, i.e., to obtain the series in F for E_0,

$$E_{\text{pert}}(N) = \sum_{i=0}^{N} E^{(i)} = \sum_{i=0}^{N} a_i F^i \tag{25}$$

parabolic coordinates are commonly used because in these coordinates the perturbation matrix is diagonal with respect to each group of mutually degenerate states.[1-3] When calculating expansion terms (25) by means of general PT, i.e., using matrix elements of the perturbation operator Fz, starting from the second term, infinite sums of complicated form need to be dealt with. Therefore, it is more convenient here to use a somewhat modified procedure.

In Eqs. (3) and (4), by changing variables ξ and η to σ and ρ

$$\sigma = (-2E)^{1/2}\xi, \qquad \rho = (-2E)^{1/2}\eta \tag{26}$$

and introducing new parameters

$$R = (-2E)^{3/2}F^{-1}, \qquad \lambda_1 = \beta_1(-2E)^{-1/2}, \qquad \lambda_2 = \beta_2(-2E)^{-1/2} \tag{27}$$

the following equations for $V(\sigma)$ and $U(\rho)$ are obtained:

$$\frac{d^2V}{d\sigma^2} + \left(\frac{1-m^2}{4\sigma^2} + \frac{\lambda_1}{\sigma} - \frac{1}{4} - \frac{\sigma}{4R}\right)V = 0 \tag{28}$$

$$\frac{d^2U}{d\rho^2} + \left(\frac{1-m^2}{4\rho^2} + \frac{\lambda_2}{\rho} - \frac{1}{4} + \frac{\rho}{4R}\right)U = 0 \tag{29}$$

where the energy is determined from the separation constants by

$$E = -\tfrac{1}{2}(\lambda_1 + \lambda_2)^{-2} \tag{30}$$

When calculating the Stark energy using PT, both Eqs. (28) and (29) are actually treated as one-dimensional Schrödinger equations with discrete bound states. Because the essential difference between (28) and (29) is the sign of R, it is sufficient to consider only one, say, Eq. (28). Supposing that

$$V(\sigma) = e^{-\sigma/2}\sigma^{(m+1)/2}f(\sigma) \tag{31}$$

we obtain the following equation for $f(\sigma)$:

$$\sigma f'' + (m + 1 - \sigma)f' + [\lambda_1 - \tfrac{1}{2}(m + 1)]f = (\sigma^2/4R)f \tag{32}$$

Assuming that F is sufficiently small, then R is a large parameter. Therefore it is natural to seek the solution of Eq. (32) in the form of a power series expansion of $1/R$.

$$f = \sum_{l=0} f^{(l)}R^{-l} \tag{33}$$

$$\lambda_1 = \sum_{l=0} \lambda_1^{(l)}R^{-l} \tag{34}$$

In zero approximation, the requirement of the finiteness of $f(\sigma)$ at $\sigma \to \infty$ leads to generalized Laguerre polynomials:

$$f^{(0)} = L^m_{n_1}(\sigma), \quad \lambda_1^{(0)} = n_1 + \tfrac{1}{2}(m + 1) \tag{35}$$

Similarly, we seek the next orders of $f(\sigma)$ in the form of linear, generalized Laguerre polynomial combinations

$$f^{(l)} = \sum_{k=-2l}^{2l} C_k^{(l)}L^m_{n_1+k} \tag{36}$$

Inserting (33), (34) and (36) into (32), we obtain the following recurrence relations:

$$\begin{aligned} kC_k^{(l)} = -\tfrac{1}{4}\{&(n_1+k)(n_1+k-1)C_{k-2}^{(l-1)} - 2(n_1+k)(2n_1+m+2k)C_{k-1}^{(l-1)} \\ &+ [6n_1^2+6n_1m+6n_1+m^2+3m+2+6k(2n_1+m+k+1)]C_k^{(l-1)} \\ &-2(n_1+m+k+1)(2n_1+m+2k+2)C_{k+1}^{(l-1)} \\ &+(n_1+m+k+2)(n_1+m+k+1)C_{k+2}^{(l-1)}\} + \sum_{i=1}^{l-j} \lambda_1^{(i)}C_k^{(l-i)} \end{aligned} \tag{37}$$

where

$$j = \begin{cases} \tfrac{1}{2}|k| & \text{even values } k \\ \tfrac{1}{2}(|k| + 1) & \text{odd values } k \end{cases}$$

From (36) it follows that $C_k^{(l)} = 0$ at $|k| > 2l$. In (37) it is convenient to choose $C_0^{(l)} = \delta_{l0}$, where δ_{l0} is the Kronecker delta.

At $k=0$, from (37) we obtain an expression allowing the definition of $\lambda_1^{(l)}$ by the coefficients $C_i^{(l-1)}$:

$$\lambda_1^{(1)} = \tfrac{1}{4}(6n_1^2 + 6n_1m + 6n_1 + m^2 + 3m + 2)$$

$$\lambda_1^{(l)} = \tfrac{1}{4}\{n_1(n_1-1)C_{-2}^{(l-1)} - 2n_1(2n_1+m)C_{-1}^{(l-1)}$$
$$- 2(n_1+m+1)(2n_1+m+2)C_1^{(l-1)}$$
$$+ (n_1+m+2)(n_1+m+1)C_2^{(l-1)}\}, \qquad l>1 \tag{38}$$

In the second order, for example, we obtain

$$\lambda_1^{(2)} = \tfrac{1}{16}(68n_1^3 + 102n_1^2 + 102n_1^2 m + 70n_1 + 102n_1 m$$
$$+ 42n_1 m^2 + 4m^3 + 21m^2 + 35m + 18) \tag{39}$$

When solving (29) we obtain a similar expression for $\lambda_2^{(l)}$:

$$\lambda_2^{(l)}(n_2,m) \leftrightarrow (-1)^l \lambda_1^{(l)}(n_1,m)$$

Substituting $\lambda_1^{(l)}$ and $\lambda_2^{(l)}$ into (30) and solving by the method of successive approximations, we obtain the series in F for the Stark energy:

$$E_{\text{pert}} = -\frac{1}{2n^2} + \frac{3}{2}n(n_1-n_2)F - \frac{n^4}{16}[17n^2 - 3(n_1-n_2)^2 - 9m^2 + 19]F^2$$
$$+ \frac{3}{32}n^7(n_1-n_2)[23n^2 - (n_1-n_2)^2 + 11m^2 + 39]F^3$$
$$- \frac{n^{10}}{1024}[5487n^4 + 35\,182n^2 - 1134m^2(n_1-n_2)^2$$
$$+ 1806n^2(n_1-n_2)^2 - 3402n^2m^2 + 147(n_1-n_2)^4 - 549m^4$$
$$+ 5754(n_1-n_2)^2 - 8622m^2 + 16\,211]F^4$$
$$+ \frac{3}{1024}n^{13}(n_1-n_2)[10\,563n^4 + 90\,708n^2 + 220m^2(n_1-n_2)^2$$
$$+ 98n^2(n_1-n_2)^2 + 772n^2m^2 - 21(n_1-n_2)^4 + 725m^4$$
$$+ 780(n_1-n_2)^2 + 830m^2 + 59\,293]F^5 + O(F^6) \tag{40}$$

2.3b. High-order PT approximation

Analytical derivation "by hand" of the first two terms in perturbation series for the Stark energy [Eq. (40)] presents no difficulties. Derivation of the third term requires some effort, and calculation of the fourth term is a tedious algebraic procedure.[37] The problem, however, can be easily computerized. In Ref. 38 Silverstone, using a method similar to that described in Sect. 2.3a, has worked out a procedure allowing the generation of "the explicit formulas for the perturbed separation constant and energy in the Stark effect in hydrogen as polynomials in the quantum numbers for arbitrary high orders" (p. 1862). With this procedure Silverstone obtained polynomial coefficients for separation constants and energy up to the seventeenth order. A simpler problem is the calculation of terms in perturbation series for the Stark energy in the case of certain states defined by a set of quantum numbers n_1, n_2, m. Using recurrence relations (37)–(38), the necessary number of terms in expansion (34) can be obtained. Some difficulties arise only when getting the series in F for the Stark energy from expansions of λ_1 and λ_2 in power series of $1/R$.[38] In Ref. 39, to calculate high-order terms of the Stark energy, the authors passed over from Schrödinger to Rikkati equations by substitutions in Eqs. (3) and (4):

$$X_1(\xi) = -\frac{2\xi^{1/2}}{U}\left(\frac{U}{\xi^{1/2}}\right)', \quad X_2(\eta) = -\frac{2\eta^{1/2}}{V}\left(\frac{V}{\eta^{1/2}}\right)'$$

Then, by substituting expansions in F for $E, \beta_1, \beta_2, X_1, X_2$ into the equations obtained, the authors obtained simple recurrence relations for calculating corresponding perturbation series for the ground state of the hydrogen atom. Another method that allowed the cumbersome procedure to be avoided when obtaining E_{pert} from expansions of λ_1 and λ_2 in powers of $1/R$ was suggested in Ref. 40. This method, in contrast to that described in Ref. 39, can be applied also for highly excited states.

2.3c. Comparison between PT and numerical results

In this subsection some conclusions will be given based on the comparison between the results of PT application and data of exact numerical calculations obtained by the method given in Sect. 2.2c. This analysis of phenomenological character rests upon extensive numerical data and will be illustrated by some typical examples.

The asymptotic nature of PT[3] gives rise to the idea that, when calculating Stark energies, perturbation series (25) should be terminated just before the term that is smallest in magnitude. For all cases considered, this rule appears to be reasonable.

The most important characteristics determining the behavior of perturbation series (25) is the electric quantum number $k = n_2 - n_1$. For $k \geqslant 0$, all terms in series (25) have a negative sign. In this case series termination before the term smallest in magnitude gives E_{pert} differing from the exact value of E_0 by not more than half of the level width Γ. Here the number of terms that need to be included in (25) for different states and electric fields corresponding to equal ionization probabilities grows with the increase of k. Therefore, in some cases when calculating Stark energies of highly excited states, it is reasonable to include a very large number of terms of the perturbation series.

For $k<0$, series (25) is oscillating. At small $|k|$, the previously mentioned procedure still enables E_0 to be determined for fields of experimental interest with an accuracy comparable to the value of Γ. However, with increase of $|k|$, the error of energy determination by using (25) increases, and for states with $|k| \simeq n \gg 1$, it may exceed the value of the level width by many orders of magnitude. Here the ratio of the error to the value of Γ for each state increases with the decrease of Γ. The error may be decreased by more than an order of magnitude if the energy is determined by

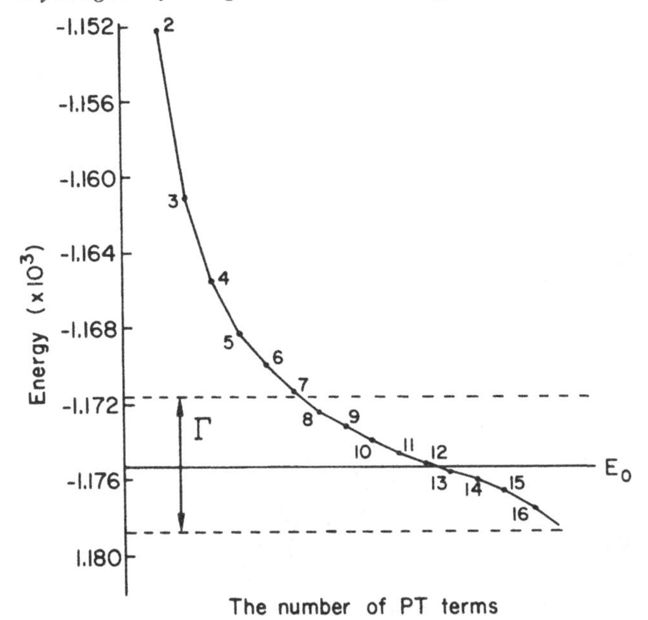

Fig. 2.1. Perturbation theory and exact numerical calculation for the hydrogen atom at $n=25$, $n_1=0$, $n_2=24$, $m=0$; $F=3.6\times10^{-7}$, $E_0=-0.001\,175\,31$, $\Gamma=7\times10^{-6}$.

$$E_{av} = \tfrac{1}{2}[E_{pert}(N-1) + E_{pert}(N)] \tag{41}$$

where N is the number of the smallest magnitude term in (25). In these cases, much more accurate results for E_0 can be achieved by using diagonal Padé approximants. To illustrate this, some examples will be given.

Figure 2.1 illustrates states with $k>0$. Perturbation terms monotonically decrease at first and then start to increase. The fourteenth term is the smallest one in series (25); $E_{pert}(13)$ lies inside the level width Γ.

A similar example for the state $n=30$, $n_1=0$, $n_2=29$, $m=0$ ($F=800$ $V\,\text{cm}^{-1}$) has been analyzed in Ref. 41. The authors terminate series (25) at $N=25$ because of roundoff error in the twenty-fifth term. The value of $E_{pert}(24)$ lies outside the level width Γ. Our calculation, however, shows that the twenty-fifth term in (25) is not the smallest one. Terms in (25) are decreasing at least up to the thirtieth term; $E_{pert}(29)$ is already in the limits of Γ. For this case, we could not attain the smallest term because of computation difficulties. The latter example illustrates a peculiarity characteristic of the high-order PT application: the necessity of high computer capacity with a large memory core. Paradoxical as it seems, the exact numerical calculation described in Sect. 2.2c has no such requirements on the computer.

For the case shown in Fig. 2.2, the smallest PT term is the thirteenth. Series terms for highly excited states with small positive and negative k at fields F that are not too small are changing nonmonotonically with increasing i. It is

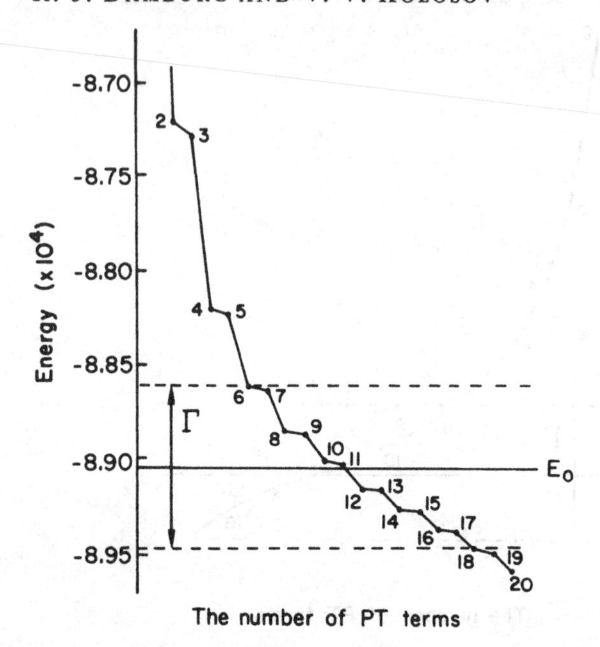

Fig. 2.2. Perturbation theory and exact numerical calculation for the hydrogen atom at $n=25$, $n_1=11$, $n_2=12$, $m=1$; $F=4.6\times10^{-7}$, $E_0=-0.000\,890\,424$, $\Gamma=8.567\times10^{-6}$.

as if series (25) consisted of two series with terms that essentially differed in magnitude. In each series taken separately, the corresponding terms are changing monotonically. This peculiarity of perturbation series behavior probably escaped Silverstone's attention.[38] The procedure of a "maximum useful field strength"[38] practically results in termination of series (25) before the first increasing term, which essentially decreases the accuracy of E_0 determination for the whole set of states.

For the case shown in Fig. 2.3, the thirteenth term is the smallest in magnitude in series (25). The value of k is negative but small, and termination of series (25) before the smallest, thirteenth term gives $E_{\text{pert}}(12)$ within the limits of Γ.

Figure 2.4 serves to illustrate a state with a large negative k. Here the fifth term in series (25) is the smallest in magnitude. Strictly speaking, the PT approximation proves only that the Stark energy is within the limits $-3.319\,273\,82\times10^{-4}<E_0<-2.976\,138\,95\times10^{-4}$. However, this estimate is too rough because Γ in this case is small. Formula (41) improves the result: $E_{\text{av}}=-3.171\,307\,29\times10^{-4}$.

2.3d. Padé approximation

The example shown in Fig. 2.4 is typical for states with large negative k. Though the coefficients a_i in (25) contain essential information on the func-

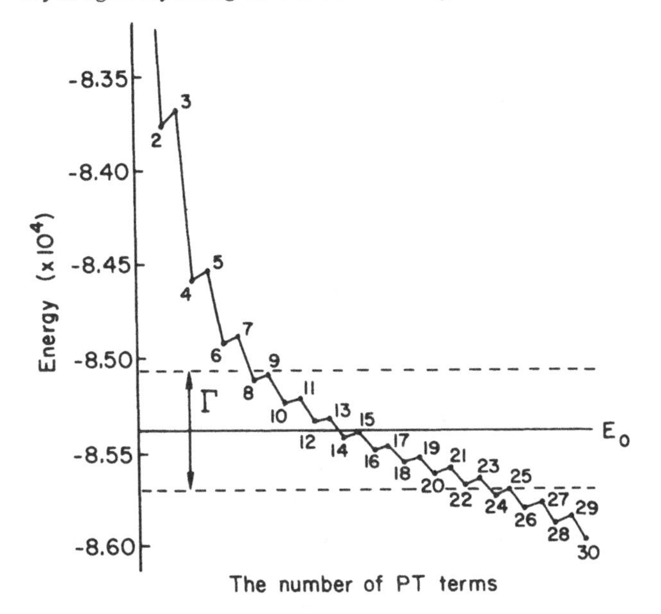

Fig. 2.3. Perturbation theory and exact numerical calculation for the hydrogen atom at $n=25$, $n_1=12$, $n_2=11$, $m=1$; $F=4.6\times10^{-7}$, $E_0=-0.000\,853\,866$, $\Gamma=6.358\times10^{-6}$.

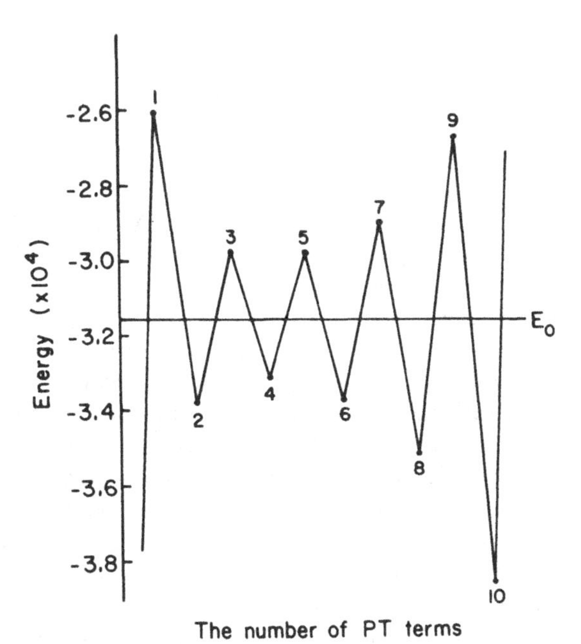

Fig. 2.4. Perturbation theory and exact numerical calculation for the hydrogen atom at $n=25$, $n_1=24$, $n_2=0$, $m=0$; $F=6\times10^{-7}$, $E_0=-0.000\,315\,641\,108$, $\Gamma=9.48\times10^{-11}$.

tion $E_0(F)$, one cannot by the direct application of (25) determine the Stark energy with an accuracy comparable to the level width. Much better results can be obtained by using diagonal Padé approximants based on the a_i coefficient.

The Padé approximant $[L,M]$ to $E_0(F)$ is the ratio of two polynomials of degrees L and M:

$$[L,M] = P^{[L]}(F)/Q^{[M]}(F)$$

whose power series agrees with (25) through order $L+M$:[42]

$$[L,M] - E_{\text{pert}}(L+M) = O(F^{L+M+1})$$

It is possible to calculate $[L,M]$ by the recurrence relations

$$[L,M+1] = [L,M]+\{([L+1,M]-[L,M])^{-1} + ([L-1,M]-[L,M])^{-1}-([L,M-1]-[L,M])^{-1}\}^{-1} \quad (42)$$

which determine $[L,M]$ via $[L,O]\equiv E_{\text{pert}}(L)$.

There is an explicit formula for the diagonal Padé approximant $[N,N]$:

$$[N,N] = \frac{\begin{vmatrix} a_1 & a_2 & \cdots & a_{N+1} \\ \vdots & \vdots & \ddots & \vdots \\ a_N & a_{N+1} & \cdots & a_{N+N} \\ E_{\text{pert}}(0) & E_{\text{pert}}(1) & \cdots & E_{\text{pert}}(N) \end{vmatrix}}{\begin{vmatrix} a_1 & a_2 & \cdots & a_{N+1} \\ \vdots & \vdots & \ddots & \vdots \\ a_N & a_{N+1} & \cdots & a_{N+N} \\ 1 & 1 & \cdots & 1 \end{vmatrix}} \quad (43)$$

The comparison between the numerical values of $E_{\text{pert}}(N)$ and $[N/2,N/2]$ for the case shown in Fig. 2.4 is given in Table 2.12. Table 2.13 illustrates another typical example. Although the Padé approximation for the two states given in Tables 2.12 and 2.13 is very effective, we would not recommend its use in all cases. Without dwelling upon the general problem of the convergence of diagonal Padé approximants in a strict mathematical sense,[43–45] we just mention that in the case when the direct application of (25) allows the determination of the Stark energy with the accuracy comparable to the level width (see Figs. 2.1–2.3), the Padé approximants (43) have no evident advantages over ordinary PT. On the contrary, in some cases, especially at small $|k|$, the convergence of the Padé approximants can be unsatisfactory. A typical example is given in Table 2.14 and in Fig. 2.5.

Thus, PT [Eq. (25)] and the Padé approximation [Eq. (43)] are, to a great extent, complementary forms that can be used for different cases in the manner just explained.

Table 2.12. Perturbation theory $E_{pert}(N)$ and diagonal Padé approximation $[N/2, N/2]$ for the energy of the hydrogen atom at $n=25$, $n_1=24$, $n_2=0$, $m=0$; $F=6\times10^{-7}$, $E_{exact}=-0.000315641108$, $\Gamma=9.48\times10^{-11}$

N	$E_{pert}(N)\times10^4$	$[N/2, N/2]\times10^4$	N	$E_{pert}(N)\times10^4$	$[N/2, N/2]\times10^4$
1	−2.6	—	15	2.4780106	—
2	−3.3836328	−3.2843254	16	−12.119789	−3.1564033
3	−2.9731550	—	17	11.269141	—
4	−3.3192738	−3.1566464	18	−26.618116	−3.1564016
5	−2.9761389	—	19	35.370724	—
6	−3.3664756	−3.1528810	20	−66.985744	−3.1564439
7	−2.8891732	—	22	—	−3.1564100
8	−3.5100561	−3.1451223	24	—	−3.1564072
9	−2.6673330	—	26	—	−3.1564096
10	−3.8526645	−3.1563710	28	—	−3.1564099
11	−2.1379858	—	30	—	−3.1564096
12	−4.6782064	−3.1562953	32	—	−3.1564100
13	−0.8387261	—	34	—	−3.1564100
14	−6.7440200	−3.1560264			

Table 2.13. Perturbation theory $E_{pert}(N)$ and diagonal Padé approximation $[N/2, N/2]$ for the energy of the hydrogen atom at $n=12$, $n_1=8$, $n_2=1$, $m=2$; $F=7.77907429\times10^{-6}$ (40 kV cm^{-1}), $E_{exact}=-0.0026570738$, $\Gamma=2.15\times10^{-8}$

N	$-E_{pert}(N)\times10^3$	$[N/2, N/2]\times10^3$	N	$-E_{pert}(N)\times10^3$	$[N/2, N/2]\times10^3$
1	2.49206	—	11	2.65356	—
2	2.67118	2.64351	12	2.66032	2.65701
3	2.63415	—	13	2.65347	—
4	2.66293	2.64299	14	2.66070	2.65700
5	2.64835	—	15	2.65301	—
6	2.66110	2.65706	16	2.66141	2.65706
7	2.65196	—	17	2.65213	—
8	2.66042	2.65655	18	2.66258	2.65706
9	2.65319	—	19	2.65068	—
10	2.66021	2.65731	20	2.66444	2.65706

2.4. Asymptotic expansion for Γ

2.4a. Boundary conditions

In Sect. 2.3a, the asymptotic expansion for the Stark energy E_0 in F was derived by PT without the explicit use of boundary conditions (7) and (8) for Eqs. (3) and (4). Equations (3) and (4) were actually treated as the problem on bound states. Obtaining an asymptotic expansion for the level width Γ is a

Table 2.14. Diagonal Padé approximants $[N,N]$ for the energy of the hydrogen atom at $n=7$, $n_1=3$, $n_2=2$, $m=1$; $F=6.5\times10^{-5}$, $E_{\text{exact}}=-0.010\,140\,55$, $\Gamma=3.597\times10^{-5}$

N	$-[N,N]\times10^2$	N	$-[N,N]\times10^2$
1	0.982 072	10	1.012 55
2	1.012 58	11	1.004 18
3	1.011 24	12	1.020 93
4	1.015 08	13	1.015 45
5	1.014 16	14	1.014 69
6	1.009 72	15	1.013 35
7	1.024 48	16	1.012 76
8	1.015 27	17	1.006 78
9	1.013 97	18	0.989 41

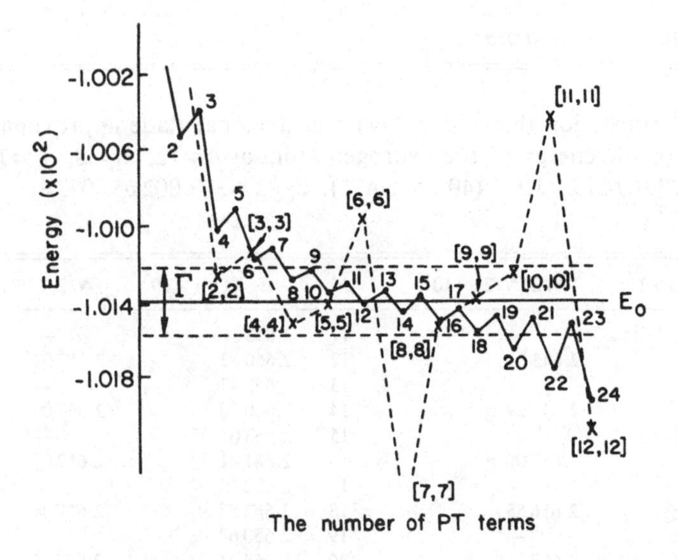

Fig. 2.5. Perturbation theory, diagonal Padé approximants, and exact numerical calculation for the hydrogen atom at $n=7$, $n_1=3$, $n_2=2$, $m=1$; $F=6.5\times10^{-5}$, $E_0=-0.010\,140\,55$, $\Gamma=3.597\times10^{-5}$. Dots: $E_{\text{pert}}(N)$; crosses: $[N/2, N/2]$.

much more delicate mathematical problem. For its solution, it is necessary to bear in mind that Eq. (4) with boundary condition (8) corresponds to the continuum spectrum problem, i.e., to the scattering problem. Therefore, the solution of Eq. (4) should be modified. The explicit use of boundary condition (7) when solving Eq. (4) would not change the eigenvalues of β_1 obtained in Sect. 2.3a. Therefore, to derive the asymptotic expansion for Γ, we may use the values of β_1 obtained by PT.

2.4b. Solution of Eq. (29)

It is convenient to seek the solution of Eq. (29), which is actually another form of Eq. (4) in two regions. The first region, $0 \leqslant \rho < \rho_1$, where $\rho_1 \ll R$, includes the origin $\rho = 0$. The second region, $\rho_0 < \rho < \infty$, where $\rho_0 < \rho_1$, includes the outer turning point. Solutions thus obtained can be linked in the matching region $\rho_0 < \rho < \rho_1$, where both of them are valid. This procedure allows us to determine the asymptotic expansion for Γ.

Region $0 \leqslant \rho < \rho_1$

We seek the solution of Eq. (29) in the region $0 \leqslant \rho < \rho_1$ in the form similar to that of Eq. (28) in Sect. 2.3a:

$$U(\rho) = Ce^{-\rho/2}\rho^{(m+1)/2}f(\rho) \tag{44}$$

where C is a constant.

Substituting (44) into (29) and imposing the requirement of the finiteness of $f(\rho)$ as $\rho \to \infty$ in zero order, we obtain

$$f^{(0)} = L_k^m(\rho), \qquad \lambda_2^{(0)} = k + \tfrac{1}{2}(m+1) \tag{45}$$

Similarly, in the next order,

$$\begin{aligned} f^{(1)} &= -\tfrac{1}{8}(k+m)(k+m-1)L_{k-2}^m + \tfrac{1}{2}(k+m)(2k+m)L_{k-1}^m \\ &\quad - \tfrac{1}{2}(k+1)(2k+m+2)L_{k+1}^m + \tfrac{1}{8}(k+1)(k+2)L_{k+2}^m \\ \lambda_2^{(1)} &= -\tfrac{1}{4}(6k^2 + 6km + 6k + m^2 + 3m + 2) \end{aligned} \tag{46}$$

The derivation of formulae (45) and (46) implies that the Schrödinger equation (29) has discrete eigenvalues and that $k = n_2$ is a positive integer or zero. As shown in Sect. 2.3a such an approach to Eqs. (28) and (29) allows the asymptotic expansion for E_0 to be obtained. However, even in the presence of an infinitesimal uniform electric field, the spectrum of eigenvalues becomes continuous. Nevertheless, when F is sufficiently small, E_0 still has a clear physical meaning as a center of the band into which the discrete atomic level is evolving in the presence of the electric field. To describe the band, we abandon the strict condition $k = n_2$ and consider k to be a variable in a narrow interval $n_2 \pm \Delta$, where, as will be shown later, $\Delta \simeq \exp(-\tfrac{2}{3}R)$. The solution given for Eq. (44) would satisfy Eq. (29) if we insert $n_2 + \delta$ (where δ is an exponentially small variable) for n_2 and replace the generalized Laguerre polynomials by confluent hypergeometric functions. The latter are further expanded in powers of δ:

$$\begin{aligned} L_{n_2}^m(\rho) &\xrightarrow[n_2 \to n_2 + \delta]{} \frac{\Gamma(n_2 + m + 1 + \delta)}{\Gamma(n_2 + 1 + \delta)\Gamma(m+1)} F(-n_2 - \delta, m+1, \rho) \\ &= L_{n_2}^m(\rho) + \delta \frac{(n_2 + m)!}{n_2!\,m!} Q(-n_2, m+1, \rho) + O(\delta^2) \end{aligned} \tag{47}$$

where

$$Q(-n_2, m+1, \rho) = \{\partial F(-n_2 - \delta, m+1, \rho)/\partial\delta\}_{\delta=0} \tag{48}$$

In the matching region $\rho_0 < \rho < \rho_1$, we are to use the asymptotic expansion for (47). The asymptotic behavior of $L^m_{n_2}(\rho)$ at large ρ is well known. The asymptotic behavior of function (48) is determined by

$$Q(-n_2, m+1, \rho) \underset{\rho\to\infty}{\simeq} (-1)^{n_2+1} \frac{m!\, e^{\rho}}{(n_2+m)!\, \rho^{n_2+m+1}} \sum_{l=0} \frac{(n_2+l)!(n_2+m+l)!}{l!\rho^l} \tag{49}$$

The use of asymptotic expansion (49) in the matching procedure results in the intrinsic error $\sim\delta^2$; therefore it is useless to consider the terms $\sim\delta^2$ in expansion (47). A more detailed discussion on this subject for a similar problem is given in Ref. 46.

Region $\rho_0 < \rho < \infty$

Introducing a new variable

$$x = 1 - (\rho/R) \tag{50}$$

into Eq. (29), we obtain

$$\frac{d^2U}{dx^2} + \left(-\frac{R^2}{4}x + \frac{\lambda_2 R}{1-x} + \frac{1-m^2}{4(1-x)^2}\right)U = 0 \tag{51}$$

To obtain the solution of Eq. (51), the comparison equations method is used.[47] An Airy equation is chosen as the comparison equation,

$$(d^2W/dy^2) - yW = 0 \tag{52}$$

We seek the solution of Eq. (51) in the form

$$U(x) = (dy/dx)^{-1/2}W(y) \tag{53}$$

Inserting (53) into (51) and taking (52) into account, we obtain, for $y(x)$,

$$(dy/dx)^2 y = \frac{R^2 x}{4} - \frac{\lambda R}{1-x} - \frac{1-m^2}{4(1-x)^2} + \frac{1}{2}\langle y;x\rangle \tag{54}$$

where

$$\langle y;x\rangle = \frac{y'''}{y'} - \frac{3}{2}\left(\frac{y''}{y'}\right)^2 \tag{55}$$

Thus the solution of Eq. (51) in the region $-\infty < x < 1$ including the outer turning point $x=0$ is

$$U = a\,(dy/dx)^{-1/2}\,\mathrm{Ai}(y) + b\,(dy/dx)^{-1/2}\,\mathrm{Bi}(y) \tag{56}$$

where y is the solution of Eq. (54), Ai and Bi the independent solutions of the

Airy equation (52), and a and b the coefficients determined by matching (56) and (44) in the region $\rho_0 < \rho < \rho_1$.

The integration of (54) yields

$$\int_{y_0}^{y} y^{1/2}\,dy = \frac{R}{2}\int_{x_0}^{x} x^{1/2}\left(1 - \frac{4\lambda_2}{Rx(1-x)} - \frac{1-m^2}{R^2x(1-x)^2} + \frac{2}{R}\langle y;x\rangle\right)^{1/2} dx \tag{57}$$

where $y_0 = y(x_0)$.

If $R \gg 16\lambda_2$, the integrand in (57) can be expanded in power series of $1/R$. By integrating the expression obtained, we determine y in the region $0<x<1$. For λ_2, we take its expansion in power series of $1/R$ obtained when deriving E_{pert}. Finally, for the region $0<x<1$, we obtain

$$\frac{2}{3}y^{3/2} = \frac{R}{3}x^{3/2} - \frac{2n_2+m+1}{2}\ln\left(\frac{1+x^{1/2}}{1-x^{1/2}}\right) + \frac{1}{2R}\left(\frac{(2n_2+m+1)^2}{x^{1/2}} - \frac{[2n_2^2+(2n_2+1)(m+1)]x^{1/2}}{1-x}\right) + O\left(\frac{1}{R^2}\right) \tag{58}$$

Using the asymptotic expansion for $\mathrm{Ai}(y)$ and $\mathrm{Bi}(y)$ for $y\to\infty$ and taking into account (58), we obtain, for $0<x<1$,

$$\begin{aligned} U(x) = {} & \frac{a}{2}x^{-1/4}\left(\frac{1+x^{1/2}}{1-x^{1/2}}\right)^{(2n_2+m+1)/2} \\ & \times \exp\left(-\frac{R}{3}x^{3/2}\right)\left[1 - \frac{1}{2R}\left(\frac{5}{12}x^{-3/2} + (2n_2+m+1)^2x^{-1/2}\right.\right. \\ & \left.\left. - \frac{2n_2+m+1}{x(1-x)} - \frac{(2n_2^2+2n_2m+2n_2+m+1)x^{1/2}}{1-x}\right) + O\left(\frac{1}{R^2}\right)\right] \\ & + bx^{-1/4}\left(\frac{1+x^{1/2}}{1-x^{1/2}}\right)^{-(2n_2+m+1)/2} \\ & \times \exp\left(\frac{R}{3}x^{3/2}\right)\left[1 + \frac{1}{2R}\left(\frac{5}{12}x^{-3/2} + (2n_2+m+1)^2x^{-1/2}\right.\right. \\ & \left.\left. + \frac{2n_2+m+1}{x(1-x)} - \frac{(2n_2^2+2n_2m+2n_2+m+1)x^{1/2}}{1-x}\right) + O\left(\frac{1}{R^2}\right)\right] \end{aligned} \tag{59}$$

Similarly, if we use the asymptotic forms[48] of the Airy functions for $y\to-\infty$, we obtain the asymptotic solution of (56) for $x\to-\infty$:

$$\begin{aligned} U(x) \simeq {} & a|x|^{-1/4}\sin\left[\tfrac{1}{3}R|x|^{3/2} + (2n_2+m+1)\tan^{-1}|x|^{1/2} + \tfrac{1}{4}\pi\right] \\ & + b|x|^{-1/4}\cos\left[\tfrac{1}{3}R|x|^{3/2} + (2n_2+m+1)\tan^{-1}|x|^{1/2} + \tfrac{1}{4}\pi\right] \end{aligned} \tag{60}$$

If the variable η is introduced, expression (60) coincides with (8) under the conditions

$$B = (a^2 + b^2)^{1/2} \tag{61}$$

and

$$\Phi = \tfrac{1}{4}(4n_2 + 2m + 3)\pi + \tan^{-1}(b/a) \tag{62}$$

Matching procedure

Function (44), with both exponentially descending and ascending parts, and function (59) are different forms of the solution of Eq. (29). Both functions are valid in the matching region $\rho_0 < \rho < \rho_1$. To match them together, we return to the variable ρ in (59) and perform the corresponding expansions under the conditions $\rho \ll R$ that allows us to express coefficients a and b in terms of c and δ. Because the matching procedure is rather cumbersome, we give here only the results:

$$b = C\frac{(-1)^{n_2}(4R)^{n_2+(m+1)/2}}{e^{R/3}n_2!} \times\left[1 - \frac{1}{8R}\left(34n_2^2 + 34n_2 m + 46n_2 + 7m^2 + 23m + \frac{53}{3}\right) + O\left(\frac{1}{R^2}\right)\right] \tag{63}$$

$$a = C\frac{(-1)^{n_2+1}e^{R/3}(n_2+m)!\,2\delta}{(4R)^{n_2+(m+1)/2}} \times\left[1 + \frac{1}{8R}\left(34n_2^2 + 34n_2 m + 22n_2 + 7m^2 + 11m + \frac{17}{3}\right) + O\left(\frac{1}{R^2}\right)\right] \tag{64}$$

Asymptotic formulae for Γ

It has been mentioned that δ is introduced to describe the band into which discrete atomic levels are evolving in the presence of the electric field. Using Eq. (40), we may find the change of the energy corresponding to the change of k by the value of δ:

$$\Delta E = \frac{\partial E_0}{\partial n_2}\delta = \delta\left(\frac{1}{n^3} - \frac{3}{2}(2n_2 + m + 1)F + O(F^2)\right) \tag{65}$$

It can be seen from (61)–(64) that, for $\delta = 0$, i.e., at $E = E_0$, the asymptotic amplitude B is a minimum but the resonance part of the phase shift $\tan^{-1}(b/a)$ goes through $\frac{1}{2}\pi$. Thus, E_0 corresponds to the resonance energy of the quasi-stationary state.

Comparing (62) to the Breit–Wigner parametrization of the phase shift [Eq. (11)], we obtain the relation for the determination of Γ:

$$b/a = \Gamma/2(E_0 - E) \tag{66}$$

Inserting expressions (63) and (64) for coefficients a and b and the value $E_0 - E = \Delta E$ from (65) into (66), we obtain the following expression for Γ:[49, 50]

$$\Gamma = \frac{(4R)^{2n_2+m+1}e^{-2R/3}}{n^3 n_2!(n_2+m)!} \times \left(1 - \frac{n^3}{4}(34n_2^2 + 34n_2 m + 46n_2 + 7m^2 + 23m + \frac{53}{3})F + O(F^2)\right) \tag{67}$$

where

$$R = (-2E_0)^{3/2}/F \tag{68}$$

Expanding R in power series of F, we obtain

$$\Gamma = \left(\frac{4}{Fn^3}\right)^{2n_2+m+1} \exp\left[3(n_1 - n_2) - \frac{2}{3Fn^3}\right] \frac{1}{n^3 n_2!(n_2+m)!} \times \left(1 - \frac{n^3}{8}\left[36n(n_1 - n_2) - 21(n_1 - n_2)^2 + 17n^2 + 68n_2^2 + 68n_2 m + 92n_2 + 5m^2 + 46m + \frac{163}{3}\right]F + O(F^2)\right) \tag{69}$$

The first term in (69) for the ground state $n_1 = n_2 = m = 0$ has been given in Ref. 3. A correct expression for the first term in the case of arbitrary quantum states was first obtained in Ref. 51 because in an earlier-published paper[52] the multiplier $\exp 3(n_1 - n_2)$ in the corresponding formula was omitted. Later this term was repeatedly derived (see, for example, Ref. 53).

A comparison with the results of exact numerical calculations shows that formula (67), where R is calculated by (68) and no additional expansion in F is performed, has a wider range of applicability in F.[50] However, the simpler formula, Eq. (69), is useful for the analysis of the ionization probability curves considered as functions of the electric field F.[54]

For the case $n_2 = m = 0$, the next term of the asymptotic expansion for Γ has been obtained:[50]

$$\Gamma = \frac{4Re^{-2R/3}}{n^3}\left[1 - \frac{53n^3}{12}F - \frac{n^6}{8}\left(48n_1^2 + 210n_1 + \frac{3221}{36}\right)F^2 + O(F^3)\right] \tag{70}$$

or, if expanded in F,

$$\Gamma = \frac{4}{Fn^6}\exp\left(3n_1 - \frac{2}{3Fn^3}\right)\left[1 - \frac{n^3}{4}\left(16n_1^2 + 35n_1 + \frac{107}{3}\right)F + \frac{n^6}{96}\left(768n_1^4 + 3900n_1^3 + 7567n_1^2 + 7346n_1 + \frac{7363}{3}\right)F^2 + O(F^3)\right] \tag{71}$$

Table 2.15. Level widths for the hydrogen atom at $n_2=m=0$

n_1, n_2, m	F	Γ_1	Γ_2	Γ_3	Γ_{num}
0, 0, 0	1.6×10^{-2}	1.8990×10^{-16}	1.7369×10^{-16}	1.7316×10^{-16}	1.7316×10^{-16}
	3.0×10^{-2}	2.6100×10^{-8}	2.2642×10^{-8}	2.2379×10^{-8}	2.2376×10^{-8}
	5.0×10^{-2}	1.038×10^{-4}	8.087×10^{-5}	7.797×10^{-5}	7.717×10^{-5}
	8.0×10^{-2}	8.29×10^{-3}	5.36×10^{-3}	4.77×10^{-3}	4.51×10^{-3}
1, 0, 0	3.0×10^{-3}	2.44×10^{-10}	2.18×10^{-10}	2.12×10^{-10}	2.12×10^{-10}
	4.0×10^{-3}	1.672×10^{-7}	1.436×10^{-7}	1.361×10^{-7}	1.364×10^{-7}
	5.0×10^{-3}	7.609×10^{-6}	6.265×10^{-6}	5.736×10^{-6}	5.729×10^{-6}
4, 0, 0	1.5×10^{-4}	1.668×10^{-11}	1.530×10^{-11}	1.406×10^{-11}	1.431×10^{-11}
	2.0×10^{-4}	5.110×10^{-8}	4.546×10^{-8}	3.868×10^{-8}	4.025×10^{-8}
	2.5×10^{-4}	4.739×10^{-6}	4.085×10^{-6}	3.103×10^{-6}	3.272×10^{-6}

$\Gamma_1, \Gamma_2, \Gamma_3$: values calculated using one, two, and three terms in expansion (70). Γ_{num}: results of exact numerical calculations.

Table 2.16. Comparison of the results for level widths obtained by using asymptotic formulae (67) and (69) with exact numerical data

n_1, n_2, m	F	Γ [Eq. (69)][a]	Γ [Eq. (67)]	Γ_{num}
3, 0, 1	1.4×10^{-4}	9.67×10^{-11}	1.55×10^{-11}	1.43×10^{-11}
	1.6×10^{-4}	8.66×10^{-9}	1.07×10^{-9}	9.61×10^{-10}
	1.8×10^{-4}	2.78×10^{-7}	2.67×10^{-8}	2.30×10^{-8}
	2.0×10^{-4}	4.35×10^{-6}	3.24×10^{-7}	2.66×10^{-7}
9, 0, 0	1.4×10^{-5}	3.17×10^{-10}	1.70×10^{-12}	1.62×10^{-12}
	1.6×10^{-5}	1.07×10^{-7}	2.75×10^{-10}	2.63×10^{-10}
	1.8×10^{-5}	9.72×10^{-6}	1.19×10^{-8}	1.13×10^{-8}
	2.0×10^{-5}	3.55×10^{-4}	2.05×10^{-7}	1.90×10^{-7}
24, 0, 0	5.445×10^{-7}	5.21×10^{-5}	2.34×10^{-13}	4.62×10^{-13}
	5.834×10^{-7}	9.03×10^{-3}	9.79×10^{-12}	2.24×10^{-11}
	6.223×10^{-7}	8.18×10^{-1}	2.10×10^{-10}	5.53×10^{-10}

[a] In formula (69), only the first term of the asymptotic expansion has been used because use of the second term results in a negative value of Γ.

Tables 2.15 and 2.16 show the results obtained using formulae (70), (67), (69) compared with the results of the exact numerical calculation. The values of R in (68) have been calculated by PT up to the fourth order in F.

As seen from the tables, asymptotic formula (67) yields good results for Γ when n_2 and m are small. At large n_2 (and m), the region of applicability of asymptotic formulae for Γ is limited to very low fields that are of no practical interest.

Table 2.17. Comparison of the results for level widths obtained by using asymptotic formula (67) and semiempirical formula (72) with exact numerical data

n_1, n_2, m	F	Γ [Eq. (67)][a]	Γ [Eq. (72)]	Γ_{num}
0, 4, 0	1.0×10^{-4}	4.6×10^{-11}	4.4×10^{-12}	4.2×10^{-12}
	1.5×10^{-4}	8.02×10^{-5}	2.43×10^{-6}	1.92×10^{-6}
	2.0×10^{-4}	5.03×10^{-2}	4.76×10^{-4}	1.78×10^{-4}
0, 9, 0	9.0×10^{-6}	2.99×10^{-7}	2.30×10^{-10}	2.20×10^{-10}
	1.0×10^{-5}	8.00×10^{-5}	2.78×10^{-8}	2.40×10^{-8}
	1.1×10^{-5}	6.56×10^{-3}	1.03×10^{-6}	7.17×10^{-7}
1, 10, 2	2.723×10^{-6}	2.69×10^{-6}	4.72×10^{-10}	2.44×10^{-10}
	2.917×10^{-6}	2.24×10^{-4}	2.11×10^{-8}	8.91×10^{-9}
	3.111×10^{-6}	9.69×10^{-3}	4.93×10^{-7}	1.55×10^{-7}
0, 29, 0	1.283×10^{-7}	3.34×10^{-3}	1.81×10^{-14}	2.59×10^{-14}
	1.322×10^{-7}	1.96×10^{-1}	4.83×10^{-13}	6.35×10^{-13}
	1.361×10^{-7}	8.70	9.77×10^{-12}	1.15×10^{-11}

[a] In formula (67), only the first term has been taken into account.

2.4c. Semiempirical formula for Γ

The analysis of the numerical results enabled us to find a useful modification of formula (67) by rewriting the multiplier $1-\alpha F+O(F^2)$ entering this formula as $\exp(-\alpha F)$

$$\Gamma = \frac{(4R)^{2n_2+m+1}}{n^3 n_2!\,(n_2+m)!} \times \exp\left(-\frac{2}{3}R - \frac{n^3F}{4}\left(34n_2^2+34n_2m+46n_2+7m^2+23m+\frac{53}{3}\right)\right) \quad (72)$$

As $F\to 0$, both formulae (72) and (67) are asymptotically correct.

In Table 2.17, the results for large n_2 obtained by using (72) and (67) are compared with exact numerical data. As before, the values of R were calculated by PT up to the fourth order in F. For all cases considered, i.e., for different n_1, n_2, m (in the table only some examples are given), for fields of practical interest, formula (72) yields results that differ from the exact ones by not more than a factor of 3.

2.4d. Stability of different sublevels

A qualitative explanation for the ionization of hydrogen atoms by an electric field, valid for the limit $F\to 0$, has been given in Ref. 2. What conditions are favorable for the ionization? First, the electron orbit should have a large radius, i.e., n should be large. Second, in parabolic coordinates wave functions of the hydrogen atom at $F=0$,

$$U_{n_1n_2m}=\frac{e^{\pm im\varphi}}{(\pi n)^{1/2}}\frac{n_1!^{1/2}n_2!^{1/2}\epsilon^{m+3/2}}{(n_1+m)!^{3/2}(n_2+m)!^{3/2}}$$

$$\times e^{-\epsilon(\xi+\eta)/2}(\xi\eta)^{m/2}L^m_{n_1+m}(\epsilon\xi)L^m_{n_2+m}(\epsilon\eta) \tag{73}$$

where $\epsilon=(-2E)^{1/2}$ is asymmetric with respect to the plane $z=0$. The coordinate η increases in the anode direction. If n is fixed, then the sublevel corresponding to the electron orbit that is more stretched in this direction, i.e., with a higher value of $k=n_2-n_1$, has a higher probability of ionization. Thus, for the same n the energetically lower sublevels are less stable [see Eq. (40)]. The experimental data of Traubenberg et al.[4-6] for some sublevels at $n=5$ are in agreement with the theory.[2]

Now we consider the asymptotic formula for Γ [Eq. (69)]. In the limit $F\to 0$, the formula is in accord with theoretical conclusions based on qualitative consideration of the problem. However, formula (69) shows that with the increase of F some of the ionization probability curves intersect. The first intersection of the ionization probability curves corresponds to the sublevels $0,0,n-1$ and $p,p-1,n-2p$, where

$$n=\begin{cases}2p & \text{even values}\\ 2p+1 & \text{odd values}\end{cases}$$

Based on (69), we can estimate the value of F^0 at which the first intersection occurs:

$$F^0=\frac{4(p-1)!}{(ne)^3(n-p)(n-p+1)\cdots(n-1)} \tag{74}$$

Using the same approach, we can analyze other intersections. Still, it should be noted that the intersections of ionization probability curves do not result in the intersections of the Stark sublevels.

Using (74) [and the asymptotic formula for Γ, Eq. (72)], it is easy to check that for $n\leqslant 4$ the first crossing of the ionization probability curves occurs in the region of such strong fields that the very notion of the quasi-stationary state loses its physical meaning. At $n=5$, the ionization probability curves corresponding to the sublevels $0,0,4$ and $2,1,1$ are the first to intersect. For this case, the asymptotic approach gives the crossing point as $F^0_{\rm as}=107\times 10^4$ V cm^{-1}. Exact numerical calculation leads to a similar value: $F^0_{\rm num}=116\times 10^4$ V cm^{-1}. The value of the ionization probability for the sublevels $0,0,4$ and $2,1,1$ in the vicinity of the crossing point is very large: $\omega\simeq 10^{12}$ s^{-1}.

With an increase in n, the number of intersections of the ionization probability curves rapidly increases; some of them occur at low electric fields where ionization probabilities are small.

The case $n=7$ has been considered in detail in Ref. 54. Ionization proba-

Table 2.18. Field strengths F^0 corresponding to crossing points of ionization probability curves for the sublevels n_1, n_2, m and n_1', n_2', m'

n	n_1, n_2, m	n_1', n_2', m'	F^0	n	n_1, n_2, m	n_1', n_2', m'	F^0
6	0, 0, 5	3, 2, 0	3.8×10^{-5}	8	0, 0, 7	4, 3, 0	3.1×10^{-6}
	0, 0, 5	2, 1, 2	6.0×10^{-5}		0, 0, 7	3, 2, 2	4.2×10^{-6}
7	0, 0, 6	3, 2, 1	1.1×10^{-5}		1, 0, 6	4, 2, 1	7.6×10^{-6}
	1, 0, 5	4, 2, 0	2.5×10^{-5}		0, 0, 7	2, 1, 4	1.1×10^{-5}
	0, 0, 6	2, 1, 2	2.6×10^{-5}		0, 1, 6	3, 3, 1	1.5×10^{-5}
	1, 0, 5	3, 1, 2	4.2×10^{-5}		2, 0, 5	5, 2, 0	1.6×10^{-5}
					1, 0, 6	3, 1, 3	1.6×10^{-5}
					2, 0, 5	4, 1, 2	2.5×10^{-5}

bility curves have been calculated numerically for all 28 sublevels in a wide range of F. The ordering of the curves and crossing points has been analyzed by the use of asymptotic formulae. At $F\simeq 70$ kV cm^{-1}, the curves corresponding to the sublevels 0, 0, 6 and 3, 2, 1 are the first to cross. At such fields the ionization probability for the said sublevels is negligibly small: $\omega \simeq 10^{-40}$ s^{-1}.

Listed in Table 2.18 are data on the ionization probability crossing points obtained by using asymptotic formulae. At present there is no experimental data on ionization probability intersections at constant n.

2.4e. Threshold field

When considering the problem of the stability of sublevels, it should be noted that in a relatively narrow range of F the ionization rate increases with F so rapidly that, for many purposes, it is treated as zero for $F<F_{cr}$ and infinite for $F>F_{cr}$. This leads to the idea of a "threshold field" F_{cr}.

To estimate F_{cr}, for want of something better, the classical value has been used:[2]

$$F_{cr} = 1/16n^4 \tag{75}$$

Formula (75), as well as other classical criteria,[55,56] yields no information on the value of Γ in the region of F_{cr}. New experimental methods require a more accurate characteristic defining the behavior of the atom in the electric field, i.e., the function $\Gamma(F)$ especially for the region where Γ starts exceeding the natural width of the corresponding sublevel. Therefore, either the results of exact numerical calculations may be used or, if there is no need for very accurate data, it is sufficient to apply, for example, formula (72).

2.5. Asymptotic relations between the energy shift and ionization probability

Analysis of the Schrödinger equation (1) for complex F leads to the derivation of a relation between PT coefficients and the value of Γ. Here, to make the

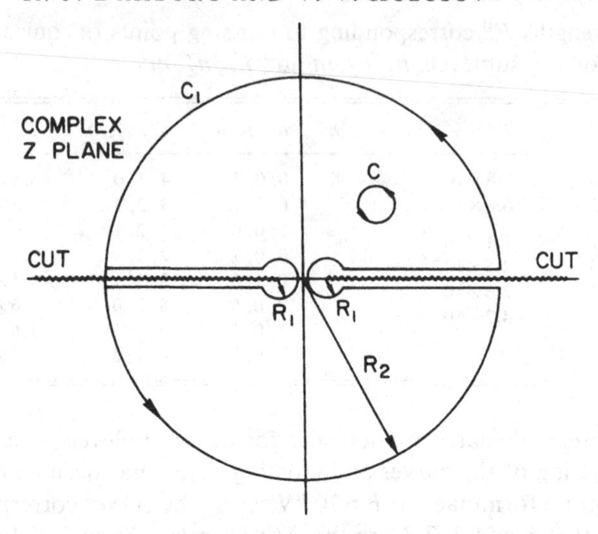

Fig. 2.6. Change of contour in the Z plane used to derive the dispersion relation in (77).

review more complete, we shall attempt to outline some ideas on this problem. For a more consistent mathematical consideration, see Refs. 57–60.

Earlier, Eqs. (3) and (4), which were obtained from (1), were treated under boundary conditions (7) and (8), where F was supposed to be real. However, Eq. (3) can be solved subject to the condition $V(\xi) \to 0$, as $\xi \to \infty$ for any value of F in the complex plane, except one branch cut running from -0 to $-\infty$. This solution selects out a discrete set of complex eigenvalues $\lambda_1(n_1, m, R)$. Similarly, the solution of Eq. (4) obtained with the condition $U(\eta) \to 0$ as $\eta \to \infty$ selects out a discrete set of eigenvalues $\lambda_2(n_2, m, R)$. Therefore, based on Eq. (30), we may conclude that the function $E(n_1, n_2, m; z)$ is analytic in the complex Z plane, except for two cuts along the real z axis, one running from -0 to $-\infty$, the other from $+0$ to $+\infty$.

We use the Cauchy theorem to write

$$E(n_1, n_2, m; F) = \frac{1}{2\pi i} \oint_C \frac{E(n_1, n_2, m; z)}{z - F} dz \tag{76}$$

where the contour C is shown in Fig. 2.6 and may be shifted to the contour C_1.

Letting $R_1 \to 0$ and $R_2 \to \infty$ and assuming that the contributions from the large and small circles vanish, we obtain[61]

$$E(n_1, n_2, m; F) = -\frac{1}{\pi} \int_0^\infty dx \left(\frac{2\Gamma(n_1, n_2, m; x)}{x - F} + \frac{2\Gamma(n_2, n_1, m; x)}{x + F} \right) \tag{77}$$

Here we note that the discontinuity of E across the branch cuts is purely imaginary and is equal in value to 2Γ.

Formula (25) may be written in the form

$$E(n_1, n_2, m; F) = \sum_{i=0}^{\infty} a_i^{n_1 n_2 m} F^i = E(n_2, n_1, m; -F) \tag{78}$$

and (69) as

$$\Gamma(n_1, n_2, m; F) = \left(\frac{4}{Fn^3}\right)^{2n_2+m+1} \times \frac{\exp\{3(n_1-n_2) - [2/(3n^3F)]\}}{n^3 n_2!(n_2+m)!} \sum_{j=0} b_j^{n_1 n_2 m}(3n^3F/2)^j \tag{79}$$

Using (77)-(79) and expanding the denominators of (77) in F/x, we obtain

$$a_k^{n_1 n_2 m} = -\left(\frac{3n^3}{2}\right)^k \frac{e^{3(n_1-n_2)} 6^{2n_1+m+1}}{2\pi n^3 n_2!(n_2+m)!} \sum_{j=0} b_j^{n_1 n_2 m}(2n_2+m+k-j)! - \left(-\frac{3n^3}{2}\right)^k \frac{e^{3(n_2-n_1)} 6^{2n_1+m+1}}{2\pi n^3 n_1!(n_1+m)!} \sum_{j=0} b_j^{n_2 n_1 m}(2n_1+m+k-j)! \tag{80}$$

Equations similar to (80) for $a_k^{n_1 n_2 m}$ were first obtained for the anharmonic oscillator by Bender and Wu.[57]

As seen from (80), when $n_1 = n_2$, only even-order terms in (78) are nonzero, and they are all negative. When $n_1 \neq n_2$, terms with the largest values of $(2n_i+m+k)!$ eventually dominate in (78). Formula (80) qualitatively explains the behavior of terms in the perturbation series described in Sect. 2.3c. This formula is asymptotic in k because of the asymptotic character of the expression for Γ, and therefore it yields sufficiently accurate numerical results only at large values of k.

Equation (80) can be applied more successfully when $n_1 = n_2 = m = 0$. Using the known coefficients b_1^{000} and b_2^{000} [see formula (71)], we obtain

$$a_k^{000} = -\frac{6}{\pi}\left(\frac{3}{2}\right)^k k! \left\{1 - \frac{107}{18k} + \frac{7363}{648k(k-1)} - \cdots\right\} \tag{81}$$

Listed in Table 2.19 are the exact numerical values of the PT coefficients a_k^{000} and their asymptotic values obtained by using one, two, or three terms in Eq. (81).

In Refs. 59 and 60, to obtain additional b_j^{000} coefficients by extrapolation using (81), exact numerical values of a_i^{000} (up to $k \simeq 150$–160) have been used. Besides, in Ref. 59, some other additional coefficients b_j^{100} and b_j^{010} have been obtained by similar calculations. The results are given in Table 2.20. Those coefficients $b_j^{000}, b_j^{100}, b_j^{010}$ that are so marked in the table have been calculated exactly [see formulae (69) and (71)]. Numerical results for b_1^{000} and b_2^{000} obtained in Refs. 59 and 60 are in good agreement with the exact results. It is difficult therefore to understand the discrepancy between the values of b_3^{000},

Table 2.19. Comparison of the exact numerical values of the PT coefficients and their asymptotic values obtained from Eq. (81)

	Number of terms used in Eq. (81)			
k	One	Two	Three	a_k^{000}(num)
6	-1.5661×10^4	-1.4503×10^2	-0.6078×10^4	$-0.490\,777\times10^3$
10	-3.9965×10^8	-1.6208×10^8	-2.1254×10^8	$-1.945\,320\times10^8$
30	$-9.714\,03\times10^{55}$	$-7.789\,21\times10^{55}$	$-7.916\,08\times10^{55}$	$-7.897\,811\times10^{55}$
150	$-2.828\,68\times10^{289}$	$-2.716\,58\times10^{289}$	$-2.718\,02\times10^{289}$	$-2.717\,978\times10^{289}$

Table 2.20. Numerical calculations for b_j^{000}, b_j^{100}, and b_j^{010} coefficients

j	b_j^{000} [a]	b_j^{000} [b]	b_j^{100} [a]	b_j^{010} [a]
1	$-\frac{107}{18}$ [c]	$-\frac{107}{18}$ [c]	$-\frac{130}{9}$ [c]	$-\frac{142}{9}$ [c]
2	$\frac{7363}{648}$ [c]	$\frac{7363}{648}$ [c]	$\frac{33\,053}{324}$ [c]	$\frac{28\,655}{324}$ [c] $\pm 0.000\,05$
3	-47.0360 ± 0.0002	-35.673	-663.45 ± 0.1	-381.832 ± 0.01
4	92.65 ± 0.2	-37.1	4450 ± 20	1455.7 ± 1.0
5	-1350 ± 20	-1.1×10^3	$-34\,000\pm3000$	$-11\,700\pm200$

[a]From Ref. 59; [b]from Ref. 60.
[c]Coefficient value was exactly calculated.

b_4^{000}, and b_5^{000} calculated in Refs. 59 and 60. Silverstone et al.[59] used the coefficients $b_j^{n_1 n_2 m}$ in formula (79) to obtain the values of $\Gamma(0,0,0;F)$, $\Gamma(1,0,0;F)$, and $\Gamma(0,1,0;F)$ at small F. When the additional coefficients are used, the agreement with the exact numerical results becomes better, especially for the state $0,1,0$ for which only two terms were known earlier.

2.6. Classical and quasi-classical methods

2.6a. Determination of the Stark energy

The quasi-classical WKB method was one of the first used together with PT to develop the theory of the Stark effect in quantum mechanics.[1-3] The use of the WKB method for the Stark problem was first employed by Wentzel,[62] who was one of the originators of the method. He used high-order WKB approximations to obtain Stark energy expansions (40) up to second order in the electric field strength.

Lanczos, to obtain the Stark energy, used quantization rules:[63]

$$\int_0^{\xi_1} [\Phi_1(\xi)]^{1/2}\, d\xi = \pi(n_1 + \tfrac{1}{2}) \tag{82}$$

$$\int_0^{\eta_1} [\Phi_2(\eta)]^{1/2}\, d\eta = \pi(n_2 + \tfrac{1}{2}) \tag{83}$$

where $\Phi_1(\xi)$ and $\Phi_2(\eta)$ are the functions standing for double kinetic energy in Eqs. (3) and (4), and ξ_1 and η_1 the corresponding turning points. Ignoring the terms $\sim(m^2-1)$ and assuming $m=0$ in expressions for $\Phi_1(\xi)$ and $\Phi_2(\eta)$, Lanczos expressed integrals in Eqs. (82) and (83) by complete elliptic integrals of the first and second kind. Using tabulated values for these integrals, he achieved good agreement with the experimental data of Gebauer and Traubenberg.[5] In the case of weak fields, the author expanded elliptic integrals in powers of small modulus and obtained the expansion for the Stark energy up to the fourth order in the field coinciding with (40) up to the first order in F.

In Ref. 64, to find the position of the level in weak fields, quantization rules derived by using ordinary WKB connection formulae were applied. In contrast to the Lanczos procedure, Rice and Good[64] made no simplifying assumptions about the values of m, and for the integrals in Eqs. (82) and (83), they obtained expansions with complete elliptic integrals. To compare their results with results obtained by PT, the authors expanded the obtained expressions and got the first three terms of the energy expansion in powers of F. The third term in this expansion differs from the term in formula (40) by the value of $(10/16)n^4F^2$. Note that the corresponding term in Ref. 63 differs from the exact one by the value of $(19-9m^2)n^4F^2/16$. To obtain the value of the Stark energy in the region of strong fields, Rice and Good used, instead of (83), the condition of minimum wave function amplitude of Eq. (4) beyond the outer turning point and obtained expressions including complete elliptic integrals. For the intermediate region of F, the authors recommend interpolation between the results for weak and strong fields.

In Ref. 65, before applying the quantization rules, the Langer transformation was made in Eqs. (3) and (4):

$$\xi=e^{2\delta x_1},\quad V(\xi)=e^{\delta x_1}y_1(x_1),\quad \eta=e^{2\delta x_2},\quad U(\eta)=e^{\delta x_2}y_2(x_2);\quad \delta>0$$

Potentials in the resulting equations have no singularities at the origin, and $y_1, y_2 \to 0$ as $x_1, x_2 \to \pm\infty$, thus satisfying the conditions for the applicability of the quantization rule. The authors determined the Stark energy by the high-order WKB approximations through terms in $\hbar^4$. In the region of strong fields, numerical integration of the expressions is required. In the limit $F\to 0$, the Stark energy expansion through terms in F^4 has been derived. The fourth-order term differs from the corresponding term in (40) by $405(n_1-n_2)^4n^{10}F^4/128$. In Ref. 37, this difference has been interpreted as the result of an error made in Ref. 65.

Quantization rules have been also considered in Ref. 66. Here, in contrast to Ref. 63, the integrals in Eqs. (82) and (83) were expressed in terms of generalized hypergeometric functions $F(-\frac{1}{2},\frac{1}{2},2,Z)$. The system of equations obtained for determining the Stark energy allows only numerical solution.

In Ref. 67, expressions obtained by integration of (82) and (83) were expanded in power series using Lagrange's formula. For weak and strong fields, the authors suggest two different expansions for the separation constants Z_1 and Z_2 with tabulated values of coefficients. The Stark energy expansion obtained in Ref. 67 coincides with (40) only up to the first order in F. The authors do not give an explicit formula for the corresponding "strong-field expansion"; however, their example for the case $n=11$, $n_1=10$, $n_2=0$, $m=0$ shows that this series converges to a wrong value for the Stark energy. Thus, for instance, at $F=1.6\times10^{-5}$, the corresponding value of the Stark energy obtained in Ref. 67 with the accuracy up to seven stabilized figures is: $E_{HY}=-0.001\,797\,659$, whereas the exact numerical calculation gives $E_0=-0.001\,806\,449\,97$ and the Padé approximation gives $[8,8]=-0.001\,806\,15$.

In Refs. 56, 68, and 69, the classical method based on the assumption of adiabatic invariance of the classical action integrals was applied in the study of the Stark effect. After separation of the classical Hamiltonian in parabolic coordinates, the problem reduces to the numerical calculation of two classical action integrals I_ξ and I_η. To present the results in compact form, classical scaling laws were used. To obtain the Stark energy, the action integrals were transformed by simple quantization rules into quantum numbers. Classical representation enables the energy to be determined only for the fields where the classical motion is bounded. Moreover, because the classical representation neglects tunneling, it provides no information on the level width Γ. On the whole, the Stark energies obtained by the classical adiabatic theory are close to those obtained by quantum-mechanical calculations, though in some cases they are worse than the PT results and do not allow determination of E_0 with the accuracy comparable to the level width.[41]

2.6b. Determination of ionization probability

It was Oppenheimer who in 1928 first considered the theory of the field ionization of hydrogen in terms of quantum mechanics.[70] His formula includes a correct exponential dependence $\exp(-2/3F)$ preceded, however, by a wrong factor.

Later, Lanczos[71] reconsidered the problem with particular attention to the disappearance of spectral lines of the Balmer series in the experiment of Traubenberg et al.[6] When considering the transition probability through a one-dimensional potential barrier, Lanczos used the WKB approximation and obtained the following expression for the decay constant δ defining the level width Γ:

$$\delta=\exp\left(-2\int_{\eta_1}^{\eta_2}|\Phi(\eta)|^{1/2}\,d\eta\right)\Big/4\int_0^{\eta_1}[\Phi(\eta)]^{-1/2}\,d\eta \tag{84}$$

where η_1 and η_2 are, respectively, inner and outer turning points. The integrals in (84) were expressed by complete elliptic integrals; the Stark energy and the separation constants were determined by use of the third-order PT. The results obtained by Lanczos agree with the experimental results.[6]

A more elaborate application of the WKB-type approximation for determining ionization probabilities for the hydrogen atom in a uniform electric field was developed by Rice and Good.[64] In contrast to Lanczos, these authors considered the full three-dimensional problem. Their formulae for level widths were also found in terms of complete elliptic integrals.

In Ref. 72, the authors, using the methods of Lanczos[71] and Rice and Good,[64] calculated ionization probabilities for the hydrogen atom in an electric field for all states up to $n=7$ and for extreme Stark components up to $n=25$. Results obtained by the Lanczos method differ in many cases from those obtained by the method of Rice and Good by an order of magnitude.[72] Though the method of Rice and Good is more consistent, still, in Ref. 64 the authors partially neglected the terms $\sim(m^2-1)$ in their approach, whereas the Lanczos method allowed these terms to be taken into account. Besides, expressions derived by Rice and Good are more complicated for practical use than those of Lanczos. However, Bailey et al.[72] gave no preference to either method.

In Table 2.21, the results obtained in Ref. 72 are compared with those of exact numerical calculations performed by the method described in Sect. 2.2c. Though the numerical results calculated by the method of Rice and Good in some cases appear to be more accurate than those obtained by the Lanczos method, both methods can lead to a wrong ordering of ionization probability curves.[54]

In Ref. 66, a quasi-classical approximation was applied to a numerical calculation of the level width Γ. In the limit $F \to 0$, an analytic formula for Γ was obtained. This formula, however, does not coincide with the first term in (69).

2.7. Some remarks on the comparison between theory and experiment

2.7a. Nonhydrogenic atoms

Because Chapters 3 and 13 in the present volume review in detail the experimental studies on the behavior of Rydberg atoms in electromagnetic fields, we shall just touch upon the problem in view of the most recent investigations. For a review of earlier studies see, for example, Ref. 73.

Precise measurements of the Stark shift and the lifetime of alkali atoms have been carried out at the Massachusetts Institute of Technology by Dr. Kleppner and co-workers.[74,75] The results of their experimental studies[75] of excited states of Na with $n=12$–15, $m=2$ appear to be in good agreement with those for the Stark energy calculated by the low-order PT. Values of lifetimes for low-lying Stark sublevels agree well with those calculated for correspond-

Table 2.21. Ionization probabilities obtained by Bailey et al.[72] using the Lanczos method and the Rice and Good method compared with exact numerical data (Sect. 2.2c)

		Ionization probability (s^{-1})		
n_1, n_2, m	F (V cm^{-1})	Lanczos method[a]	Rice and Good method[b]	Exact data
0, 4, 0	6.280×10^5	4.146×10^7	3.088×10^8	2.938×10^8
	7.142×10^5	1.744×10^9	1.298×10^{10}	1.226×10^{10}
	8.283×10^5	5.240×10^{10}	3.873×10^{11}	3.499×10^{11}
4, 0, 0	9.675×10^5	4.896×10^8	5.611×10^8	3.870×10^8
	1.133×10^6	1.637×10^{10}	1.959×10^{10}	1.338×10^{10}
	1.302×10^6	2.091×10^{11}	2.513×10^{11}	1.666×10^{11}
0, 9, 0	5.178×10^4	1.425×10^8	1.389×10^9	1.310×10^9
	5.814×10^4	7.560×10^9	7.479×10^{10}	6.707×10^{10}
	6.710×10^4	1.934×10^{11}	1.379×10^{12}	1.265×10^{12}
9, 0, 0	9.134×10^4	8.525×10^7	4.668×10^8	3.170×10^8
	1.021×10^5	1.802×10^8	9.894×10^9	6.585×10^9
	1.252×10^5	9.821×10^{10}	6.266×10^{11}	3.596×10^{11}
24, 0, 0	2.973×10^3	8.206×10^6	8.623×10^5	5.17×10^5
	3.199×10^3	3.438×10^8	3.379×10^7	2.253×10^7
	3.432×10^3	7.080×10^9	7.031×10^8	4.628×10^8

[a] Lanczos.[71]
[b] Rice and Good.[64]

ing hydrogen states. However, for higher sublevels, there is a disagreement with the hydrogen theory that the authors attribute to level mixing.

In Ref. 76, the excited products of the reaction Na(nd) + Xe → Na(nl) + Xe were investigated for $n=37$ and $n=42$ by use of selective field ionization. To calculate ionizing fields, hydrogenic theory, namely formula (72), was used. Good agreement between the theory and the experimental data was observed.

2.7b. Hydrogen atom

Studies of the hydrogen atom in single, well-defined Rydberg states in a uniform electric field are of a quite recent origin.[7,8] In Ref. 7, the first results on precise measurements of Stark energies for $n=25$, $n_1=21$, $n_2=2$, $m=1$ were given. The results agree well with those of exact calculation. Here, to calculate the Stark energy, either the high-order PT or the Padé approximants obtained based on PT can be applied. It should be noted that a thorough development of these methods requires a certain comparison with the results of exact numerical calculations (see Sect. 2.3c).

At the VII ICAP, Koch, in his review report,[8] presented the first, still unpublished, results on measuring the level width Γ for a number of states with $n = 30$–40. His data were in very good agreement with the results of exact numerical calculations.

References and notes

1. E. V. Condon and G. H. Shortley, *The Theory of Atomic Spectra* (Cambridge University Press, 1963).
2. H. A. Bethe and E. E. Salpeter, *Quantum Mechanics of One- and Two-Electron Atoms* (New York: Academic Press, 1957).
3. L. D. Landau and E. M. Lifschitz, *Quantum Mechanics* (New York: Pergamon Press, 1958).
4. H. R. Traubenberg and R. Gebauer, *Z. Phys. 54,* 307 (1929); *56,* 254 (1929).
5. R. Gebauer and H. R. Traubenberg, *Z. Phys. 62,* 289 (1930); *71,* 291 (1931).
6. H. R. Traubenberg, R. Gebauer, and G. Lewin, *Naturwissenschaften 18,* 417 (1930).
7. P. M. Koch, *Phys. Rev. Lett. 41,* 99 (1978).
8. P. M. Koch, in *Proceedings, VII ICAP,* (New York: Plenum Press, 1980), p. 181.
9. M. H. Alexander, *Phys. Rev. 178,* 34 (1969).
10. M. Hehenberger, H. V. McIntosh, and E. Brändas, *Phys. Rev. A 10,* 1494 (1974).
11. P. Froelich and E. Brändas, *Phys. Rev. A 12,* 1 (1975).
12. P. Froelich and E. Brändas, *Int. J. Quantum. Chem. Symp. 10,* 353 (1976).
13. E. Brändas and P. Froelich, *Phys. Rev. A 16,* 2207 (1977).
14. W. P. Reinhardt, *Int. J. Quantum Chem. Symp. 10,* 359 (1976).
15. C. Cerjan, R. Hedges, C. Holt, W. P. Reinhardt, K. Scheibner, and J. J. Wendoloski, *Int. J. Quantum Chem. 14,* 393 (1978).
16. J. N. Bardsley, *Int. J. Quantum Chem. 14,* 343 (1978).
17. W. P. Reinhardt, *Proceedings, XI ICPEAC* (Amsterdam: North-Holland Publ., 1980), p. 729.
18. J. O. Hirschfelder and L. A. Curtiss, *J. Chem. Phys. 55,* 1395 (1971).
19. N. A. Guschina and V. K. Nikulin, *Chem. Phys. 10,* 23 (1975).
20. H. J. Ralph, *J. Phys. C 1,* 378 (1968).
21. J. D. Dow and D. Redfield, *Phys. Rev. B 1,* 3358 (1970).
22. D. F. Blossey, *Phys. Rev. B 2,* 3976 (1970).
23. J. Fauchier and J. D. Dow, *Phys. Rev. A 9,* 98 (1974).
24. R. J. Damburg and V. V. Kolosov, *J. Phys. B 9,* 3149 (1976).
25. E. Luc-Koenig and A. Bachelier, *J. Phys. B 13,* 1743 (1980).
26. I. W. Herbst and B. Simon, *Phys. Rev. Lett. 41,* 67 (1978).
27. L. Benassi, V. Grecchi, E. Harrell, and B. Simon, *Phys. Rev. Lett. 42,* 67 (1979).
28. S. Graffi and V. Grecchi, *Commun. Math. Phys. 62,* 83 (1978).
29. I. W. Herbst, *Commun. Math. Phys. 64,* 279 (1979).
30. L. Benassi and V. Grecchi, *J. Phys. B 13,* 911 (1980).
31. E. Luc-Koenig and A. Bachelier, *Phys. Rev. Lett. 43,* 921 (1979).
32. E. Luc-Koenig and A. Bachelier, *J. Phys. B 13,* 1769 (1980).
33. R. J. Damburg and V. V. Kolosov, Abstracts, *V ICAP* (New York: Plenum Press, 1977), p. 202.
34. R. J. Damburg and V. V. Kolosov, *Phys. Lett. 61A,* 233 (1977).
35. R. J. Damburg and V. V. Kolosov, *J. Phys. B 12,* 2637 (1979).
36. J. Killinbeck, *Rep. Prog. Phys. 40,* 963 (1977).

37. S. P. Alliluyev and I. A. Malkin, *Zh. Eksp. Teor. Fiz. 66,* 1283 (1974), [*Sov. Phys. JETP 39,* 627 (1974)].
38. H. J. Silverstone, *Phys. Rev. A 18,* 1853 (1978).
39. S. P. Alliluyev, V. L. Eletsky, and V. S. Popov, *Phys. Lett. 73A,* 103 (1979).
40. R. J. Damburg and V. V. Kolosov, *Izv. Akad. Nauk. Latv. SSR., Ser. Fiz. Techn. Nauk 3,* 3 (1980).
41. H. J. Silverstone and P. M. Koch, *J. Phys. B 12,* L537 (1979).
42. G. A. Baker, Jr., *Essentials of Padé Approximants* (New York: Academic Press, 1975).
43. B. Simon, *Ann. Phys.* (*N.Y.*) *58,* 76 (1970).
44. S. Graffi and V. Grecchi, *Lett. Math. Phys. 2,* 335 (1978).
45. S. Graffi, V. Grecchi, S. Levoni, and M. Maioli, *J. Math. Phys. 20,* 685 (1979).
46. R. J. Damburg and R. Kh. Propin, *J. Phys. B 1,* 681 (1968).
47. T. M. Cherry, *Trans. Am. Math. Soc. 68,* 224 (1950).
48. J. C. P. Miller, "The Airy Integral," in *An Index of Mathematical Tables,* eds. A. Fletcher, J. C. P. Miller, and L. Rosenhead (London: Scientific Computing Service, 1946).
49. R. J. Damburg and V. V. Kolosov, *An Asymptotic Approach to the Stark Effect for the Hydrogen Atom* (in Russian)(Riga: Zinatne, 1977).
50. R. J. Damburg and V. V. Kolosov, *J. Phys. B 11,* 1921 (1978).
51. S. Yu. Slavjanov, *Problemi Matematicheskoi Fiziki* (Leningrad: Leningrad State University, 1970), p. 125.
52. B. M. Smirnov and M. I. Chibisov, *Zh. Eksp. Teor. Fiz. 49,* 841 (1965) [*Sov. Phys. JETP 21,* 624 (1965)].
53. T. Yamabe, A. Tachibana, and J. J. Silverstone, *Phys. Rev. A 16,* 877 (1977).
54. R. J. Damburg and V. V. Kolosov, *J. Phys. B 12,* 2637 (1979).
55. D. R. Herrick, *J. Chem. Phys. 65,* 3529 (1976).
56. D. Banks and J. G. Leopold, *J. Phys. B 11,* L5 (1978).
57. C. M. Bender and T. T. Wu, *Phys. Rev. D 7,* 1620 (1973).
58. I. W. Herbst and B. Simon, *Phys. Rev. Lett. 41,* 67 (1978).
59. H. J. Silverstone, B. G. Adams, J. Cizek, and P. Otto, *Phys. Rev. Lett. 43,* 1498 (1979).
60. V. L. Eletsky and V. S. Popov, *Dok. Akad. Nauk SSSR 250,* 74 (1980) [*Sov. Phys. Dokl. 25,* 27 (1980)].
61. Strictly speaking, instead of (77) the *N*-times-substracted dispersion relation would have to be used, but this would not, however, change the final result.[59]
62. G. Wentzel, *Z. Phys. 38,* 518 (1926).
63. C. Lanczos, *Z. Phys. 65,* 431 (1930).
64. M. H. Rice and R. H. Good, *J. Opt. Soc. Am. 52,* 239 (1962).
65. J. D. Bekenstein and J. B. Krieger, *Phys. Rev. 188,* 130 (1969).
66. G. F. Drukarev, *Zh. Eksp. Teor. Fiz. 75,* 473 (1978) [*Sov. Phys. JETP 48,* 237 (1978)].
67. F. T. Hioe and H. I. Yoo, *Phys. Rev. A 21,* 426 (1980).
68. D. Banks and J. G. Leopold, *J. Phys. B 11,* 37 (1978).
69. D. Banks and J. G. Leopold, *J. Phys. B 11,* 2833 (1978).
70. J. R. Oppenheimer, *Phys. Rev. 31,* 66 (1928).
71. C. Lanczos, *Z. Phys. 68,* 204 (1931).
72. D. S. Bailey, J. R. Hiskes, and A. C. Riviere, *Nucl. Fusion 5,* 41 (1965).
73. R. N. Il'in, in *Atomic Physics,* eds. S. J. Smith and G. K. Walters (New York: Plenum Press, 1973), vol. 3.

74. M. G. Littman, unpublished thesis, Massachusetts Institute of Technology, 1977.
75. M. G. Littman, M. L. Zimmerman, and D. Kleppner, *Phys. Rev. Lett. 37,* 486 (1976).
76. F. G. Kellert, T. H. Jeys, K. A. Smith, F. B. Dunning, and R. F. Stebbings, *Phys. Rev. A 23,* 1127 (1981).

3

Rydberg atoms in strong fields

D. KLEPPNER, MICHAEL G. LITTMAN,
AND MYRON L. ZIMMERMAN

3.1. Introduction

The development of techniques to create and study Rydberg atoms in the laboratory has generated renewed interest in the interactions of atoms with electric and magnetic fields. The scale of the interactions is so alien to the usual situation, however, that the subject is in many respects completely fresh. Laboratory fields, though insignificant compared with atomic fields, are not necessarily small compared with the much diminished Coulombic field experienced by an electron far from the nucleus of a highly excited atom. The applied fields can dominate the system, resulting in a new class of atomic phenomena that offers new insights and new challenges to atomic theory.

In this chapter we shall describe experimental and theoretical progress in research on Rydberg atoms in applied fields. We shall limit the discussion to static, or quasi-static, fields and to "one-electron" atoms, that is, atoms that are accurately described by single-electron states. The problem of hydrogen in electric fields is treated elsewhere in this volume and the electric field discussion will emphasize the behavior of alkali-metal atoms. The theoretical situation for atoms in magnetic fields is far less satisfactory than for electric fields; there is no comprehensive theory for hydrogen, much less for other atoms. We shall review the theoretical situation for hydrogen but draw on data from alkali-metal atoms to provide the experimental illustrations.

We can gain insight into the relative strengths of atom–field interactions in Rydberg atoms from simple scaling arguments. We shall use atomic units ($\hbar = m = e = 1$) for theoretical discussions and spectroscopic units for presenting experimental results. The relations between atomic and practical units are shown in Table 3.1.

The term energy of hydrogen is

$$W_n = -1/2n^2 \tag{1}$$

where n is the principal quantum number. The characteristic dimension of the atom is

$$\langle r \rangle \simeq n^2 \tag{2}$$

Table 3.1. Relations between atomic and practical units

Parameter	Practical unit	Atomic unit
Energy	2.2×10^{5} cm^{-1}	1 hartee
Distance	0.53×10^{-8} cm	1 bohr
Electric field	5.1×10^{9} V cm^{-1}	—
Magnetic field	1.7×10^{3} T	—

The contribution to the Hamiltonian arising from the interaction of the electron with a field **F** is **F•r**. The interaction energy thus has a characteristic value

$$W_{\mathrm{E}} \simeq n^2F \tag{3}$$

In a strong magnetic field, the dominant interaction is the diamagnetic term $\frac{1}{8}\alpha^2B^2r^2\sin^2\theta$, where α is the fine-structure constant. The characteristic value of the diamagnetic interaction is

$$W_{\mathrm{B}} \simeq \tfrac{1}{8}\alpha^2n^4B^2 \tag{4}$$

We immediately obtain the following scaling relations:

$$W_{\mathrm{E}}/W_0 \simeq n^4F \tag{5a}$$

$$W_{\mathrm{B}}/W_0 \simeq n^6B^2 \tag{5b}$$

The high n-dependence of these ratios is responsible for the qualitatively different behavior of Rydberg atoms in strong fields compared with the behavior of "ordinary atoms," i.e., atoms in low-lying states.

The fundamental effect of an electric field is to change the boundary conditions for the electron from closed to open, converting the bound states into continuum states. In principle, this occurs in an arbitrarily small field; in practice, the effect starts to be important at a characteristic field $F_c=1/(16n^4)$. For $n=15$, the value of F_c is about 6 kV cm^{-1}, a field strength that is easily achievable in the laboratory.

Magnetic fields, in contrast, do not ionize atoms; they tend to compress the charge to the nucleus and increase the binding. As the field is increased, the electron's motion becomes dominated by the magnetic force, though the Coulomb interaction continues to play an important role even at infinite fields. The most interesting and least understood region is where $W_{\mathrm{B}} \simeq W_0$. For $n=40$, this occurs at approximately 10 T, again a field that is easily reached in the laboratory.

In this chapter we shall summarize recent experimental and theoretical advances on the structure of Rydberg atoms in strong fields. The illustrations are drawn largely from work at the Massachussetts Institute of Technology. Some of the material was presented in a series of lectures by one of the authors.[1]

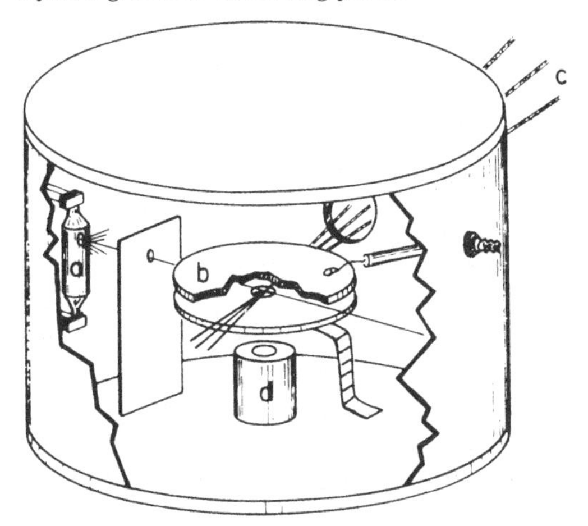

Fig. 3.1. Apparatus for observing Stark structure of Rydberg atoms: a, the atomic beam source; b, the electric field plates; c, the pulsed laser beams; and d, the electron multiplier.

3.2. Experimental methods

The study of Rydberg atoms in strong fields has been made possible by the development of tunable laser light sources. The high power of these sources is needed because of the small oscillator strengths for optical transitions to Rydberg states, particularly in strong fields where considerable l mixing occurs, and their wide tuning range is useful for generating panoramic displays of high-field structure.

Atomic beam methods are particularly well suited to strong-field spectroscopy. They largely eliminate collisional effects and Doppler broadening and greatly facilitate application of the electric or magnetic fields. Because the density of Rydberg atoms must be kept low to avoid collisions and because their radiative lifetimes tend to be prohibitively long, absorption or fluorescence spectroscopy is not usually practical for studying high-field structure. On the other hand, field ionization works extremely well for detecting resonance transitions to Rydberg states. Following the laser pulse, a high electric field is applied, which tears the highly excited electron away from the core ion. The electron or the ion is detected with an electron multiplier or other charge-sensitive device. The technique is efficient, selective, and relatively simple.

A schematic diagram of an apparatus used at the Massachusetts Institute of Technology to study Stark structure of alkali-metal atoms is shown in Fig. 3.1. We give herewith some experimental details.

Source. Of the many different atomic beam source designs in common use, we have found a "tube oven" to be particularly convenient. The alkali metal is encased in a stainless-steel tube, approximately 12 mm in diameter by 0.5 mm in wall thickness, which is crimped at each end. The vapor effuses from a 0.5-mm aperture in the center of the tube. Heat is generated by a current of several hundred amperes, which passes through the tube via water-cooled electrodes. The atomic density in the laser interaction region is approximately 10^9 atoms cm^{-3}.

Laser excitation. Rydberg states of alkali-metal atoms can be populated in two or three steps using optical photons from pulsed dye lasers. The dye lasers are of the grazing incidence type[2] and are pumped by the second or third harmonic of a Nd:YAG laser. Typical characteristics are: pulse duration, 5 nsec; repetition rate, 10 Hz; peak power, 3 kW; linewidth, 0.1 cm^{-1}; and scan range, 100 cm^{-1}.

The angular momentum of the final state is determined by the laser polarization. In the case of an electric field where m_l is generally a good quantum number for Rydberg states, transitions with $\Delta m_l = -1, 0, +1$ are given by left-hand, linear, and right-hand polarization, respectively. For low-lying levels where fine structure (and hyperfine structure) may be important, the selection rules are governed by the usual angular momentum coupling arguments.

Narrowband cw dye lasers are also useful as excitation sources. These provide much higher resolution, though generally at the expense of a reduced spectral sweep range. Pulsed and cw laser sources are both needed for a complete picture of high-field structure because panoramic maps and detailed high-resolution studies offer complementary views.

Field plates. A pair of field plates is used to apply the pulsed ionizing field. The interaction region, defined by the crossing of the atomic and laser beams, is at the center of the plates. The ionization fragments, electrons or ions, emerge through an array of small holes in one of the plates. Typically, the plates are 10 cm in diameter and 1 cm apart. The plates can also be used to apply the static electric field in Stark structure studies.

3.3. Stark structure of one-electron atoms

In this section we shall discuss methods for calculating the Stark structure of one-electron atoms and compare the results with experimental observations. We shall consider relatively low fields where the states are effectively stationary, leaving high-field structure and field ionization for a separate discussion. To introduce the subject we start by summarizing briefly the solution for hydrogen.

3.3a. Stark structure of hydrogen

We shall neglect relativistic and radiative interactions, assume that the usual field-free reduced-mass correction applies, and adopt the simplified Hamiltonian

$$H = -\tfrac{1}{2}\nabla^2 + V(r) + Fz \tag{6}$$

where $V(r)$ is a central potential and the last term describes the interaction of the electron with an electric field directed along the z axis. In the case of hydrogen, the Hamiltonian is

$$H = -\tfrac{1}{2}\nabla^2 - (1/r) + Fz \tag{7}$$

Solutions to this Hamiltonian are discussed in the literature[3] and in Chap. 2 of this volume, so we shall review only the essential features of the solution.

Equation 7 is separable in parabolic co-ordinates $r=(\xi+\eta)/2$; $z=(\xi-\eta)/2$; $x=(\xi\eta)^{1/2}\sin\phi$; $y=(\xi\eta)^{1/2}\cos\phi$. The eigenfunctions are of the form $[1/(\xi\eta)^{1/2}]f(\xi)g(\eta)\exp(im\phi)$, where f and g satisfy

$$f'' + k_1(\xi)^2 f = 0, \qquad g'' + k_2(\eta)^2 g = 0 \tag{8}$$

and

$$k_1(\xi)^2 = \frac{E}{2} + \frac{Z_1}{\xi} + \frac{1-m^2}{4\xi^2} - \frac{F}{4}\xi$$

$$k_1(\eta)^2 = \frac{E}{2} + \frac{Z_2}{\eta} + \frac{1-m^2}{4\eta^2} + \frac{F}{4}\eta \tag{9}$$

Z_1 and Z_2 are separation parameters that are related by $Z_1+Z_2=Z$ ($Z=$ atomic number). In zero field, f and g are Laguerre polynomials of order n_1 and n_2, respectively, which are known as the parabolic quantum numbers and are related to the principal quantum number n and the "magnetic" quantum number m by

$$n_1 + n_2 + |m| + 1 = n \tag{10}$$

Solutions to Eq. (8) can be carried out to arbitrary order in the field.[4] The energy through second order is

$$W = -\frac{1}{2n^2} + \frac{3}{2}n(n_1-n_2)F - \frac{1}{16}n^4[17n^2 - 3(n_1-n_2)^2 - 9m^2 + 19]F^2 \tag{11}$$

An energy-level diagram for some states in the vicinity of $n=19$ is shown in Fig. 3.2. Although curvature of the levels due to the second- and higher-order terms is clearly visible, the dominant feature is the linear Stark effect.

The zero-field parabolic eigenfunctions are contrasted with the familiar spherical eigenfunctions in Figs. 3.3 and 3.4. The nodal lines of the spherical states are seen to be concentric circles and radii, whereas those of the parabolic states are intersecting parabolas, where n_1 and n_2 are the number of nodes along the lines of constant ξ ($\equiv r+z$) and η ($\equiv r-z$), respectively. The most striking feature of the parabolic functions is their lack of parity. The center of mass of the charge distribution is displaced from the origin, giving rise to a permanent electric dipole moment. Such a lack of inversion symmetry with a central potential is possible only if the angular momentum

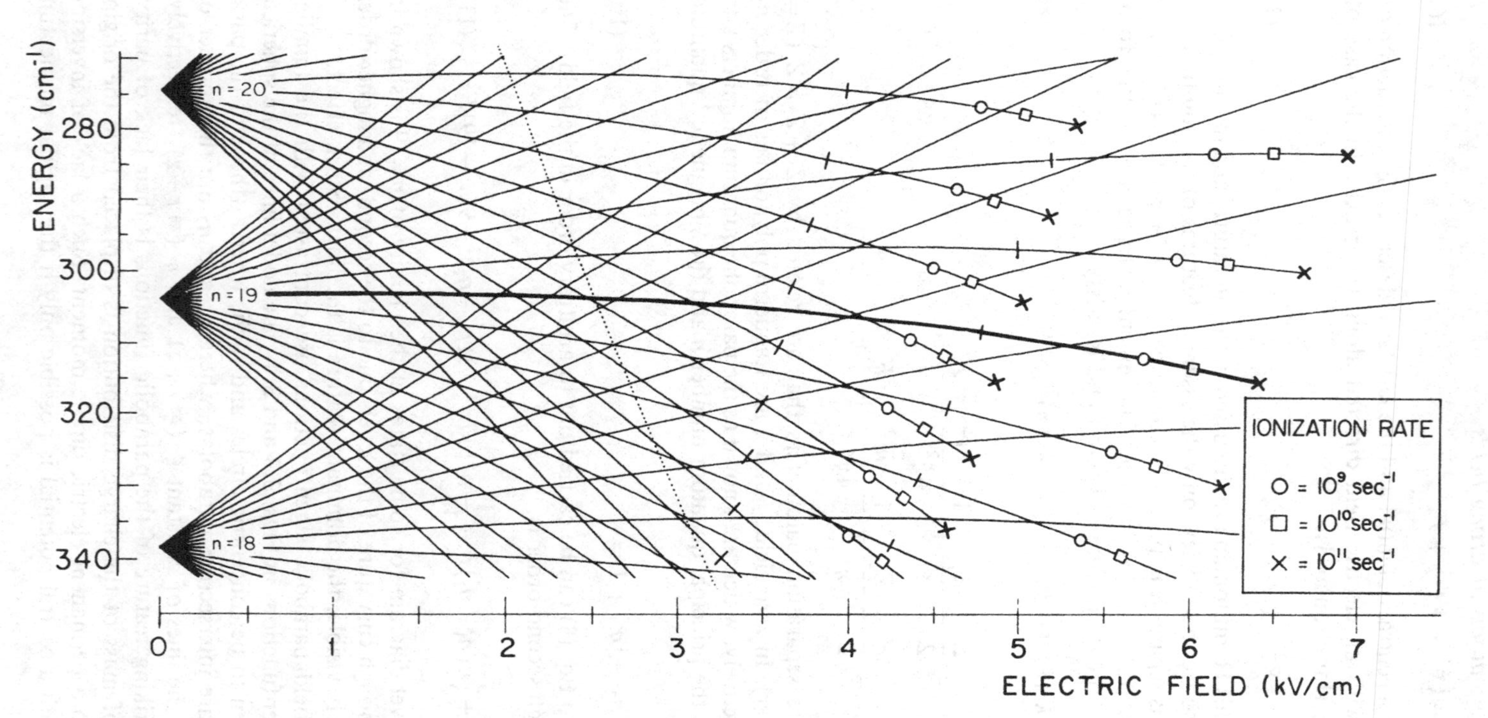

Fig. 3.2. Stark structure of $|m|=1$ levels of hydrogen in the vicinity of $n=19$. The ionization rates are discussed in Sect. 3.4a. The dotted line is the classical threshold for ionization, explained in Sect. 3.4b. (From Littman, Kash, and Kleppner[31])

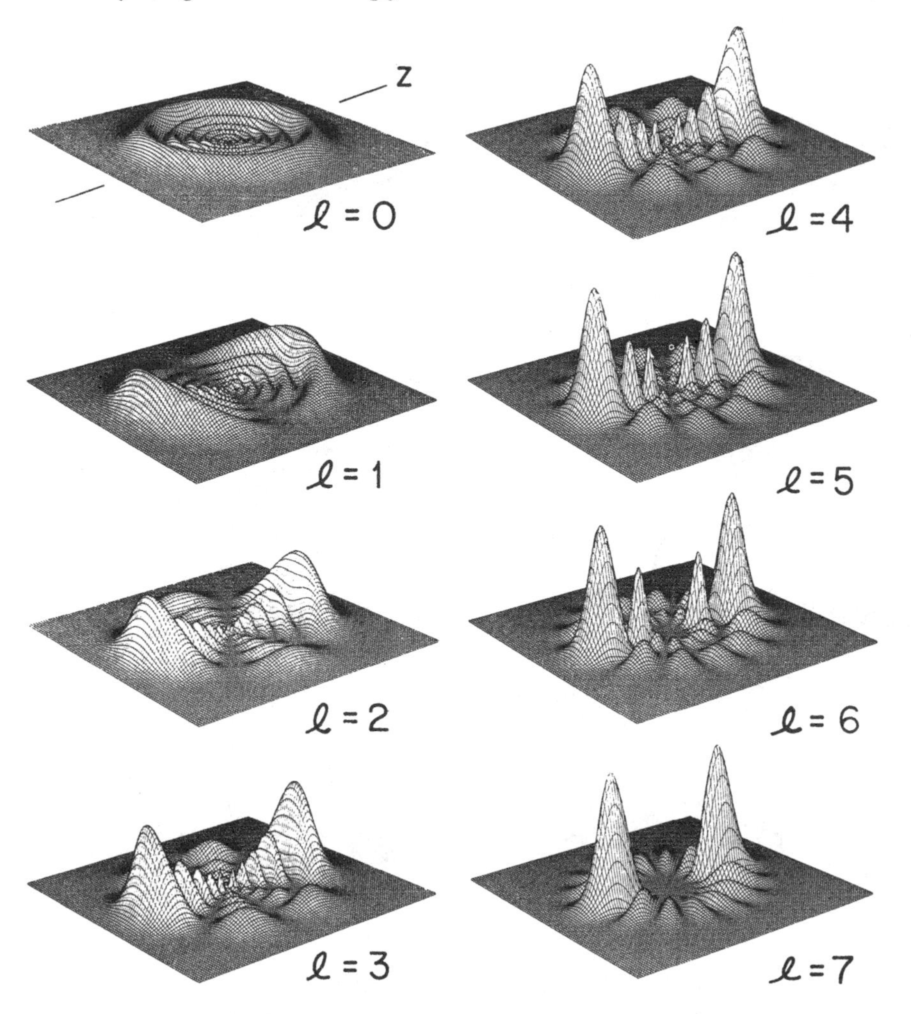

Fig. 3.3. Charge distribution for hydrogen in a plane containing the z axis. The charge density has been multiplied by r^2 to aid visibility. The states are $n=8$, $m=0$, $l=0 \rightarrow 7$. (Courtesy of W. P. Spencer)

states are degenerate. In hydrogen the degeneracy is a consequence of a special symmetry of the Coulomb field, dynamic in nature, which is accompanied by a unique constant of motion, the Lenz–Pauli vector.[5] The symmetry is preserved in an electric field[6] and the constant of motion – the z component of a generalized Lenz–Pauli vector – corresponds physically to the permanent dipole moment.[7] The dominant feature of hydrogenic Stark structure, the linear Stark effect, is thus a unique property of the Coulomb central potential. There are other consequences of the symmetry that will also be important to our study. For example, levels of the same m cross,[8] contrary to the usual case where the "no-crossing" theorem applies.[9] In high electric fields where field

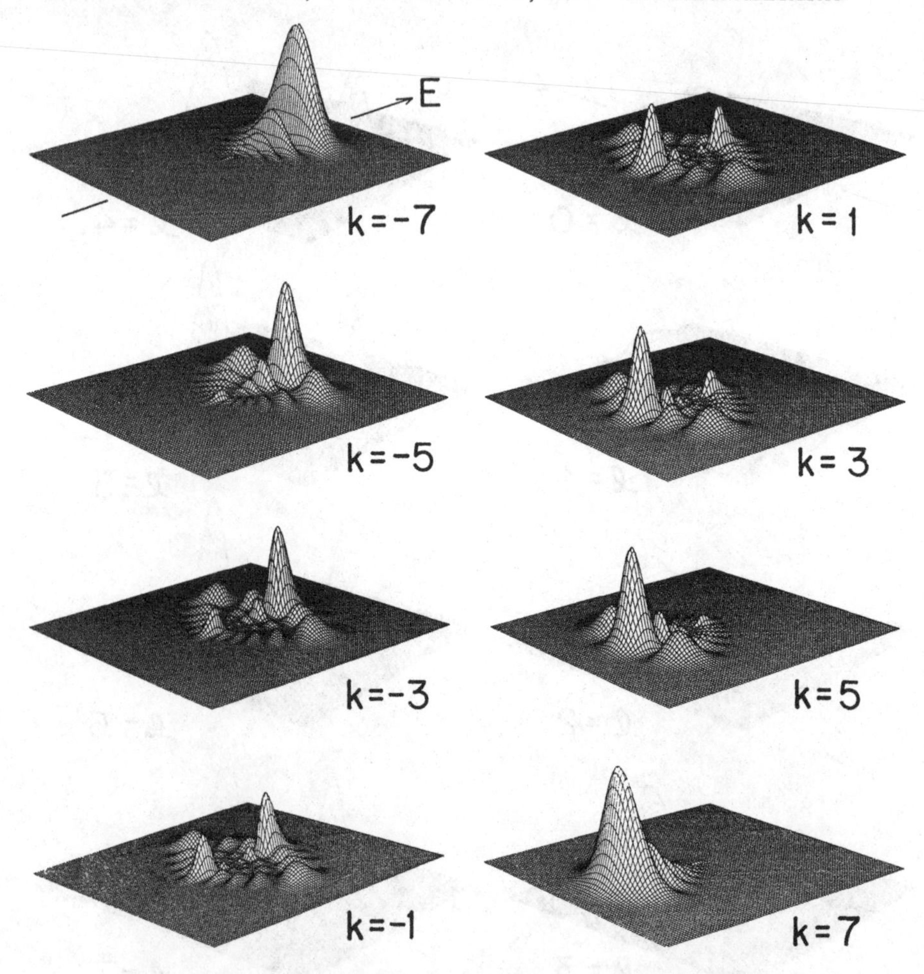

Fig. 3.4. Charge distribution for hydrogen, displayed as in Fig. 3.3, for "parabolic" eigenstates: $n=8$, $m=0$, $k=-7 \rightarrow +7$ $(k=n_1-n_2)$. The dipole moments that give rise to the first-order Stark effect are conspicuous. Note that the nodal lines form families of parabolas. (Courtesy of W. P. Spencer)

ionization is important, this symmetry allows stable states to coexist with a continuum of rapidly ionizing states.

3.3b. Stark structure of the alkali-metal atoms

The theoretical apparatus that works so admirably for hydrogen fails for other "one-electron" atoms because the Hamiltonian, Eq. (6), is separable only for a pure Coulomb central potential. Fortunately, accurate approximation schemes are available based on the following observation: The potential experienced by the outer electron in any neutral atom approaches $-1/r$ for r

sufficiently large. For atoms with a core of closed-shell electrons, such as the alkali-metal atoms, "sufficiently large" means larger than the core radius, r_c, typically a few atomic units. Highly excited electrons spend most of their time at large distance ($r \simeq n^2$), where the wave functions can be accurately represented by Coulombic functions. This observation is the starting point of the "Coulomb approximation" method of Bates and Damgaard[10] for calculating oscillator strengths of the alkalis; it also underlies quantum-defect theory.

Quantum defects were discovered empirically in the early days of spectroscopy by Rydberg. The energies of the alkalis can be written as

$$W_{n,l} = -1/2(n - \delta_l)^2 \tag{12}$$

where δ_l, the quantum defect, is a constant for given l or a slowly varying function of energy. Quantum defects are large for states that penetrate the core and small for states in which the centrifugal barrier prevents penetration. Empirically, the defects are large ($\delta_l \geqslant 1$), when l is less than l_c, which is the maximum angular momentum of a core electron. When l is greater than l_c, the defects are small, typically 10^{-2} or less.

A rigorous theory of quantum defects was developed by Seaton,[11] who showed that $\delta_l = \phi_l/\pi$, where ϕ_l is the low-energy electron-scattering phase shift. Alternatively, it can be shown by a WKB argument[1] that

$$\delta_l = (1/\pi)\left[\int^{r_c} k\,dr - \int^{r_c} k_c\,dr\right] \tag{13}$$

where $k = [E - V_{\mathrm{eff}}(r)]^{1/2}$, V_{eff} is the central potential (including the centrifugal term) and k_c is similar to k, but for a Coulomb potential. The lower limits of the integrals are the zeros of the integrands; the upper limits are the core radius. Because $|V(r)| \gg |E|$ in the core region, the energy can be taken as zero in Eq. (13) without introducing any large error. Alternatively, the integrands can be expanded in powers of E/V_{eff}, giving the usual energy expansion of the quantum defect.

There is yet another approach to quantum defects, one that will be most useful for our purposes and to which we shall return. A quantum defect can be viewed as the measure of any short-range perturbation on a hydrogenic system. The diagonal matrix element gives rise to changes in the zero-field term energies. In the presence of a field, the off-diagonal elements of the perturbation play an important role in determining the Stark structure.

Let us now turn to the problem of calculating the Stark structure of the alkali-metal atoms, that is, of finding the eigenvalue of Eq. (6). In the absence of a general solution, we resort to computational methods. One approach is to start with a basis of zero-field Coulombic functions $|W, l, m\rangle$, where $W = -1/2(n - \delta_l)^2$, and to diagonalize the Hamiltonian using some suitably selected set of basis states. Methods for calculating matrix elements and solving the secular equation are described by Zimmerman, Littman, Kash, and Kleppner;[12] we shall present here only a brief summary of this work.

3.3c. Computational methods

The problem is to diagonalize the Hamiltonian for which the Stark matrix elements are given by

$$\langle W',l',m'|Fr\cos\theta|W,l,m\rangle = \delta_{m,m'}\delta_{l,l+1}F\langle l',m|\cos\theta|l,m\rangle\langle W',l'|r|W,l\rangle \tag{14}$$

The angular matrix element is

$$\langle l+1,m|\cos\theta|l,m\rangle = [(l+1)^2 - m^2]/(2l+3)(2l+1)^2 \tag{15}$$

For generality, we shall write the radial matrix element in the form

$$R^j = \langle W',l'|r^j|W,l\rangle \tag{16}$$

where j is a nonnegative integer. A number of techniques are available for evaluating the radial matrix elements, all of which exploit the Coulombic properties of the wave functions. Among these are semiclassical methods, polynomial expansion techniques,[12] and numerical integration.[13,14] Unfortunately, semiclassical methods are not well suited to evaluating matrix elements (they are dominated by contributions from turning-point regions where the approximations are least accurate), whereas polynomial expansion methods tend to be numerically unstable because of the importance of small differences between large terms. The most satisfactory technique we have found is based on the numerical integration of the radial wave function. Although inelegant in concept, the method turns out to be fast and precise. Furthermore, unlike the other methods, it can easily accommodate long-range perturbations to the central potential, for instance, fine structure ($\sim r^{-3}$) or core polarization ($\sim r^{-4}$).

The radial integration can take advantage of the Numerov algorithm.[15] This algorithm, which is accurate to sixth order in the step size, allows speedy integration of any equation that can be put in the form $y''=f(x)y$. Because Coulombic wave functions are quasi-periodic with a wavelength that increases rapidly with distance, a change of scale is required to keep the number of integration points in each period approximately constant. A logarithmic scale is adequate, but the Numerov algorithm can be adapted to any scale of the form $x=r^k$, $k>0$. For a Coulombic potential, the value $k=\frac{1}{2}$ is optimal.

The major source of numerical error in a matrix element is caused by the normalization error that results from truncating the radial integration near the core. In the vicinity of $n=15$, the major fractional error in the matrix element due to this is typically 10^{-4}; the average numerical error is far smaller. The limiting error usually turns out to be due to experimental uncertainty in the quantum defects. Because only a small number of states have significant quantum defects, this error is effectively diluted throughout the Stark manifold. Thus the Stark structure can generally be calculated to a precision somewhat higher than the spectroscopic determination of the unperturbed energies.

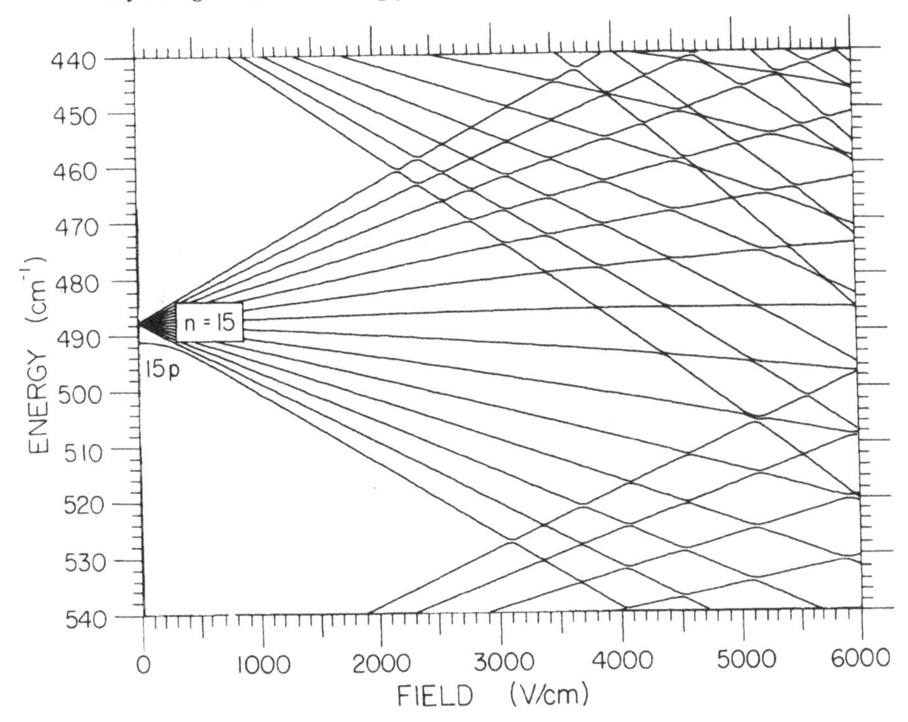

Fig. 3.5. Calculated Stark map of Li, $|m|=1$, in the vicinity of $n=15$. (From Zimmerman et al.[12])

Once the radial matrix elements are evaluated, diagonalizing the energy matrix is a straightforward task, limited only by computer capabilities. The results that we shall present were computed with a basis of 130 states using a small laboratory computer. The energies are generally accurate to 0.01 cm^{-1}; the relative energy (for instance, at an avoided crossing) is considerably more accurate.

3.3d. Results

The Stark structure of the alkali-metal atoms, or any Rydberg atom for that matter, is too complex to be characterized by a series of parameters or a list of eigenenergies. The most effective method we have found to display the properties of the atom–field system and to compare them with experiment is to generate a "Stark map," an energy-level diagram obtained by diagonalizing the energy matrix at a large number of field values (typically 100) and connecting the points to generate a plot of energy vs. field.

Theoretical and experimental results for $m=1$ states of lithium in the vicinity of $n=15$ are shown in Figs. 3.5 and 3.6, respectively, and the two are

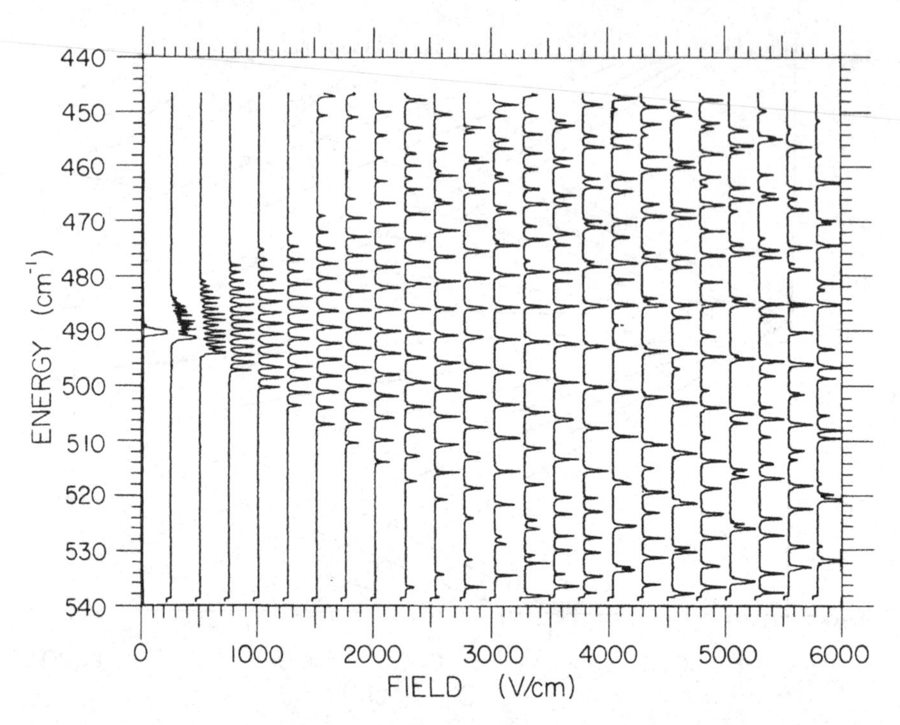

Fig. 3.6. Experimental Stark map for Li, $|m|=1$. Each vertical line represents a scan of the laser in the applied field indicated. The horizontal peaks are generated by pulsed field-ionization signals.

compared in Fig. 3.7. (Results in this section are from Ref. 12). Only one level, the p state, has a significant quantum defect, and it is so small ($\delta_1=0.05$) that Fig. 3.5 looks generally hydrogenic. At very low fields, the p state is slightly displaced from the term manifold and displays a second-order Stark effect. In spite of their similarities, however, the lithium and hydrogen Stark structures fundamentally differ from each other. None of the lithium's levels cross. Most of the avoided crossings are too small to be observed in Fig. 3.5 (the apparent avoided crossings are artifacts of the plotting procedure), but their effect can be important.

The Stark structure for lithium, $m=0$, is shown in Fig. 3.8. The structure, which is typical of one-electron atoms, differs dramatically from the hydrogenic structure displayed in the $m=1$ manifold. The s state, which has a quantum defect of 0.4, is displaced close to midway between the hydrogenic terms. As a result, the Stark levels show strong repulsions, causing them to lose completely their low-field character. This means, for instance, that knowledge of the energy and slope of a high-field state is not sufficient to identify the low-field quantum numbers. The complete Stark map is needed to

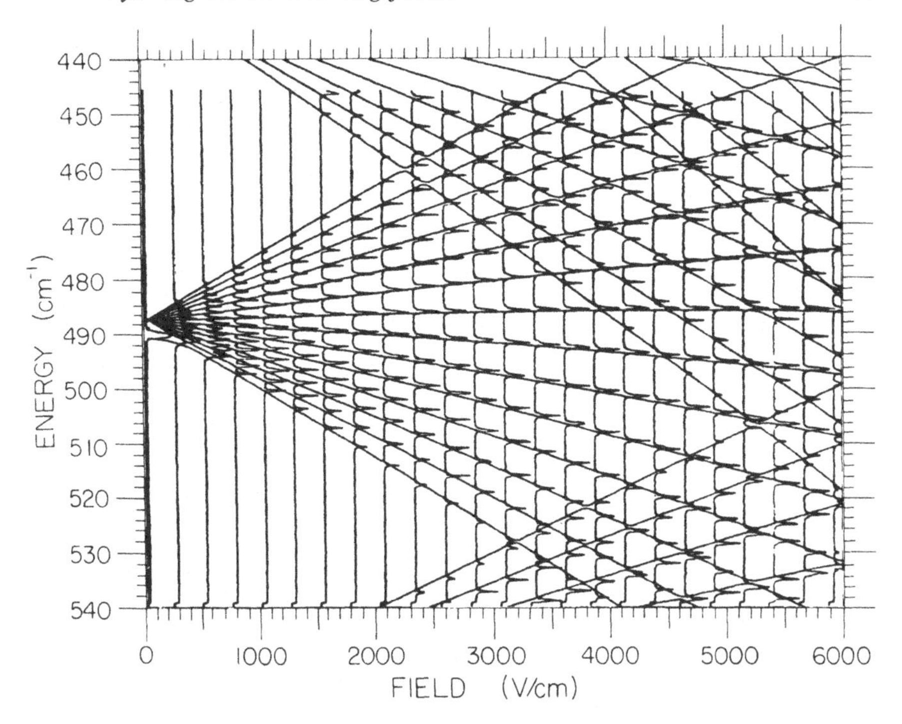

Fig. 3.7. Comparison of theory and experiment for Li, $|m|=1$ levels· (Figs. 3.5 and 3.6 superimposed)

establish the correspondence between low- and high-field states. There is no way to assign quantum numbers by counting the nodes. Occasional degeneracies - apparent level crossings - seem to occur more or less randomly. An experimental map of the $m=0$ states of lithium is shown in Fig. 3.9 and the theory and experiment are compared in Fig. 3.10. The signal intensity varies markedly across the map. The variations in oscillator strength that are responsible for this can be calculated by generating the eigenvectors from the diagonalized energy matrix. The oscillator strengths, like the energies, are in good agreement with experiment.

Fine structure becomes increasingly prominent for the heavier alkalis and must be included in the calculations. A straightforward method is to calculate the matrix elements of the Stark interaction in the J basis, where $j=l\pm\frac{1}{2}$. With the appropriate modifications to the angular matrix elements, the procedure is the same as before. However, the number of states needed for the diagonalization has doubled. This problem can be avoided by carrying out separate diagonalizations for $l=j\pm\frac{1}{2}$, ignoring fine structure (the "center-of-gravity" energies are used). The two separate results are then mixed by the fine-structure operator to give the Stark energies for the two values of j.

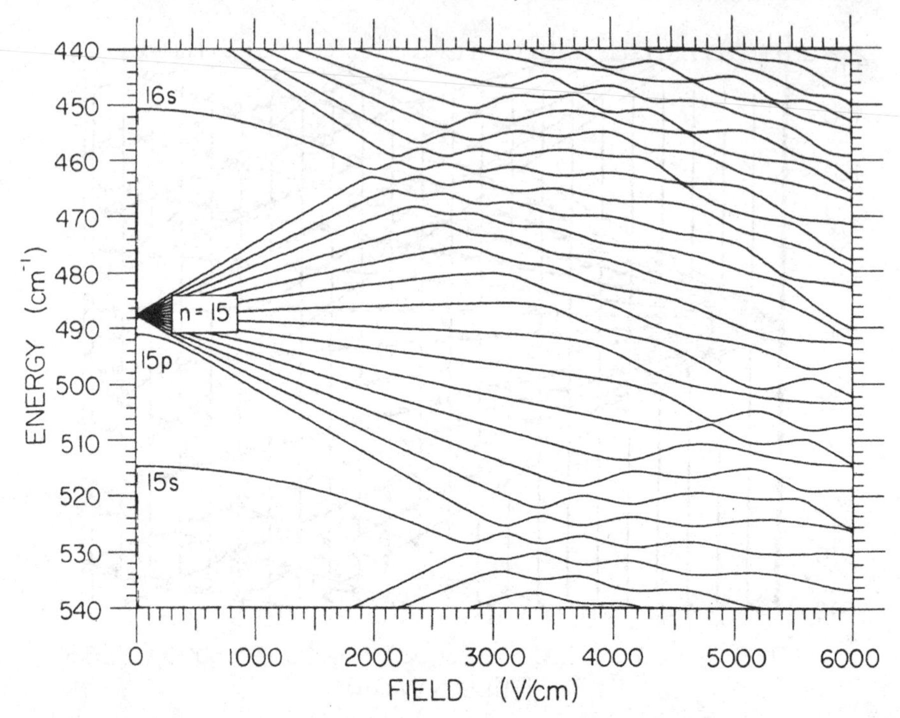

Fig. 3.8. Calculated Stark map of lithium, $|m|=0$, in the vicinity of $n=15$. Compare with Fig. 3.5.

3.3e. Search for a systematic solution

The computational method for calculating Stark structure that we just described suffers from a number of limitations. Although straightforward, the method is cumbersome and tedious. As n increases, the method becomes increasingly less tractable, and at high electric fields, it loses accuracy. More seriously, the method does not reflect the inherent simplicity of the problem. If we consider that the Stark structure of alkali-metal atoms differs from hydrogen Stark structure, which is known exactly, only because of a small number of nonvanishing quantum defects, perhaps two or three, then the enormous apparatus required to solve the secular equation seems inappropriate. A theory is needed to obtain the Stark structure of any one-electron atom directly from the hydrogen solution, providing a transformation generated by the quantum defects. A notable advance toward such an approach was made recently by Komarov, Grosdanov and Janev,[16] who developed a general method for calculating the size of avoided crossings.

Komarov et al. took the core interaction to be some short-range, spherically symmetric potential $V(r)$. They assumed that the exact energy of the alkali-

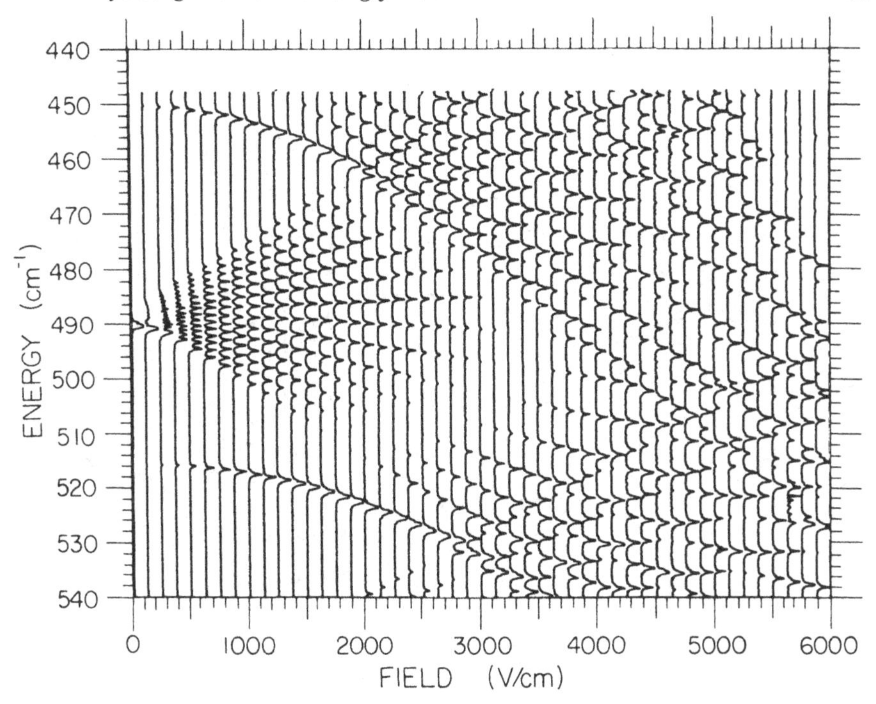

Fig. 3.9. Experimental Stark map of Li, $m=0$.

metal Rydberg states at zero field was given by first-order perturbation theory:

$$W_{nl} = (-1/2n^2) + \langle nl|V(r)|nl\rangle \tag{17}$$

For $n \gg 1$, hydrogen radial wave functions can be expanded as[17]

$$R_{nl}(r) = n^{-3/2}[f_l(r) + n^{-2}g_l(r) + O(n^{-4})] \tag{18}$$

which is valid for $r \ll n^2$ and $l \ll n$. Thus

$$\langle nl|V(r)|n'l\rangle = (nn')^{-3/2}[\delta_l^0 + \tfrac{1}{2}\delta_l^1(n^{-2}+n'^{-2}) + O(n^{-4})] \tag{19}$$

where

$$\delta_l^0 = \langle f_l(r)|V(r)|f_l(r)\rangle, \qquad \delta_l^1 = \langle g_l(r)|V(r)|f_l(r)\rangle \tag{20}$$

Then

$$W_{n,l} = \tfrac{1}{2}n^{-2} + n^{-3}[\delta_l^0 + \delta_l^1 + O(n^{-4})] \tag{21}$$

where δ_l^0 can be identified as the energy-independent quantum defect. (The quantum defects are determined from spectroscopic data and are constrained to the range $-0.5 \leqslant \delta \leqslant 0.5$).

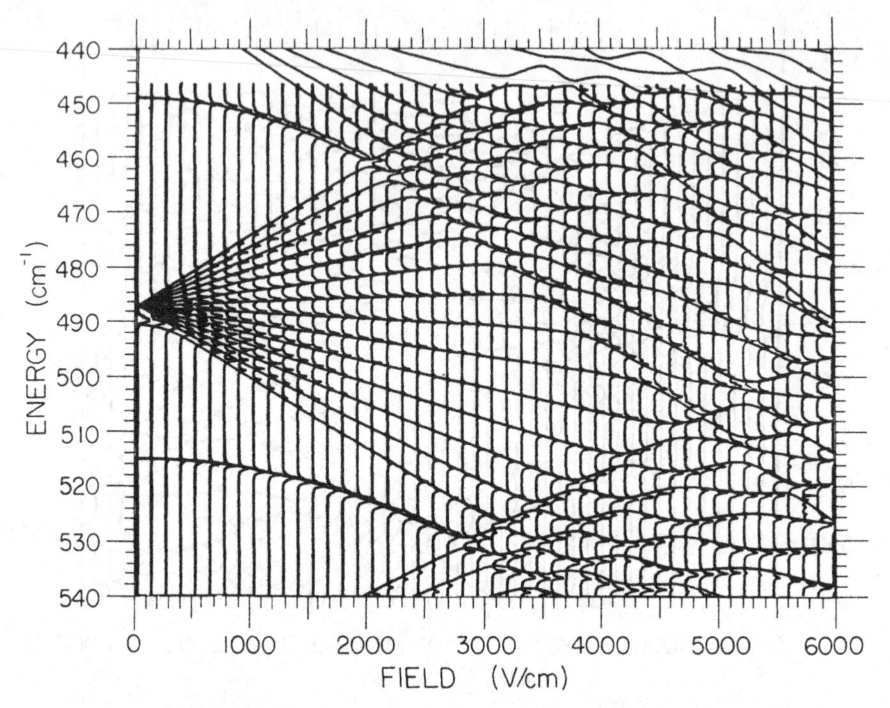

Fig. 3.10. Comparison of theory and experiment for Li, $|m|=0$ levels. (Figs. 3.8 and 3.9 superimposed)

The zero-field parabolic basis is assumed to describe the states of hydrogen in the electric field. Although this basis is not "good" in the presence of the field, as evidenced by the second- and higher-order terms in the Stark effect, it provides a good approximation to the wave functions near the origin where the Coulombic interaction dominates the electric field. The eigenstates of the alkalis in the field can be created from a superposition of the parabolic states, but, in contrast to hydrogen, the eigenstates are drastically modified by the field. The wave functions do not change near the origin, however, where the Coulomb field dominates. Thus the matrix elements of the core potential are essentially unchanged by the field. The leading error in a matrix element arises from the change in normalization at the origin as a function of field, which is largest for the levels with the largest Stark slope, the "extreme" components.

A unitary transformation relates the hydrogenic spherical and parabolic levels:[18]

$$|nkm\rangle = \sum_{l=|m|}^{n-1} C\left(\frac{n-1}{2}, \frac{m-k}{2}, \frac{n-1}{2}, \frac{m+k}{2}, l, m\right)|nlm\rangle \tag{22}$$

where $C(\cdots)$ is a Clebsch–Gordan coefficient and $k=n_2-n_1$. The matrix elements of the core interaction can now be evaluated in the parabolic basis:

$$\langle n'k'm|V(r)|nkm\rangle = \sum_{l=|m|}^{N} [\delta_l^0 + \tfrac{1}{2}\delta_l^1(n^{-2}+n'^{-2})]C(n'k'lm)C(nklm) \tag{23}$$

where N is the lesser of $n-1$ and $n'-1$. In the case of a well-isolated avoided crossing (e.g., when the quantum defects are small), degenerate perturbation theory predicts that the size of the avoided crossing will be twice the matrix element coupling the two states.

In one case where an avoided crossing was measured accurately,[19] the crossing between (18, 16, 0, 1) and (19, 2, 15, 1) in lithium at 1008 V cm^{-1} (the crossing is encircled in Fig. 3.18), the value from Eq. (23) was 7.87×10^{-2} V cm^{-1}, which agrees to within 3% with the measured value $7.6820(12)\times10^{-2}$ V cm^{-1}.[20] The difference appears to be due to second-order effects that were neglected in the calculation.

If we consider the behavior of avoided crossings as the quantum defects increase, the problem reaches the point where we can no longer treat them as in isolated two-level systems. A multitude of matrix elements may affect the energies. In such a case, a numerical diagonalization is required. However, the task involves diagonalizing the core perturbation using a parabolic basis, a task far simpler than diagonalizing the electric field interaction in a spherical basis. For example, because the average core interaction in a term decreases as n^{-4}, whereas the density of states increases as n^4, the number of states needed for a given precision is independent of n. In contrast, the electric field interaction increases as n^2, and the size of the calculation in a spherical basis increases rapidly with n.

This approach gives good agreement with numerical calculations and experiments for small quantum defects but remains to be tested for large quantum defects. A generalization to multielectron systems should be possible but remains to be developed.

3.4. Field ionization

Hydrogen has no true stationary states in an electric field; the electron will inevitably tunnel through the Coulomb barrier and be carried away by the field, a process known as *field ionization*. At low fields, however, tunneling occurs so slowly that the states are effectively stationary. (This approximation underlies the perturbative approach to the Stark energy described in the last section.) Field ionization starts to become observable when the tunneling rate becomes comparable to the natural radiative decay rate. For fields in this region, the energy can be described by a complex quantity whose imaginary part is the ionization rate. In very strong fields, the states ionize so rapidly that they lose the discrete character and their decay no longer follows a simple exponential law. Under such conditions, the problem is most naturally approached from the point of view of scattering theory and the eigenstates are replaced by resonances in the continuum.

The problem of hydrogen in arbitrarily large electric fields has attracted renewed theoretical interest and a considerable body of literature has been devoted to it. Chapter 2 of this volume discusses theory and Chapter 13 discusses experiment for hydrogen (see also Ref. 21). The theoretical background was reviewed in some detail by Luc-Koenig and Bachelier.[22] Consequently, we shall focus our attention on field ionization in other one-electron atoms. As we shall see, their ionization behavior differs markedly from that of hydrogen.

Although the primary interest in field ionization stems from its role as an elementary atomic process, field ionization also needs to be understood because of practical applications. Field ionization has proven to be an invaluable experimental technique for detecting Rydberg atoms. It has played a pivotal role in numerous experiments on Rydberg atoms, including both spectroscopic and scattering studies, and it promises to be useful in applications ranging from infrared detection[23] to isotope separation.[24] Thus there are important practical reasons for understanding the ionization process in some detail.

A related topic that we shall also discuss in this section is the discovery of positive energy resonances in an electric field. Although their existence was not anticipated theoretically, it is now recognized that they represent a characteristic feature of atoms in strong fields.

3.4a. Field ionization in hydrogen

As we have seen, the Stark structure of hydrogen provides a useful background for understanding Stark structure in other one-electron atoms. Similarly, the field-ionization behavior of hydrogen serves as a natural point of departure for understanding field ionization in other one-electron atoms. For this purpose, we shall briefly review some of the qualitative features of field ionization. Detailed discussions can be found in Refs. 4, 21, and 22.

The energy levels of hydrogen vary uniformly with the field, and every state can be identified by its zero-field quantum numbers $(n, n_1, n_2, |m|)$. In a moderate electric field, each level ionizes at some uniform rate Γ, which is a rapidly increasing function of the field. In this regime, each state undergoes simple exponential decay with a lifetime $\tau = 1/\Gamma$. The width of each energy level is $\Delta W = \Gamma$. Ionization rates for a few levels from the $n = 14$ manifold are shown in Fig. 3.11. (These rates were calculated by Bailey, Hiskes, and Riviere[25] using a WKB technique developed by Rice and Good.[26]) The rapid increases of Γ with field, a factor of 10^6 for a 20% change in field, is characteristic of the tunneling process. Note that the state $(n_1 = 0, n_2 = 13)$, which is the lowest member of the linear Stark manifold, ionizes at the lowest field, and the level with the largest energy $(n_1 = 13, n_2 = 0)$ ionizes at the highest field. The experimental points, from work by Littman et al.,[27] are from measurements on $|m| = 2$ levels of sodium: this particular state is expected to be accurately described by hydrogenic theory.

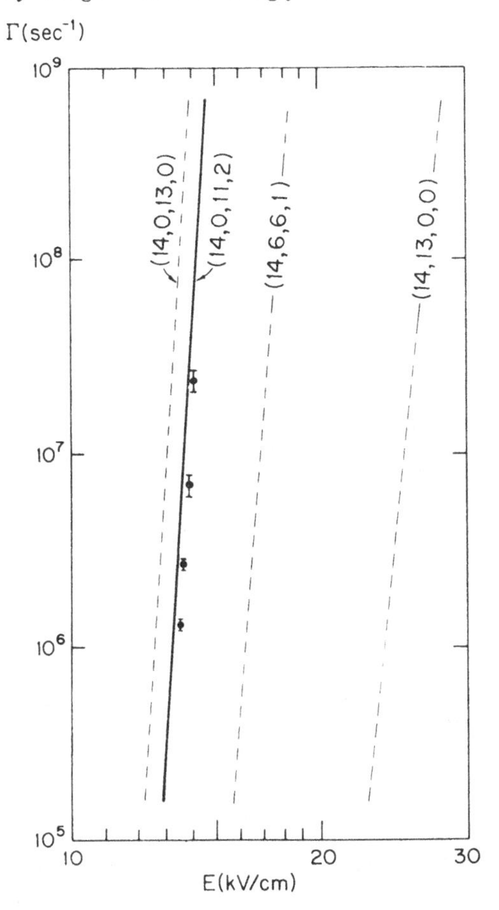

Fig. 3.11. Experimental ionization rates for a hydrogenlike state of sodium ($n=14$, $n_1=0$, $n_2=11$, $|m|=2$). Dashed lines are calculated by Bailey, Hiskes, and Riviere[25]; solid line is interpolated.

Figure 3.2 provides a somewhat different picture of the characteristic field-ionization behavior of hydrogen. The tick mark on each level indicates the field for which the ionization rate equals the radiative decay rate. Other marks indicate the fields for higher ionization rates. When the rates exceed about 10^{11} s^{-1}, the levels become so broad that they overlap, effectively forming a continuum. From an experimental point of view, an eigenstate effectively terminates when its ionization rate reaches about 10^{11} s^{-1}.

3.4b. Field ionization in alkali-metal atoms

Accurate theory for field ionization of a one-electron atom exists only for hydrogen. If the central potential departs from $-1/r$, the problem is not

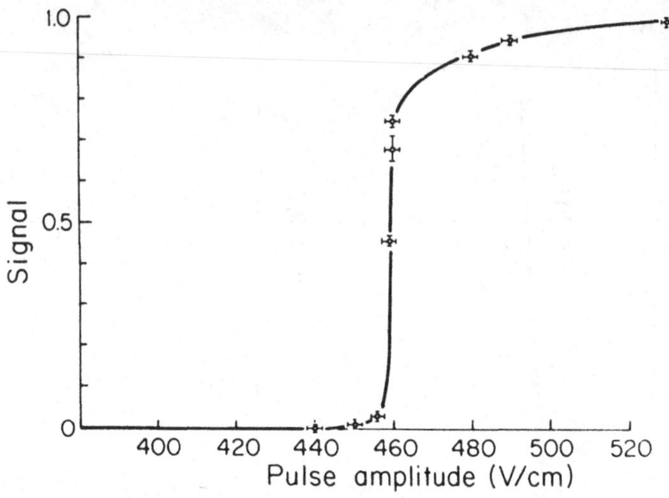

Fig. 3.12. Ionization signal vs. amplitude of electric field pulse for the 31s state of sodium.

separable. Although no entirely adequate theoretical approach has been developed for the general problem, we can understand many of the important qualitative features from simple energy arguments.

The potential for a single electron in a Coulomb field and an applied field **F** along the z axis is

$$V_F(r) = -(1/r) + Fz \tag{24}$$

where V_F has a maximum on the z axis at $z=-1/F^{1/2}$. This maximum is actually a saddlepoint because V increases off the z axis. The saddlepoint potential is

$$V_{\text{sp}} = -2F^{1/2} \tag{25}$$

For a state to be bound, its energy must be below the V_{sp} or

$$W_{\text{th}} = -2F^{1/2} \tag{26}$$

Thus, in any given applied field **F**, there is a "threshold" energy for ionization. Alternatively, if the field applied to a Rydberg atom is slowly increased, the atom will ionize at a critical value

$$F_{\text{c}} = W^2/4 \tag{27}$$

For a rough estimate of the critical field, we can neglect the Stark effect and take $W=-1/(2n^2)$. This yields

$$F_{\text{c}}^* = 1/16n^4 \tag{28}$$

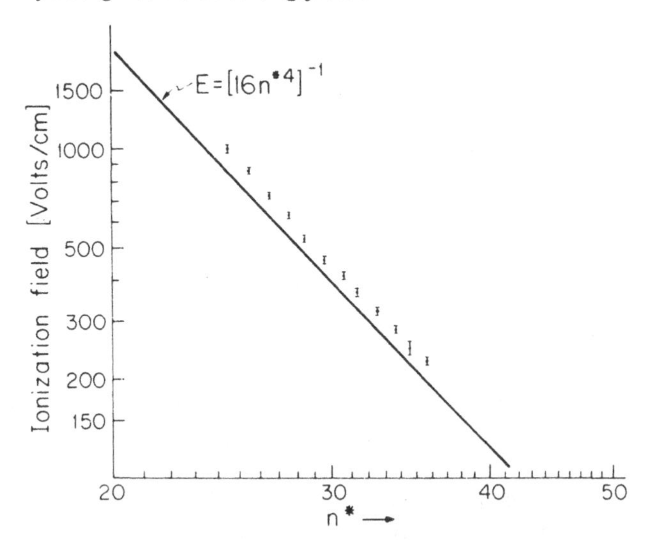

Fig. 3.13. Threshold field vs. effective quantum number for a sequence of sodium s states. (Data from Ducas et al.[28])

In laboratory units, $F_c^* = 3.2 \times 10^8/n^4$ V cm^{-1}; for $n=30$, $F_c^* = 400$ V cm^{-1}. If a Rydberg state is populated and an electric field pulse is then applied, the atoms often start to ionize abruptly when the pulse amplitude reaches a field close to F_c^*. An example of this threshold behavior is shown in Fig. 3.12 and a plot of the measured threshold field for a number of s states in sodium is shown in Fig. 3.13. For an alkali in an s state, we might expect to replace the principle quantum number by the effective principal quantum number $n^* = n - \delta_0$, so that Eq. (28) becomes $F_c^* = 1/(16n^{*4})$. The data show the predicted $1/n^{*4}$ dependence. The displacement between the experimental and theoretical plots is not surprising in view of the fact that we neglected the shift in energy due to the Stark effect.

The data in Fig. 3.13 show that, in this case at least, field ionization obeys a simple well-defined scaling law. In fact, the threshold field is so well defined that by measuring it we can often determine the state of the atom.

A graphical view of ionization at the saddlepoint potential, sometimes called "saddlepoint" ionization, is obtained by plotting the locus of F_c vs. energy. The stable region lies below the parabola $W_{th} = -2F^{1/2}$. Other graphical methods for visualizing field ionization were presented by Jaquinot.[29]

In deriving the criterion for stability, we neglected the atom's angular momentum. Angular momentum can be introduced into our classical model by a simple argument.[30] Although total angular momentum is not a constant of motion because of the noncentral field F, the component L_z of angular momentum about the z axis is conserved. Denoting the magnitude of L_z by m, we have an effective potential

$$V_{\text{eff}} = -(1/r) + Fz + (m^2/2\rho^2) \tag{29}$$

where $\rho^2 = x^2 + y^2$. The effect of the last term, the centrifugal potential, is to displace the saddlepoint from the z axis and raise it. The result is a slight correction to Eq. (27), which becomes

$$W = -2F^{1/2} + |m|F^{3/4} + \tfrac{3}{16}m^2F + \cdots \tag{30}$$

For low-$|m|$ states, the last two terms represent a small correction that shifts the threshold field by a few percent at most. The ionization threshold is shown in Fig. 3.2 by a dotted line.

These ideas were demonstrated experimentally by panoramic studies of field ionization in lithium.[31] Rydberg states were excited in a dc field, using the methods described earlier. To show the onset of spontaneous field ionization, the high-voltage ionizing pulse was delayed a few microseconds and a timing gate was used to reject signals that appeared prior to that delay. With this scheme, stable Rydberg atoms produced signals, but atoms that spontaneously ionized during the delay period did not. When the signals are plotted to produce a Stark structure map, the levels simply vanish at the critical field. An example is shown in Fig. 3.14. The states are $|m| = 1$ levels of lithium in the vicinity of $n = 19$. (These levels are shown in Fig. 3.2.) The dotted line is the locus given by Eq. (30). The agreement of experiment with the predicted value for the threshold field is striking. There seems no reason to doubt that the simple classical model is, in some sense, realistic. Nevertheless, as we shall see, serious difficulties remain.

It should be pointed out that we can actually do a rigorous classical analysis. The two-dimensional Kepler problem in an applied field was solved by Banks and Leopold[32] to yield values for the critical field at which the motion becomes unbound. The results disagree markedly with those of our simple one-dimensional treatment, as do the quantum-mechanical results. In Fig. 3.2, for instance, the saddlepoint threshold, Eq. (30), can be compared with the points at which $n = 19$ levels ionize by tunneling. The saddlepoint threshold has little to do with the tunneling fields. The levels ionize in reverse order according to the two descriptions, and the ionization fields can differ by a factor larger than three.

3.4c. Resolution of the discrepant views

The difficulty arises because the "saddlepoint" model answers the question "at which field is ionization energetically possible?" but has essentially nothing to say about the ionization rate. The data in Fig. 3.14, as well as a great many other observations, show that the rate grows abruptly at the threshold field, but these observations are on alkali metals or other simple atoms. The fact that the atoms are not exactly hydrogenic is crucial.

Let us consider the broadening of a particular Stark level of hydrogen, as

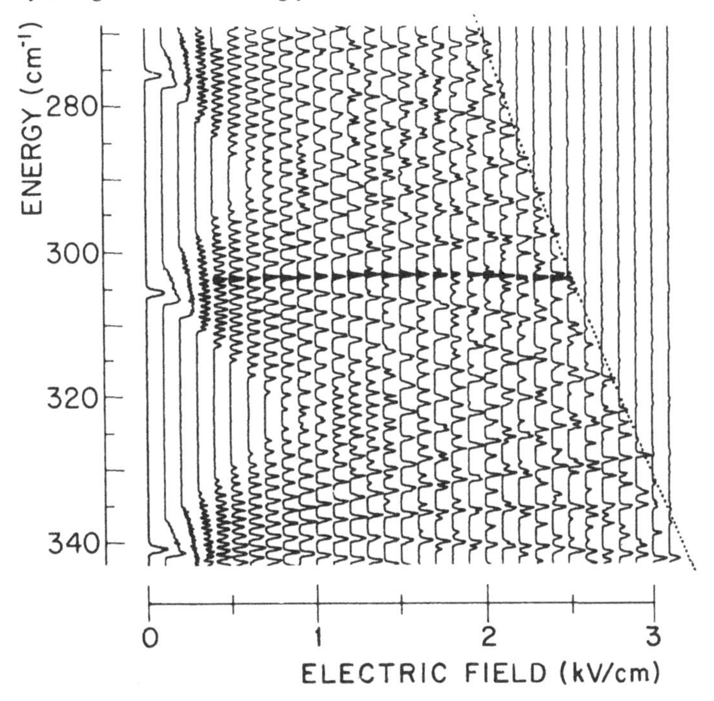

Fig. 3.14. Threshold ionization for $|m| = 1$ states of lithium in the vicinity of $n=19$. Disappearance of the signal occurs when the ionization rate exceeds $3 \times 10^5\ \mathrm{s}^{-1}$. Dotted line is the locus of stability given by Eq. (30). Calculated levels are shown in Fig. 3.2. (From Littman, Kash, and Kleppner[31])

shown in Fig. 3.15. Note that many levels from higher terms cross *it* and that these levels may be extremely broad because of field ionization. Thus, in the presence of an electric field, each Stark level is actually embedded in a "sea" of higher-lying levels, both discrete and continuous.

The degenerate "sea" of levels has no effect on the dynamics of hydrogen because the Stark states do not mix as a result of the dynamic symmetry discussed in Section 3.3a. The symmetry is only exact, however, for a pure Coulomb potential. Any perturbation to the potential, for instance, the core interaction in an alkali-metal atom, upsets the symmetry and causes the hydrogenic Stark levels to mix. If a stable state mixes with a decaying state, it also decays.

We can illustrate these ideas with a simple example based on a two-level system. Energies W_a and W_b, respectively, depend on the field and, in the absence of any perturbation levels, are degenerate at some value of the field F_0. We assume that level a decays by ionization at rate Γ_a (which can also depend on the field) but that the decay rate of state b is negligible. The states are coupled by a perturbation V, for instance, the core perturbation in an alkali-metal atom.

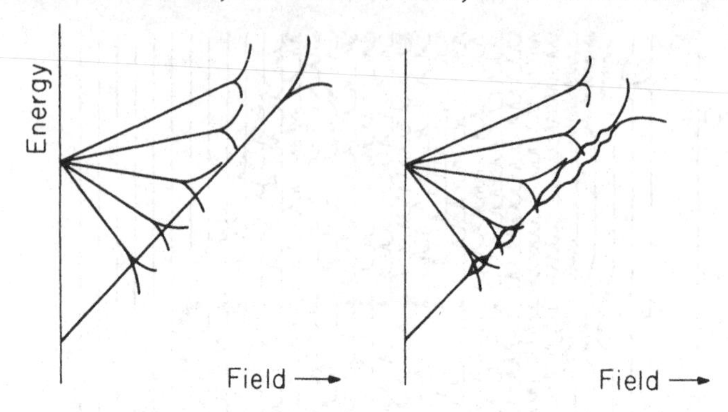

Fig. 3.15. Sketch of Stark levels of hydrogen in strong electric fields (*left*) and of an alkali-metal atom (*right*). Hydrogenic Stark levels do not interact; in an alkali, the levels can mix, causing an otherwise stable level to decay. Level widths indicate the decay rates.

The eigenvalue equation for the complex energy is

$$\begin{pmatrix} W_a - \frac{1}{2}i\Gamma_a - W & V \\ V^* & W_b - W \end{pmatrix} = 0 \tag{31}$$

For simplicity we consider the solution at the level crossing, $W_a = W_b = W_0$, for the case $\Gamma_a \gg V$. The energies are, approximately,

$$W_a' = W_0 - \tfrac{1}{2}i\Gamma_a, \qquad W_b' = W_0 + (2i|V|^2/\Gamma_a) \tag{32}$$

The real parts of the energy are equal, in violation of the "no-crossing" theorem. Such behavior was first pointed out by Lamb[33] in his discussion of the effect of radiative damping on the fine structure of hydrogen. More importantly, state b has an imaginary component of energy. The damping rate $\Gamma_b' = 4V^2/\Gamma_a$ is small compared with Γ_a, but in many experimental situations, it can be significant.

The situation is generally more complicated than this model suggests because of the simultaneous interaction of many levels. Nevertheless, in one case where only a few levels are important,[27] the analysis gives good agreement with the observation of the decay rate, as shown in Fig. 3.16.

3.4d. Field ionization and autoionization

For fields between the "saddlepoint" threshold and the point where hydrogen ionizes (i.e., where the ionizing rate exceeds the radiative decay rate), the ionizing process is similar to autoionization. In each case, degenerate stable and continuum levels are coupled by a perturbation that mixes some of the continuum character into the stable state. If the perturbation is known, the decay rate can be calculated.

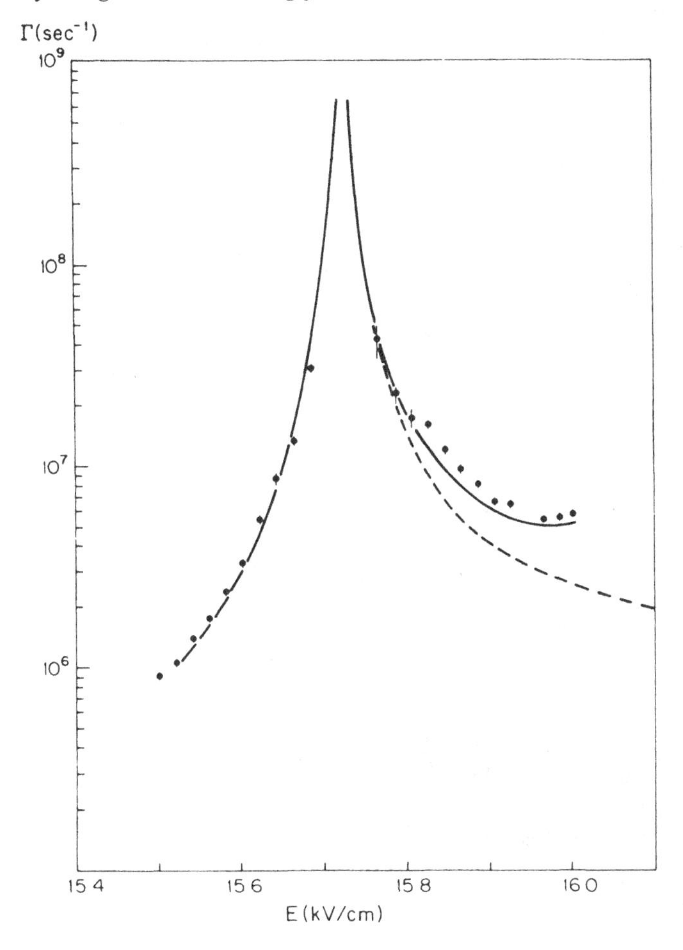

Fig. 3.16. Field ionization at a level crossing, as observed in sodium. Dots are experimental points; solid lines are calculated values; dash lines are calculated neglecting another level crossing at higher fields.

Autoionization is generally studied in frequency, rather than in time, by measuring the energy dependence of a photoionization. Autoionizing lines have a characteristic shape, the Fano profile,[34] which arises from interference between the discrete and continuum states. Feneuille et al.[35] observed a Fano profile in the photoionization cross section of rubidium in the saddlepoint region, providing further evidence that field ionization in this region is indeed an autoionization process.

The two mechanisms of field ionization, tunneling and autoionization, are qualitatively different. Autoionization commences at the saddlepoint field, but the autoionization rate does not increase exponentially with the field. As a

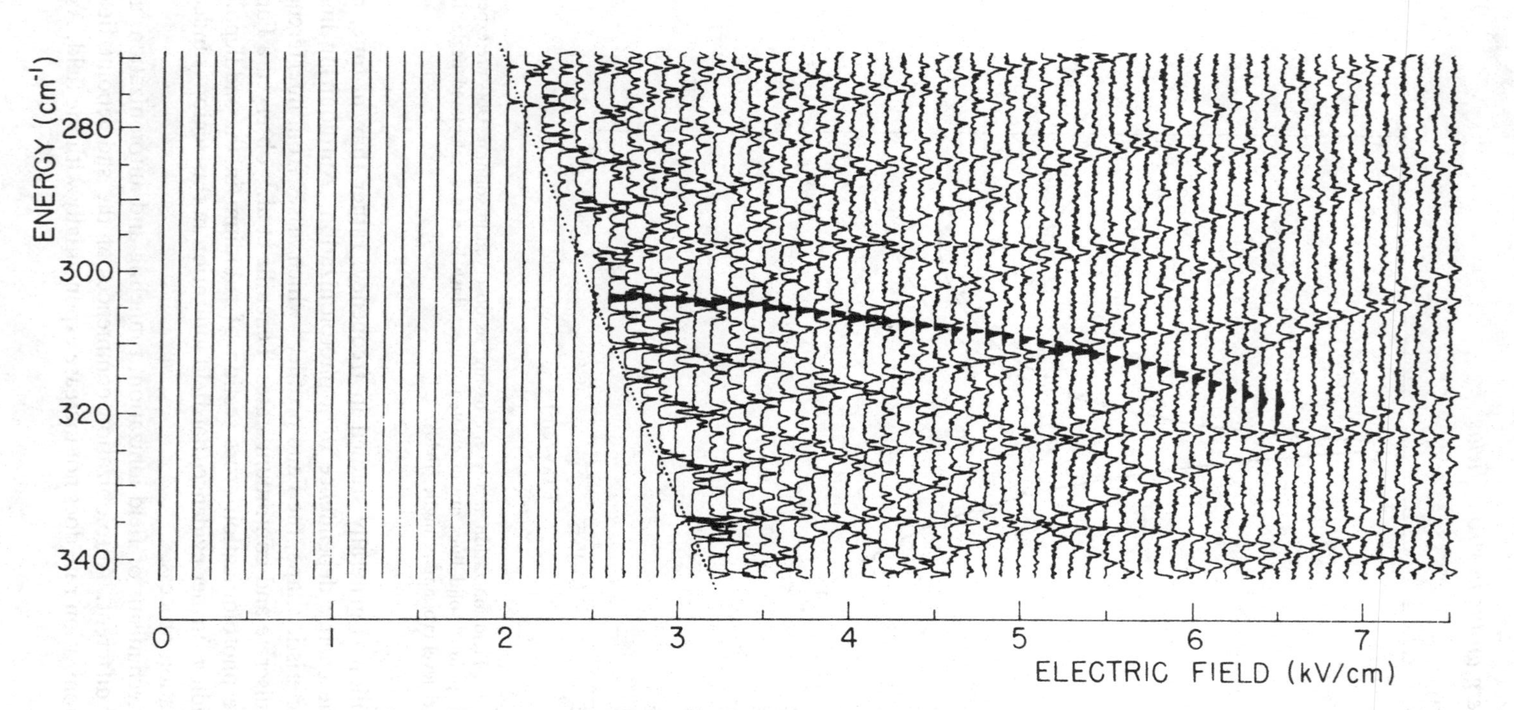

Fig. 3.17. Autoionizing states of lithium, $|m| = 1$, above the classical threshold. This is a continuation of the map for nonionizing states (Fig. 3.14). One level has been darkened for clarity.

result, levels in the autoionizing region retain their character for a considerable range of field, in contrast to the situation of tunneling, where the levels rapidly merge into the continuum as the field is increased. The spectra in Fig. 3.17 confirm these features. This is a continuation of Fig. 3.14, but the signals arise entirely from ionization in the static field (no ionizing pulse is applied). The levels first appear at the saddlepoint threshold and retain their discrete character to the tunneling region, where they abruptly disappear. Because the system, lithium ($m=1$), is essentially hydrogenic, the Stark structure is similar to hydrogen and the levels tunnel in agreement with hydrogenic theory (Fig. 3.2).

The basic principles of field ionization appear to be well understood, but a general theory for calculating the ionization rates for one-electron atoms is lacking. Although major features of field ionization can often be predicted with confidence, many applications require a detailed understanding of the ionization rates, frequently for adjacent levels. In such cases, there is, at present, no alternative to careful experimental observations; better theoretical approaches are needed.

3.4e. Experimental aspects of field ionization

Field ionization has been widely adopted for detecting Rydberg atoms because of its simplicity and efficiency and because it can be used to discriminate the Rydberg state. In its simplest form, a pulsed field is applied and the ion or electron signal is measured. By controlling the pulse amplitude, the final state can be distinguished.[28,36-41] Gallagher, Hill, and Edelstein[42] used pulsed field ionization to detect microwave transitions between fine-structure states of sodium Rydberg atoms; the change in m_j in zero field altered the value of m_l in the ionizing field and changed the ionization characteristics. Fabre, Goy, and Haroche[43] used a controlled voltage ramp for the ionizing field. A gated detector then provided a time signature for the individual Rydberg substates. Leuchs, Smith, and Walther[44] applied pulsed field ionization to detect quantum beats in Rydberg states.

A central problem in detecting a low-field Rydberg state by pulsed field ionization is understanding how the state evolves as the field is applied. If the field increases slowly, the state follows an adiabatic path, which means that none of the levels cross. If the field increases rapidly, the atom can traverse the avoided crossing diabatically: the atom "jumps" across the avoided crossing. The threshold fields can differ significantly for the two cases. Observations of nonadiabatic effects were reported by Gallagher et al.[45] and Jeys et al.[46]

The probability of traversing a simple avoided crossing diabatically is given by the Landau–Zener formula[47]

$$P = \exp[-2\pi|V|^2/\hbar\,(dW/dt)] \tag{33}$$

where V is the matrix element of the perturbation that gives rise to the avoided crossing ($2V$ is the separation of the levels at the crossing) and dW/dt the rate

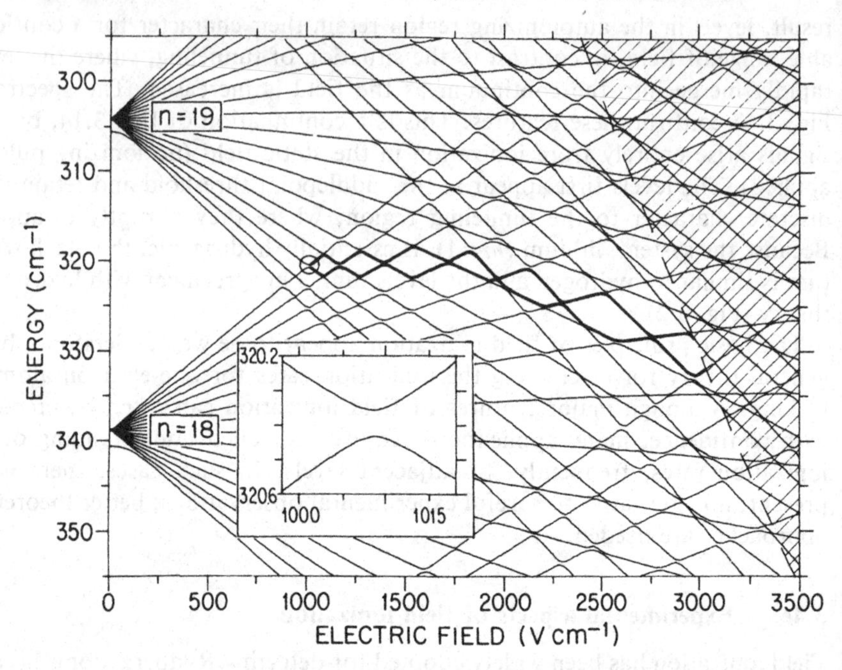

Fig. 3.18. Adiabatic routes followed by two levels to the ionization threshold (dashed). Inset shows details of the encircled avoided crossing. (From Rubbmark et al.[19])

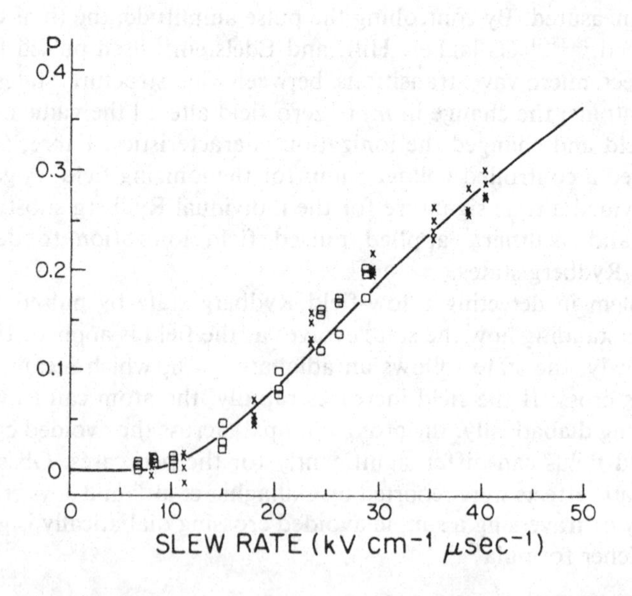

Fig. 3.19. Diabatic transition probability vs. electric field slew rate for the avoided crossing shown in Fig. 3.18. The solid line is the Landau–Zener result, Eq. (33). The crosses and squares represent data for two different experimental conditions. (From Rubbmark et al.[19])

of change of the level separation as the avoided crossing is approached. In an electric field, $dW/dt=(dW/dF)(dF/dt)$, where dF/dt is the slew rate of the field.

The probability for a diabatic transition in a rising electric field was measured by Rubbmark et al.[19] for the avoided crossing shown in Fig. 3.18. The level separation was determined by radio-frequency spectroscopy and all the other parameters in Eq. (33) were individually measured. The results, shown in Fig. 3.19, are in good agreement with theory.

To predict how a Rydberg state will field ionize, we need to understand the level structure in sufficient detail to predict the path it will follow to the threshold. In practice, we often make a reasonable guess and hope for the best. To predict the ionization rates in the threshold region is an even more difficult task, but this is the critical problem in experimentally resolving two levels that are close in energy. The rates are by no means discontinuous at the threshold, though they can vary rapidly (and irregularly), as Fig. 3.16 shows. Unfortunately, these can be calculated only in the simplest cases and we generally have to investigate experimentally each level of interest.

3.4f. Positive energy resonances

According to the saddlepoint model for ionization, the threshold field vanishes at zero energy and no bound positive energy states are possible. Simple scaling arguments based on hydrogenic tunneling theory likewise preclude the possibility of long-lived positive energy states. It came, therefore, as something of a surprise when Freeman and his colleagues[48] observed resonances for the photoionization cross section for rubidium in an electric field that extend into the positive energy region. Although the detailed theory remains to be tested, the basic properties appear to follow from rather simple arguments.[49]

The resonances can be explained in terms of a semiclassical model for motions in the separated potentials for η and ξ. However, some insight can be obtained by making the *ansatz* that the problem is one dimensional, involving motion along the z axis. This is most plausible for $m=0$ states, the only states for which the resonances have been observed.

If we consider the highest-energy Stark state for a particular term, for instance the $k=7$ state shown in Fig. 3.4, then it is apparent that the electron spends most of its time along the positive z axis, the "uphill" side of the potential well. The motion is essentially confined to the region $z>0$, so that the penetration of the electron cloud beyond the origin is minimal in this state. Thus the electron is effectively trapped in a potential well $-(1/z)+Fz$, and it is also confined to the region $z>0$. For the positive energy states, the electron spends most of its time in the linear potential region, and the eigenstates are separated by the characteristic separation of a linear potential well.

These ideas can be developed semiclassically using either the one-dimensional approximation[49] or the separated two-dimensional (parabolic) potentials.[50] The result is a characteristic $F^{3/4}$ law for the spacing between resonances, in good agreement with the observations. The scaling law was also obtained by Rau[51] from his general theory for motion in a "strong mixing" regime.

A fully quantum mechanical treatment was developed by Luc-Koenig and Bachelier[52] in which the wave functions are numerically generated for the positive energy region. The resonances arise from variations in the density of states and in the oscillator strengths for the transition. In contrast to the semi-classical theories, the quantum-mechanical theory predicts resonances for $m=1$ states. Such resonances have not yet been observed. However, the theory was developed for hydrogen, and the experiments were carried out on an alkali. The role of symmetry in the problem is not clearly understood. Experiments on hydrogen and extension of the theory to deal with alkali atoms are both needed.

3.5. Atoms in strong magnetic fields

3.5a. Background

Magnetic fields have played a useful and sometimes important role throughout the history of atomic physics. The Zeeman effect, for example, provided the first direct evidence of the electromagnetic origin of light, and the Stern-Gerlach experiment proved the reality of spatial quantization. Over the years, studies of atom–field interactions have provided important keys to understanding spin-dependent interactions and angular momentum coupling schemes in atoms. The Zeeman effect is intrinsically small compared with the electrostatic interaction, however, so that the magnetic interaction in these early studies was always feeble. There are two reasons for this. First, laboratory-size magnetic fields are intrinsically small. The atomic unit of field is $e/a_0^2 = 1.7\times10^3$ T (1 T $=10^4$ G), whereas laboratory fields are generally less than 30 T. Second, the coupling constant for the paramagnetic interaction, the Bohr magneton, is also small, $\alpha/2$ ($\simeq 4\times10^{-3}$) atomic units.

At very high magnetic fields, however, the diamagnetic interaction, which increases quadratically with field, is important. This interaction scales as n^4B^2; in high excited atoms, it can exceed the Coulomb interaction, providing an opportunity to study atomic structure under strong field conditions. Much of the recent interest in such studies stems from the discovery by Garton and Tomkins[53] that the absorption spectrum of barium in a strong magnetic field can exhibit periodic resonances that extend above the ionization limit. The structure is now well understood, but it is recognized that this reflects only one aspect of a much broader problem. Numerous experimental and theoretical attacks on the problem are in progress. Undoubtedly a major attraction of this work is that, in spite of its fundamental simplicity, the problem of hydrogen in a magnetic field does not have a general solution. In fact, it can be

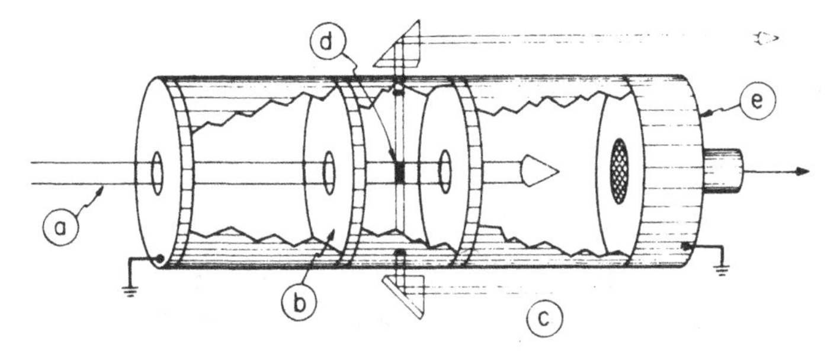

Fig. 3.20. Interaction region for the study of Rydberg atoms in magnetic fields. The apparatus is on the axis of a superconducting solenoid, not shown: (a) atomic beam, (b) field-ionization electrode, (c) laser beam, (d) interaction region, (e) surface-barrier diode. (From Castro[55])

argued that this is the principal remaining problem in the elementary quantum mechanics of a one-electron atom.

A comprehensive article by Garstang[54] reviewed the subject of atoms in strong magnetic fields and summarized progress up to a few years ago. In this section we shall briefly review the theory and go on to discuss recent developments.

3.5b. Experimental methods

Techniques for studying highly excited atoms in strong fields include ultraviolet absorption spectroscopy and various forms of laser spectroscopy. One class of experiments uses a gas cell; the excited atoms are detected by collisional ionization in a space-charge-limited diode. The results we shall present here were obtained with atomic beam methods, following the general principles described in Sect. 3.2. The atomic beam moves parallel to the magnetic field, which largely eliminates the motional electric field. This is extremely helpful in studying high-field spectra because motional Stark shifts can be large and motional fields can mix states with different values of m_l, further complicating a complicated problem.

A diagram of the experimental interaction region is shown in Fig. 3.20. A superconducting solenoid, not shown, provides a field up to 10 T. The atoms are ionized with a pulsed electric field. The electrons are accelerated to 7 kV and detected with a surface-barrier diode. For studies of sodium, a two-step excitation scheme is used: 3s → 4p followed by 4p → Rydberg state. Even-parity states with $m_l = 0, \pm 1, \pm 2$ can be excited, depending on the laser polarization.

3.5c. Basic Hamiltonian

An electron moving in a potential $V(r)$ and a magnetic field $\mathbf{B} = \nabla \times \mathbf{A}$, where $\mathbf{A}$ is the vector potential, is governed by the Hamiltonian (neglecting spin)

$$H = \tfrac{1}{2}(\mathbf{p} + \alpha\mathbf{A})^2 + V(r) \tag{34}$$

The Coulomb gauge ($\nabla \cdot \mathbf{A} = 0$) is most convenient for our problem. For a uniform field along the z axis, $\mathbf{A} = (\mathbf{B} \times \mathbf{r})/2$, and Eq. (34) becomes

$$H = H_0 + \tfrac{1}{2}\alpha\mathbf{L} \cdot \mathbf{B} + \tfrac{1}{8}\alpha^2 B^2 r^2 \sin^2\theta \tag{35}$$

The first term is the Hamiltonian for the free atom, the second the orbital paramagnetic interaction, and the last the diamagnetic interaction H_d. Our goal is to understand the structure in fields where the diamagnetic interaction plays the principal role, but to set the problem in a more familiar context, we shall display the Hamiltonian with all the major spin-dependent terms included:

$$H = H_0 + H_p + H_d + H_s + H_{hf} \tag{36}$$

where

$$H_p = \tfrac{1}{2}\alpha(\mathbf{L} + g_e\mathbf{S}) \cdot \mathbf{B} \tag{37a}$$

$$H_d = \tfrac{1}{8}\alpha^2 B^2 r^2 \sin^2\theta \tag{37b}$$

$$H_s = \xi(r)\mathbf{L} \cdot \mathbf{S} \tag{37c}$$

$$H_{hf} = a(\mathbf{I} \cdot \mathbf{J}) + \tfrac{1}{2}\alpha g_I' \mathbf{I} \cdot \mathbf{B} \tag{37d}$$

where H_p, H_d, H_s, and H_{hf} are, respectively, the paramagnetic, diamagnetic, spin–orbit, and hyperfine terms, g_e and g_I' are the electronic and nuclear g factors, $\mathbf{S}$ and $\mathbf{I}$ their spins, $\xi(r)$ the radial spin–orbit operator, and a the hyperfine constant.

3.5d. Low-field solution

At low magnetic fields, the spin–orbit interaction couples $\mathbf{L}$ and $\mathbf{S}$ to form total angular momentum $\mathbf{J}$, and the hyperfine interaction couples $\mathbf{J}$ and $\mathbf{I}$ to form the total angular momentum $\mathbf{F}$. As the field is increased, the electron-field interaction H_p first exceeds the hyperfine interaction H_{hf} and then exceeds the spin–orbit interaction H_s (Paschen–Back effect). The electronic energy is not significantly altered by any of these interactions, however, and analyzing their effects is chiefly a matter of properly recoupling the angular momentum. Such problems have been extensively studied, and because our goal is to understand high-field behavior where these interactions are unimportant, we shall simply neglect electronic and nuclear spin and deal with a hypothetical hydrogen atom governed by the Hamiltonian

$$H = \tfrac{1}{2}p^2 - (1/r) + \tfrac{1}{2}\alpha\mathbf{L} \cdot \mathbf{B} + \tfrac{1}{8}\alpha^2 B^2 r^2 \sin^2\theta \tag{38}$$

If we omit the last term and treat the third as a perturbation, we obtain the familiar result

$$E(n,l,m) = -\tfrac{1}{2}n^{-2} + \tfrac{1}{2}\alpha mB \tag{39}$$

In this region, n, l, and m are all good quantum numbers. Each l manifold is separated into $2l+1$ sublevels split by the linear Zeeman energy $\frac{1}{2}\alpha B$.

The diamagnetic interaction $H_d = \frac{1}{8}\alpha^2 B^2 r^2 \sin^2\theta$ couples states with $\Delta l = 0, \pm 2$ and Δn unrestricted. The only good quantum numbers are m and parity. The matrix elements are[55]

$$\langle nlm|H_d|n'lm\rangle = \frac{\alpha^2 B^2}{8}\, 2\left[\frac{(l^2+l-1+m^2)}{(2l-1)(2l+3)}\right]\sigma(nl;n'l) \tag{40a}$$

$$\langle nlm|H_d|n'(l+2)m\rangle$$
$$= -\frac{\alpha^2 B^2}{8}\left[\frac{(l+m+2)(l+m+1)(l-m+2)(l-m+1)}{(2l+5)(2l+3)^2(2l+1)}\right]^{1/2}\sigma(nl;\, n'(l+2)) \tag{40b}$$

where $\sigma(nl;n'l')$ is the radial matrix element of r^2 between the two states.

If we consider, for instance, the 4p, $m=1$, states of hydrogen, the diamagnetic and paramagnetic terms are equal at $B = 2\times 10^3$ T, which sets the scale for what we might call "strong diamagnetism." The H_β line ($n=4 \rightarrow 2$) was actually observed at 2×10^4 T in the spectrum from a white-dwarf star;[56] 2×10^4 T cannot really be considered a strong field as far as $n=4$ states are concerned, however, because the diamagnetic interaction remains small compared with the Coulomb interaction. For true strong field behavior, the diamagnetic energy must be comparable to or greater than the Coulomb energy.

At low fields, n remains an approximate quantum number because the diamagnetic n-mixing perturbations are second order. Within each term, H_d mixes states of the same parity and m. For the hypothetical hydrogen atom described by the Hamiltonian Eq. (38), H_d can be diagonalized by some operator that transforms the spherical representation into what we shall call the magnetic representation. The situation is reminiscent of the transformation from the spherical to the parabolic representation, except that the parabolic transformation was known [cf. Eq. (22)], whereas the magnetic transformation is not. Nevertheless, we can generate the transformation by diagonalizing H_d, using a spherical basis set. The result of one such computation is shown in Fig. 3.21, which shows the charge density for the $n=8$, $m=0$ "magnetic states." We can label the states within the manifold (n,m) by an index k, which has the values $0, 1, 2, \ldots, n-|m|-1$ (where k plays the role of a quantum number, but the operator for which it is an eigenvalue is so far unknown). Because parity is also a good quantum number, the submanifold can be divided into two groups with k even or odd. The diamagnetic energy for these states is sketched in Fig. 3.22. If we write the diamagnetic energy as

$$\langle H_d\rangle = C(k)B^2 \tag{41}$$

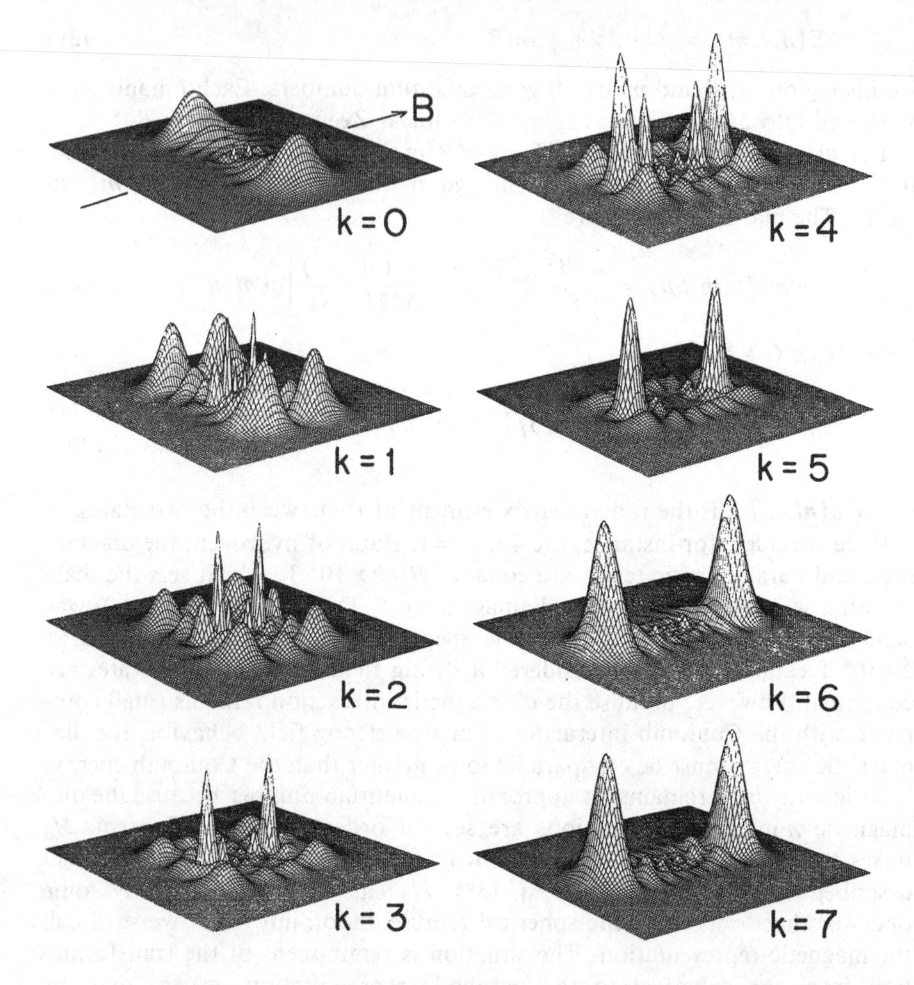

Fig. 3.21. Plots of charge density for $n=8$, $|m|=0$, states of hydrogen in a low magnetic field, drawn as in Fig. 3.3. The index k serves to label the states but it is not the eigenvalue of any known operator. (Courtesy of W. P. Spencer)

then it can be seen that $C(k)$ is largest for the state most spread out in the xy plane and smallest for the state localized near the z axis. These are, respectively, labeled $k=0$ and $k=7$ in Fig. 3.21. Physically, the $k=0$ state has the largest diamagnetic interaction because $\langle x^2+y^2 \rangle$ is a maximum, whereas the $k=7$ state has the smallest interaction. Recall that in the spherical representation the low-l states have the largest values of $\langle r^2 \rangle$. Consequently, low-l states tend to be correlated with the low-k states and high-l states tend to be correlated with high-k states. Because the oscillator strength distribution reflects the distribution of a particular l state, this correlation plays an important role in the appearance of diamagnetic spectra.

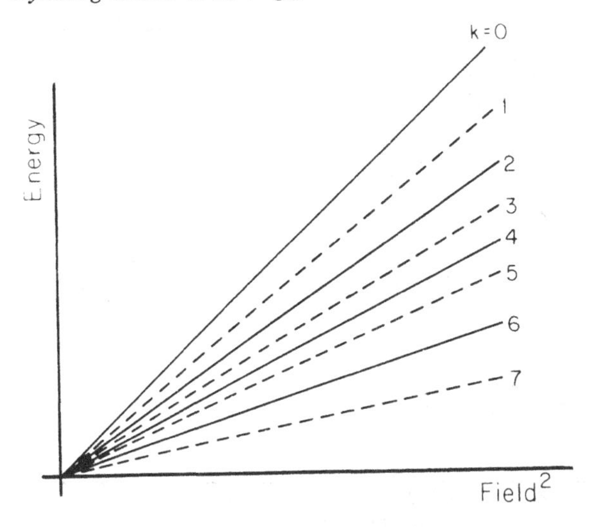

Fig. 3.22. Sketch of the magnetic field dependence of the energy levels for hydrogen, $n=8$, $m=0$. Solid and dashed lines represent even and odd parity states, respectively. The states are shown in Fig. 3.21.

3.5e. Solution in an intermediate field

As the magnetic field is increased, $\langle H_d \rangle$ becomes comparable to the term separation and the matrix elements of H_d, which are off-diagonal in n, become important and need to be included in any calculation. Lacking an analytical solution for the eigenstates and energies, we can resort to numerical computation by expanding the basis set to include adjacent terms with the same parity. The result of one such computation, in the vicinity of even-parity, $n=8$, states for hydrogen, is shown in Fig. 3.23. As the terms start to overlap, the levels repel strongly and the regularities are lost.

Somewhat unexpectedly, the picture becomes simpler at higher values of n. Figure 3.24 shows level structure in the vicinity of $n=12$ for even-parity, $m=1$, states of hydrogen. The apparent sharp level crossings are reminiscent of the hydrogen Stark problem, except that in the case of an electric field the level crossings reflect a symmetry in the Hamiltonian that has no counterpart in the magnetic case. However, there is an approximate symmetry. The size of the avoided crossings in hydrogen decreases exponentially with n.[57] This suggests that if the diamagnetic Hamiltonian is written as $H_d = H_s + V$, where H_s is an approximate Hamiltonian that obeys the unknown symmetry and $V = H_d - H_s$, then V can be treated as a perturbation. For high-n states, the effect of V should be negligible, i.e., less important, for instance, than radiative decay. Identification of such a symmetry could lead to a complete solution for the magnetic field problem.

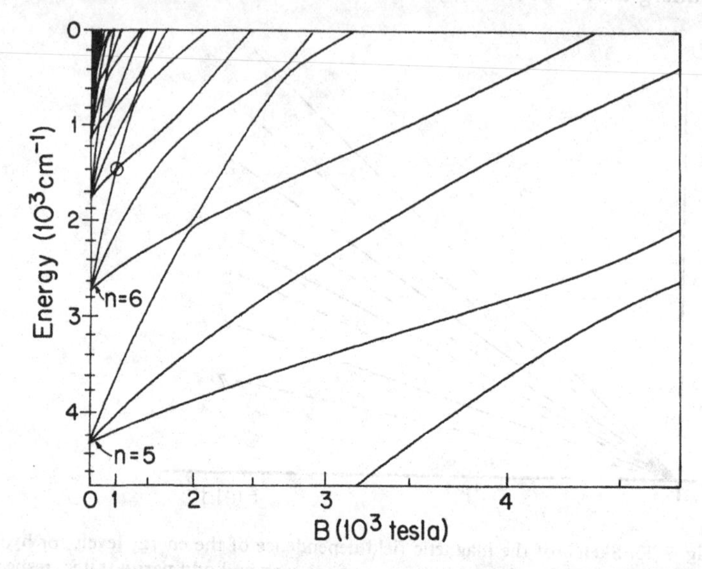

Fig. 3.23. Energy levels of even-parity, $|m|=0$, states of hydrogen in the vicinity of $n=6$, plotted on a scale quadratic in B. Note the strong repulsions between many levels.

There are other strands of evidence pointing to the existence of an approximate symmetry. Numerical studies by Clark and Taylor[58] on the oscillator strengths for hydrogen in a strong magnetic field show degeneracies that are highly suggestive of the symmetry, and experimental studies to be described next confirm the fundamental regularities of diamagnetic level structure.

3.5f. Very high field solution

In very high magnetic fields, it is tempting to treat the Coulomb interaction as a perturbation and to take the free-electron problem as the starting point. The theory for a free electron in a magnetic field, initially developed by Landau, is described in many quantum mechanics texts.[47,59] The Hamiltonian is

$$H_0 = \tfrac{1}{2}(\mathbf{p} + \alpha\mathbf{A})^2 \tag{42}$$

For a uniform field along the z axis, the energy can be written as

$$W = W_z + W_N \tag{43}$$

where

$$W_N = [N + \tfrac{1}{2}(m + |m|) + \tfrac{1}{2})\alpha B \tag{44}$$

N an integer and αB the cyclotron frequency (eB/mc is the cyclotron frequency in laboratory units). Then $W_z = \frac{1}{2}p_z^2$ is the energy of motion along the field.

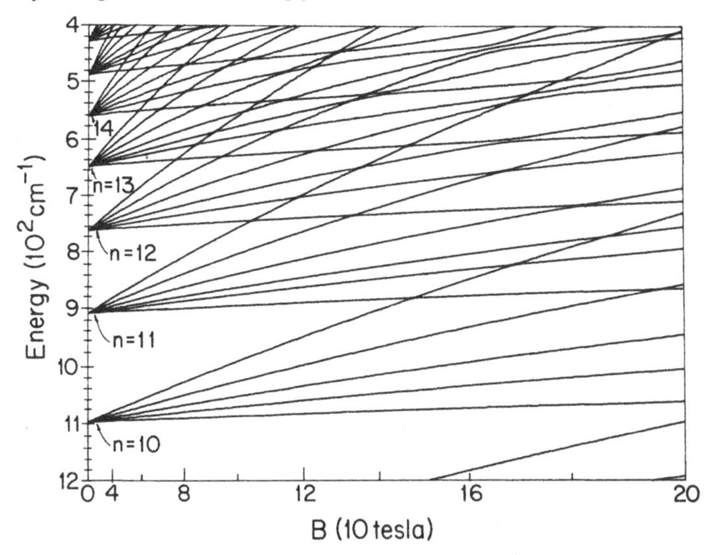

Fig. 3.24. Energy of even-parity, $m=1$ states of hydrogen as a function of magnetic field (plotted on a squared scale) for terms in the range $n=10$–14.

The wave function for the free electron is, in cylindrical coordinates,

$$\Psi_f(N, m, p_z) = \Phi_{N,m}(\rho) \exp(im\phi) \exp(ip_z z) \tag{45}$$

where m is the magnetic quantum number and

$$\Phi_{N,m}(\rho) = A\rho^{|m|} \exp(-\rho^2/4R^2) L_{N+|m|}^{|m|}(\rho^2/2R^2) \tag{46}$$

with A a normalizing constant and $R=(\alpha B)^{-1/2}$ the cyclotron radius.

When the Coulomb interaction is included, the Hamiltonian becomes

$$H = \tfrac{1}{2}p^2 + \tfrac{1}{2}\alpha \mathbf{L} \cdot \mathbf{B} + \tfrac{1}{8}\alpha^2 B^2 \rho^2 - (\rho^2 + z^2)^{-1/2} \tag{47}$$

This equation is not separable and we lack any sort of general solution.

One promising approach to the problem is the so-called adiabatic approximation,[60,61] which rests on the assumption that the frequency of transverse motion is high compared with the frequency of longitudinal motion. The z motion takes place in an effective potential that is evaluated by averaging the longitudinal Coulomb interaction over many cycles of the cyclotron frequency. The separated wave function can be written as

$$\Psi_{Nm\mu} = \Phi_{Nm}(\rho) \exp(im\phi) f_{Nm\mu}(z) \tag{48}$$

where μ is a quantum number associated with the z motion. If we write the energy as

$$E = E_{Nm} + \epsilon_{Nm\mu} \tag{49}$$

then Schrödinger's equation becomes

$$H\Psi_{Nm\mu} = (E_{Nm} + \epsilon_{Nm\mu})\Psi_{Nm\mu} \tag{50}$$

Substituting Eq. (48), left-multiplying by $\Phi^*_{Nm}(\rho)\exp(-im\phi)$, and integrating over ρ and ϕ yields

$$f''(z) + 2[\epsilon_{Nm\mu} - V_{Nm}(z)]f(z) = 0 \tag{51}$$

where $V_{Nm}(z)$ is an effective potential defined by

$$V_{Nm}(z) = -\int |\Phi_{n,m}(\rho)|^2(\rho^2 + z^2)^{-1/2}\rho\, d\rho \tag{52}$$

An alternative version of the adiabatic approximation involves retaining a Coulomb potential term $-1/\rho$ in the ρ equation, whose solutions are modified Landau states. The effective z potential is then $-\langle(\rho^2+z^2)^{-1/2}+(1/\rho)\rangle$, where the average is over the transverse motion.[55]

In principle, the adiabatic approximation offers a systematic approach to the high-field problem. For instance, it should lend itself to an iterative procedure in which the solution for the z motion is used to refine the ρ potential, and so on. In practice, its success has been limited to highly localized states in the xy plane that give rise to the quasi-Landau resonances, however, and for these states the adiabatic approximation works remarkably well.

3.5g. Quasi-Landau resonances

In 1969, Garton and Tomkins[53] discovered a dramatic modulation in the absorption spectrum of barium in a field of 2.5 T. The modulation pattern, now known as the quasi-Landau resonances, extends above the zero-field ionization limit of the atom and has a period of close to $1.5\omega_c$ at the limit (ω_c is the cyclotron frequency αB). At higher energies, the period approaches ω_c, as might be expected for a free electron. The quasi-Landau resonances were successfully explained by Edmonds[61] and Starace[62] using a semiclassical argument based on the adiabatic approximation. O'Connell[63] explained the basic features of the resonances from very direct arguments based on elementary consideration of energy, and Rau[51] showed that the quasi-Landau resonances display the periodic structure that is characteristic of motion in the "strong-mixing" region.

An important clue to the nature of the quasi-Landau resonances is that they are observed in barium's σ spectrum ($\Delta m = \pm 1$) but not in the π spectrum ($\Delta m = 0$). The σ lines arise from states that tend to be localized in the xy plane, whereas the π line comes from a state with a node in the xy plane. (The conditions for an antinode in the xy plane are even parity and even m or odd parity and odd m). Edmonds and Starace argued that a reasonable approximation for the σ states is to neglect the z motion and treat the problem as two dimensional. Writing $g(\rho)=\rho^{-1/2}\phi(\rho)e^{im\phi}$, where $\phi(\rho)e^{im\phi}$ is the transverse wave function, the Schrödinger equation becomes

$$g''(\rho)+2\left[E-\left(-\frac{1}{\rho}+\frac{m^2-\frac{1}{4}}{2\rho^2}+\frac{\alpha}{2}mB+\frac{1}{8}\alpha^2B^2\rho^2\right)\right]g(\rho)=0 \tag{53}$$

The energy levels can be found numerically or from a WKB argument:

$$\int_{\rho_1}^{\rho_2}[2E-2V(\rho)]^{1/2}\,d\rho=(n+\tfrac{1}{2})\pi \tag{54}$$

where $V(\rho)$ is the effective potential in Eq. (53) and the integration extends between the turning points. (In the semiclassical limit $m^2-\frac{1}{4}\to m^2$.) The separation between levels is found by evaluating

$$dE/dn=\pi\left\{\int_{\rho_1}^{\rho_2}[2E-2V(\rho)]^{1/2}\,d\rho\right\}^{-1} \tag{55}$$

At $E=0$, the result is $dE/dn=1.5\omega_c$. Numerical evaluation for other energies shows that the spacing slowly diminishes as the energy increases.

Several groups studied quasi-Landau structure experimentally[64-66] and were able to account quantitatively for the field dependence and spacing of the resonances by using the semiclassical arguments or by finding the solutions of Eq. (53) numerically. Gay, Delande, and Burabin[67] studied positive and negative energy resonances for hydrogenic states of cesium ($m=3$) and obtained excellent agreement with Eq. (54) over a wide range of fields and energies.

The quasi-Landau resonances have attracted wide attention because they demonstrate dramatically that motion in the strong-mixing region can be understood by simple dynamic arguments. Nonetheless, the semiclassical explanation is by no means complete. It cannot account for line intensities and it fails to show how high- and low-field states are generally related. More seriously, because it applies only to a single level from each diamagnetic manifold, it cannot come to grips with the general problem of atomic structure in a strong magnetic field. Recent experimental and computational advances suggest that the general problem may be tractable.

3.5h. Structure in the strong-mixing regime

One reason for the prominence of the quasi-Landau resonances in absorption spectra is that Stark mixing due to the motional electric field tends to average together nearby levels, helping to emphasize general patterns in line strength. Nonetheless, the resonance can be observed in fully resolved spectra.[57,68] Figure 3.25, for instance, shows spectra for even-parity, $m=-2$, states of sodium. (These states are expected to exhibit hydrogenic behavior because the largest quantum defect δ_2 is only 0.015.) Lines have been drawn through the uppermost level of each manifold starting at low field; these levels evolve into

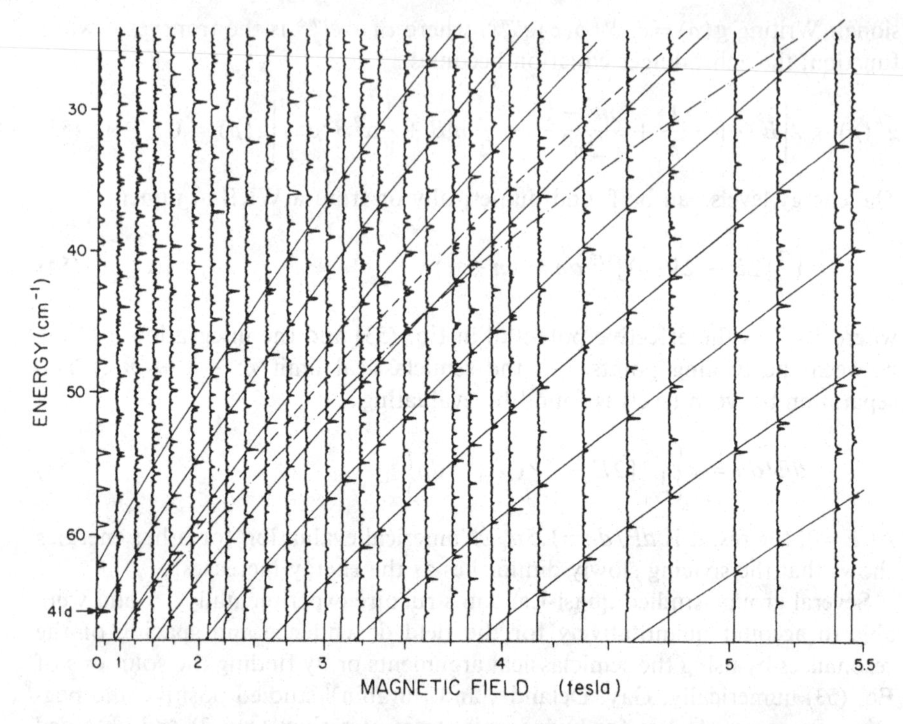

Fig. 3.25. Experimental energy-level map for even-parity, $m=-2$, states of sodium. The magnetic field is plotted on a squared scale. Solid lines indicate the evolution of the highest level for several n manifolds. Dashed and dash-dot lines show the second and third highest levels of the $n=40$ manifold, respectively. The evolution is most clearly seen by sighting along the drawn lines.

the quasi-Landau resonances. In Fig. 3.26, a single sweep is displayed magnified; the dominant peaks show an energy separation approaching $1.5\omega_c$ at $E=0$. Under conditions of low resolution or in the presence of Stark mixing, the spectrum would show the smooth periodic fluctuations that are the signature of quasi-Landau resonance.

Close inspection of Fig. 3.25 shows how the quasi-Landau resonances originate. They arise by a concentration of oscillator strength in the highest level of the manifold. The concentration is not 100%, however, because the second highest level is also visible (dashed line). Other intermediate levels are also visible, but at low intensity.

In Fig. 3.27, we show the diamagnetic spectrum of even-parity, $m=-1$, states of sodium. These states have a node in the xy plane and are not expected to show the quasi-Landau resonances. Instead, we observe a multitude of lines of approximately equal intensity. Under low-resolution conditions, the spectrum would be featureless. Figure 3.28 displays the oscillator

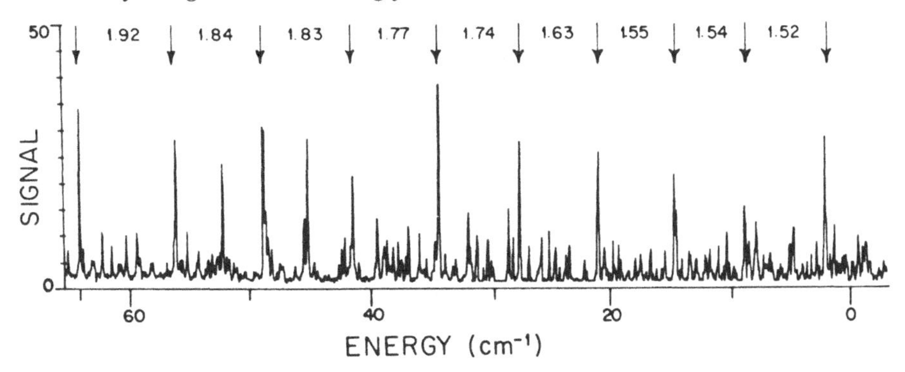

Fig. 3.26. Quasi-Landau spectrum, taken from data in Fig. 3.25. The arrows indicate the quasi-Landau levels. The numbers between the arrows give the level separation in units of $\hbar\omega_c$. The spacing increases with binding energy from the value of 1.5 at $E=0$, in agreement with a WKB analysis. (From Castro et al.[68])

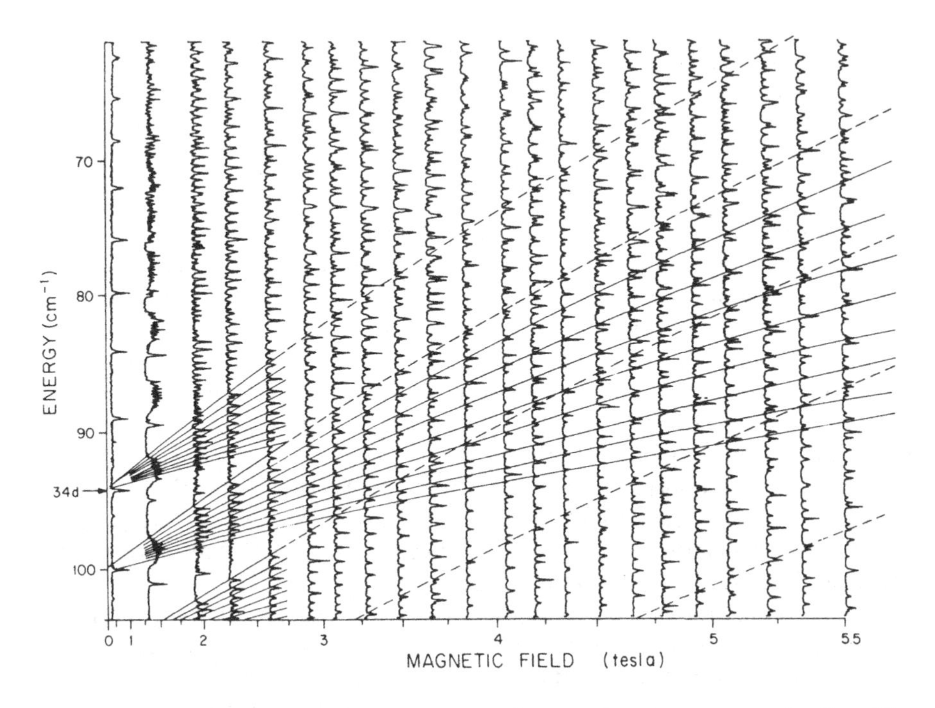

Fig. 3.27. Experimental energy-level map of even-parity, $m=-1$, states of sodium. The dashed lines are to indicate the evolution of the highest level for several n manifolds. Note that the corresponding levels of adjacent n manifolds are equally spaced. The levels are most clearly seen when viewed close to the plane of the paper, along the dashed line. Note contrast with Fig. 3.24.

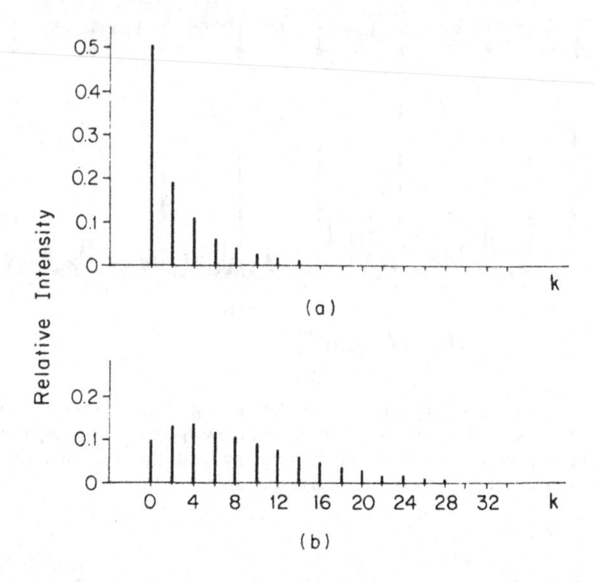

Fig. 3.28. Oscillator-strength distribution for 3p→nd transition for even-parity levels of the $n=35$ manifold: (a) $m=-2$; (b) $m=-1$. The visibility of quasi-Landau resonances is due to the concentration of oscillator strength shown in (a). (From Castro[55])

strength for the $n=8$, even-parity states of hydrogen calculated at low field.[55] The difference in oscillator strength between even and odd m states is dramatic. For even m states, over half of the strength is concentrated in the $k=0$ level, whereas for odd m states, the oscillator strength is broadly distributed. Figure 3.27 shows that the diamagnetic levels evolve smoothly, crossing with no apparent repulsions, as would be expected with the approximate symmetry described in Sect. 3.5e. Thus each state preserves its low-field character into the strong-mixing regime, exactly as in the case of the Stark structure of hydrogen. Another surprising feature is that corresponding members of each manifold (highest, second highest, etc.) all show the *same* periodicity. Thus the quasi-Landau structure appears to be a general property of *all* states of the atom, not merely those localized in the xy plane.

Similar observations were made by Clark and Taylor[58] on the basis of numerical calculations. Structure in the strong-mixing regime appears to be characterized by simple regularities, a rather unexpected finding in view of the apparent complexity of the problem. The accumulating evidence that an approximate symmetry governs the structure of hydrogen in strong magnetic fields gives hope that the last elementary problem in the quantum mechanics of the one-electron atom may be nearing a solution.

References and notes

1. D. Kleppner, in *Les Houches Summer School,* eds. J. C. Adam and R. Ballian (New York: Gordon & Breach, 1981), Session 28.
2. M. G. Littman and H. J. Metcalf, *Appl. Opt. 17,* 2224 (1979).
3. H. A. Bethe and E. E. Salpeter, *Quantum Mechanics of One- and Two-Electron Atoms* (Springer-Verlag, Berlin, 1957).
4. H. J. Silverstone, *Phys. Rev. A 18,* 1853 (1978).
5. L. I. Schiff, *Quantum Mechanics,* 3rd ed. (New York: McGraw-Hill, 1968).
6. K. Helfrich, *Theor. Chim. Acta 24,* 271 (1972).
7. P. J. Redmond, *Phys. Rev. B 133,* 1352 (1964).
8. G. J. Hatton, *Phys. Rev. A 16,* 1347 (1977).
9. E. Merzbacher, *Quantum Mechanics* (New York: Wiley, 1970), pp. 428–9.
10. D. R. Bates and A. Damgaard, *Philos. Trans. R. Soc. London 242,* 101 (1949).
11. M. J. Seaton, *Comments At. Mol. Phys. 2,* 37 (1970).
12. M. L. Zimmerman, M. G. Littman, M. M. Kash, and D. Kleppner, *Phys. Rev. A 20,* 2251 (1979).
13. C. Froese, *Can. J. Phys. 41,* 1895 (1963).
14. A. Lindgard and S. E. Nielsen, *J. Phys. B. 8,* 1183 (1975).
15. J. M. Blatt, *J. Comput. Phys. 1,* 378 (1967).
16. I. V. Komarov, T. P. Grosdanov, and R. K. Janev, *J. Phys. B 13,* L573 (1980).
17. A. Erdélyi, ed., *Higher Transcendental Functions* (New York: McGraw-Hill, 1953), vol. I.
18. D. Park, *Z. Phys. 159,* 155 (1960).
19. J. Rubbmark, M. M. Kash, M. G. Littman, and D. Kleppner, *Phys. Rev. A, 23,* 3107 (1981).
20. M. M. Kash (private communication).
21. P. A. Koch, in *Atomic Physics,* eds. D. Kleppner and F. M. Pipkin (New York: Plenum, 1981), vol. 7.
22. E. Luc-Koenig and A. Bachelier, *J. Phys. B 13,* 1743 (1980).
23. T. W. Ducas, W. P. Spencer, A. G. Vaidyanathan, W. H. Hamilton, and D. Kleppner, *Appl. Phys. Lett. 35,* 382 (1979).
24. G. I. Bekov, V. S. Letokhov, O. I. Matveev, and V. I. Mishin, *Zh. Eksp. Teor. Fiz. 75,* 2092 (1978) [*Sov. Phys. JETP 48,* 1052 (1978)].
25. D. S. Bailey, J. R. Hiskes, and A. C. Riviere, *Nucl. Fusion 5,* 41 (1965).
26. M. H. Rice and R. H. Good, Jr., *J. Opt. Soc. Am. 52,* 239 (1962).
27. M. G. Littman, M. L. Zimmerman, and D. Kleppner, *Phys. Rev. Lett. 37,* 486 (1976).
28. T. W. Ducas, M. G. Littman, R. R. Freeman, and D. Kleppner, *Phys. Rev. Lett. 35,* 366 (1975).
29. P. Jaquinot, *Laser Spectroscopy,* eds. H. Walther and K. W. Rothe (New York: Springer-Verlag, 1979), vol. IV, p. 236.
30. W. Cooke and T. Gallagher, *Phys. Rev. A 17,* 1276 (1978).
31. M. G. Littman, M. M. Kash, and D. Kleppner, *Phys. Rev. Lett. 41,* 103 (1978).
32. D. Banks and J. G. Leopold, *J. Phys. B 8,* 1 (1977).
33. W. E. Lamb, *Phys. Rev. 85,* 259 (1952).
34. U. Fano, *Phys. Rev. 124,* 1866 (1961).
35. S. Feneuille, S. Liberman, J. Pinard, and A. Taleb, *Phys. Rev. Lett. 42,* 1406 (1979).

36. R. F. Stebbings, C. J. Latimer, W. P. West, F. B. Dunning, and T. B. Cook, *Phys. Rev. A 12,* 1453 (1975).
37. R. V. Ambartsumyan, G. I. Bekov, V. S. Letokhov, and V. I. Mishin, *JETP Lett. 21,* 279 (1975).
38. K. C. Smith, J. A. Schiavone, and R. S. Freund, *J. Chem. Phys. 59,* 5225 (1973).
39. A. F. J. Van Raan, G. Baum, and W. Raith, *J. Phys. B 9,* L173 (1976).
40. T. F. Gallagher, L. M. Humphrey, W. E. Cooke, R. M. Hill, and S. A. Edelstein, *Phys. Rev. A 16,* 1098 (1977).
41. S. Liberman and J. Pinard, *Phys. Rev. A 20,* 507 (1979).
42. T. F. Gallagher, R. M. Hill, and S. A. Edelstein, *Phys. Rev. A 14,* 744 (1976).
43. C. Fabre, P. Goy, and S. Haroche, *J. Phys. B 10,* L183 (1977).
44. G. Leuchs, S. J. Smith, and H. Walther, in *Laser Spectroscopy,* eds. H. Walther and K. W. Rothe (New York: Springer-Verlag, 1979), vol. IV.
45. T. F. Gallagher, L. M. Humphrey, R. M. Hill, and S. A. Edelstein, *Phys. Rev. Lett. 37,* 1465 (1976).
46. T. H. Jeys, G. W. Foltz, K. A. Smith, E. J. Beiting, F. G. Kellert, F. B. Dunning, and R. F. Stebbings, *Phys. Rev. Lett. 44,* 398 (1980).
47. L. D. Landau and E. M. Lifshitz, *Quantum Mechanics: Non-Relativistic Theory,* 3rd ed. (Elmsford, N.Y.: Pergamon Press, 1977).
48. R. R. Freeman, N. P. Economu, G. C. Bjorklund, and K. T. Lu, *Phys. Rev. Lett. 41,* 1463 (1978).
49. R. R. Freeman and N. P. Economu, *Phys. Rev. A 20,* 2356 (1979); also R. R. Freeman, in *Atomic Physics,* eds. D. Kleppner and F. M. Pipkin (New York: Plenum, 1981), vol. 7.
50. A. R. P. Rau and K. T. Lu, *Phys. Rev. A 21,* 1057 (1980).
51. A. R. P. Rau, *Phys. Rev. A 16,* 613 (1977).
52. E. Luc-Koenig and A. Bachelier, *J. Phys. B 13,* 1769 (1980).
53. W. R. S. Garton and F. S. Tomkins, *Astrophys. J. 158* 839 (1969).
54. R. H. Garstang, *Rep. Prog. Phys. 40,* 105 (1977).
55. J. C. Castro, unpublished thesis, Massachusetts Institute of Technology, 1981.
56. J. R. D. Angel, *Astrophys. J. 216,* 1 (1977).
57. M. L. Zimmerman, M. M. Kash, and D. Kleppner, *Phys. Rev. Lett. 45,* 1092 (1980).
58. C. W. Clark and K. T. Taylor, *J. Phys. B 13,* L737 (1980).
59. C. Cohen-Tannoudji, B. Diu, and F. Laloë, *Quantum Mechanics* (New York: Wiley, 1977).
60. L. I. Schiff and H. Snyder, *Phys. Rev. 55,* 59 (1979).
61. A. R. Edmonds, *J. Phys.* (*Paris*) *31* (C4), 71 (1970).
62. A. F. Starace, *J. Phys. B 6,* 585 (1973).
63. R. F. O'Connell, *Astrophys. J. 187,* 275 (1974).
64. R. J. Fonck, F. L. Roesler, D. H. Tracy, and F. S. Tomkins, *Phys. Rev. A 21,* 861 (1980).
65. H. Crosswhite, U. Fano, K. T. Lu, and A. R. P. Rau, *Phys. Rev. Lett. 43,* 963 (1979).
66. N. P. Economou, R. R. Freeman, and P. F. Liao, *Phys. Rev. A 18,* 2506 (1979).
67. J. C. Gay, D. Delande, and F. Burabin, *J. Phys. B 13,* L279 (1980).
68. J. C. Castro, M. L. Zimmerman, R. Hulet, D. Kleppner, and R. R. Freeman, *Phys. Rev. Lett. 45,* 1780 (1980).

4

Spectroscopy of one- and two-electron Rydberg atoms

C. FABRE AND S. HAROCHE

4.1. Introduction

The spectroscopy of Rydberg atoms has played a very large part in the recent renewal of the studies of highly excited states. Within only a few years, an impressive number of term values, term intervals, and fine and hyperfine splittings have been measured in these states in a great range of atomic species from helium to uranium. The availability of these data has, in turn, stimulated new theoretical calculations.

Spectroscopy is often the first step in the investigation of new physical objects. It allows us to measure physical constants and to analyze the dominant processes in the dynamics of these systems. In Rydberg states, these processes appear very simple at first, because it is well known that highly excited states of any atom behave, to a large extent, like hydrogen atoms with a very large principal quantum number n. All physical quantities vary rapidly with n, of course, so the scale and relative importance of the physical properties are completely different from those of moderately excited levels. This is revealed by quite unusual spectroscopic properties: electronic transitions falling in the microwave range; very large dipole moments; fine and hyperfine interactions stronger than electrostatic interactions; Stark and Zeeman effects of the order of, or more important than, the coupling with the atomic core. This explains the great interest in these states among spectroscopists.

Another feature of the spectroscopy of Rydberg states is that it is actually performed on a long series of states. It permits the systematic study of atomic properties over a large range of n (and sometimes l) values and, thus, the study of how atomic properties evolve when the electron energy approaches nearer and nearer to the ionization limit. Energy splittings can quite generally be fitted by simple analytic formulas expressed as functions of n (or l). The pioneering work of Rydberg,[1] whose name has been given to these highly excited states, is of course the most famous example of studies of such series. Considerations of regularities in spectra close to the ionization limit led him to propose the simple "Rydberg formula":

$$\sigma_n = \sigma_0 - \frac{R_\infty}{(n-\delta)^2} \tag{1}$$

for the wave number σ_n of the lines, where R_∞ is the universal Rydberg constant, n an integer, and σ_0 and δ the characteristic constants of the atom and the series. Classic spectroscopy of highly excited states thus provided the first great step in the understanding of line spectra, which was in turn decisive in the birth of quantum mechanics. Nowadays, the basic laws of the atomic structure are, of course, well known. It is, nevertheless, very important to have precise information about long series of states. This allows us to obtain deeper insights into the various kinds of approximations that have to be made to compute atomic spectra and a better understanding of the physical processes involved in the interaction of the valence electron with the atomic core.

Indeed, we shall see in the following that the simple picture in terms of hydrogenlike atoms is not always valid and that some properties of Rydberg atoms are completely different from those of hydrogen, even when n tends to infinity. The discrepancy from the hydrogen model leads then to a deeper knowledge of the properties of the atomic core and of its interaction with the valence electron. The very excited electron appears, in fact, as a probe of the nonpointlike character of the core, in the same way that the differences between the actual orbit of an earth satellite and the Keplerian orbit give information on the mass distribution inside the earth.

The purpose of this chapter is to describe the most striking features of Rydberg state spectroscopy, without any claim to completeness. For the sake of simplicity, we shall restrict our subject to the spectroscopy of one- and two-electron atoms (i.e., hydrogen, the alkalis, helium, and the alkaline earths), which display some of the basic physical features common to the spectroscopy of all very excited species (including more complex atoms or molecules).

Because the spectroscopy of atoms perturbed by external fields (electric or magnetic) is covered in other chapters of this book, we shall deal here only with free-atom spectroscopy. We shall describe only energy interval measurements, without considering in detail other spectroscopy-related experiments such as lifetime or oscillator-strength determinations.

The first section of this chapter will give the minimal theoretical framework necessary to understand most characteristics of Rydberg atom spectroscopy. The second section will describe the most widely used experimental techniques. The last sections will give a brief review of the data obtained in one- and two-electron systems, along with possible applications of these studies to fields outside the domain of pure spectroscopy.

4.2. Characteristics of Rydberg atoms relevant to spectroscopy

4.2a. The quantum-defect model for Rydberg series

Atomic Rydberg states appear to be quite simple physical objects: a small set of physical parameters is enough to account for the energies of a great number of Rydberg states. This property, which does not exist for low-lying states, is one of the most attractive features of very excited states. The aim of this section is to outline briefly the simple theoretical framework that allows us to explain the experimental spectroscopic data now available on these states.

For highly excited states, the immensely difficult problem of finding the energy levels of $N+1$ electrons interacting with the nucleus and with each other is drastically simplified. First, the separation into a core of N electrons and a distant electron (valence electron) is quite natural because of the great difference between their binding energies. The problem is mainly a one-electron problem, with some corrections. Second, because the core wave function is limited to a small region of space (or radius r_0) around the nucleus, the effective potential $\mathcal{V}(r)$ "seen" by the outer electron is mainly a Coulomb potential:

$$\mathcal{V}(r) \simeq -e^2/r \qquad \text{for} \quad r \geqslant r_0 \tag{2}$$

To a first approximation, Rydberg states appear, therefore, as simple hydrogenic excited states for which energies and wave functions are known.

Furthermore, because we are dealing with high quantum numbers, we approach the classical limit of quantum mechanics. As a consequence, semiclassical pictures (WKB), or even classical pictures, can provide an accurate description of highly excited states in terms of electron orbits and can yield simple physical pictures of their dynamics.

The hydrogenlike approximation (point-core model)

The first-order approximation in the description of Rydberg states is to suppose that the overlap of the core with the outer electron wave function is negligible. This wave function is then exactly known and can be written as

$$R_{nl}(r) = f(n, l; \rho)/r \qquad \text{with} \quad \rho = r/a_0 \tag{3}$$

where $f(n, l; \rho)$ is the regular Coulomb function according to the notations of Ref. 2 and a_0 the Bohr radius. We shall restrict ourselves to states with low-l values, which are usually observed in spectroscopy. In this case, the wave function exhibits many oscillations $(n-l-1)$ (dashed line in Fig. 4.1), and the usual polynomial expansion is not very convenient. The WKB semiclassical theory provides a simpler expression, especially in the region far from the classical turning points r_1 and r_2 of the radial motion $(r_1 \ll r \ll r_2)$:[3]

$$R_{nl}(r) = (-1)^l (2/\pi^2)^{1/4} n^{-3/2} \rho^{-3/4} \cos[(8\rho)^{1/2} - \tfrac{3}{4}\pi] \tag{4}$$

When $n \to \infty$, r_1 and r_2 are given by

$$r_1 \simeq (l + \tfrac{1}{2})^2 a_0; \qquad r_2 \simeq 2n^2 a_0 \tag{5}$$

Classically, the Rydberg electron moves on a very eccentric elliptic orbit, with a "perihelion" and an "aphelion" that are, respectively, equal to r_1 and r_2. It can be easily shown that, near the perihelion r_1, the orbit depends mainly

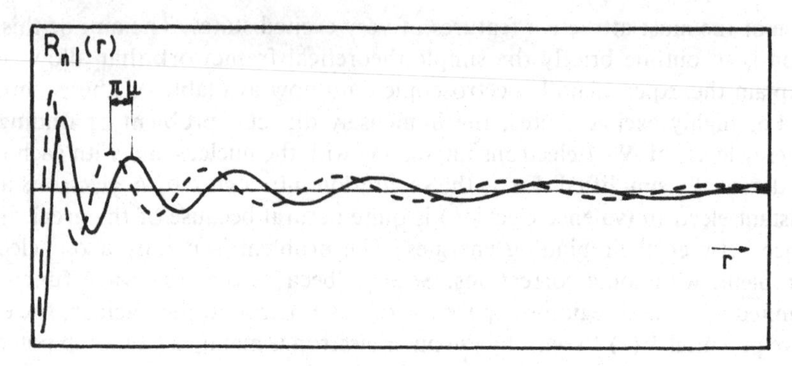

Fig. 4.1. Oscillatory part ($r_1 \ll r \ll r_2$) of the electron radial wave function in a Rydberg state (WKB approximation). Dashed line, point-core hydrogen case; solid line, extended-core case with a quantum defect of 0.25. The phase shift $\pi\mu$ responsible for the quantum defect is indicated by the arrow.

on l and is energy independent, whereas the outer part of the same orbit depends mainly on n. One can then analyze a Rydberg state in terms of a low-energy collision between an outer electron and the core, the impact parameter being fixed by l and the frequency of collisions by n (the frequency is, more precisely, proportional to n^{-3}).

It is evident that the point-core model can only be a crude approximation: if l is low enough so that $r_1 < r_0$, where r_0 is the core radius, the outer electron will collide with the core at each revolution, whatever the value of n, and this collision will strongly affect the atomic properties. We must then make a distinction between penetrating orbits (small-l values) and nonpenetrating orbits (larger-l values), for which the point-core model holds.

Long-range effects

Let us first consider the nonpenetrating orbits. They correspond to angular quantum number l greater than a given value l_0, where l_0 is, in general, a small integer ($l_0=1$ for Li, 2 for Na and K, 3 for Rb and Cs, for example). In that case, the wave functions of the core and the outer electron do not appreciably overlap. The core is then considered as only a source of an electrostatic potential $\mathcal{V}(r)$, which can be developed in a multipolar expansion:

$$\mathcal{V}(r) = -(e^2/r) + \delta\mathcal{V}(r) \tag{6}$$

where $\delta\mathcal{V}(r)$ mainly contains the induced dipole and quadrupole contributions[4,5]

$$\delta\mathcal{V}(r) \simeq -\frac{1}{2}\left(\frac{\alpha'_{\mathrm{d}}}{r^4} - \frac{\alpha'_{\mathrm{Q}}}{r^6}\right) \tag{7}$$

with α'_{d} and α'_{Q}, respectively, the effective dipole and quadrupole polarizabilities of the core. $\delta\mathcal{V}$ can be treated as a perturbation with respect to the

Coulomb term, and the energy corrections are readily calculated. They appear to be proportional to $1/n^3$ when $n \to \infty$ and the total binding energy can be written as[5-7]

$$E_{nl} = -\frac{R_\infty}{n^2} + \frac{2R_\infty \mu_l}{n^3} \simeq -\frac{R_\infty}{(n-\mu_l)^2} \tag{8}$$

where R_∞ is the Rydberg constant and

$$\mu_l \simeq \frac{3\alpha'_d}{4l^5} + \frac{35\alpha'_Q}{16l^9} \tag{9}$$

We must then correct the principal quantum number n in the Balmer formula by a small quantum defect μ_l, which decreases very rapidly with l.

Short-range effects: the quantum-defect method

The problem is more complicated for the penetrating orbits that strongly interact with the core and are more sensitive to the details of the core wave function. Hartree,[8] Ham,[9] and mainly Seaton[2] and his co-workers have developed a theory, the quantum-defect method (QDM), which is well suited to the description of such Rydberg series.

The theory rests on the following point: the interaction between the electron and the core is limited to a small region of space around the nucleus (of radius r_0). For $r > r_0$, i.e., in the largest region of space explored by the electron, the potential is Coulombic. As a consequence, the wave function for $r > r_0$ is exactly known: it is a Coulomb wave function satisfying specific boundary conditions for $r = r_0$ and $r \to \infty$ (no longer for $r = 0$ and $r \to \infty$ as in the hydrogen atom). The value of the wave function at $r = r_0$ is determined by the detailed properties of the short-range electron–core interaction.

Furthermore, by assuming some simple regularity conditions on the characteristics of this short-range interaction, Ham[8] showed that the wave function must be an *analytic function of the energy* in the whole space, which restricts the size of the Hilbert space in which solutions are to be found. A possible basis of the space of analytic Coulomb wave functions is the set of two functions $f(\nu, l; \rho)$ and $g(\nu, l; \rho)$ given by Eq. (2.22) of Ref. 10. The function f is regular at $r = 0$ and has the following asymptotic form when $r \to \infty$:[11]

$$\begin{aligned} f(\nu, l; \rho) \simeq{} & (-1)^l \nu^{1/2} [\Gamma(\nu - l)\Gamma(\nu + l + 1)]^{1/2} \\ & \times [(1/\pi)(2r/\nu)^{-\nu} e^{r/\nu} \sin \pi\nu - (2r/\nu)^\nu e^{-r/\nu} e^{i\pi\nu}] \end{aligned} \tag{10}$$

The function g is irregular at the nucleus and, when $r \to \infty$, is in quadrature with f:

$$\begin{aligned} g(\nu, l; \rho) \simeq{} & (-1)^l \nu^{1/2} [\Gamma(\nu - l)\Gamma(\nu + l + 1)]^{1/2} \\ & \times \{-(1/\pi)(2r/\nu)^{-\nu} e^{r/\nu} \cos \pi\nu + (2r/\nu)^\nu e^{-r/\nu} \exp[i\pi(\nu + \tfrac{1}{2})]\} \end{aligned} \tag{11}$$

Functions $f(\nu, l; \rho)$ and $g(\nu, l; \rho)$ correspond to the same energy:

$$E = -R_\infty / \nu^2 \tag{12}$$

The effect of the core for the wave function outside the core reduces to the determination of the coefficients in the linear combination of functions f and g.

Single-electron model. Let us first assume that the core is simply described by a wave function Φ that does not depend on the outer electron properties. The wave function ψ of the $N+1$ electrons is then simply given by $\psi = \Phi R_{nl}(r)$, where the outer electron wave function R_{nl} can be written as

$$R_{nl}(r) = \cos \pi\mu\, f(\nu, l; \rho) - \sin \pi\mu\, g(\nu, l; \rho), \qquad r \geqslant r_0 \tag{13}$$

where μ, called the quantum defect, depends on l and n and must be an analytical function of the energy; thus μ is determined by the boundary conditions at the core surface $r = r_0$. (If the core reduces to a point charge, then $\mu = 0$.)

The energy quantization will be given by the boundary condition $R_{nl}(r) \to 0$ for $r \to \infty$. Using the asymptotic expansion of Eqs. (10) and (11), we see that ν and μ are linked by the following relation:

$$\sin \pi(\nu + \mu) = 0 \tag{14}$$

so that $\nu = n - \mu$, where n is an integer. The energy levels are then given by the well-known Rydberg formula

$$E_n = -\frac{R_\infty}{(n - \mu)^2} \tag{15}$$

where the quantum defect μ, introduced by Rydberg[1] as a phenomenological parameter, has now a very precise and simple physical meaning.

The quantum-defect method can be used in two different ways. First, by making a model of the core wave function, it is possible to calculate ab initio the outer electron wave function for $r = r_0$ and, hence, to calculate the quantum defect μ. Second, from the experimental knowledge of Rydberg state energies, an experimental quantum defect μ can be determined. Expression (13) then gives an exact expression of the wave function outside the core, which allows us to calculate other physical properties of these states.

The semiclassical (WKB) expression for QDM wave functions is also very simple. For $r_1 \ll r \ll r_2$, it is given as

$$R_{nl}(r) = (-1)^l (2/\pi)^{1/4} \nu^{-3/2} \cos[(8\rho)^{1/2} - \tfrac{3}{4}\pi + \pi\mu] \tag{16}$$

The quantum defect appears here as a *phase shift* of the wave function oscillations outside the core (solid line in Fig. 4.1). This is the same phase shift that appears in collision theory, confirming the strong relation between a highly excited state and a collision state between a low-energy electron and the core.[12] In fact, all physical properties have a continuous behavior when crossing the ionization limit from the Rydberg states ($E \lesssim 0$) to the low-energy continuum states ($E \gtrsim 0$).

We can easily show by WKB arguments that μ appears as the difference between two action integrals corresponding, respectively, to the Coulombic potential $-e^2/r$ and to the non-Coulombic potential $\mathcal{V}(r)$ of the core:

$$\pi\mu = \int_0^{r_0} \left[\frac{2m}{\hbar^2}\left(E + \frac{e^2}{r}\right)\right]^{1/2} dr - \int_0^{r_0} \left\{\frac{2m}{\hbar^2}[E - \mathcal{V}(r)]\right\}^{1/2} dr \qquad (17)$$

Because the binding energy E is very small compared with $-e^2/r$ and $\mathcal{V}(r)$ near $r=0$, it appears clearly from Eq. (17) that μ is nearly energy independent (i.e., $\nu = n - \mu$ independent). More precisely, expression (17) can be expanded as a power series in E, i.e., as a power series in $1/\nu^2$, whose leading term is the ν-independent one:

$$\mu = \mu_0 + \frac{\mu_1}{\nu^2} + \frac{\mu_2}{\nu^4} + \cdots \qquad (18)$$

with the μ_0, μ_1, μ_2 factors being only l dependent.

The interpretation of μ as a phase shift of the atomic wave function allows us to give a simple physical analogy to the concept of quantum defect in the single-electron case. In fact, the perturbation to the Coulombic potential induced by the core near the nucleus is very similar to a change of index of refraction produced by a dielectric medium placed near a mirror in a resonant electromagnetic cavity: it induces in the same way a shift in the eigenmode phases and changes the corresponding eigenfrequencies, these effects being accounted for by the replacement of n by $n - \mu$.

Finally, let us consider classically the effect of the core. The orbit is elliptic outside the core and strongly modified inside the core. In any case, as $\mathbf{L}$ is conserved, the orbit is always in the same plane: at each revolution, the major axis of the ellipse *precesses* by an angle $\Delta\theta$ (Fig. 4.2). Thus $\Delta\theta$ vanishes when $\mathbf{L}=0$ and is maximum when the orbit is "tangent" to the core. This precession, which is the classical counterpart of the quantum defect, is simply linked to it by the formula

$$\Delta\theta = -2\pi\,\partial\mu/\partial l \qquad (19)$$

Multichannel quantum-defect theory. The single-electron model that we just outlined is only a first approximation because we considered the atomic core as a fixed source for the potential experienced by the outer electron, which is not perturbed by it and stands as a "spectator" of the valence electron dynamics. This simple model holds only if the minimum excitation energy of the core is much larger than the ionization energy of the outer electron. This is the case for alkali atoms but not for more complex atoms, for which the outer electron can temporarily excite the core during their mutual "collision." This kind of coupling obviously strongly modifies the valence

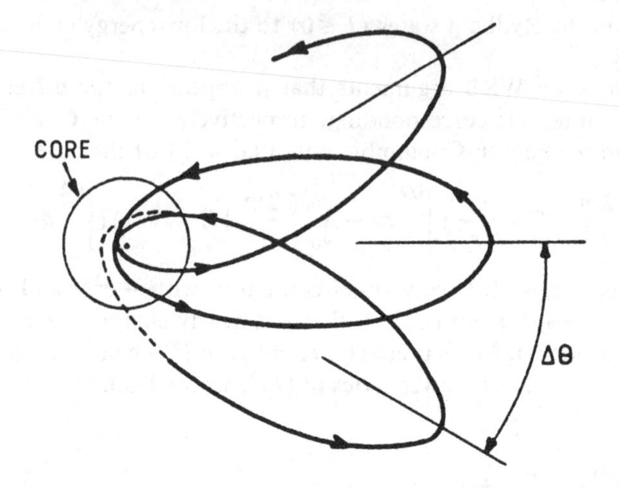

Fig. 4.2. Classical orbit of a Rydberg electron exhibiting the precession $\Delta\theta$ of the ellipse axis.

electron dynamics and results in a drastic modification of the quantum defect as a function of energy, for example, for levels in a Rydberg series that lies close to a state of another series converging to an excited core state. In general, the situation is even more complicated because one is faced with several levels and several ionization limits, without any possibility of unambiguously assigning each level to a series converging to a given limit. We cannot speak of a series perturbed by a single level but, rather, of different series mutually interacting.

Multichannel quantum-defect theory (MQDT), initiated by Seaton and co-workers[10,13] and further developed by Fano and co-workers,[12,14] is an extension of single-channel QDM, which provides a convenient tool to treat this complex problem by reducing it to the determination of a minimum set of physical parameters. As the outer electron and the core interact over a wide range of mutual distances, corresponding to different physical situations, MQDT uses two different bases for the spatial wave functions:

1. The basis of "close-coupling channels" ψ_α, which diagonalizes the $N+1$ electron Hamiltonian describing the short-range interaction (in which the electron–electron interaction plays a dominant role). This basis is well suited to the description of tightly bound states.
2. The basis of "collision channels" defined by a definite state of the N electron core ϕ_i and by the angular momenta of core and outer electron (along with the specification of their coupling). This basis is well suited to the description of the collision between an electron and the core.

The unitary matrix $\{U_{i\alpha}\}$ allows transformation from the first basis to the second.

Let M be the number of channels necessary to describe the coupling between the electron and the core and N the number of ionization limits I_i (i.e., of core states; $N \leqslant M$). Each observed level of energy E has N effective quantum numbers ν_i defined by

$$E = I_i - (R_\infty / \nu_i^2) \tag{20}$$

The corresponding state may be described in terms of close-coupling channels inside the core ($r \leqslant r_0$) and in terms of collision channels outside the core, by the following expressions:

$$\psi = \sum_{\alpha=1}^{M} A_\alpha \psi_\alpha \tag{21}$$

$$= \sum_{i=1}^{M} Z_i \phi_i h_i(r) \tag{22}$$

where $h_i(r)$, the wave function of the outer electron, is a linear combination of the analytic Coulomb wave functions f and g outside the core. It is related to the close-coupling channels ψ_α by the boundary conditions at $r = r_0$. More precisely, the wave function for $r \geqslant r_0$ can be written as

$$\psi = \sum_{i,\alpha} A_\alpha \Phi_i U_{i\alpha} [\cos \pi\mu_\alpha f(\nu_i, l; \rho) - \sin \pi\mu_\alpha g(\nu_i, l; \rho)] \tag{23}$$

This expression, which is the extension of the single-channel expression (13), gives exactly the wave function outside the core when the M quantum defects μ_α corresponding to the M close-coupling channels and the matrix elements $U_{i\alpha}$ are known. The A_α coefficients are determined by the nondivergence condition when $r \to \infty$, i.e., according to asymptotic expressions (10) and (11), by the relation:

$$\sum_\alpha A_\alpha U_{i\alpha} \sin \pi(\nu_i + \mu_\alpha) = 0, \qquad i = 1, \ldots, N \tag{24}$$

The solution of this set of linear homogeneous equations is nonzero only when the following relation among ν_i, μ_α, and $U_{i\alpha}$ is satisfied:

$$\det | U_{i\alpha} \sin \pi(\nu_i + \mu_\alpha) | = 0 \tag{25}$$

The theoretical values for the effective quantum numbers ν_i are then given by the intersection of the $N-1$ dimensional surfaces $\mathcal{L}$ defined by relation (25) and the line $\mathcal{E}$ defined by relations

$$I_i - (R_\infty / \nu_i^2) = I_j - (R_\infty / \nu_j^2), \qquad i \neq j \tag{26}$$

Here again, it is possible, using this technique, to calculate the ν_i from an ab initio determination of parameters $U_{i\alpha}$ and μ_α,[15,16] or to determine the parameters $U_{i\alpha}$ and μ_α by fitting relations (25) from the experimental deter-

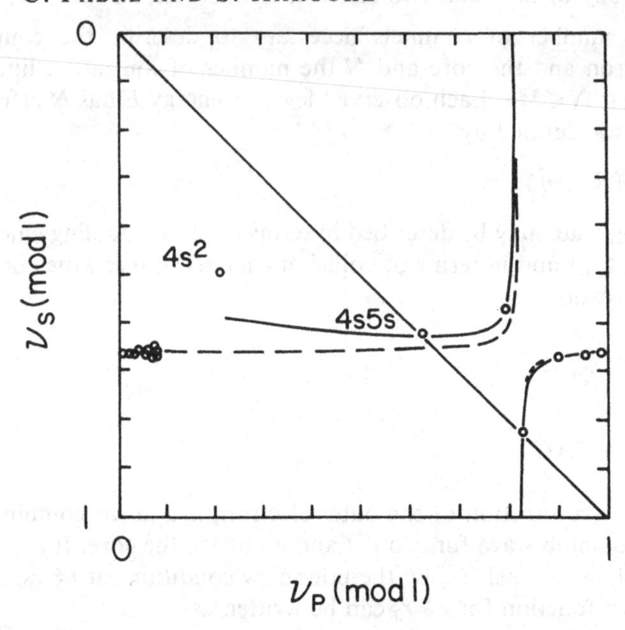

Fig. 4.3. Lu–Fano diagram of the $^1S_0$ spectrum in calcium. (From Armstrong, Esherick, and Wynne[17])

minations of effective quantum numbers ν_i. This last procedure is now used very often to analyze the complex spectra of two-electron Rydberg atom series. A convenient way is to display the "Lu–Fano diagrams,"[14] i.e., to plot ν_i (mod 1) as a function of ν_j (mod 1) for each experimental level. If there is no interaction between the series converging to limits I_i and I_j, ν_i (or ν_j) is roughly constant for levels belonging to series i (or j) and the points form two intersecting straight lines parallel to the axes. The interaction between the series appears in the Lu–Fano diagram as an "anticrossing" between the lines, and the strength of the interaction is related to the distance between the branches. When all the physical parameters $U_{i\alpha}$ and μ_α are determined, it is possible, by the set of equations (24), to calculate the A_α (or Z_i) coefficients and hence to know the mixing of the different channels. Figure 4.3 displays as an example the Lu–Fano diagram of the $^1S_0$ spectrum in calcium (taken from Ref. 17), which is a simple two-channel–two-limit problem ($N=M=2$). The two interacting series are the 4sns and 4pnp series, respectively, converging to the 4s limit ($I_s = 49\,305.99$ cm^{-1}) and the 4p limit ($I_p = 74\,609$ cm^{-1}). Each measured level, characterized by two quantum numbers ν_s and ν_p, is then represented by a point in Fig. 4.3. Except for the most bound states of the series ($4s^2$ and 4s 5s), all points belong to the two branches of an hyperbola shown by dashed lines in Fig. 4.3 and correspond to the following values of the parameters: $\mu_1 = 0.3471(3)$, $\mu_2 = 0.171\,756(7)$, $U_{11} = 0.9810(3)$, $U_{12} = -0.194(2)$.

A better fit, corresponding to the solid line of Fig. 4.3, is obtained by taking into account the linear dependence of μ_i on energy. The very good agreement that is finally obtained shows that the hypothesis of a simple two-channel coupling is good and that there are no other interacting series.

4.2b. Fine and hyperfine structures

The radial part of the fine- and hyperfine-structure Hamiltonians is proportional to $1/r^3$ [except for the Fermi contact term in the hyperfine structure, which is proportional to $\delta(\mathbf{r})$]. As a result, the fine- and hyperfine-structure energy corrections will be mostly sensitive to the value of the electronic wave functions in a limited region around the nucleus.

As far as the nonpenetrating orbits are concerned, we can use the hydrogen wave functions to compute the splittings. An important property of those wave functions is that for $r \simeq 0$, the following factorization between the n and l dependencies occurs:

$$R_{nl}(r) \simeq \phi_l(r)/n^{3/2} \tag{27}$$

This is simply the quantum translation of the property, already mentioned, that the classical orbits do not depend on the energy at small radial distances. Therefore, the mean value of an operator depending mainly on small values of r (i.e., r^{-k} with $k>1$) in such states will be proportional to n^{-3}. This is the case for the core polarization correction of Sect. 4.2a and also for fine- and hyperfine-structure corrections. More precisely, the fine-structure splitting of the nonpenetrating orbits is given by[3]

$$\Delta E_{\text{fs}} = \alpha^2 R_\infty / n^3 l(l+1) \tag{28}$$

and the hyperfine structure splitting (separation between outermost components when $j \geqslant I$) by[3]

$$\Delta E_{\text{hfs}} = \frac{m}{M_{\text{p}}} \frac{8\alpha^2 I(j+\frac{1}{2})}{n^3(2l+1)j(j+1)} R_\infty \tag{29}$$

Note that the l dependence of ΔE_{fs} is only l^{-2}, whereas the polarization term dependence given by Eq. (9) is mainly l^{-5}. This means that for high-l values the fine-structure splitting is larger than the shift due to the core polarization quantum defect.

The calculation of such splittings is much more complex in penetrating orbits because the main contribution comes from the values of wave functions inside the core, which are not given by QDM. This method cannot yield the exact values of the structures but provides information on their n dependence. The quantum-defect method shows that the wave function for small r, even in the core, can be factorized in a way that is analogous to the hydrogen wave functions:[2]

$$R_{nl}(r) = [\nu^{-3}\zeta(\nu)]^{-1/2}\phi_l(1/\nu^2, r) \tag{30}$$

where $\zeta(\nu) = 1 + (\partial\mu/\partial\nu)$ and ϕ_l is an analytical function of energy, i.e., of $1/\nu^2$. The ν dependence of the fine and hyperfine splittings can therefore be written as

$$\Delta E_{nl} = \frac{\zeta(\nu)}{\nu^3}\left[A_l + \frac{B_l}{\nu^2} + \frac{C_l}{\nu^4} + \cdots\right] \tag{31}$$

As $\zeta(\nu) \simeq 1$ in Rydberg states, the ν dependence of fine and hyperfine structures can be fitted by a polynomial in odd powers of $1/\nu$, starting from ν^{-3}. Because the quantum defect μ, as previously mentioned, is itself an analytic function of energy, it can be easily shown that any analytic function of $1/\nu^2$ is analytic in $1/n^2$. Therefore, the structures can also be expanded (neglecting the $\zeta(\nu)$ factor) as[18]

$$\Delta E_{nl} \simeq \frac{1}{n^3}\left(A_l' + \frac{B_l'}{n^2} + \frac{C_l'}{n^4} + \cdots\right) \tag{32}$$

These two kinds of energy expansions have been used in the literature.[18-20] The precise ab initio computation of such structures is difficult and has been undertaken for only a few Rydberg levels, on helium,[21] and on more complex alkali atoms using higher-order perturbation theory,[22,23] many-body formalism,[24] or pure relativistic central field approach.[25]

For some alkali Rydberg series (D series of Na and K, F series of Rb and Cs), the fine structure appears to be inverted (the maximum j level lies below the minimum j level). This is very puzzling because it can be readily shown that in the usual model of one electron moving in an effective potential, fine structure is never inverted. This effect received a lot of experimental and theoretical attention. It means that higher-order effects (correlations and exchange effects) are in this case so important during the collision between the electron and the core that they play a dominant role. Sternheimer and co-workers[23] pointed out that the main contribution comes from a crossed term between exchange and fine-structure effects, and several different theoretical approaches were able to account for this inversion.[24-26] Finally, let us mention that expansion (31) holds even in the case of inverted structure with $A_l < 0$. This is not surprising because all the higher-order effects taken into account are only important when the valence electron is inside the core, where factorization (30) is valid.

4.2c. Other parameters relevant to spectroscopy

Oscillator strengths

Oscillator strengths involving Rydberg levels are important parameters in our discussion because they give the transition probabilities between levels. They are proportional to the square of the electric dipole matrix elements $D_{nl}^{n'l'}$:

$$D_{nl}^{n'l'} = qR_{nl}^{n'l'} = q\int_0^\infty r^3 R_{nl}(r) R_{n'l'}(r)\, dr \tag{33}$$

where $R_{nl}(r)$ is the radial part of the $|nl\rangle$ state wave function ($n \gg l \simeq 1$). Because of the r^3 factor, such matrix elements are primarily sensitive to values of wave functions far from the nucleus outside the core, where the form of Eq. (13) for wave functions is valid. As pointed out by Bates and Damgaard,[27] QDM is therefore able to give a good approximation to the actual matrix elements:

$$R_{nl}^{n'l'} \simeq \int_{r_0}^\infty R_{nl}^{(\mathrm{QDM})}(r) R_{n'l'}^{(\mathrm{QDM})}(r) r^3\, dr \tag{34}$$

r_0 being the core radius. Such matrix elements, which can be numerically computed,[28-30] depend therefore on the core properties only through the value of the quantum defect, which is a great simplification compared with the fine and hyperfine calculations. We shall now focus on two important limiting cases.

First, when $|nl\rangle$ and $|n'l'\rangle$ are both neighboring Rydberg states ($n \simeq n' \gg 1$), it can be shown that[28]

$$R_{nl}^{n'l'} = \frac{3}{2} n_c^2 a_0 \left(1 - \frac{l_c^2}{n^2}\right)^{1/2} \left[g_0(\Delta\nu) + \frac{l_c \Delta l}{n_c} g_1(\Delta\nu) + \left(\frac{l_c \Delta l}{n_c}\right)^2 g_2(\delta\nu) + \cdots \right] \tag{35}$$

where $n_c = 2\nu\nu'/(\nu+\nu')$, $l_c = \max(l, l')$, $\Delta l = l - l'$, and g_0, g_1, and g_2, are oscillatory functions of $\Delta\nu$ tabulated in Ref. 28 [$g_0(0) = 1$, $g_i(0) = 0$ for $i \geqslant 1$]. In first approximation, this reduces to

$$R_{nl}^{n'l'} \simeq \tfrac{3}{2} \nu^2 a_0 g_0(\Delta\nu) \qquad \text{when} \quad n \simeq n' \gg 1 \tag{36}$$

For a given $\Delta\nu = (n-\mu) - (n'-\mu')$, $R_{nl}^{n'l'}$ is proportional to $a_0 \nu^2$, which is the spatial extension of the Rydberg state wave functions. When $\Delta\nu$ increases, $g_0(\Delta\nu)$, displayed in Fig. 4.4, oscillates with a period tending to 2 and decreases with a general $\Delta\nu$ dependence of $\Delta\nu^{-5/3}$.[31] This is due to the alternately constructive and destructive interference between the two oscillatory wave functions.[32]

In the second case, when $|nl\rangle$ is a Rydberg state and $|n'l'\rangle$ a deeply bound state, the two wave functions overlap in a region of extension $n'^2 a_0$, i.e., in a limited sphere around the nucleus, where factorization is valid;[26] as a result, the n dependence of such a matrix element is simply $\nu^{-3/2}$, whereas the n' dependence has not a simple form:

$$R_{nl}^{n'l'} \propto \nu^{-3/2} \qquad \text{when} \quad n \gg n' \simeq 1 \tag{37}$$

The conclusion of this study is as follows: for a given flux of resonant electromagnetic radiation, the transition probability between two levels $|nl\rangle$ and $|n'l'\rangle$ is (1) much larger than between low-lying states when both states are neighboring Rydberg states (the probability increases as ν^4) and (2) much

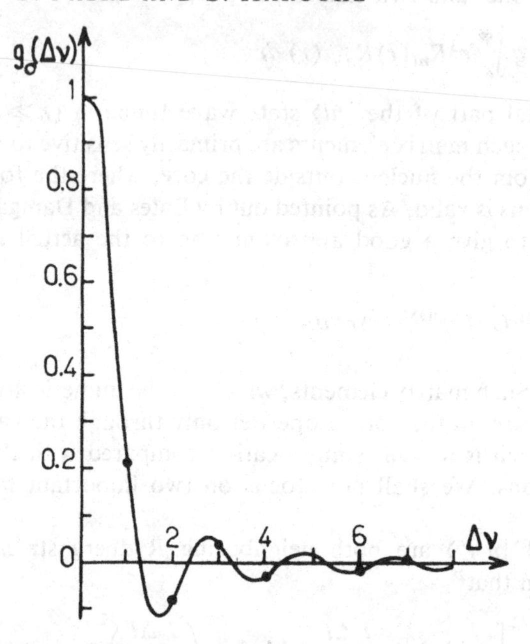

Fig. 4.4. Plot of the function $g_0(\Delta\nu)$ exhibiting the oscillatory dependence of the electric dipole matrix element $R_{nl}^{n'l'}$ versus $\Delta\nu=(n-\mu)-(n'-\mu')$. (From Picart et al.[28])

weaker than between low-lying states when the transitions link a Rydberg state and a low-lying state (the probability decreases as ν^{-3}). These two points will be very important in the following discussion of spectroscopy experiments.

Finally, relativistic effects can play an important role in determining oscillator strengths in the heavy alkalis. For example, the ratio ρ between oscillator strengths of transitions linking the ground state to the $j=\frac{3}{2}$ and $\frac{1}{2}$ fine-structure sublevels of the P series in cesium appears experimentally to depart strongly from the hydrogenic value of 2 (the ratio of the statistical weights of the two lines). In fact, it increases so quickly with n that it reaches a value of 1170 for level 30P.[33,34] The theoretical calculation of Norcross, which takes into account contamination of wave functions via the fine-structure interaction, is in good agreement with experimental data.[35] The same effect exists in rubidium but is not so important (ρ saturates to the value 5.9).[36]

Lifetimes

The spontaneous decay rate γ_{nl} of a level $|nl\rangle$ is given by the following well-known formula:

$$\gamma_{nl}=\sum_{n'l'}\frac{4}{3}\alpha\frac{\max(l,l')}{2l+1}\left(\frac{E_{nl}-E_{n'l'}}{\hbar}\right)^3\left(\frac{R_{nl}^{n'l'}}{c}\right)^2 \tag{38}$$

We see that the contribution of each $n'l'$ level depends on two factors that behave differently when the difference $n-n'$ increases: the energy factor increases rapidly, the squared matrix element decreases. For low angular momentum states ($l \ll n$), the net effect can be shown to be a fast increase with $n-n'$. The spontaneous decay of an n, $l \ll n$, state can populate all the n' levels (such that $l<n'<n$) compatible with the $\Delta l = \pm 1$ selection rule. In each term of Eq. (38), the n dependence of $R_{nl}^{n'l'}$ is simply $\nu^{-3/2}$, whereas the energy transition is close to $E_{n'l'}$ and does not depend on n. The dominant partial decay rates will then vary as ν^{-3}, and therefore the total lifetime of a Rydberg level will be proportional to ν^3 to a first approximation. This ν dependence has a very simple classical explanation: the classical electron only radiates when it is strongly accelerated, i.e., near the perihelion, where the orbit is independent of energy. The energy radiated per revolution is therefore constant and the radiative decay constant reflects only the variation with ν of the frequency of the electronic rotation.

The lifetime of sodium nS Rydberg states, for instance, is given by the formula[30]

$$\tau_{nS} = 1.38 \times (n - 1.348)^3 \qquad \text{(in ns)} \tag{39}$$

This corresponds to a lifetime of 32 μs for $n=30$. Rydberg states can be considered as metastable states, with very narrow intrinsic linewidths. This characteristic is very important for spectroscopic studies because it opens the way to very high resolution studies, with metrological implications. We must mention, however, that we have supposed thus far that Rydberg states interact with the photon vacuum. In an actual experiment, this is not the case because of the existence of blackbody radiation background. Gallagher and Cooke[37] pointed out that such thermal effects may appreciably affect Rydberg state lifetimes. This point is discussed in detail in Chap. 5.

Finally, let us mention that the previous analysis is not valid for high-l Rydberg states, for example, for the "circular" Rydberg states ($l=n-1$), which are observed in the interstellar medium (whose lifetimes are much longer than those of the low-l states). In this case, the spontaneous decay can only populate the nearby $|n'=n-1, l'=n-2\rangle$ state by emission of a long-wavelength photon whose frequency varies as n^{-3}. The energy term in Eq. (38) then varies as n^{-9} and the $(R_{nl}^{n'l'})^2$ factor as n^4 so that the decay rate can be shown to be proportional to n^{-5} ($\tau \simeq 2.3$ ms for $n=30$). This microwave spontaneous decay is actually what is used to monitor the population of interstellar Rydberg states.[38]

Polarizabilities

In order to obtain reliable spectroscopic data on Rydberg states, one must be sure that the studied atoms are not perturbed by some other interaction. The main reason for such broadening and shifting effects on Rydberg levels is electrostatic perturbation. In the low-field regime, an electrostatic field F parallel to the z axis produces an energy shift on level $|nlm\rangle$ given by

$$\Delta E_{nlm} = -\frac{1}{2}(\alpha_{nl})_0 F^2 - \frac{1}{2}(\alpha_{nl})_2 \frac{3m^2 - l(l+1)}{l(2l-1)} F^2 \tag{40}$$

where $(\alpha_{nl})_0$ and $(\alpha_{nl})_2$ are, respectively, the scalar and tensor polarizabilities of the levels, which are related to matrix elements $R_{nl}^{n'l'}$ by

$$(\alpha_{nl})_i = -q^2 \sum_{l'=l\pm1} a_{ll'}^{(i)} \sum_{n'} \frac{|R_{nl}^{n'l'}|^2}{E_{nl} - E_{n'l'}} \tag{41}$$

where $a_{ll'}^{(i)}$ is an angular coupling factor on the order of 1. The main contribution to $(\alpha_{nl})_i$ arises from levels with $n' \simeq n$. In this case, $E_{nl} - E_{n'l'}$ is proportional to n^{-3} and $R_{nl}^{n'l'}$ to n^2. The polarizabilities are therefore varying like n^7, which implies that Stark shifts in Rydberg states are orders of magnitude larger than in low-lying states. For example, the ground state (3s) of a sodium atom is shifted by 1 MHz when $F \simeq 7$ kV cm^{-1}. A field F of 1 V cm^{-1} is enough to shift the 30S level by the same amount. Polarizabilities increase also with l because levels of opposite parities lie closer for high-l values.

If we now turn to the dynamical Stark effect, expression (40) holds when replacing the static polarizabilities by the dynamical ones, which are given by the same expression (41) with $E_{nl} - E_{n'l'} \pm \hbar\omega$ instead of $E_{nl} - E_{n'l'}$. If $\hbar\omega$ is on the order of the energy splitting between Rydberg levels, the dynamical polarizabilities are also proportional to n^7: line shifts due to quasi-resonant irradiation may be important, even for low impinging fluxes. On the other hand, it has been shown theoretically[39] and experimentally[36] that high-frequency nonresonant irradiation has very little effect on Rydberg level positions.

4.3. Experimental techniques

We shall summarize in this section the main experimental methods used to prepare and to detect Rydberg atoms and to measure their energy intervals. Some of these techniques are mere generalizations to very excited states of methods already well known in studies of moderately excited species and we shall review them here just for the sake of completeness. Sometimes these standard atomic physics methods are not well suited for the study of Rydberg atoms. New methods, taking specific advantage of the peculiar properties of these states (i.e., their great sensitivity to electric fields or their large collision cross sections) have been developed to replace less-efficient or less-sensitive excitation or detection procedures. We shall review these new methods in greater detail in this section.

Most experiments can be divided into three phases: (1) a "preparation stage," during which Rydberg states are populated; (2) an "evolution stage," during which the Rydberg atoms evolve freely or are subjected to some kind of perturbation; (3) a "detection stage," during which the evolution undergone in the previous stage is observed. Let us first review the main features of stages 1 and 3.

4.3a. Preparation of Rydberg atoms

Rydberg atoms can be prepared either from the positive energy continuum states by electron–ion recombination or from the atomic ground state by electron or light excitation. All these methods are effectively used in experiments and have their specific advantages and drawbacks.

Electron recombination

The process of electron recombination is the one by which Rydberg atoms are created in the interstellar medium (binding of photoionized ions and electrons). This process can also be used in the laboratory. An ion beam in the kilo electron volt energy range – prepared by a plasmatron-type source – impinges on a solid (foil) or gas target and captures electrons onto very highly excited orbitals. The resulting Rydberg atoms can be studied in the beam downstream. The method has been used for hydrogen[40] (proton beam) and for helium[41] Rydberg atoms. It is convenient when selective excitation of a given Rydberg level is not required because capture results in a mixture of levels with different excitation energies. However, because it requires fast beams, it cannot lead to extremely high resolution (transit time problem).

Electron excitation

Electron impact on ground-state atoms can also be used to prepare excited Rydberg atoms. The experiment is performed in a low-pressure discharge in a cell or an atomic beam system. The method has been extensively used, in particular to prepare Rydberg states of He atoms.[42] It is very simple and convenient but, again, lacks selectivity. Furthermore, unless the spectroscopic measurement is performed in a carefully controlled afterglow, the discharge conditions induce perturbations on the levels under study (electric fields, collisions) whose effects have to be extrapolated to zero pressure.

Light excitation

Light excitation is by far the most versatile and refined technique for preparing Rydberg atoms. Starting from a ground state (or from an electronically prepared metastable level or even from a molecular ground state[43]), the atomic system is excited with one or several laser beams. Usually, the energy interval between the ground state (or metastable state) and the ionization limit falls in the UV part of the spectrum. The excitation, which is sketched on Fig.

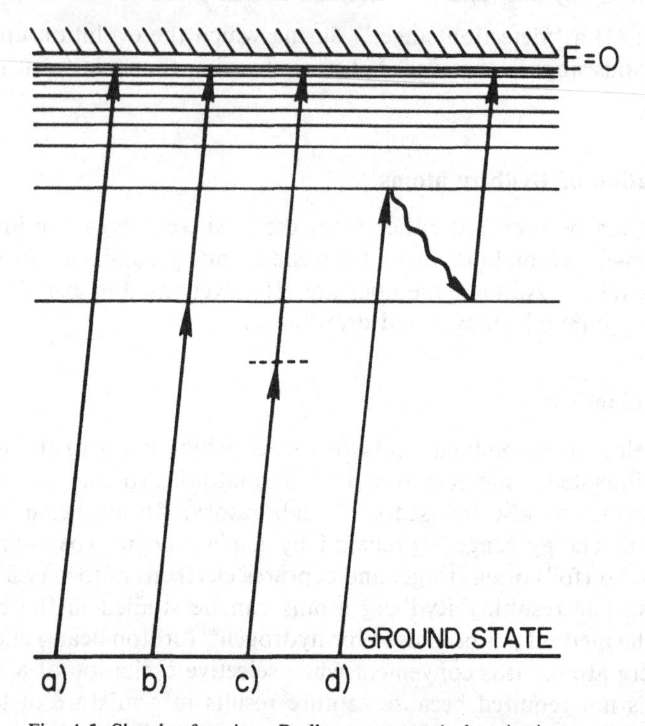

Fig. 4.5. Sketch of various Rydberg state optical excitation processes: (a) direct one-photon pumping, (b) two-photon stepwise excitation, (c) double quantum excitation with an intermediate virtual state, (d) combination of stepwise optical excitation and spontaneous decay.

4.5 for various cases, then requires a frequency-doubled visible dye laser[44] (Fig. 4.5*a*) or a multistep process involving a single frequency or several different frequencies. The multistep process can be resonant [stepwise excitation (Fig. 4.5*b*)][19] or nonresonant [virtual intermediate state with an energy mismatch between the first photon and the intermediate levels[45] (Fig. 4.5*c*)].

There exists a very large number of variants of such excitation processes. In some of them, one of the photons can be provided by the atomic system itself. For example (Fig. 4.5*d*), a first laser photon can be used to bring the atom to an excited level. Then the system decays to a lower metastable state by spontaneous emission. A second laser is then used to excite from this latter level up to the Rydberg series.[46]

The common feature of all these excitation mechanisms is the requirement of a lot of light intensity at least for the last step. As pointed out in Section 4.3c, the transition connecting a strongly bound state to a Rydberg level has a very small oscillator strength. This explains why many experiments have been made possible only with the advent of tunable, powerful dye lasers. A very convenient source for these experiments is the Hänsch-type pulsed laser,[47] pumped either by a N_2 laser or by the second or third harmonic of a Yag laser. The pulsed excitation is very convenient for all experiments involving time-resolved techniques. Some experiments also make use of cw lasers. In some instances, the tuning to the Rydberg level is achieved through the use of the Doppler effect and a fast beam.[48]

Light excitation has the big advantage of energy and polarization selectivity. With appropriate choices of laser frequencies and polarizations, it is possible to prepare levels of well-defined principal quantum number n, l, and magnetic quantum number m. The angular momentum l depends, through parity conservation, on the number of steps used to excite the system: starting from an s level, one excites a p state in a single step, an s or a d in two steps, an f or a p in three steps. The method does not seem to be well adapted for the preparation of high angular momentum levels, which would require a very high order excitation process. In fact, it is possible to excite such levels by taking advantage of the high polarizability of Rydberg levels: a relatively small electric field mixes the levels of opposite parity and breaks the parity conservation laws for the excitation process. In the case of a pulsed laser excitation, the electric field can then be reduced to zero after the light pulse and the atomic system are allowed to relax adiabatically toward the field-free levels.[49] All angular momentum states can be populated in this way. The electric field used for these experiments might, in some instances, be the motion field seen in the rest frame of atoms moving fast in an external magnetic field.[50] Let us mention also that the methods of electron recombination or excitation and light excitation can be combined to prepare Rydberg states via moderately excited states.[48]

4.3b. Detection of Rydberg atoms

Optical methods

A first method consists in analyzing the characteristics of the fluorescence light emitted by Rydberg atoms cascading down to lower bound states. This method is in general used to study low-lying excited states. In the case of Rydberg atoms, it suffers from the fact that the fluorescence yield corresponding to transitions in the visible or UV part of the spectrum is quite small, as explained in Sect. 4.2c. The intensity of fluorescence is thus very weak, and averaging procedures have to be used. This method is nevertheless convenient for moderately excited Rydberg atoms ($n \simeq 10$–20).[19,51,52]

Field-ionization method

The field-ionization method is certainly the most sensitive and most widely used detection method for Rydberg atoms.[44,53-56] It takes advantage of the very low ionization threshold of the atoms in dc electric fields. As soon as a

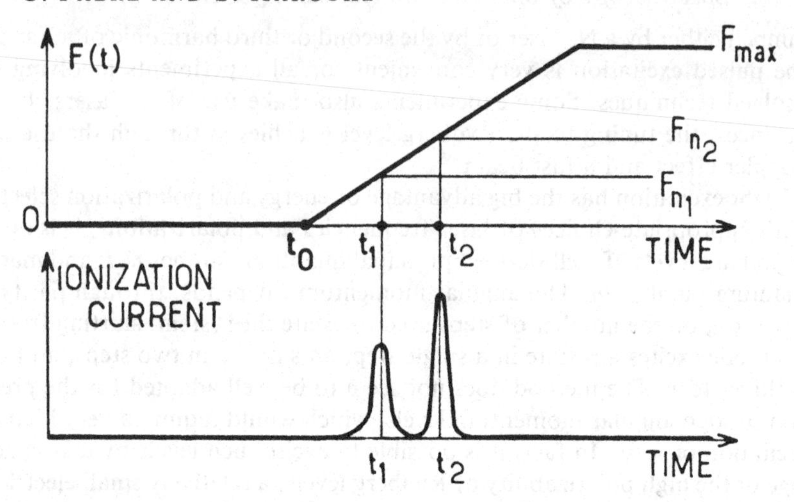

Fig. 4.6. Scheme of field-ionization detection of Rydberg states. *Top,* Ramped electric field $F(t)$ starts after a delay t_0 following the Rydberg state preparation and reaches at times t_1 and t_2 the ionization thresholds F_{n_1} and F_{n_2} for levels n_1 and n_2. *Bottom,* corresponding ion-current peaks proportional to the population of levels n_1 and n_2 at time t_0 (the system evolution between time t_0 and times t_1 and t_2 can be neglected).

field on the order of the internal field (i.e. $\sim 1/n^4$) is applied, Rydberg atoms break up into an ion and an electron that can be accelerated and detected with a channeltron or an electron multiplier. Selectivity in the detection results from the fact that the ionization threshold depends on the binding energy, i.e., on n and l, and also varies for a given level between different m_l substates.[55] Small changes in the various Rydberg state populations or orientations, either spontaneous or induced by lasers or rf fields, can thus be easily detected. The method is used either in a threshold procedure (with a square-shaped pulse of electrostatic field) or in a time-resolved procedure, as can be seen in Fig. 4.6. At a given time t_0 after the pulsed excitation of Rydberg atoms, an electric field ramp $F(t)$ (typically a few hundred volts per centimeter) is applied to them.[56] This field reaches at different times t_1 and t_2 the thresholds F_{n_1} and F_{n_2} for ionization of Rydberg states $|n_1\rangle$ and $|n_2\rangle$ for example. Each state thus appears as a time-resolved peak in the ionization signal (Fig. 4.6, *bottom*). The method is very simple, energy selective, and extremely sensitive. A single Rydberg atom in a well-defined state can be detected in this way.[57]

Space-charge-diode (thermionic) method

Another method that can be applied in atomic cell experiments is the method of thermionic diode detection.[58,59] The cell in which the Rydberg atoms are studied contains electrodes between which a small voltage is applied. When no Rydberg states are present, the current in this diode is space-charge limited.

When a Rydberg atom is produced, it has a probability of being ionized through collisions with other atoms. The resulting ions and electrons modify the space charge in the cell and the diode starts to pass some current. This current is detected, which results in a strong amplification of the initial ion or electron. This method is both very simple and elegant. It has, though, some drawbacks when compared with the previous one: perturbations due to the necessary collisions with the background gas and to the applied voltage and no obvious selectivity for Rydberg levels of different n and m_l. It can be used, however, inside a cell, which is a simpler setup than an atomic beam device.

Optogalvanic method

A variant of the previous technique is the optogalvanic method,[60] which involves the existence of a high electron temperature plasma in the cell: a relatively high current is driven through the atomic cell and the $V(I)$ characteristic curve of the discharge is monitored with an oscilloscope. When laser excitation excites some of the atoms to Rydberg states, the electric properties of the discharge change and the current through the cell is modified. In a pulsed experiment, this entails a transient change in the optogalvanic signal, which is easily detected. This method has been shown to be very versatile and sensitive. It has been used in particular to detect Ba Rydberg levels.[60] For very high resolution work, it would, however, suffer the same kind of limitations as the previous method.

4.3c. The methods of optical spectroscopy

The spectroscopy of Rydberg states can be divided into two main types. In the first, optical spectroscopy, energy intervals between a tightly bound state (in general the ground state) and a highly excited state, corresponding to transitions falling in the optical part of the spectrum, are measured. In the second type of spectroscopy, energy intervals between Rydberg states, corresponding to transitions of longer wavelengths (infrared, microwave), are directly measured. We shall discuss these techniques in Sects. 4.3c and 4.3d, respectively.

In optical spectroscopy, a frequency-tunable optical source (usually a laser) is used to irradiate the atoms. The population of Rydberg atoms is then monitored as a function of the frequency of the source using one of the techniques described in the previous section. The resolution of these measurements then depends on the spectral purity of the laser source and on the various perturbations acting on the Rydberg atoms (see Sect. 4.3e).

Doppler-limited techniques

When only moderate resolution is required, the lasers are operated multimode and no attempt is made to cancel the motional Doppler effect. Under these circumstances, which are common to a very large number of experiments, a resolution on the order of a fraction of an angstrom, good enough to resolve successive Rydberg levels up to $n \simeq 50$, can be achieved. This is, for example, the resolution needed in MQDT analysis experiments. The lasers used are generally pulsed dye lasers.

Laser–atomic beam methods

When higher resolution is required, single-mode frequency-stabilized lasers are used, and at the same time, the Doppler effect, which is usually the main factor of atomic line broadening, has to be eliminated. The simplest method to achieve this result is to cross the laser and the atomic beam at right angles.[36,61] The spectrum obtained by tuning the single-mode laser frequency is then Doppler free except for a small residual Doppler effect due to beam divergence. Resolutions on the order of several megahertz using cw dye lasers or tens of megahertz using pulsed dye lasers can be obtained by this method. Figure 4.7 shows as an example an optical spectrum obtained in this way on rubidium 48P states,[36] displaying four fine- and hyperfine-structure lines for each isotope, with a resolution of 60 MHz. In this experiment, Rb atoms are excited by a single-mode cw laser pulse-amplified in a Nd–Yag-pumped cell and detected by field ionization.

Two-photon Doppler-free methods

Another way of eliminating the Doppler effect consists of inducing a two-photon transition using a standing wave: the Doppler effects of counterpropagating photons exactly cancel each other and Doppler-free lines are observed in this way.[43,62–64] Higher light intensities have to be used to induce these nonlinear two-photon processes. Rydberg spectra of rubidium at megahertz and kilohertz resolutions were obtained in this way. In some of these experiments, the resolution was in fact limited by transit time (see Sect. 4.3e).

4.3d. Double resonance and related techniques

When one is interested, not in optical frequencies corresponding to transitions between Rydberg atoms and low excited states, but rather in small energy intervals between neighboring Rydberg atoms, the methods of optical spectroscopy are generally not the most convenient and sensitive. In these methods, the fine-structure or small-term energy splittings are obtained as differences between large energy intervals and the resolution is limited by the optical linewidth. The methods of double resonance, level crossings, or quantum beats are then, in general, preferable.

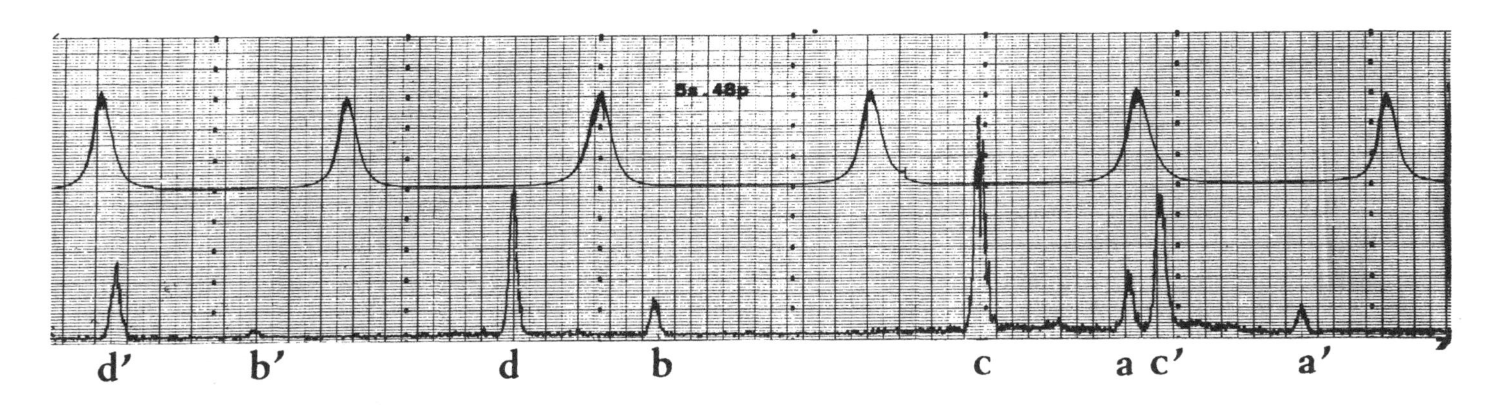

Fig. 4.7. Optical spectrum of the 5S–48p transition in Rb, with a resolution of about 60 MHz. The upper trace is a reference Fabry–Perot fringe pattern (spacing 1500 MHz). Components *a*–*d* refer to the fine and hyperfine components of ^{85}Rb (the fine structure of the 48p and the hyperfine structure of the 5S level are resolved). Components *a'*–*d'* refer to the corresponding lines in ^{87}Rb. (From Liberman and Pinard[36])

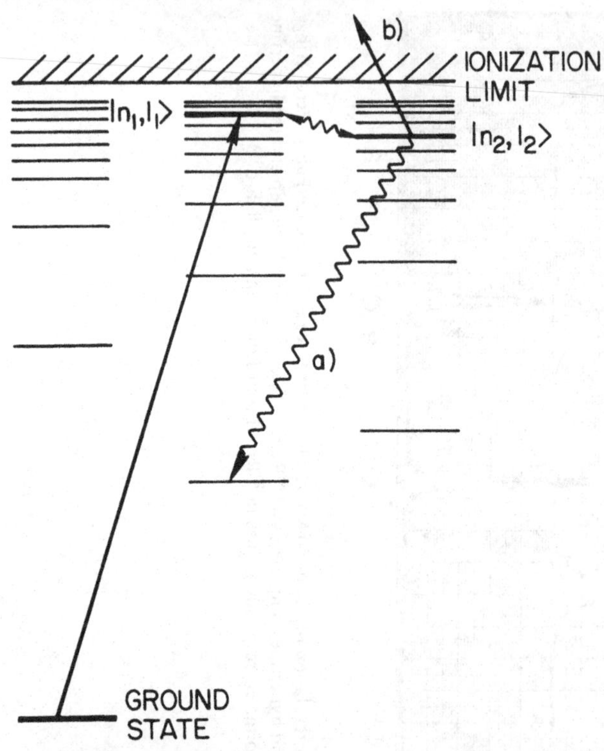

Fig. 4.8. Sketch of the various transitions involved in a typical double-resonance experiment involving Rydberg atoms. The successive arrows refer to the optical excitation, the microwave transition, and the detection process – either fluorescence (arrow a) or ionization (arrow b).

Microwave, millimeter-wave, and submillimeter-wave spectroscopy

A typical double-resonance experiment on Rydberg atoms is presented in Fig. 4.8. An atomic beam is irradiated by laser pulses preparing the atoms in a well-defined Rydberg state $|n_1 l_1\rangle$. The atoms thus excited undergo a microwave or a rf resonance transition to a nearby Rydberg level $|n_2 l_2\rangle$; this can be either a single-photon or a multiphoton transition. The resonance is then detected by a change of a specific atomic property depending on the state of the atom. In can be either a change in the fluorescence pattern (arrow a in Fig. 4.8) or a change in the field ionization current (arrow b). Resonance lines are recorded in most cases by sweeping the microwave or rf field frequency across the resonance.[42, 65, 66] In several other experiments,[67–69] the frequency of the low-frequency field is kept constant and a resonant signal is obtained when sweeping an external field acting on the atoms (magnetic[68, 69] or electrostatic[67] field). In this case, zero-field structures are then obtained by fitting the experimental data with theoretical calculations of the Zeeman or Stark pattern.

The resolution is now limited not by the optical frequency source jitter or

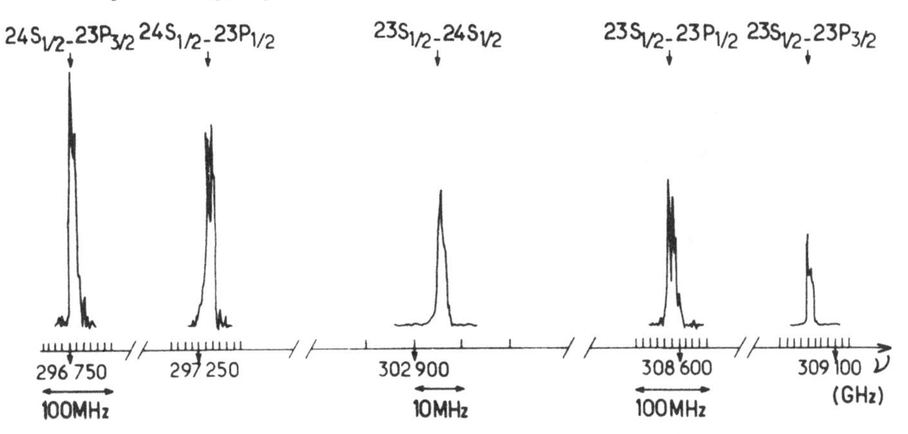

Fig. 4.9. Double-resonance spectrum in Na Rydberg atoms exhibiting the single-quantum transitions 24S–23P and 23S–23P (with their fine structure) and the double-quantum 23S–24S transition. (From Fabre, Haroche, and Goy[66])

optical Doppler effect but by the microwave Doppler effect. Source-frequency jitter can be reduced to a few hertz in the gigahertz domain by frequency-locking klystrons or carcinotrons.[70] The Doppler effect is, on the other hand, much smaller than in the optical domain (a few hundred kilohertz for a millimeter transition instead of a few hundred megahertz). It can be further suppressed by similar techniques as in the optical domain (for example, by using the now standard two-photon absorption in a standing wave).

A very interesting characteristic of these double-resonance experiments performed on Rydberg atoms is the extremely low level of microwave or millimeter radiation needed to saturate the transitions.[66,67] As explained in Sect. 4.2c, the electric dipole matrix elements between neighboring Rydberg levels ($\Delta n \simeq 1$) increase as n^2. As a result, the flux of resonant electromagnetic radiation necessary to saturate a transition between such states is (for $n \simeq 30$) reduced by a factor 10^6 with respect to a transition between low-lying states: picowatts of power are thus generally enough to saturate single-photon lines and microwatts to induce two-photon transitions. Very weak sources of radiation, such as the high-order harmonic output of a Shottky diode frequency multiplying a milliwatt power klystron are then powerful enough to perform spectroscopy on these atoms.

Figure 4.9 gives as an example a spectrum obtained by double resonance techniques on sodium 23S–23P, 24S–23P, 23S–24S lines,[66] using a millimeter-wave source around 300 GHz (back-wave oscillator) and the selective field ionization technique.

Maser emission spectroscopy

In fact, the coupling of Rydberg atoms to microwave radiation is so strong that the threshold for maser action can very easily be reached on transitions

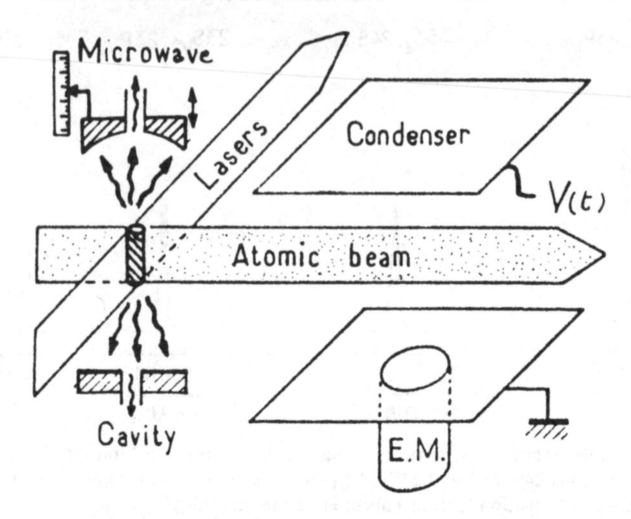

Fig. 4.10. Sketch of a Rydberg atom maser system. The atomic beam is pumped in a tunable microwave cavity and maser action is monitored either through the field-ionization signal or directly by detecting the small microwave emission escaping through the waveguide. Spectroscopy is performed by tuning the cavity length.

connecting nearby levels. Figure 4.10 illustrates the principle of a Rydberg atom maser.[71] The atomic beam under study is prepared by a laser pulse in a well-defined Rydberg state. It then crosses a millimeter-wave cavity tuned to a transition linking the initially prepared state to a lower state. The system mases above a threshold on the order of a few thousand inverted atoms. Maser action is monitored either indirectly by analysis of the level populations after the atoms have left the cavity[71] or directly by detecting the small microwave radiation bursts.[72] By tuning the cavity across the Rydberg transition frequencies, small energy intervals such as Rydberg state fine structure can be measured. For example, the maser action on the ns-$(n-1)p$ transition in sodium does occur at two slightly different cavity positions corresponding to the ns-$(n-1)P_{1/2}$ and ns-$(n-1)P_{3/2}$ transition frequencies. By measuring the change in cavity lengths corresponding to these two lines, the fine-structure intervals of the sodium Rydberg nP states can be determined.

In these maser-emission spectroscopy experiments, the maser action is triggered by the blackbody radiation background. Thus, these experiments can be described as double resonance induced by the broad blackbody background filtered by the cavity and amplified by the atoms.

Level-crossing spectroscopy

The determination of small energy intervals in excited atomic states can also be performed by level-crossing spectroscopy. This well-known technique has been successfully applied to a variety of alkali atom Rydberg states for the measurement of fine or hyperfine structures.[69] Let us recall that the method consists of preparing, by convenient continuous optical or electronic bombardment excitation, a linear superposition of excited substates. This superposition evolves at the Bohr frequencies corresponding to the energy splittings and is damped by spontaneous emission. A steady state is reached under the effect of these competing processes. By applying a magnetic field to the atoms, the various atomic sublevels are shifted, and for some specific magnetic field value, crossings of energy levels are obtained. For the field value corresponding to these crossings, the linear superposition of crossing states evolves at zero frequency and the steady-state fluorescence signals, sensitive to this linear superposition, undergo a resonant change when the field is swept around the crossing point. It is thus possible to determine with accuracy the values of the magnetic field corresponding to the crossing points and, from these values, to deduce the zero-field atomic parameters such as fine- and hyperfine-structure constants.

A variant of the method is the so-called level-anticrossing spectroscopy method:[73] it might indeed happen that either an internal or an external perturbation couples the levels that would otherwise cross each other, thus inducing in the energy diagram an avoided crossing or anticrossing situation. Around the anticrossing point, the atomic fluorescence also undergoes a resonant change. The width in magnetic (or electric) field of these anticrossing signals as well as their amplitude yields interesting information about the internal atomic structure and such signals can also be used for fine- or hyperfine-structure determinations.

The level-crossing experiments are very similar to the double-resonance ones. The basic difference lies in the fact that in the latter the preparation and detection are performed on atomic populations in which changes are induced by an external rf field, whereas in the former, preparation and detection are made on superposition states whose free evolution around crossing points provides the spectroscopic information.

When, in a level-crossing experiment, the preparation is performed by light irradiation, the polarization of the exciting beam has to be chosen in order to prepare a superposition of states. The fluorescence detection has to be performed also with the proper choice of polarizations in order to be able to detect a linear superposition of crossing substates. Because they generally rely on fluorescence detection, these methods become increasingly difficult to use when the principal quantum number n increases. They have thus been principally developed for moderately excited Rydberg atoms ($n \simeq 10$–20). It is, however, worth noting that level-crossing signals do not necessarily have to be detected by fluorescence. Any other method sensitive to state superposition in the Rydberg level of interest can, in principle, be used. This is the case for the field-ionization or the photoionization method, which will be discussed later.

Quantum-beat spectroscopy

Quantum-beat experiments[74] are the transient counterpart of the level-crossing ones. The superposition of atomic substates is prepared by a short-pulse excitation and the evolution of this superposition is directly monitored at the atomic system Bohr frequencies. This evolution entails a modulation of conveniently polarized fluorescence signals.[19,75]

Quantum beats can also be viewed as a quantum interference effect: the atomic system can follow several different paths through the various sublevels of excited states. To each path corresponds a quantum-mechanical amplitude. Because the final state reached by the fluorescence decay is common to all paths, all these amplitudes interfere with each other and the interfering term exhibits the modulation of interest. This modulation can be analyzed by Fourier transformation to yield the structures under study (fine or hyperfine structures in general). Here again, the detection of fluorescence limits the method to moderately excited Rydberg levels. Even so, averaging methods have to be used to extract the signal from the noise.

An interesting variant of the fluorescence detection consists in using the field-ionization method:[76,77] the rate of field ionization depends on the respective direction of the excited-state alignment and of the ionizing electric field. When varying the delay between the excitation and the detection electric field pulse, the angle between these two directions periodically changes, with a period corresponding to the beats to be studied. As a result, a modulation of the ion current is observed as a function of the detection delay. Figure 4.11 gives the modulations observed in the ionization signal as a function of this delay for the levels 21D–31D of sodium.[76] The actual mechanism of field ionization in such an experiment is rather complicated. In order to describe it, one has to know how the field-free energy levels are modified when a pulse of electric field is applied on the atoms and how the atomic levels "follow" these modified levels from the field-free situation up to the ionizing field.

The existence of anticrossings of Stark levels in the field-dependent energy diagram plays a very important role in this process (see Fig. 4.12). It is clear that here again the modulation will arise from a quantum-interference effect, which can occur if the various substates whose coherence is being studied are allowed to evolve under the effect of the electric field switching into a *single* ionizing level. Only in this case are the channels corresponding to the various substates truly indistinguishable and the quantum-interference effect observable. To achieve this result, the field-ionization process should not have a purely adiabatic behavior: the system must have a nonzero probability of jumping from one anticrossing branch to the other one when the field is swept across it. In the adiabatic case, on the contrary, the different zero-field levels would follow distinct paths, each one along its own anticrossing branch, and would end up into different high-field states with no possible interference effects. One way of obtaining such a nonadiabatic evolution is to use a very fast ionizing electric field, rising in a time much shorter than the Bohr period

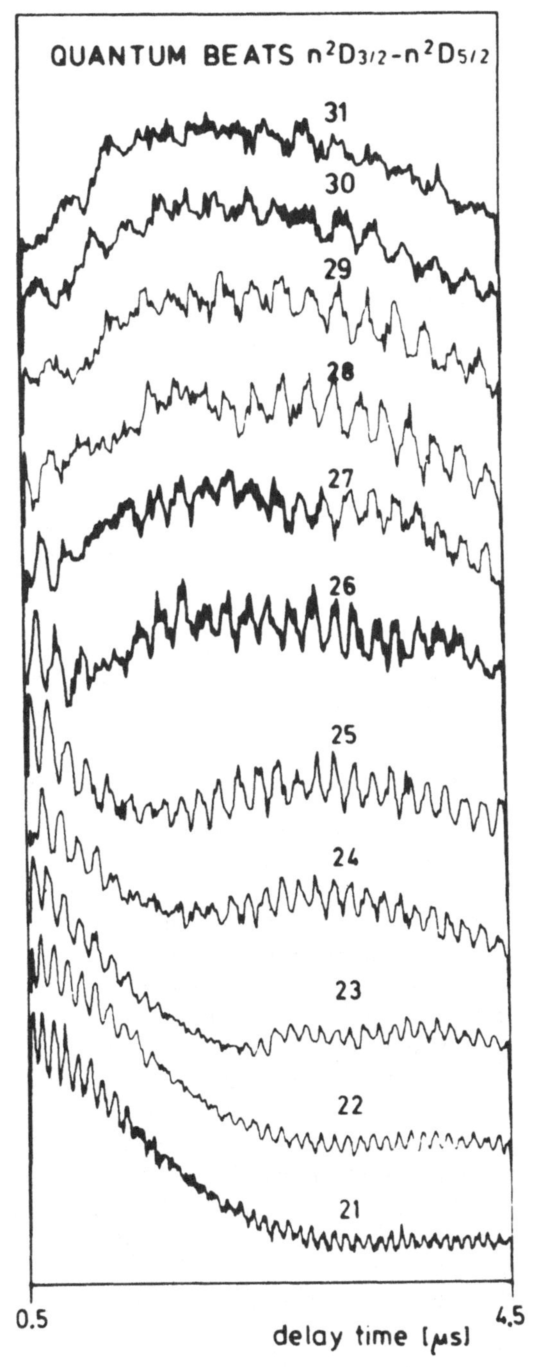

Fig. 4.11. Quantum-beat signal of high-lying ^{2}D states of sodium detected by delayed field ionization. (From Leuchs, Smith, and Walther[76])

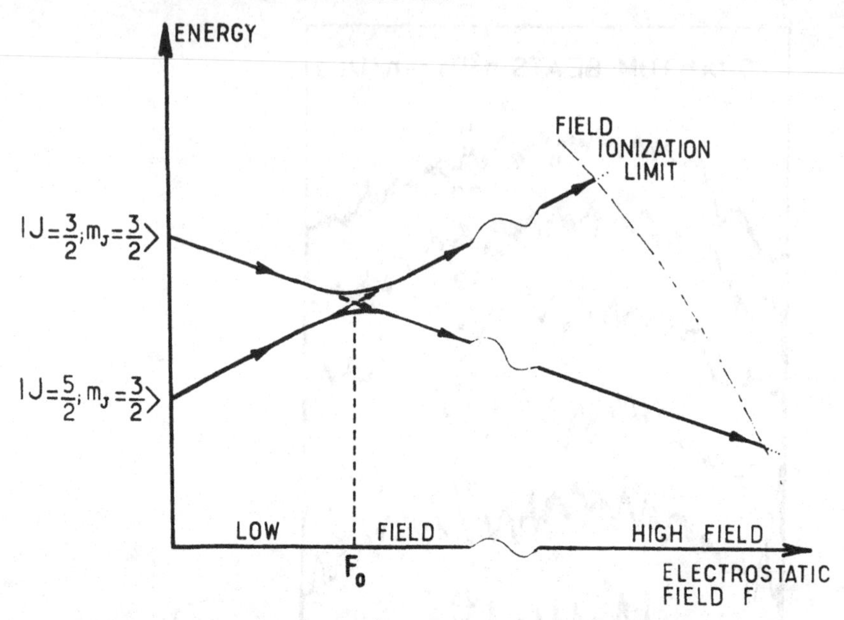

Fig. 4.12. Simplified Stark diagram showing the transition between the zero field and the ionization region for two sublevels that anticross for a given low-field value F_0. In the adiabatic case (slow electric field increase), each level follows its own anticrossing branch and ionizes in different fields. No quantum interference is expected in this case. In the nonadiabatic situation (fast electric field increase in the low-field region), each zero-field sublevel evolves into a mixture of the anticrossing states so that the ionization process cannot distinguish anymore between the two initial sublevels. Quantum beats can then be observed.

associated with the anticrossing gap (practically, nanosecond-long pulses are short enough for sodium *n*D states).[76] Such pulses are much faster than the one used in ordinary field-ionization detection (where microsecond pulses are used) and are rather difficult to produce because they should also have a very stable pulse-to-pulse amplitude (the beats being monitored with a sampling method involving repetition of the experiment with increasing delays). A variant of this technique[77] consists in applying, after a variable delay, a very small electric field of a few volts per centimeter, rising in a very short time (on the order of 1 ns), which is not large enough to ionize the atoms but is fast enough to bring them nonadiabatically through the low-field region where the level mixing necessary for the quantum-beat observation does occur. After this nonadiabatic mixing has been secured, the atoms are easily ionized by a subsequent much larger electric field pulse, which does not need to be fast anymore. This method seems to be more convenient than the previous one because it does not require the difficult realization of fast, intense, and stable electric field pulses.

The field-ionization quantum-beat technique has been successfully used to determine fine structure splittings for highly excited sodium *n*D states

($n \simeq 20$–40). It has indeed proven to be more sensitive in this case than the ordinary fluorescence quantum-beat method, which could not be used above $n=17$.[19]

4.3e. General limitations to resolution

As stated in Sect. 4.2c, Rydberg states have very long lifetimes when compared with low atomic excited states. As a result, the ultimate resolution can be very high, and metrological applications of the spectroscopy of such states may be considered. On the other hand, Rydberg atoms appear to be very perturbable species. Therefore, all the sources of line broadening and shifting must be carefully studied in order to eliminate them and to obtain reliable spectroscopic data on such states. We shall review here the main physical processes that restrict the experimental resolution.

Perturbation by static fields

We shall deal first with magnetic perturbations. The paramagnetic Zeeman effect, which does not depend on the principal quantum number n, has therefore no specific properties in low-l Rydberg states. This is not the case for the diamagnetic Zeeman effect, which depends on the mean value of x^2+y^2 (if the field $\mathbf{B}_0$ is parallel to the $0z$ axis): diamagnetic shifts turn out to be proportional to $\nu^4 B_0^2$ and increase rapidly with n. Nevertheless, for commonly studied Rydberg states ($10<n<40$), the diamagnetic perturbation is negligible in usual stray magnetic fields (1 G shifts the $n=30$ level by only 200 Hz).

On the contrary, the electrostatic perturbation is very important in Rydberg states because of their very large dimensions. The static polarizabilities increase as n^7 and are also proportional to l^5 for nonpenetrating orbits. In most experimental setups, the stray electrostatic inhomogeneous field cannot be reduced to less than about 50 mV cm^{-1}, which induces a level broadening on the order of 2 kHz, 25 kHz, 500 kHz, and 5 MHz, respectively, for levels 30S, 30P, 30D, and 30F in sodium. Electrostatic perturbation is therefore a strong limitation to the resolution of spectroscopic experiments, especially for nonpenetrating orbits (or for experiments performed on hydrogen). To eliminate it, very careful electrostatic shielding should be used. On the contrary, if only low-l Rydberg states are studied (penetrating orbits), the Stark line shifting and broadening can be negligible in the usual stray fields. This is especially true in double-resonance experiments between analogous Rydberg states (for example nS–$(n+1)$S double-quantum transitions), which are only sensitive to differential Stark shifts.

Perturbation by radiation fields

Rydberg states are very sensitive to electromagnetic waves whose frequencies are on the order of the characteristic frequency of the electronic rotation around the nucleus ($\sim$300 GHz for $n=30$). A minute power of resonant radiation is enough to saturate a transition between Rydberg states and hence to broaden the observed line (100 pW cm^{-2} MHz^{-1} saturates a 30S–30P transition in 10 μs). When the radiation is nonresonant, the levels are shifted by the ac Stark effect, which is proportional to $n^7 l^5$, as is the dc Stark effect. Double-resonance experiments must then be performed with extremely low powers so as not to saturate the line and to reduce the line shifts due to quasi-resonance of the impinging wave with nearby Rydberg levels.

On the other hand, Rydberg states are quite insensitive to optical fields: there is therefore no problem with radiative line shifting and broadening in the optical spectroscopy of such states.

Let us briefly mention here the strong effect of a peculiar radiation field – the blackbody radiation field – that shortens the lifetimes of Rydberg states. This reduces not only the ultimate resolution of experiments, but also slightly shifts the Rydberg levels (24 kHz for $T=300$ °K).[37] This shift, which is roughly independent of n, cannot be seen in double-resonance experiments, but only in optical spectroscopy. These effects are described in Chap. 5.

Collisions

The effect of collisions will be mentioned here only briefly for the sake of completeness because it will be discussed in great detail in Chap. 7. Let us say simply that, because collisional cross sections with Rydberg states first increase with n as n^4 (geometrical cross section) for $n \lesssim 10$ and then decrease with n because of the transparency of the very large Rydberg atoms, their effect is not as dramatic as it might appear at first sight. Collisional line-shifting and broadening effects are more important than in low-lying states, but in most experimental situations (with atomic beams or low-pressure cells), they do not limit resolution.

Velocity-dependent effects

As in the case of transitions between low-lying states, the Doppler effect limits the resolution to about 10^{-6} for atoms with thermal velocities. For higher resolution, the usual techniques of Doppler-free spectroscopy have been used for Rydberg states, as explained in Sect. 4.3c and 4.3e (laser-beam technique or two-photon spectroscopy). To our knowledge, the saturated absorption technique has not been used because of the difficulty in saturating the optical transitions to Rydberg states.

Another velocity-dependent limitation to resolution is due to the finite interaction time between atoms and radiation, which is, in most cases, less than the natural lifetime of the level. This time can be increased by cooling the atoms, by extending the electromagnetic wave beam, or even by using the

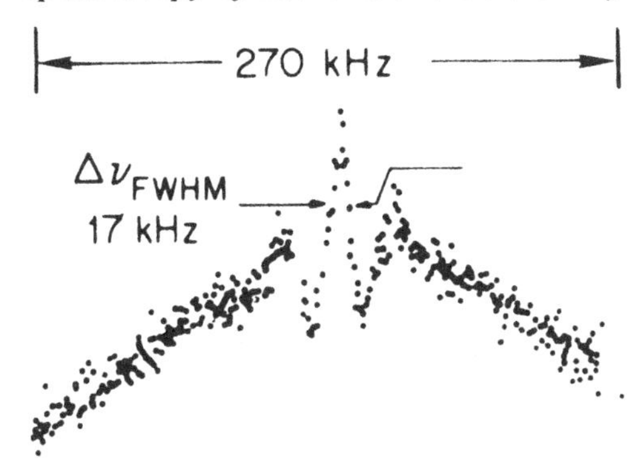

Fig. 4.13. Optical two-photon Ramsey fringes observed on the 5S–32S transition in ^{85}Rb. The atomic beam crosses two regions of laser standing waves separated by 4.2 mm. (From Lee, Helmcke, and Hall[64])

Ramsey technique of separated fields. Transit times on the order of 100 μs, corresponding to a few kilohertz linewidths, are currently achieved.

Ultimate resolution

To illustrate the previous discussion, let us take two experimental examples (one optical, one double resonance) in which the goal was the highest possible resolution.

In the optical resonance experiment,[64] double-quantum optical transitions were induced between ground and nS states of a rubidium atomic beam ($n \simeq 30$), with an ultrahigh-stability single-mode dye laser (2-kHz linewidth). The stray field (150 mV cm^{-1}) is responsible for a line shift of less than 10 kHz. The Doppler effect is suppressed by two-photon absorption inside an optical Fabry–Perot cavity. Transit-time broadening is reduced by folding this cavity, which yields two interacting regions separated by 4 mm. When sweeping the dye laser frequency, Ramsey fringes were recorded with a width of 17 kHz, corresponding to a resolution of 10^{-10} (Fig. 4.13). This very high resolution (close to the highest ever achieved in optical measurements) makes Rydberg states good candidates for metrological applications. However, the problem of shifts and distortions of the Ramsey fringes has not yet been completely solved.

In the double-resonance experiment (Fig. 4.14) double-quantum millimeter-wave transitions were induced between Rydberg nS and $(n+1)$S states of a sodium atomic beam ($30 < n < 40$) with a phase-locked back-wave oscillator around 100 GHz.[70] The stability of the millimeter source was better than 1 kHz, i.e., less than the natural linewidth (2 kHz). The stray electrostatic field (~30 mV cm^{-1}) shifted the line by less than 1 kHz. The Doppler effect is

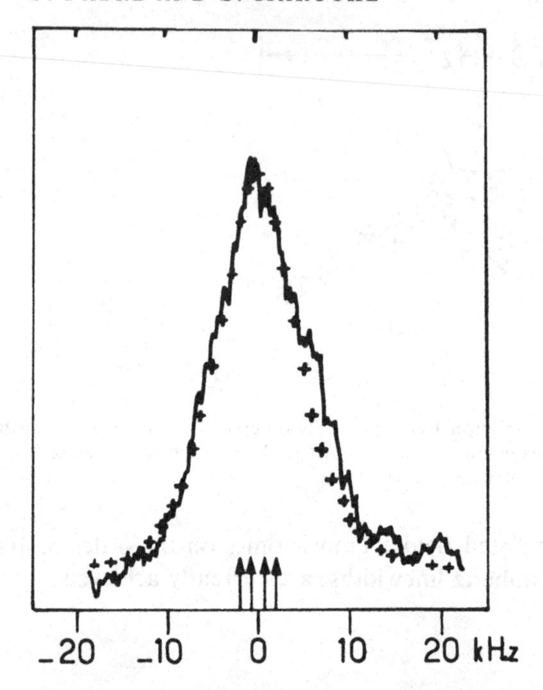

Fig. 4.14. Millimeter-wave two-photon resonance observed on the 39S→40S transition in Na. The atomic beam crosses a millimeter cavity with a waist of 2 cm. The transition frequency is $2\nu = 118\ 520\ 443\ (4)$ kHz. The four arrows indicate unresolved hyperfine components. The crosses give a theoretical line shape computed without any adjustable parameter (except height and line center). (From Goy et al.[70])

suppressed by two-photon absorption inside a millimeter Fabry–Perot cavity. The linewidth is then limited by the transit time inside the cavity mode (2 cm wide at the waist) to 11 kHz, which allows a line-center determination of 2 kHz (4×10^{-8} resolution). Such an experiment could open the way to a new measurement of the Rydberg constant if it could be repeated on hydrogen Rydberg states.

This quick description of two high-resolution experiments shows strong similarities between the techniques in different frequency domains. The optical experiment may reach a higher relative resolution but yields wavelengths in terms of the length standard, whereas the millimeter-wave experiment directly yields transition frequencies in terms of the more precise time standard.

4.4. One-electron Rydberg atom spectroscopy

We shall review in this section the various spectroscopy experiments performed on Rydberg atoms with a closed-shell core (one-electron atoms) for which the single-electron quantum-defect model is valid.

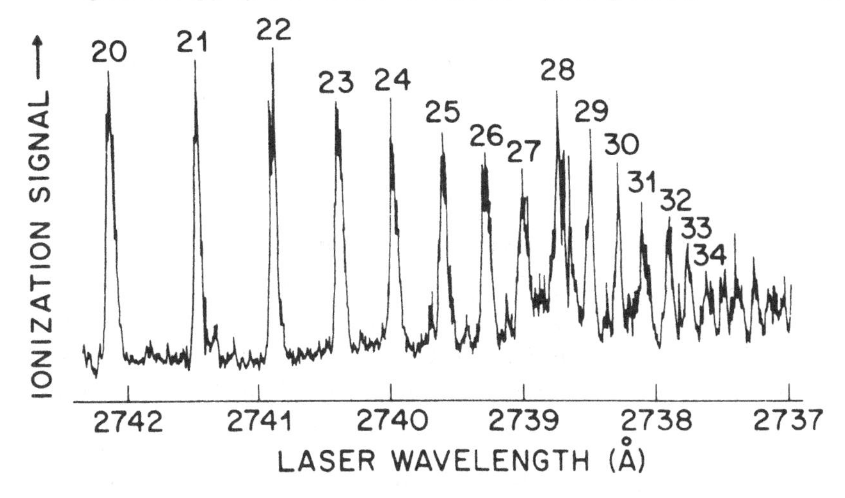

Fig. 4.15. Three-photon excitation of hydrogen Rydberg n states ($20<n<34$) recorded on the ionization signal as a function of UV laser wavelength. (From Bjorklund, Freeman, and Storz[79])

4.4a. Hydrogen

Paradoxically, there have been, up to now, almost no precise spectroscopic studies of the simplest of all Rydberg atoms. The reason stems from the experimental difficulties in the excitation of hydrogen (large ionization energy), from the great sensitivity of these atoms to electric perturbations, and from the fact that any measurement, to be of value, should be extremely precise because the theoretical position of the levels is known with the present accuracy of the Rydberg constant (3 parts in 10^9). In fact, the spectroscopy of hydrogen Rydberg states would be extremely interesting if it could reach a resolution high enough to yield an improved value for this constant. Several groups are presently planning experiments to achieve this goal.

The excitation of hydrogen Rydberg states will either be a combination of electron capture and infrared laser excitation (according to the technique developed by Koch, Gardner, and Bayfield[78] for the study of various microwave or electric field–Rydberg atom interaction processes) or a direct excitation from the ground state by multiphoton optical pumping according to a method recently developed by Bjorklund, Freeman, and Storz.[79] These authors made use of a frequency-doubled Yag-laser-pumped pulsed dye laser to induce a three-photon 1s–np transition in hydrogen atoms obtained by an electric discharge in a molecular hydrogen stream. By tuning the laser frequency, they obtained an optical absorption spectrum of the hydrogen Rydberg series (see Fig. 4.15) and showed that this method (or another optical variant making use of intense laser excitation) could be used with profit in a spectroscopic experiment involving hydrogen Rydberg states.

The sources that could be used to induce transitions between these states could be either in the millimeter-wave domain (transition between adjacent levels) or a frequency-stabilized infrared laser (CO or CO_2) because coincidences do exist between hydrogen Rydberg transitions and these laser frequencies.

4.4b. Term-interval and quantum-defect measurements in alkalis

The spectroscopy of the alkali Rydberg states (Li, Na, K, Rb, Cs) is much more developed than that of hydrogen. Laser excitation of these states is indeed much easier and the perturbing effects of stray electric fields are smaller. Hence, a very large number of experiments have been performed on these species, leading to a considerable increase in available spectroscopic data on these atoms in the last several years. We shall review in this section the term-interval measurements, which will amount to a determination of the quantum defects.

Direct optical spectroscopy measurements - two-photon s–s or s–d transitions - have yielded precise measurements of quantum defects in rubidium,[62] potassium,[63] and cesium.[43] Double-resonance experiments using klystrons, back-wave oscillators, or CO_2 lasers have also yielded precise quantum-defect values in Na (s, p, d, f, and h states),[65-67,70,80,81] in Li,[82] and in Cs.[52]

For core-nonpenetrating states, these quantum-defect measurements amount to the determination of the core dipolar and quadrupolar polarizabilities. These quantities have been determined with precision in Li,[82] Na[65] and Cs.[52]

For core-penetrating states (s in Li; s and p in Na and K; s, p, and d in Rb and Cs), the quantum defects are much larger and reflect the complex distribution of electrons within the core. Systematic variations of quantum defects with n within a given series of Rydberg states have been observed in various experiments,[62,70,81] the relative variation of the quantum defect being on the order of 10^{-5} to 10^{-6}. As shown in Sect. 4.2c, these variations can be expanded in powers of the binding energy, i.e., along even powers of $1/\nu$. Empirical polynomial expressions have been found that fit the experimental data very well.[62,70] Figure 4.16 from Ref. 70 shows as an example the quantum-defect variations in Na S states as a function of the binding energy. The precision of these experiments is well beyond the accuracy of ab initio calculations of these quantum defects. In fact, one value in performing such experiments lies in possible metrological applications (see Sect. 4.6).

4.4c. Fine- and hyperfine-structure measurements in the alkalis

All the experimental techniques described in the previous section (optical, double resonance, quantum beats, level crossings) have been used to measure the fine-structure intervals for a lot of alkali Rydberg states: d, f, and g states of Li;[82] p,[66,80,81] d,[19,49,76,77,83] f,[84] and g and h[65] states of Na; d states of K;[85] p

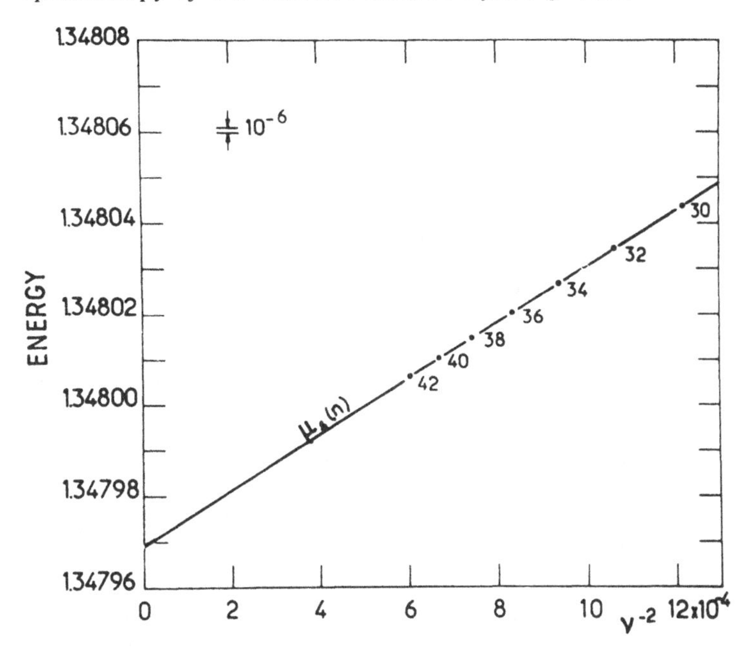

Fig. 4.16. Plot of Na S state quantum-defect variations measured by double resonance. (From Goy et al.[70])

states[36] and d states[62] of Rb; and f states of Cs.[52,61] Long series of levels have been investigated in these experiments: systematic variations of the fine-structure interval as a function of n have been obtained in this way and have always been found to be in good agreement with the semiempirical law in odd powers of $1/\nu$ discussed in Sect. 4.2b. For core-penetrating states, the fine-structure interval appears to be much larger than the corresponding interval in hydrogen, whereas for nonpenetrating states, the interval can be shown to tend quickly toward its hydrogenic value. For series corresponding to the threshold-l value between the two regimes ($l=2$ for Na and K and 3 for Rb, Cs), the structure appears to be inverted, except in Li: such an inversion, as already mentioned in Sect. 4.2b, is highly nonhydrogenic. This very interesting problem has been extensively studied by several groups, and a great deal of experimental data on those "strange" series has been gathered. A typical example involves the inverted fine structure of the Na D series. Fine-structure intervals of most levels from $n=3$ to $n=40$ (ranging from 1.5 GHz to 1.5 MHz) have been measured using optical methods,[86] level crossings,[83] and optical[19] and field-ionization[76,77] quantum beats by various groups. Figure 4.17 shows the variation of this interval with n. The solid line corresponds to the following expansion in odd powers of $\nu=n-\mu_d$:

$$\Delta E_{fs} = \frac{A}{\nu^3} + \frac{B}{\nu^5} \tag{42}$$

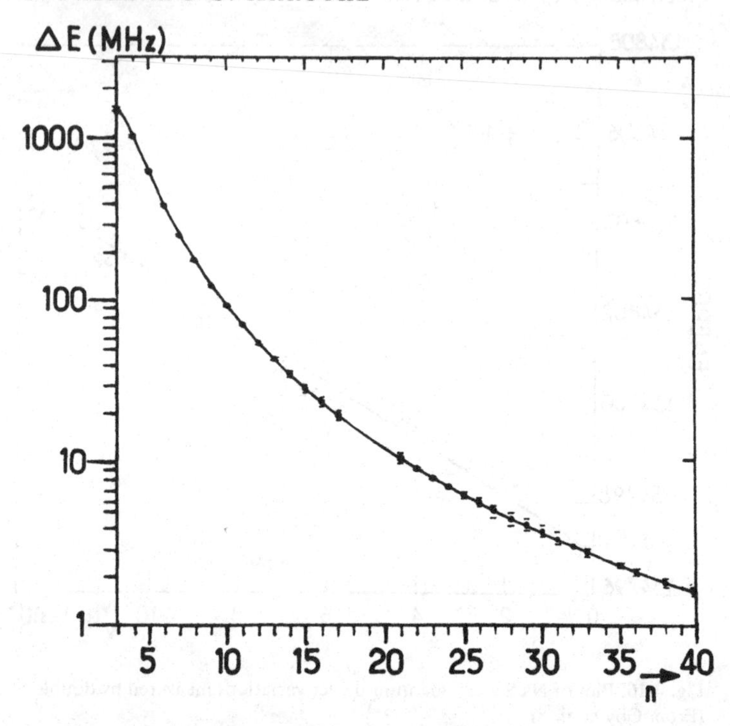

Fig. 4.17. Semilogarithmic plot of fine-structure interval of sodium nD states as a function of principal quantum number n $(3<n<40)$. The points give the experimental data taken from several experiments and the solid line gives the semiempirical fit by the expression $(A/\nu^3)+(B/\nu^5)$.

with A = −97 800 (1000) MHz and B = 52 (2) × 10^4 MHz and the points are the experimental results. It is interesting to note that 31 experimental values ranging over three orders of magnitude can be very precisely fitted by the simple formula (42), depending only on two adjustable constants. Similar results, though less extensive, have been obtained for the inverted series of the other alkalis.

The usually smaller hyperfine structures have been much less studied in Rydberg series, though some structures have been measured for moderately excited states ($n \lesssim 10$) in Na S,[69] Rb S,[68] P,[69] cesium S,[68,87] and D series.[69,75] Isotope shifts of potassium nS and nD states ($n \leqslant 18$) have also been measured.[88]

4.5. Two-electron Rydberg atom spectroscopy

The spectra of Rydberg atoms with two electrons outside the closed shells of the core are much more complex than those of the alkalis (see Sect. 4.2a). The experiments on these states are now advancing very rapidly. Let us consider successively the case of single-electron excitation in helium (which is, in many respects, similar to a single-electron atom), then in alkaline earths, and finally doubly excited two-electron atoms (so-called planetary atoms).

4.5a. Helium Rydberg state spectroscopy

Because helium is the simplest multielectron atom, an extensive experimental effort has been made in the exploration of its Rydberg series. A very important series of experiments in this field is that of Wing and MacAdam.[42, 89-91] They used electron bombardment excitation and microwave double resonance with fluorescence detection and measured a very large number of fine structure and electrostatic term intervals for n varying between 6 and 18 and l between 0 and 4. A submegahertz precision was obtained by these authors for most transitions lying in the 7-50-GHz range. Using similar techniques, Cok[41] recently measured some F–G and G–H intervals. Using the magnetic field anticrossing technique with a small electric field to mix singlet and triplet states, Beyer and Kollath[73] also measured, with somewhat less precision, several D to higher-l-state intervals for n between 7 and 10 and l between 3 and 7. Panock et al.[50] also used a field-mixing technique to perform motional Stark effect spectroscopy on helium. The experiment was performed on an He beam placed in a very high magnetic field. The motional electric field mixed levels of different l, thus allowing several transitions between S and high-l states. These transitions were driven by a CO_2 laser. Several high-field spectra involving high-l (up to $l=8$ with $n=9$) states were obtained. With the magnetic field in the experiment known, values for the zero-field energies of these states were obtained for the first time.

From these and other experiments, a very large body of precise experimental data is now available on moderately excited Rydberg states of He. These data are well fitted by semiempirical formulae involving expansions in odd powers of $1/n$,[88] which show that the one-electron quantum-defect model is still a good approximation in the case of He (the excitation energy of the ion He^+ core is very high). The agreement with ab initio calculations is less good. In particular, systematic discrepancies on the order of 2% exist between the measured term F–G intervals and the theoretical calculations.[41] These discrepancies should stimulate further theoretical calculations.

4.5b. Singly excited alkaline-earth atoms

Extensive studies of the singly excited alkaline-earth atom Rydberg spectra have recently been made using laser excitation. In order to reach Rydberg levels or to study levels of the same parity as the ground state, a multiphoton process was necessary. In a first type of experiment, a dye laser was used to populate odd-parity levels and the broadband background of a second dye cell was used in absorption to monitor the spectrum of the transitions linking the

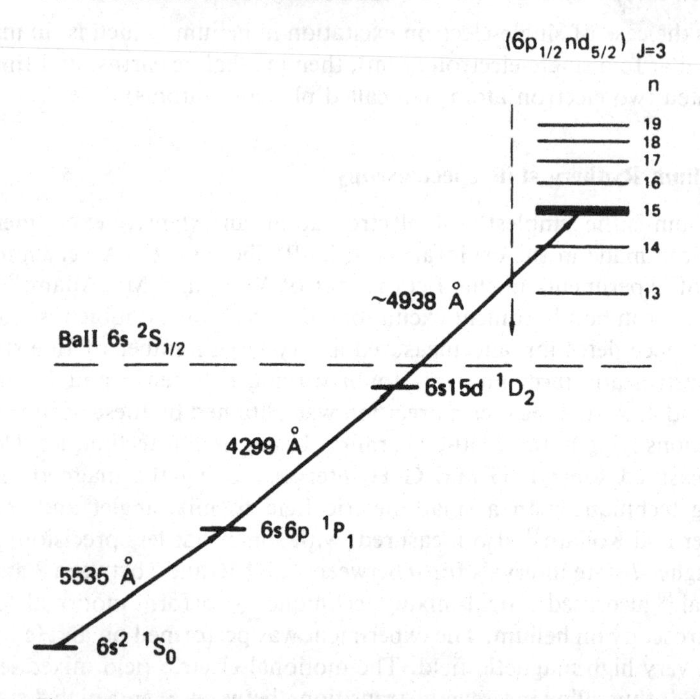

Fig. 4.18. Level diagram showing the excitation process for two-electron-excited autoionizing states in the case of Ba. (From Gallagher, Cooke, and Safinya[98])

initially populated level to the Rydberg even-parity states.[92,93] In more recent experiments, a direct multiphoton or multistep process using two or more tunable dye lasers was used to populate Rydberg states that are monitored[59] by thermionic or optogalvanic detection methods.[94,95] The atomic species under study (Ca, Sr, Ba) were generally prepared in a heat pipe oven. The precision of the experiment was usually only moderate because pulsed-Yag laser-pumped or nitrogen-laser-pumped dye lasers were used. This resolution is, however, generally sufficient to identify series of levels and to allow a multichannel quantum-defect analysis to be performed. Figure 4.3 presents the result of such an analysis on calcium Rydberg states.[17] Similar graphs were discussed by various authors for a large number of Rydberg series and the perturbing series were identified by this method. Other MQDT analyses include those in Be, Mg np^2 terms by Lu;[96] in Sr series by Rubbmark and Borgström;[93] in $J=1$ spectra of Ca, Sr, and Ba by Armstrong et al.;[94] and in the $J=0$ spectrum of Ba by Aymar et al.[95]

The method of double resonance was also used to study Sr Rydberg states in order to determine D and F state quantum defects.[97] More precise spectroscopic measurements, involving fine- and hyperfine-structure studies, will now certainly be developed in alkaline-earth atoms to study the effect of perturbing levels on these structures.

4.5c. Doubly excited alkaline-earth atoms

If the two outer electrons of an alkaline-earth atom are excited, the corresponding energy level lies above the ionization limit for the single-excitation Rydberg series. As a result, the level is autoionizing. If at least one of the principal quantum numbers is large, a Rydberg autoionizing state is obtained. The study of this peculiar type of Rydberg atom has recently been initiated, in particular, by Gallagher et al.[98] Several dye lasers (at least three) had to be used for the excitation process illustrated on Fig. 4.18 (in the case of Ba). Two lasers were used to promote one electron to a Rydberg level ($6s^2$–6s 6p–6s nd), then a third laser excited the remaining 6s electron to an excited $6p_{1/2}$ or $6p_{3/2}$ state, the system being excited in a $6p_{1/2,3/2}$ nd autoionizing Rydberg state. Several studies of these levels were performed, including an analysis of decay channels and branching ratios for autoionization. These studies are beyond the scope of this chapter, although the analysis of the position of the levels in a Rydberg series is interesting because the resulting quantum defects depend on the excitation level of the inner electron (quantum defects for the outer d electron are different when the core is in the 6s and in the 6p level, for example).

A very elegant way to obtain immediately the whole spectrum of autoionizing levels corresponding to a Rydberg series consists of using a very intense laser excitation for the last excitation step promoting the inner electron. The excitation is then allowed, not only for the strong single-electron 6s nd–6p nd transition, but also for the weak "two-electron" transition 6s nd–6p n'd with $n' \neq n$. The Rydberg orbitals nd (with an s electron in the core) and n'd (with a p electron in the core) are indeed not orthogonal to each other and the two electron dipole electric transition matrix element $\langle 6s\ nd|D|6p\ n'd\rangle = \langle 6s|D|6p\rangle\langle nd \text{ core in } 6s|n'd \text{ core in } 6p\rangle$ is nonzero. As a result, Safinya and Gallagher[99] obtained, when tuning this third laser frequency, the spectrum of autoionizing states shown in Fig. 4.19. The interference of the strong and weak transitions gives rise to the well-known Beutler–Fano asymmetric profiles for the weak lines.[100]

The change in quantum defects for the two different core configurations can be measured from the comparison between the singly and doubly excited Rydberg series. Knowing this change, the overlap $\langle nd|n'd\rangle$ integrals can be calculated and thus all the relevant parameters for the Beutler–Fano profiles of Fig. 4.19 can be measured. This type of measurement is a very good illustration of the simplicity and predictability of the quantum-defect model for Rydberg atoms. Extension of these kinds of measurements to more excited cores is now under consideration. The goal is, of course, to be able to prepare and study atoms in which both outer electrons will be in large orbitals. Such "planetary" atoms should behave as giant heliumlike species and the correlation properties of the electrons in Rydberg orbitals should be very interesting to study.

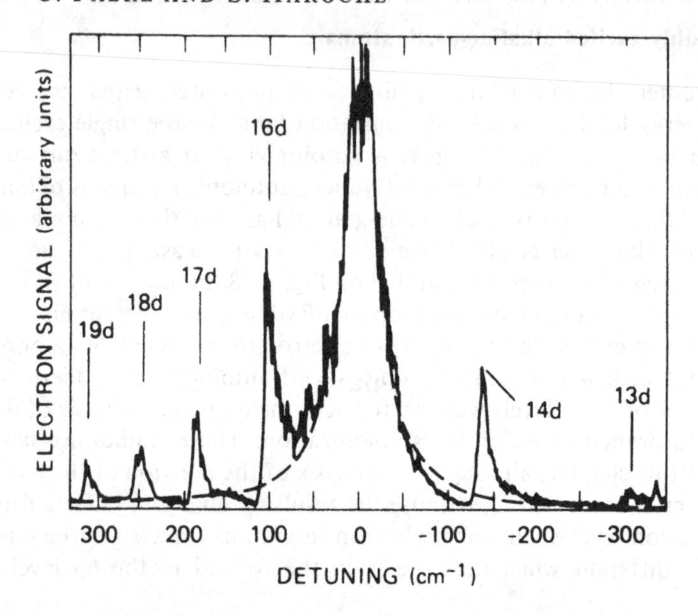

Fig. 4.19. Power-broadened spectrum of the 6s 15d $^1D_2$–$(6P_{1/2}\ 15d_{5/2})_{J=3}$ line in Ba showing asymetric side bands corresponding to the two-electron 6s 15d–6p $n \neq 15$d transitions. (From Gallagher, Cooke, and Safinya[98])

We have restricted our description here to free atomic spectroscopy experiments involving only the determination of Rydberg state energy intervals. In fact, techniques identical or very similar to the ones described in this chapter have been used to measure other Rydberg atom parameters in free or perturbed atomic systems: lifetimes;[75, 101-103] oscillator strengths;[33,34] polarizabilities,[82, 104, 105] and Stark[32] and Zeeman[106-108] diagrams. Details about some of these experiments can be found in other chapters of this book.

4.6. Conclusion: connections with other fields

Beyond the purely fundamental interest of atomic energy level investigation, the spectroscopic studies described in the previous sections have or may have promising applications in the various practical fields that we shall review in this last section.

4.6a. Possible applications to metrology

The precise determination of quantum defects either in the optical or in the millimeter-wave domain amounts to measuring with the same precision the absolute position of a ladder of states converging to the ionization limit. As indicated in Sect. 4.3e, under "Ultimate resolution," such measurements have already been performed in wavelength units in rubidium[64] and in frequency units in sodium.[70] They provide us with accurate frequency or wavelength markers throughout a very large spectroscopic domain. In particular, the millimeter-wave experiments (performed thus far in the $n=20$–40 range) can be used to determine, with a 10^8 resolution, the frequency of any laser line connecting, in the far infrared, two levels whose principal quantum numbers are included between these limits (50-μm to 1-mm wavelength). Extension to the 10-μm range can be obtained by extending the frequency domain of the double-resonance experiments with frequency-known CO_2 laser lines connecting $n \simeq 10$ states to Rydberg levels.[67] The ladder of Rydberg states with $n \gtrsim 10$ could thus be used as a secondary frequency standard, allowing us to determine the frequency of any laser source in the infrared with $\lambda \gtrsim 10$ μm. One would merely have to use this laser to induce, if necessary with the help of one or several additional milimeter-wave photons, transitions between two well-defined Rydberg levels. Such a frequency calibration procedure would certainly not compete in precision with the direct frequency-multiplying method involving the mutual locking of stabilized sources of increasing frequencies. Rydberg levels are indeed very sensitive to perturbations of various kinds, and frequency shifts would limit the practical resolution to about 10^8 or 10^9. However, at this moderate resolution level, the method could be useful because it is simpler to use than the direct frequency-comparison method.

4.6b. Microwave and infrared detection technology

We already emphasized the fact that double-resonance experiments on Rydberg atoms require only tiny amounts of micro- or millimeter-wave powers. Because the number of resonant frequencies in Rydberg atoms is very large (due to the large density of states near the ionization limit) and because these states can be energy shifted by using the Stark effect for example, there certainly exists the potential for developing very sensitive tunable detectors in the far-infrared or millimeter-wave domain. Here other types of detectors are lacking or have a limited sensitivity. Several groups have been investigating this possibility in various types of experiments.[109-111]

A first class of experiments made use of the very sensitive field-ionization method. It was shown, in various experiments, that field ionization of Rydberg atoms is very sensitive to the blackbody radiation background at room temperature[110-112] or at higher[113] or lower[114-116] temperatures. Ducas et al.[117] demonstrated that, instead of tuning the light frequency across a Rydberg transition as is done in an ordinary spectroscopy experiment, it is possible to tune the Rydberg levels across the light frequency and thus to record, in the far infrared, the spectrum of the radiation of interest. In a second class of experiments,[72] the Rydberg atoms were used as amplifiers of small microwave signals that are subsequently detected by other devices (Schottky diodes, for example). The Rydberg masers described in Sect. 4.3d can thus be used as preamplifiers and extremely small millimeter-wave signals can be detected in this way.[118] The sensitivity of these detectors is already competitive with the best of the other types of detectors available ($\sim 10^{-17}$ W Hz$^{-1/2}$ around 100 GHz), and there is a good reason to hope that ultimate sensitivities on the order of 10^{-19} W Hz$^{-1/2}$, corresponding to the detection of single-microwave photons in the bandpass of the detector (~ 1 MHz), could be achieved in the near future.

4.6c. Isotope separation

Because optical spectroscopy of Rydberg atoms involves selective excitation of atomic levels very near to the ionization limit, there certainly exists the possibility of using such experiments to develop isotope-separation procedures. A simple scheme consists in selectively exciting (in a multiphoton or multistep process) the level of a given isotope and subsequently field-ionizing and electrostatically separating this isotope from the other. Obviously, interest in such a method (or other related methods involving Rydberg atoms) depends on the nature of the isotope and on the energy efficiency of the process. Because the species of practical interest are not one or two electron atoms but more complex atoms (actinides) whose spectroscopy[119] is beyond the scope of this chapter, we shall not discuss this problem further.

4.6d. Rydberg states in astrophysics

For the sake of completeness, let us mention that Rydberg state microwave spectroscopy is a very active field in radioastronomy. In fact, numerous radio emission lines of Rydberg atoms produced by recombination in the interstellar medium have been detected in the last 20 years.[38,120] Not only one- and two-electron atoms (H, He) but also more complex atoms such as C have been detected in Rydberg states in various parts of the sky. Knowing the exact frequencies of the line allows us to distinguish the various atomic species (through the reduced mass factor) and gives us information about isotope abundances.

Although the environmental conditions are rather different between space and the laboratory, it is likely that the spectroscopy experiments in the laboratory will be useful to the interpretation of the interstellar spectra.

References and notes

1. J. R. Rydberg, *Z. Phys. Chem.* (*Leipzig*) *5*, 227 (1890).
2. M. J. Seaton, *Mon. Not. R. Astron. Soc. A 118*, 501 (1958).
3. A. Bethe and E. Salpeter, *Quantum Mechanics of One- and Two-Electron Atoms* (New York: Springer-Verlag, 1957).
4. L. Spruch and E. Kelsey, *Phys. Rev. A 18*, 1055 (1978).

5. R. Freeman and D. Kleppner, *Phys. Rev. A 14,* 1614 (1976).
6. M. Born and W. Heisenberg, *Z. Phys. 23,* 388 (1924).
7. C. Jaffé and W. Reinhardt, *J. Chem. Phys. 66,* 1285 (1977).
8. D. Hartree, *Proc. Cambridge Philos. Soc. 25,* 310 (1929).
9. F. Ham, *Solid State Phys. 1,* 127 (1955).
10. M. J. Seaton, *Proc. Phys. Soc. London 88,* 801 (1966).
11. U. Fano, *Phys. Rev. A 2,* 353 (1970).
12. U. Fano, *J. Opt. Soc. Am. 65,* 979 (1975).
13. M. J. Seaton, *Proc. Phys. Soc. London 88,* 815 (1966).
14. K. Lu and U. Fano, *Phys. Rev. A 2,* 81 (1970).
15. H. Saraph and M. Seaton, *Philos. Trans. R. Soc. London, Ser. A 271,* 1 (1971).
16. U. Fano and C. M. Lee, *Phys. Rev. Lett. 31,* 1573 (1973).
17. J. Armstrong, P. Esherick, and J. Wynne, *Phys. Rev. A 15,* 180 (1977).
18. T. Chang and F. Larijani, *J. Phys. B 13,* 1307 (1980).
19. C. Fabre, M. Gross, and S. Haroche, *Opt. Commun. 13,* 393 (1975).
20. G. Leuchs, S. Smith, E. Khawaja, and H. Walther, *Opt. Commun. 31,* 313 (1979).
21. T. Chang and R. Poe, *Phys. Rev. A 14,* 11 (1976).
22. H. Foley and R. Sternheimer, *Phys. Lett. 55A,* 276 (1975).
23. R. Sternheimer, J. Rodgers, and T. Das, *Phys. Rev. A 14,* 1595 (1976).
24. L. Holmgren, I. Lindgren, J. Morrison, and A. Mårtensson, *Z. Phys. A276,* 179 (1976).
25. E. Luc-Koenig, *Phys. Rev. A 13,* 2114 (1976).
26. R. Sternheimer, J. Rodgers, and T. Das, *Phys. Rev. A 17,* 505 (1978).
27. D. Bates and A. Damgaard, *Philos. Trans. R. Soc. London, Ser. A 242,* 101 (1949).
28. J. Picart, A. Edmonds, Tran Minh, and R. Pullen, *J. Phys. B 12,* 2781 (1979).
29. H. Van Regemorter, Hoang Binh Dy, and M. Prud'homme, *J. Phys. B 12,* 1053 (1979).
30. J. F. Gounand, *J. Phys. (Paris) 40,* 457 (1979).
31. C. Fabre, *Ann. Phys. (Paris) 7,* 5 (1982).
32. M. Zimmerman, M. Littman, M. Kash, and D. Kleppner, *Phys. Rev. A 6,* 2251 (1979).
33. C. J. Lorenzen and K. Niemax, *J. Phys. B 11,* L723 (1978).
34. J. M. Raimond, M. Gross, C. Fabre, S. Haroche, and H. Stroke, *J. Phys. B 11,* L765 (1978).
35. D. Norcross, *Phys. Rev. A 7,* 606 (1973).
36. S. Liberman and J. Pinard, *Phys. Rev. A 20,* 507 (1979).
37. T. Gallagher and W. Cooke, *Phys. Rev. Lett. 42,* 835 (1979).
38. B. Hoglund and P. Mezger, *Science 150,* 339 (1965).
39. P. Avan, C. Cohen-Tannoudji, J. Dupont-Roc, and C. Fabre, *J. Phys. (Paris) 37,* 993 (1976).
40. J. Bayfield and P. Koch, *Phys. Rev. Lett. 33,* 258 (1974).
41. D. Cok, unpublished thesis, Harvard University, 1980.
42. W. Wing and K. MacAdam, in *Progress in Atomic Spectroscopy,* eds. H. Hanle and W. Kleinpoppen (New York: Plenum, 1978), vol. A, chap. 11.
43. L. Pendrill, D. Delande, and J. C. Gay, *J. Phys. B 12,* L603 (1979).
44. D. Tuan, S. Liberman, and J. Pinard, *Opt. Commun. 18,* 533 (1976).
45. C. Harper, S. Wheatley, and M. Levenson, *J. Opt. Soc. Am. 67,* 579 (1977).
46. S. Svanberg, *Z. Phys. A284,* 429 (1978).

47. T. W. Hänsch, *Appl. Opt. 11,* 895 (1972).
48. J. Bayfield, *Rev. Sci. Instrum. 47,* 1450 (1976).
49. T. Gallagher, L. Humphrey, R. Hill, W. Cooke, and S. Edelstein, *Phys. Rev. A 15,* 1937 (1977).
50. R. Panock, M. Rosenbluh, B. Lax, and T. Miller, *Phys. Rev. Lett. 42,* 172 (1979).
51. S. Svanberg and G. Belin, *J. Phys. B 7,* L82 (1974).
52. G. Ruff, K. Safinya, and T. Gallagher, *Phys. Rev. A 22,* 183 (1980).
53. R. Stebbings, C. Latimer, W. West, F. Dunning, and T. Cook, *Phys. Rev. A 12,* 1453 (1975).
54. T. Ducas, M. Littman, R. Freeman, and D. Kleppner, *Phys. Rev. Lett. 35,* 366 (1975).
55. T. Gallagher, L. Humphrey, R. Hill, and S. Edelstein, *Phys. Rev. Lett. 37,* 1465 (1976).
56. C. Fabre, P. Goy, and S. Haroche, *J. Phys. B 10,* L183 (1977).
57. R. Ambartsumyan, G. Bekov, V. Letokhov, and V. Mishin, *JETP Lett. 21,* 279 (1975).
58. K. Harvey and B. Stoicheff, *Phys. Rev. Lett. 38,* 537 (1977).
59. P. Camus and C. Morillon, *J. Phys. B 10,* L133 (1977).
60. P. Camus, M. Dieulin, and C. Morillon, *J. Phys. Lett. (Paris) 40,* L513 (1979).
61. K. Frederiksson, H. Lundberg, and S. Svanberg, *Phys. Rev. A 21,* 241 (1980).
62. B. Stoicheff and E. Weinberger, in *Laser Spectroscopy,* ed. H. Walther (New York: Springer-Verlag, 1979), vol. IV, p. 264.
63. C. Harper, S. Wheatley, and M. Levenson, *J. Opt. Soc. Am. 67,* 579 (1977).
64. S. Lee, J. Helmcke, and J. Hall, *Laser Spectroscopy,* ed. H. Walther (New York: Springer-Verlag, 1979), vol. IV, p. 130.
65. T. Gallagher, R. Hill, and S. Edelstein, *Phys. Rev. A 14,* 744 (1976).
66. C. Fabre, S. Haroche, and P. Goy, *Phys. Rev. A 18,* 229 (1978).
67. T. Ducas and M. Zimmerman, *Phys. Rev. A 15,* 1523 (1977).
68. J. Farley, P. Tsekeris, and R. Gupta, *Phys. Rev. A 15,* 1530 (1977).
69. S. Svanberg, in *Laser Spectroscopy,* eds. J. Hall and J. Carlsten (New York: Springer-Verlag, 1977), vol. III, p. 183; and references therein.
70. P. Goy, C. Fabre, M. Gross, and S. Haroche, *J. Phys. B 13,* L83 (1980).
71. M. Gross, P. Goy, C. Fabre, S. Haroche, and J. M. Raimond, *Phys. Rev. Lett. 43,* 343 (1979).
72. L. Moi, C. Fabre, P. Goy, M. Gross, S. Haroche, P. Encrenaz, G. Beaudin, and B. Lazareff, *Opt. Commun. 33,* 47 (1980).
73. H. Beyer and K. Kollath, *J. Phys. B 10,* L5 (1977).
74. S. Haroche, in *High Resolution Laser Spectroscopy,* ed., K. Shimoda (New York: Springer-Verlag, 1977); and the references therein.
75. J. Deech, R. Luypaert, L. Pendrill, and G. Series, *J. Phys. B 10,* L137 (1977).
76. G. Leuchs, S. J. Smith, and H. Walther, in *Laser Spectroscopy,* ed., H. Walther (New York: Springer-Verlag, 1979), vol. IV, p. 255.
77. T. Jeys, K. Smith, F. B. Dunning, and R. F. Stebbings, *Phys. Rev. A 23,* 3065 (1981).
78. P. Koch, L. Gardner, and J. Bayfield, in *Beam Foil Spectroscopy,* eds. J. Pegg, I. Sellin (New York: Plenum, 1976).
79. G. Bjorklund, R. Freeman, and R. Storz, *Opt. Commun. 31,* 47 (1979).
80. W. Cooke, T. Gallagher, R. Hill, and S. Edelstein, *Phys. Rev. A 16,* 2473 (1977).
81. C. Fabre, S. Haroche, and P. Goy, *Phys. Rev. A 22,* 778 (1980).

82. W. Cooke, T. Gallagher, R. Hill, and S. Edelstein, *Phys. Rev. A 16,* 1141 (1977).
83. K. Frederiksson and S. Svanberg, *J. Phys. B 9,* 1237 (1976).
84. T. Gallagher, W. Cooke, S. Edelstein, and R. Hill, *Phys. Rev. A 16,* 273 (1977).
85. T. Gallagher and W. Cooke, *Phys. Rev. A 18,* 2510 (1978).
86. M. Levenson and M. Salour, *Phys. Lett. 48A,* 331 (1974).
87. G. Belin, L. Holmgren, and S. Svanberg, *Physica Scripta 14,* 39 (1976).
88. K. Niemax and L. Pendrill, *J. Phys. B 13,* L461 (1980).
89. K. MacAdam and W. Wing, *Phys. Rev. A 12,* 1464 (1975).
90. K. MacAdam and W. Wing, *Phys. Rev. A 13,* 2163 (1976).
91. K. MacAdam and W. Wing, *Phys. Rev. A 15,* 678 (1977).
92. J. Rubbmark, S. Borgström, and K. Bockasten, *J. Phys. B 10,* 421 (1977).
93. J. Rubbmark and S. Borgström, *Physica Scripta 18,* 196 (1978).
94. J. Armstrong, J. Wynne, and P. Esherick, *J. Opt. Soc. Am. 69,* 211 (1979).
95. M. Aymar, P. Camus, M. Dieulin, and C. Morillon, *Phys. Rev. A 18,* 2173 (1978).
96. K. Lu, *J. Opt. Soc. Am 64,* 706 (1974).
97. W. Cooke and T. Gallagher, *Opt. Lett. 4,* 173 (1979).
98. T. Gallagher, W. Cooke, and K. Safinya, in *Laser Spectroscopy,* ed., H. Walther (New York: Springer-Verlag, 1979), vol. IV, p. 273.
99. K. Safinya and T. Gallagher, *Phys. Rev. Lett. 43,* 1239 (1979).
100. U. Fano, *Phys. Rev. 124,* 1866 (1961).
101. T. Gallagher and W. Cooke, *Phys. Rev. A 20,* 670 (1979); and references therein.
102. H. Lundberg and S. Svanberg, *Z. Phys. A290,* 127 (1979); and references therein.
103. M. Hugon, F. Gounand, and P. Fournier, *J. Phys. B 11,* L605 (1978).
104. K. Frederiksson, L. Nilsson, and S. Svanberg, to be published in *Z. Phys.*
105. C. Fabre and S. Haroche, *Opt. Commun. 15,* 254 (1975).
106. J. C. Gay, D. Delande, L. Pendrill, and F. Biraben, *Abstracts of VII ICAP* (New York: Plenum Press, 1982), p. 212.
107. M. Zimmerman, J. Castro Nuevo, and D. Kleppner, *Phys. Rev. Lett. 40,* 1083 (1978); and references therein.
108. M. Economou, R. Freeman, and P. Liao, *Phys. Rev. A 18,* 256 (1979); and references therein.
109. D. Kleppner and T. Ducas, *Bull. Am. Phys. Soc. 20,* 1458 (1975).
110. T. Gallagher and W. Cooke, *Appl. Phys. Lett. 34,* 369 (1979).
111. E. Beiting, G. Hildebrandt, F. Kellert, G. Foltz, K. Smith, F. Dunning, and R. Stebbings, *J. Chem. Phys. 70,* 3551 (1979).
112. S. Haroche, C. Fabre, P. Goy, M. Gross, and J. M. Raimond, in *Laser Spectroscopy,* ed., H. Walther (New York: Springer-Verlag, 1979), vol. IV, p. 244.
113. P. Koch, H. Hieronymus, A. van Raan, and W. Raith, *Phys. Lett. 75A,* 273 (1980).
114. H. Figger, G. Leuchs, R. Straubinger, and H. Walther, *Opt. Commun. 33,* 37 (1980).
115. G. F. Hildebrandt, E. J. Beiting, C. Higgs, G. J. Hatton, K. A. Smith, F. B. Dunning, and R. F. Stebbings, *Phys. Rev. A 23,* 2978 (1981).
116. W. Spencer, A. Vaidyanathan, T. Ducas, and D. Kleppner, in *Abstracts of VII ICAP* (New York: Plenum Press, 1980), p. 228.
117. T. Ducas, W. Spencer, A. Vaidyanathan, W. Hamilton, D. Kleppner, *Appl. Phys. Lett. 35,* 382 (1980).

118. L. Moi, P. Goy, S. Haroche, C. Fabre, and M. Gross, *Phys. Rev.* (1982), in press.
119. J. Paisner, R. Solarz, and E. Worden, in *Laser Spectroscopy,* ed. J. Hall (New York: Springer-Verlag, 1977), vol. III, p. 160.
120. A Dupree and L. Goldberg, *Annu. Rev. Astron. Astrophy. 8,* 231 (1970).

5

Interaction of Rydberg atoms with blackbody radiation

T. F. GALLAGHER

5.1. Introduction

The interaction of blackbody radiation with atoms is by no means a new discovery. For example, absorption spectroscopy in the visible part of the spectrum is routinely performed using incandescent lamps. However, the realization that the omnipresent, room-temperature 300-K radiation is important is recent.[1-3] It is interesting to note that previously Pimbert observed blackbody-radiation-induced transitions in low-lying Rydberg states at slightly higher temperatures.[4] In this chapter we shall concentrate on the effects of the usually ignored 300-K radiation.

Blackbody radiation has two effects: to induce transitions and to produce energy shifts of the atomic levels. The magnitude of these effects on any atom depends on the match of the atomic frequencies with the blackbody frequencies and the strength of the coupling of the atoms to the blackbody radiation. The effects of blackbody radiation on Rydberg states are so dramatic because the low transition frequencies between the Rydberg states fall within the blackbody spectrum for all but the lowest temperatures and the large electric dipole moments of these transitions afford excellent coupling to the radiation. The effects of blackbody radiation are not restricted to Rydberg atoms but, as we shall see, in no other case are the effects expected to have the same magnitude.

5.2. Theory

5.2a. Blackbody radiation

A good starting point for this chapter is the characterization of blackbody radiation in terms of its energy density (the square of the electric field) as a function of frequency. This is the familiar Planck's radiation law, which is given by,[5]

$$p(\nu)\,d\nu = \frac{8\pi h\nu^3}{c^3}\,\frac{1}{[\exp(h\nu/kT)] - 1} \tag{1}$$

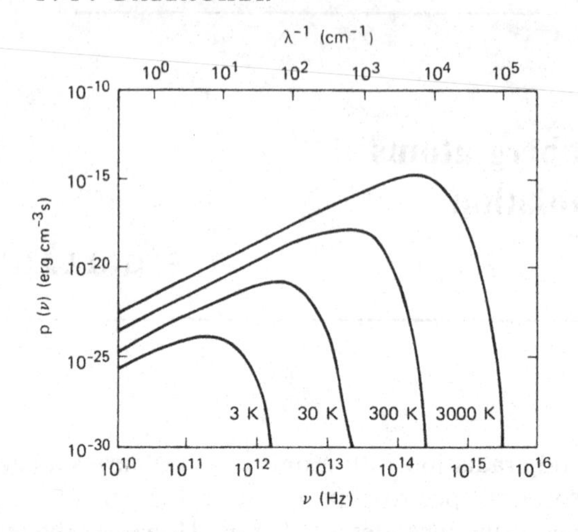

Fig. 5.1. Logarithmic plot of energy density $p(\nu)$ vs. ν for several temperatures.

where k is Boltzmann's constant, h Planck's constant, ν the frequency of the blackbody radiation, T the temperature, and $p(\nu)$ the energy density. By plotting Eq. (1) logarithmically (Fig. 5.1), we can see quite clearly why Rydberg states with their low transition frequencies, $\sim 10^{11}$ Hz, are much more strongly affected by thermal radiation, even at low temperatures, than are optical transitions at frequencies $\sim 10^{14}$ Hz.

For the calculation of blackbody-radiation-induced ac Stark shifts, it is convenient to use Eq. (1) to characterize the radiation. However, to show more clearly the spectral distribution vs. the relevant atomic frequencies, a linear plot of Eq. (1) is useful. Figure 5.2 is a plot of $p(\nu)$ with both frequency (hertz) and wave number (reciprocal centimeters) as abscissae. A typical electric dipole transition from the ground state of an atom has $\nu = 3 \times 10^{14}$ Hz and a transition between two Rydberg states has $\nu \simeq 3 \times 10^{11}$ Hz. Thus it is apparent from Fig. 5.2 that to a ground-state atom the blackbody radiation appears as a slowly varying, nearly dc field, whereas to a Rydberg atom it appears to be a rapidly varying field.

The blackbody radiation also induces transitions in the atoms by absorption or stimulated emission, and for calculating these transition rates it is more convenient to express the radiation field in terms of the number of photons per mode of the radiation field, i.e., the photon occupation number $\bar{n}$ of each mode. We may express the blackbody radiation in terms of the photon occupation number $\bar{n}$ as[5]

$$\bar{n} = \frac{1}{[\exp(h\nu/kT)] - 1} \tag{2}$$

It is useful to note that in the important case $h\nu \ll kT$, Eq. (2) reduces to

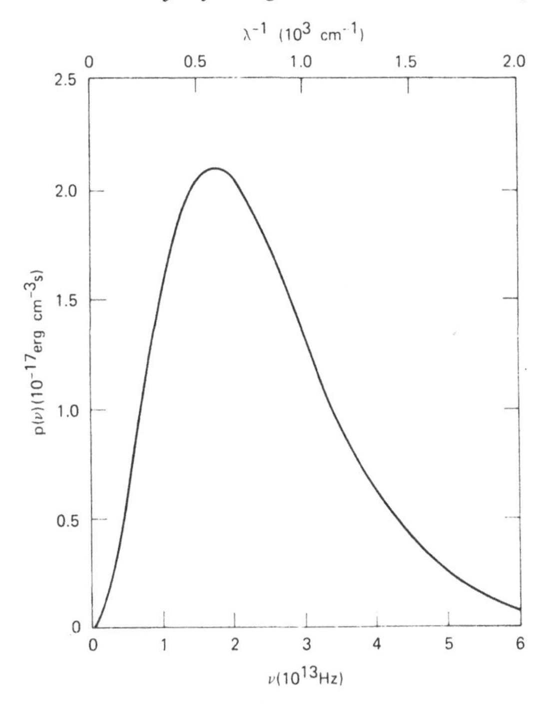

Fig. 5.2. Energy density $p(\nu)$ as a function of frequency ν and wave number λ^{-1} at 300 K.

$$\bar{n} = kT/h\nu \tag{3}$$

Figure 5.3 shows the dependence of $\bar{n}$ on ν for $T = 300\,\mathrm{K}$, again with both frequency (hertz) and wave number (reciprocal centimeters) used as abscissae. It is useful to recall that at 300 K, $kT/h \simeq 6 \times 10^{13}$ Hz or $kT/hc \simeq 200\ \mathrm{cm}^{-1}$. Because the vacuum fluctuations, which lead to spontaneous emission, are given by $\bar{n} = \frac{1}{2}$, blackbody radiation at frequencies greater than kT/h, where $\bar{n} \ll 1$, leads to insignificant effects.

For an atom in its ground state with transitions at $10^4\ \mathrm{cm}^{-1}$, blackbody-induced transitions are unimportant because $\bar{n} \ll 1$. However, for a Rydberg state with transitions at $10\ \mathrm{cm}^{-1}$, where $\bar{n} \simeq 10$, the blackbody-induced transition rates can be an order of magnitude larger than the spontaneous emission rates.

To calculate the effects of blackbody radiation on atoms, it is easier to use atomic units than the units in Eqs. (1)-(3). Therefore, we can reexpress Eq. (1) as

$$p(\omega)\,d\omega = \frac{2\alpha^3\omega^3\,d\omega}{\pi\{[\exp(\omega/kT)] - 1\}} \tag{4}$$

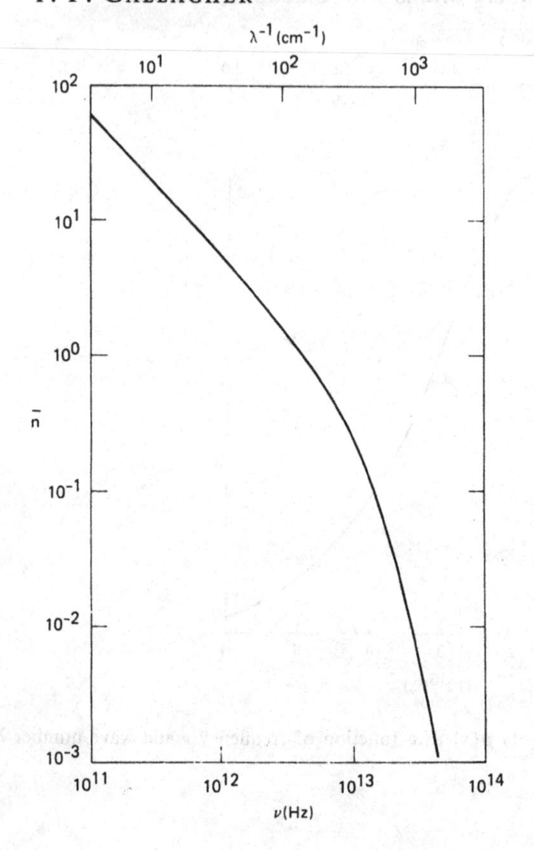

Fig. 5.3. Occupation number $\bar{n}$ as a function of frequency ν and wave number λ^{-1} at 300 K.

where α is the fine-structure constant and ω the energy in atomic units. Similarly, we may reexpress Eq. (2) as

$$\bar{n} = \frac{1}{[\exp(\omega/kT)] - 1} \tag{5}$$

and Eq. (3) for $\omega \ll kT$ as

$$\bar{n} \simeq kT/\omega \tag{6}$$

5.2b. Transitions driven by blackbody radiation

In this section we shall consider the transition rates for absorption of and stimulated emission induced by the blackbody radiation and compare them with the spontaneous emission rates. This comparison will provide a reasonable indication of the importance of the effects of blackbody-radiation-induced transitions, at least for Rydberg atoms. The development follows that used by Cooke and Gallagher.[6]

The spontaneous emission rate for a level n to n' is given by its Einstein coefficient $A_{nn'}$:[7]

$$A_{nn'} = -2\alpha^3 \omega_{nn'}^2 f_{nn'} \tag{7}$$

where the oscillator strength $f_{nn'}$ is given by

$$f_{nn'} = \frac{2}{3}\omega_{n'n} r_{nn'}^2 \frac{g_>}{2g_n + 1} \tag{8}$$

where

$$\omega_{nn'} = W_n - W_{n'} \tag{9}$$

and W_n and $W_{n'}$ are the energies of the states n and n', $r_{nn'}$ the electric dipole matrix element between the two states, g_n and $g_{n'}$ the degeneracies of states n and n', and $g_>$ the larger of g_n and $g_{n'}$. The radiative decay rate, which is the inverse of the radiative lifetime of the state n, is given by the sum of the transition rates to all lower states n':

$$1/\tau_n = -2\alpha^3 \sum_{n'} f_{nn'} \omega_{nn'}^2 \tag{10}$$

It is useful to note the variation of τ with n and l. The n dependence of τ for Rydberg states of low l is approximately given by n^{-3}. This occurs because the largest term, by far, in the sum of Eq. (10) is the highest-frequency term, corresponding to the transition to the lowest-lying state. For states of high n (and the same l), the highest frequency cannot change much with n; however, the dipole matrix element decreases with n as $n^{-3/2}$, reflecting the normalization of the Rydberg state wave function at the small radius of the low-lying state, which leads to the n^{-3} decrease in the radiative decay rate.

For the states of the same n, as l is increased, the lifetimes increase mainly because the frequencies of the available transitions to lower states decrease rapidly with increasing l, where only the states with $n \geqslant l$ are available. For example, for the $n=12$ state of hydrogen, the 12p state has a lifetime of 321 ns, whereas the $n=12$o ($l=11$) state has a lifetime of 21.3 μs, a difference of two orders of magnitude.[8]

We can express the blackbody-radiation-induced rates $K_{nn'}$ for stimulated emission and absorption from level n to n' in terms of the oscillator strength as

$$K_{nn'} = 2\alpha^3 \bar{n}_{\omega_{nn'}} \omega_{nn'}^2 |f_{nn'}| \tag{11}$$

where $\bar{n}_{\omega_{nn'}}$ is the photon occupation number of the frequency $\omega_{nn'}$. Because the frequency dependence of $K_{nn'}$ is quite different from that of $A_{nn'}$, these two processes favor different final states. This is indicated graphically in Fig. 5.4, which is a plot of $A_{18sn'}$ and $K_{18sn'}$ vs. n' for the Na 18s state and $T=300$ K. As shown in Fig. 5.4, blackbody radiation favors transitions to

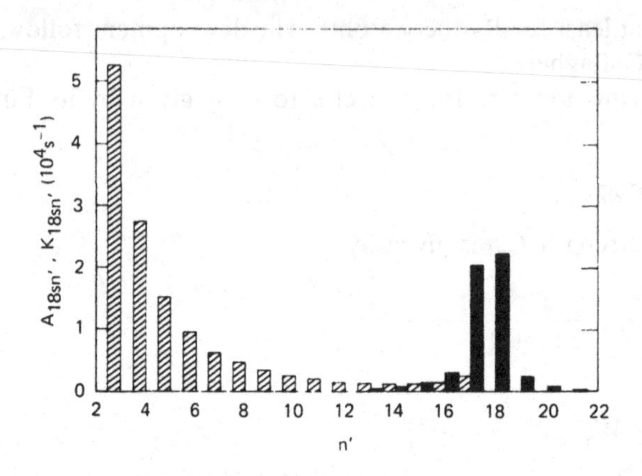

Fig. 5.4. Spontaneous emission rate $A_{18s\,n'}$ (striped bars) of the Na 18s state to n'p states and the 300 K blackbody transition rate $K_{18s\,n'}$ (black bars) as a function of n'.

nearby states and spontaneous emission favors transitions to the lowest-lying states.

For many applications it is useful to write the blackbody-radiation-induced decay rate $1/\tau_n^{\mathrm{b}}$ as

$$1/\tau_n^{\mathrm{b}} = 2\alpha^3 \sum_{n'} \bar{n}_{\omega_{nn'}} \omega_{nn'}^2 |f_{nn'}| \tag{12}$$

and the total decay rate $1/\tau_n^{\mathrm{T}}$ as

$$\frac{1}{\tau_n^{\mathrm{T}}} = \frac{1}{\tau_n} + \frac{1}{\tau_n^{\mathrm{b}}} \tag{13}$$

From Fig. 5.4 it is apparent that the most important contributions to the blackbody radiation decay rate are from transitions to nearby states for which $\omega_{nn'} \ll kT$, in which case we may substitute Eq. (6) for $\bar{n}_{\omega_{nn'}}$ and write $1/\tau_n^{\mathrm{b}}$ as

$$1/\tau_n^{\mathrm{b}} = 2\alpha^3 kT \sum_{n'} \omega_{n'n} f_{nn'} \tag{14}$$

For a one-electron atom, we may use the sum rule[7]

$$\sum_{n'} \omega_{n'n} f_{nn'} = 2/3n^{*2} \tag{15}$$

which we may use to write

$$1/\tau_n^{\mathrm{b}} = 4\alpha^3 kT/3n^{*2} \tag{16}$$

For $T = 300\,\mathrm{K}$, Eq. (16) is accurate to 30% for $n > 15$ and is thus useful for many Rydberg states under current study. In addition, it serves to bring out a

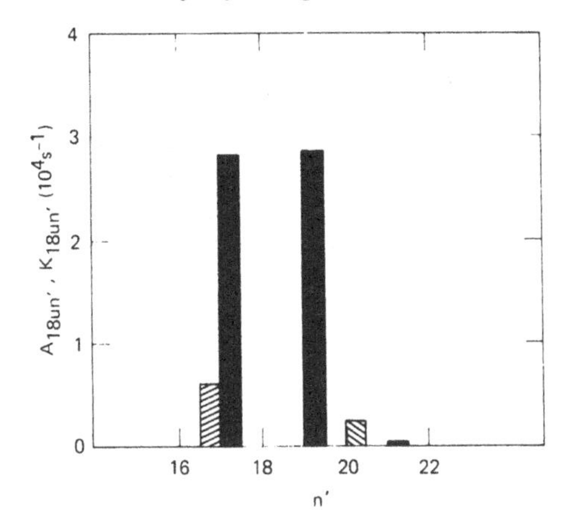

Fig. 5.5. Spontaneous emission rate $A_{18un'}$ (striped bars) of the Na 18u state to $n'l$ states and the 300 K blackbody transition rate $K_{18un'}$ (black bars) as a function of n'.

very important result: the total blackbody decay rate $1/\tau^{\mathrm{b}}$ does not depend on l. Because the radiative decay rate decreases rapidly as l is increased, for high l the blackbody-induced decay rate is much larger than the spontaneous emission rate. This is shown quite clearly in Figs. 5.4 and 5.5, where the spontaneous emission rates and blackbody emission and absorptions are shown. In Fig. 5.4, the areas under the curves for $A_{nn'}$ and $K_{nn'}$ yield the spontaneous and blackbody decay rates, and it is clear that $1/\tau^{\mathrm{b}} \simeq 0.2/\tau$. In Fig. 5.5, the analogous plots for $A_{nn'}$ and $K_{nn'}$ are given for the 18u state ($l=17$), which clearly illustrates that the blackbody decay rate is 10 times larger than the spontaneous decay rate.

It is useful to compare transition rates induced by 300-K blackbody radiation to those due to collisional processes. For $n \simeq 17$, the blackbody radiative transition rate is $\sim 10^4\ \mathrm{s}^{-1}$. For a typical background pressure of 10^{-6} Torr, a cross section of $5 \times 10^4\ \text{Å}^2$ would be required to produce the same collision rate. In fact, most small molecules normally present as background gases have cross sections that are evidently considerably smaller.[9] Thus, in many experiments, the blackbody transition rate is the most significant cause of population redistribution.

5.2c. Blackbody radiation ac Stark shift

The largest blackbody-radiation-induced energy shift for a Rydberg atom in the state n comes, not surprisingly, from the ac Stark shift ΔW_n, which can be expressed as[10]

$$\Delta W_n = \sum_{n'} \int_0^\infty \frac{r_{nn'}^2 E_{\omega_{\mathrm{b}}}^2 \omega_{nn'}}{2(\omega_{nn'}^2 - \omega_{\mathrm{b}}^2)}\, d\omega_{\mathrm{b}} \tag{17}$$

where E_{ω_b} is the electric field of the blackbody radiation in a bandwidth $d\omega_b$ at frequency ω_b. Here, not surprisingly, the approaches used by Cooke and Gallagher[6] and Gallagher et al.[11] are followed. For atoms in Rydberg states, it is generally true that the frequencies of the strong transitions are much lower than the frequency of the blackbody radiation, i.e., $\omega_{nn'} \ll \omega_b$, in which case we can ignore $\omega_{nn'}$ in the denominator of Eq. (17) and rewrite ΔW_n as

$$\Delta W_n = \left(\sum_{n'} f_{nn'}\right)\left(\int_0^\infty \frac{E_{\omega_b}^2\, d\omega_b}{4\omega_b^2}\right) \tag{18}$$

Using the oscillator-strength sum rule,[7]

$$\sum_{n'} f_{nn'} = 1 \tag{19}$$

and integrating over ω_b in Eq. (18), we may express Eq. (18) as

$$\Delta W_n = \tfrac{1}{3}\pi\alpha^3(kT)^2 \tag{20}$$

Equation (20) is accurate to $\sim 10\%$ for $n > 15$ at $T = 300$ K and indicates that all Rydberg states experience the same energy shift, which is $+2.2$ kHz at 300 K. Note that Eq. (20) also corresponds to the energy shift of a free electron. This is not surprising because the physical effect of the ac blackbody field, at a high frequency compared with the orbital frequency of the electron, is to superimpose a fast, $\sim\omega_b$, wiggle on the orbital motion of the electron, the energy of this motion being independent of the orbital velocity.

For states of low principal quantum number, it is evident that $\omega_{nn'} \gg \omega_b$, and we can ignore ω_b in the denominator of Eq. (16), in which case the Stark shift is equal to that produced by a dc field and can be expressed as

$$\Delta W_n = \left(\sum_{n'} f_{nn'}/\omega_{nn'}^2\right)\left(\int_0^\infty \tfrac{1}{4}E_{\omega_b}^2\, d\omega_b\right) \tag{21}$$

For a one-electron Rydberg atom we may use the hydrogenic sum rule[12]

$$\sum_{n'} f_{nn'}/\omega_{nn'}^2 = \tfrac{9}{2} \tag{22}$$

so that we may write Eq. (21) as

$$\Delta W_n = \int \tfrac{9}{8}E_{\omega_b}^2\, d\omega_b = -\tfrac{3}{5}(\alpha\pi)^3(kT)^4 \tag{23}$$

Evaluation of Eq. (23) gives a shift of approximately -0.036 Hz at 300 K, which is negligible for all practical purposes.

Finally, we must remember that for some states it will be true that $\omega_{nn'} \simeq \omega_b$, in which case neither of the preceding approximations is valid and Eq. (17) must be evaluated explicitly, as has been done by Farley and Wing.[13] At 300 K, this occurs for $n \simeq 8$.

Focusing our attention for a moment on Rydberg states of $n>10$, we see from Eq. (20) that ΔW_n is proportional to T^2, that is, the shift in the atomic frequency is a direct measure of the absolute temperature. This is a potentially interesting characteristic that leads to the question of how it can be observed. In a one-electron atom in which all the Rydberg states experience the same shift, +2 kHz at 300 K, and the low-lying states are unshifted, it appears that it is easiest to observe the shift of the transition from the ground state to a Rydberg state. Thus we are faced with the proposition of observing a 2-kHz shift of an optical transition frequency of 10^{15} Hz ($\Delta\nu/\nu \simeq 10^{-12}$), which is a formidable task. One of the most serious problems is simply generating the optical radiation with the necessary frequency stability. This has led to the consideration of other atomic systems in which the effect might be more pronounced, and it appears that alkaline-earth atoms, in which valence states exist in the midst of the Rydberg series, might be good candidates.[11] In such atoms the states that are predominantly Rydberg in character will experience a shift given by Eq. (20) to the extent that they have Rydberg character. The valence states, however, have strong transitions only to other valence states, transitions that are necessarily high frequency. Thus the shift of these levels is given by Eq. (23) and is negligible. It is clear that the exact size of the shift of the transition frequency between the Rydberg state and the valence state will depend on how much Rydberg character is in the valence state and vice versa; however, the magnitude of the shift will be ~1 kHz at 300 K. The principal advantage is, of course, that the frequency is much lower and falls in the radio-frequency range where signal generators with more than adequate frequency stability can be used. For example, in a specific case occurring in Ba, the frequency ν is 900 MHz and the shift $\Delta\nu$ is 0.6 kHz so that $\Delta\nu/\nu = 10^{-7}$, a factor of 10^5 larger than $\Delta\nu/\nu$ for an optical transition from the ground state to a Rydberg state.

Even without the technical problems of generating a stable frequency, the problems posed by the atoms themselves are formidable. If we use the 30d state of Rb, for example, the total decay rate for this state is $\sim 3\times10^4\ \mathrm{s}^{-1}$ and its polarizability is ~10 MHz $(V\,\mathrm{cm}^{-1})^{-2}$ so that the observation of the frequency shifts will be difficult.

5.3. Experimental observations

Experiments have been undertaken to observe both the blackbody-radiation-induced frequency shifts and stimulated emission and absorption. As might be expected from the magnitudes of the effects, most of the work has been concentrated on the latter. Nevertheless, an experiment to observe the blackbody radiation shift of the optical transition from the ground state to a Rydberg state is currently under way.[14] In this experiment it is possible to determine line centers to 100 Hz, thus it appears quite feasible to make a good measurement of the shift, which is ~1 kHz. At present, however, the only observations are of the substantially larger population redistribution effects.

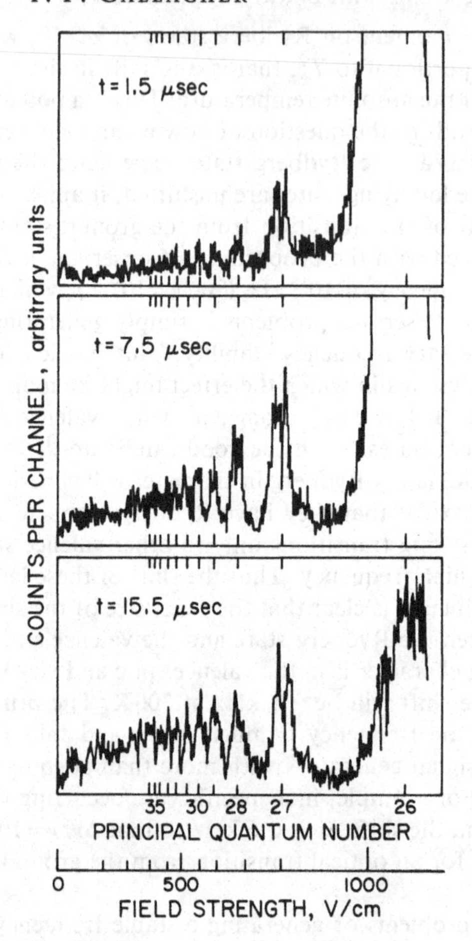

Fig. 5.6. Field-ionization spectra at different times after the laser pulse excites the Xe 26f level, showing the blackbody-radiation-induced transfer to higher levels. (From Beiting et al.[1])

5.3a. Initial verification

The initial experimental observations that called attention to the effects of blackbody radiation on Rydberg atoms were made using Rydberg states of Xe and Na at room temperature.[1-3] Two kinds of experiments were performed. In the first type, a pulsed laser was used to populate a single state and the population in higher-lying states (including the continuum) increased as a function of time after the exciting laser pulse. For example, in Fig. 5.6, the field-ionization signals show the population distribution in the Rydberg states of Xe at several times after the initial population by the laser.[1] At first glance the data of Fig. 5.6 suggest that a collision process is transferring the atoms from the initially populated Xe 27f state to the higher-lying states. However, sys-

Table 5.1. Calculated and observed populations in higher-lying *n*p states 5 μs after the initial population of the Na 18s state expressed in terms of the initial 18s population

Final state	Calculated yield (%)	Observed yield (%)
18p	4.2	5.0
19, 20p	1.6	1.6
>20p, ϵp	1.0	1.2
Total	6.8	7.8

Source: Gallagher and Cooke.[2]

Table 5.2. Na p state lifetimes

State	τ (calc.) (μs)	τ^T (calc.) (μs)	τ^T (obs.) (μs)
17p	48.4	15.5	$11.4^{+5.0}_{-1.4}$
18p	58.4	17.9	$13.9^{+8.8}_{-2.9}$

Source: Gallagher and Cooke.[3]

tematic studies in the experiments indicated no dependence of the signal on the density of any of the possible collision partners. As shown by Fig. 5.6, all values of l are not included in the final-state distributions, for in that case Fig. 5.6 would not have discrete peaks but a continuous distribution of signal intensity vs. ionizing field. In fact, in the study of the initially populated Na 18s state, the final states were all found to be p states, consistent with the $\Delta l=1$ electric dipole selection rule. Furthermore, the fraction of the atoms that underwent transitions to each of the higher-p states was in nearly perfect agreement with the values calculated on the basis of 300-K blackbody-radiation-induced transitions, as shown by Table 5.1.[3]

The second type of experiment involved the observation of the dramatic decrease in the lifetimes of Na *n*p levels caused by blackbody radiation.[2] The lifetimes of the readily accessible Rydberg p levels of alkali atoms are an order of magnitude larger than those of the s or d states, and the p states therefore are much more dramatically affected by blackbody radiation. As indicated by Table 5.2, the observed lifetimes of the Na p states are about three times shorter than the radiative lifetimes and are in reasonably good agreement with the values calculated when the effects of blackbody radiation are taken into account.

Although the nearly perfect agreement between the experimental results and those calculated for blackbody radiation makes it difficult to doubt seriously that blackbody radiation is responsible for the observed results, the initial experiments leave room for improvement. Specifically, as shown by Eq. (13), the transition rates are proportional to T, and a verification of the correct temperature dependence is clearly of interest. In principle, this is a straightforward proposition; however, at the wavelengths of interest, $\sim 300\ \mu$m, almost any material is a good reflector. Consequently, it is necessary to take care that the apparatus is constructed in such a way that the temperature in the cooled or heated interior region is known (and not equal to 300 K). For example, the classic approach of cooling a copper jacket around the interaction region to 77 K would accomplish nothing. Nevertheless, experiments were recently reported in which the final-state distribution was significantly altered by changing the temperature of the surroundings.[15-17] As an example, Fig. 5.7 shows the final-state distribution after the initial population of the Cs 38p state at temperatures of 300 and 700 K.[15] There is roughly a twofold increase in the number of atoms excited to states lying above the 38p state when the temperature is raised from 300 to 700 K, as expected.

5.3b. Effect on superradiance

The phenomenon of superradiance is very easily observed in Rydberg states, which is not particularly surprising because the threshold for superradiance is usually expressed in terms of the gain of an optical medium along its length. For superradiance to occur between two levels (crudely speaking, laser action without mirrors), the gain must be one along the length of the sample. This may be expressed as[18]

$$NL\sigma \geqslant 1 \tag{24}$$

where N is the difference in population densities of the upper and lower levels, L the length of the sample, and σ the optical cross section, given by[19]

$$\sigma = g_l \lambda^2 A_{ul} / 8\pi g_u \Delta \tag{25}$$

where g_l and g_u are the degeneracies of the upper and lower states, A_{ul} the Einstein coefficient for the transition, and Δ the linewidth of the transition, to which the main contributions are usually radiative and Doppler broadening. Because of the long wavelength of the transitions, it is apparent that the criterion for superradiance can be met with a very low density of atoms, hundreds of atoms in a volume of 10^{-3} cm^{-1}, and is easily met even in a low-density atomic beam. However, because of the very low spontaneous emission rates of these long-wavelength transitions, as shown by Fig. 5.4, superradiance would be initiated very slowly were it not for the presence of blackbody radiation. The blackbody radiation provides about 10 photons mode^{-1} compared with one-half from the vacuum fluctuations.

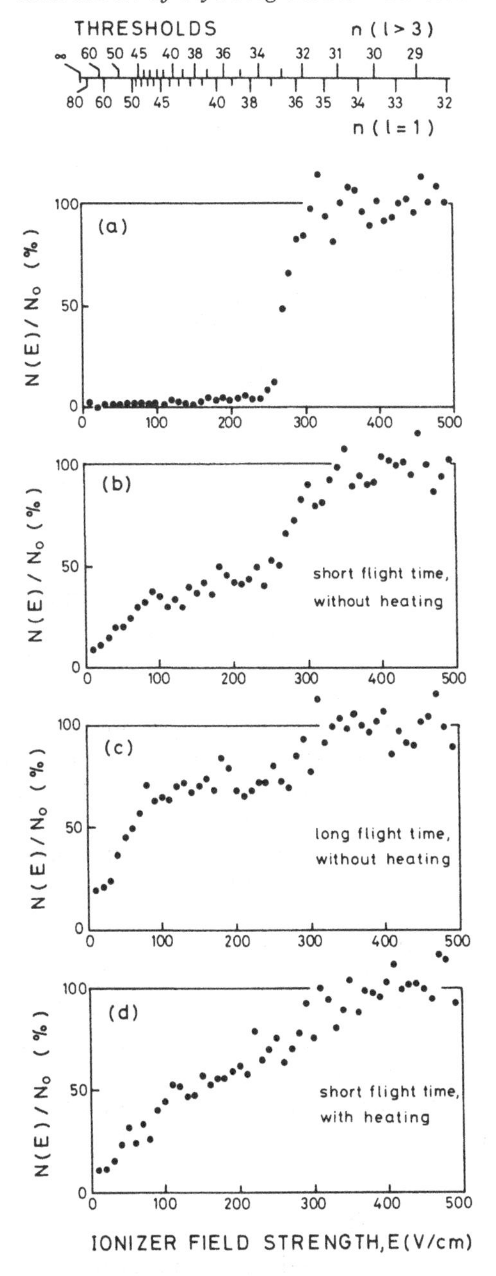

Fig. 5.7. Field-ionization spectra after the initial population of the Cs 38p state. The scale on the top gives the field-ionization thresholds of the Cs *n* levels. Spectra are shown with the detector (*a*) at the excitation region (*b–d*) 11 cm downstream. Short flight time is 206 μs; long flight time 310 μs. Without heating = 300 K; with heating = 700 K. (From Koch et al.[15])

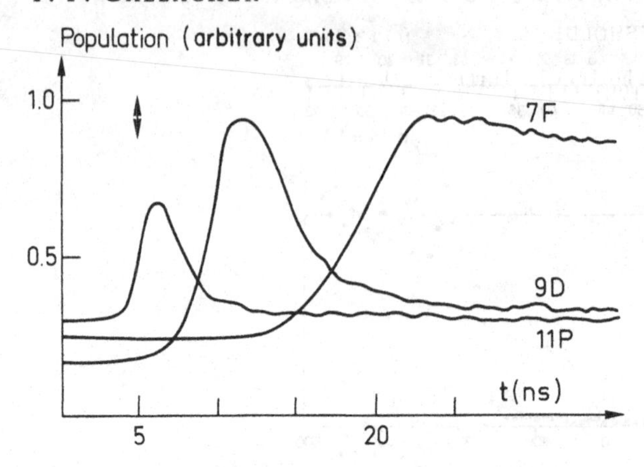

Fig. 5.8. Time dependence of the population in Rb Rydberg states after the initial population of the 12s state showing the rapid superradiant cascades. (From Gounand et al.[20])

The result of superradiance is to very rapidly transfer population from the upper to the lower state, and in Rydberg states, this has been observed in several ways under a variety of conditions. For example, in vapor-cell experiments, the population in Rydberg states was observed to undergo several very rapid superradiant cascades that populate lower lying levels and lead to characteristic fluorescence at different wavelengths, in times far shorter than they would be populated by simple radiative decay.[20] Figure 5.8 shows the population in various Rb Rydberg states after the initial population of the 12s states. Note that normal radiative decay would require ~3 μs to populate the 7f state, but as shown in Fig. 5.8, only 20 ns is required when superradiant cascades occur.

As previously mentioned, superradiance was also observed in a dilute atomic beam,[18,21] where the density requirements are somewhat lower because all the atoms are moving in the same direction, leading to smaller Doppler widths, especially in the direction perpendicular to beam travel. Finally, it is interesting to note that, as the theory of superradiance was originally formulated by Dicke,[22] only two atoms separated by a distance of less than a wavelength were considered. Only by using Rydberg atoms is it possible to realize this.

5.3c. Systematic measurements of lifetimes

As mentioned earlier, one effect of blackbody radiation is to shorten the radiative lifetimes of Rydberg states. In fact, the lifetimes of a great many alkali Rydberg states have been measured, and in some of these states there is clear evidence for the effect of blackbody radiation. The alkali atoms are of particular interest because their lifetimes can be calculated with confidence

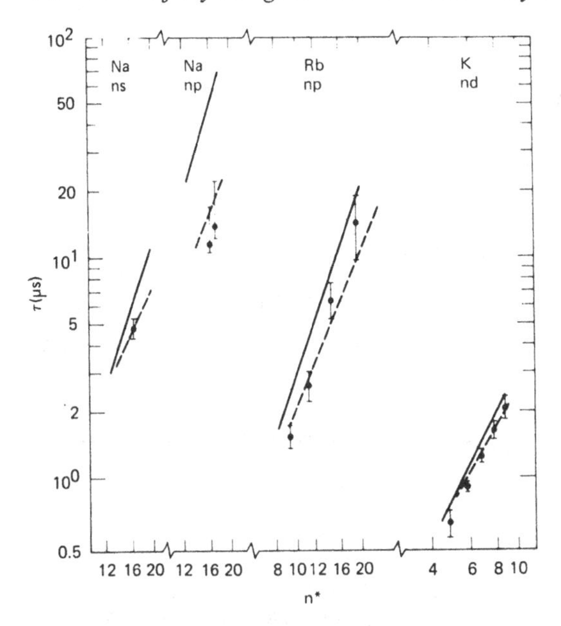

Fig. 5.9. Radiative lifetimes of alkali s, p, and d states. Solid lines, calculated values of τ; broken lines, calculated values of τ^T; circles, observed lifetimes Na *n*s (from Ref. 6), Na *n*p (from Ref. 3), Rb *n*p (from Ref. 25), and K *n*d (from Ref. 26).

using the Coulomb approximation.[23,24] Figure 5.9 shows the values of τ^T and τ and the measured lifetimes for several alkali series for which the lifetimes are reduced by factors of 1.1 to 3.[3,6,25,26] One interesting study would be the measurement of the higher-*l* lifetimes, which should be as much as 10 times shorter than the radiative lifetimes.

5.3d. Effects on Rydberg atom experiments

The rapid redistribution of population by blackbody radiation must be taken into account in experiments with Rydberg states. To appreciate fully the inconvenience and chagrin caused by this, it is necessary first to recall that many experiments with Rydberg states were performed using laser excitation with the result that only one state was populated. Because the spontaneous emission rates to other nearby states were very low (see Fig. 5.3), it was reasonable to look for collisional transfers of only small fractions (<1%) of the initially populated state to other nearby states. Naturally, the rapid redistribution of the population by 300-K blackbody radiation severely limited the sensitivity of such an approach. The problem may be minimized by cooling, but liquid He is required to reduce the blackbody transition rates by more than an order of magnitude.

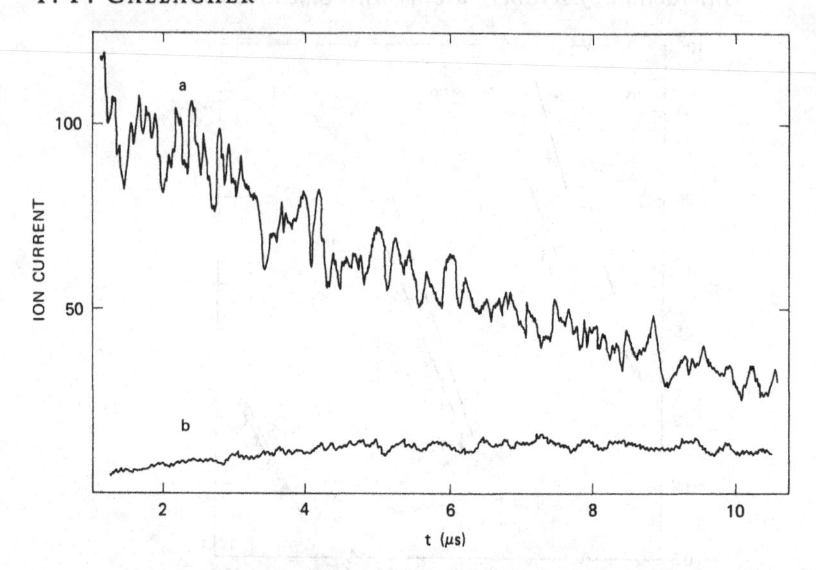

Fig. 5.10. Measurement of the lifetime of the Na 18s state using field ionization: (*a*) population in the 18s, 17p, and higher-lying levels as a function of time after the laser pulse; (*b*) population of the 18p and higher-lying states as a function of time after the laser pulse. (From Cooke and Gallagher[6])

A less serious problem is that the long lifetimes (5 μs) of Rydberg atoms appear to allow the use of delays of ~1 μs in the application of ionizing fields or other manipulations. Such latitude makes the experiments technically less demanding and therefore somewhat easier to do. In the absence of any apparent reason not to relax the technical requirements, it seems only reasonable to do so, and of course, the blackbody radiation drives a small fraction (3%) of the population into other states. The result was always a small, relatively insignificant discrepancy that was easily misinterpreted. Perhaps the best examples are experiments in which field-ionization signals at fields below a theoretically very sharp threshold were mistakenly attributed to photoionization for lack of any better explanation.[27,28]

Possibly the best way to illustrate the problems arising from blackbody radiation is to consider the measurement of the lifetime of a Rydberg state using selective field ionization, an apparently simple experiment that can lead to bizarre results.[6] The 18s state of Na is calculated[24] to have a radiative lifetime of 6.37 μs, which is reduced to 4.87 μs by the 300-K blackbody radiation. No doubt, if the lifetime were measured by observing the time-resolved 18s → 3p fluorescence, one would obtain a value for τ^{T} of precisely 4.87 μs. However, for n as high as 18, measuring the lifetime using field-ionization techniques is vastly easier. If atoms in the 18s state are created at time $t=0$, the population remaining at any later time can be determined by applying an electric field high enough to ionize the atoms in the 18s state. This measurement yields a lifetime of 7.8 μs, as shown in Fig. 5.10*a*.

At first this value is puzzling because it is even longer than the radiative lifetime. The problem is that a field high enough to ionize the 18s state will also ionize all the long-lived, higher-lying *n*p states to which population is transferred by blackbody radiation. In fact, in this particular experiment, the field-ionization threshold of the 17p state, which lies below the 18s state, is so close to that of the 18s state that the field was deliberately set high enough to ionize the 17p state as well, rather than have any ambiguity. Thus, the signal of Fig. 5.10*a* represents the population in the 17p and higher states. A direct measure of the population transferred to the p states is obtained by reducing the ionizing field so that only the 18p and higher states are ionized and then observing the population as a function of time after the initial population of the 18s state. The result, shown in Fig. 5.10*b*, clearly demonstrates that the p states constitute a major contribution to the signal at later times. The 18s population is given by the difference of Fig. 5.10*a* and 156% of Fig. 5.10*b* (the "extra" 56% accounts for the fact that in Fig. 5.10*b* the 17p state is not observed but in Fig. 5.10*a* it is). The net result is a lifetime of 4.78 μs for the 18s state, which is in excellent, and probably fortuitous, agreement with the calculated value of 4.87 μs.

5.3e. Possible observations in non-Rydberg atom systems

Having surveyed most of the observations of the interaction of blackbody radiation with Rydberg atoms, it is interesting to return to the question of whether such effects can be observed in other systems. If we consider atoms in their ground or low-lying excited states, the frequencies of the electric dipole transitions are $\sim 10\,000$ cm^{-1}. Thus $\omega \gg kT$ and $\bar{n} \simeq 0$, so there will be negligible blackbody-radiation-induced stimulated emission or absorption. Similarly, because of the high frequencies of the transitions, the Stark shift of the levels is essentially a dc Stark shift proportional to the polarizability. The largest scalar polarizabilities of the ground states of most atoms lead to shifts of ~ 0.1 Hz at 300 K. However, it may be possible to observe a shift in the hyperfine-structure frequency because of the analogous ac Zeeman effect from the blackbody radiation.[2] The shift is only 1 part in 10^{16} and is therefore currently lost in other temperature-dependent corrections such as the wall shifts in a hydrogen maser.[29] All in all, it is not surprising that the effects of blackbody radiation have not been noticed previously in low-lying states of atoms, nor does it seem likely that such effects will be observed in the near future.

We are thus led to consider molecules, specifically the rotational transitions in polar diatomic molecules, such as HF, that have low frequencies ~ 20 cm^{-1} and large dipole moments (2 D). For HF with a $J=0 \rightarrow 1$ rotational frequency of 41.8 cm^{-1}, $\bar{n}=5$, and it is immediately apparent that the blackbody-radiation-induced decay rate vastly exceeds the radiative decay rate. Unfortunately, the radiative decay rate is only $\sim 10^{-1}$ s, so that, in one radiative lifetime (10 s), a 300-K HF molecule could travel $\sim 10^{6}$ cm. It is obviously impos-

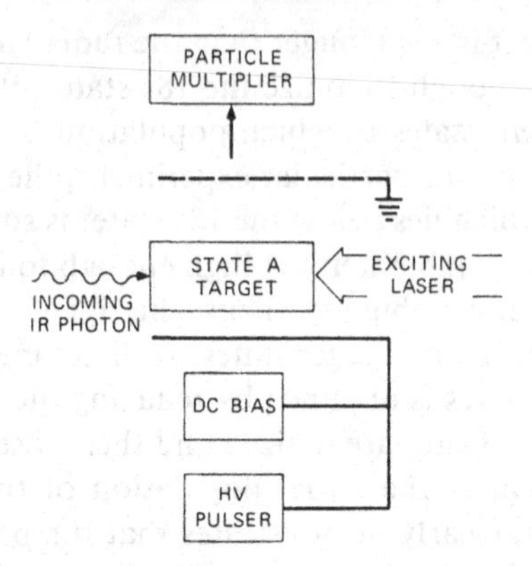

Fig. 5.11. A possible configuration for a Rydberg atom infrared detector.

sible to observe a molecule in free flight for that distance. Under practical laboratory conditions, the molecules collide with walls or other molecules in times that are orders of magnitude shorter than the radiative lifetime. Consequently, the redistribution effect of the blackbody radiation is not noticeable. Similarly, the ac Stark shifts for the rotational transitions are ~5 Hz, which is but a small part of the 40-cm^{-1} transition frequency. This shift is difficult to observe because the highest-resolution molecular-beam resonance apparatus has a linewidth of ~300 Hz, which is not achieved at 40 cm^{-1}. It is worth noting though that the long storage times available in ion traps may make molecular ions attractive candidates for such studies. Thus it appears that the interaction with blackbody radiation is evident only in Rydberg atoms, which offer the unique combination of low frequencies and large (1000-D) dipole moments.

5.4. Applications

Although we can consider the possible application of Rydberg atoms as an atomic thermometer, by far the most promising potential application is as a far-infrared (or microwave) detector, a notion first suggested by Kleppner and Ducas.[30] The basic idea of the most straightforward type of device is to use a laser to make a target of Rydberg state atoms in one state, A. This target is exposed to the radiation to be detected at the infrared frequency ω_{AB}, which is equal to the frequency of the atomic transition from A to a higher state B. If the density of atoms in state A in the target is high enough that the target is optically thick to the radiation at frequency ω_{AB}, then an incoming photon with frequency ω_{AB} will be absorbed and one atom will undergo the transition to state B. The fact that the atom has undergone the transition can be detected by selectively field-ionizing atoms in state B but not in state A. Thus each incoming photon is converted into an ion and an electron that may be easily detected by particle multipliers. A possible configuration for the proposed detector is shown in Fig. 5.11.

An alternative approach is to detect the optical fluorescence from state B but not from state A so that each incoming photon at frequency ω_{AB} is converted into an optical photon that may be easily detected with a photomultiplier.[31] The net result in either case is a photon detector for the far infrared. Because the frequencies of atomic Rydberg states are easily tuned with modest electric fields, the Rydberg atom infrared detector has many desirable features as a filter–detector combination: high sensitivity, high spectral resolution, and wide entrance apertures. The only feature it does not have is high speed; the response time would be $\sim 1\ \mu s$.

The principle for such a device was first established in radio-frequency resonance experiments in electron-beam-excited Rydberg states of He.[32] In these experiments, very low rf powers were needed to drive the microwave transitions between Rydberg states, which were detected by observing a change in the resulting optical fluorescence. Since these initial experiments, the use of laser excitation has made it possible to populate a single state efficiently, allowing the use of lower pressures (10^{-6} Torr instead of 10^{-3} Torr) and reducing collisional effects. In these laser-excitation experiments, optical detection was used to observe successfully transitions from 0.6 to 10^3 cm^{-1} at powers as low as 10^{-9} W cm^{-2}.[33,34] Perhaps more important, the efficiency of a laser has made it possible to observe transitions between Rydberg states in an atomic beam using selective field ionization. These resonance experiments covered the range from 7×10^{-4}–10^3 cm^{-1} (10 μm).[35,36]

Because it is clear that the basic idea and techniques of the proposed detector are sound, the ultimate sensitivity of the detector needs to be determined. There are several technical obstacles to overcome, such as minimizing the ambient temperature and efficiently coupling the radiation to the atoms. However, the basic physical limitation is the rate at which collisions drive the target atoms from state A to state B, a process indistinguishable from the absorption of a photon. The most likely collision partners are atoms in the ground state and the target state A, background gas molecules, ions, and electrons. Let us consider these one at a time.

In a sensibly designed system, ions and electrons should not normally be present but might be created by photoionization of the target atoms by out-of-band radiation coming in with that to be detected. However, if the detector were operating on a $\Delta n = 1$ transition, the frequency required for photoionization would be a factor of n higher than the frequency to be detected; thus removal of the photoionizing radiation would require only a coarse filter. Because collisions with small molecules that would be likely constituents in a vacuum system do not have extraordinarily high cross sections,[9] collisions with background gas molecules are not likely to cause problems for any reasonable background pressure ($\sim 10^{-6}$ Torr).

Finally, we come to two unavoidable species, atoms in the ground state and

Table 5.3. Detection sensitivities

Reference	λ^{-1} (cm^{-1})	NEP ($W\ cm^{-1}\ Hz^{-1/2}$)
Gallagher and Cooke (Ref. 2)	22	10^{-14}
Figger et al. (Ref. 38)	3.3	10^{-17}
Ducas et al. (Ref. 39)	20	5×10^{-15}

the target state A. Because we can, in principle, pump all the atoms to the Rydberg state A, let us only consider Rydberg atom–Rydberg atom collisions. Recent calculations indicate that such collisions have large cross sections of $\sim n^4$ Å^2 and high branching ratios for the dipole-allowed transitions to nearby states.[37] To better understand the magnitude of the effect, let us consider a detector that is to operate on the Na $n=20s \rightarrow n=20p$ transition at 16.2 cm^{-1}. Such a detector, using a beam of Na to form an active volume 1 cm^3, would require a 20s density of $\sim10^4$ cm^{-3}, which, with a cross section of 10^5 Å^2 for 20s–20s collisions producing a 20p state, would lead to a rate of 10^{-2} s^{-1} for collisional transitions to the 20p state. Thus one collision would occur during each observation time, which is assumed to be 1 μs. This is equivalent to NEP$=3\times10^{-19}$ W cm^{-1} $Hz^{-1/2}$, which is certainly consistent with recent observations of the sensitivity of the proposed detector.

The first experimental test of the sensitivity of Rydberg atoms as an IR detector was to compare the effect of the known 300-K background radiation with collision rates for the possible collision partners by varying the number density of the possible collision partners.[3] This led to a value of $<10^{-16}$ W cm^{-1} $Hz^{-1/2}$ for the NEP for the detection of 23-cm^{-1}. Curiously, the major limitation was the large transition rate induced by the 300-K blackbody radiation, which made it impossible to measure the collision rates. Since this experiment, more sophisticated measurements have been taken using both blackbody radiation[38] and a strongly attenuated far-infrared laser.[39] The reported sensitivities are given in Table 5.3 as a function of frequency. Note that the laser attenuation measurement at 20 cm^{-1} is less subject to errors than either of the blackbody radiation measurements.

The method just described is straightforward and, in some sense, proven. However, there is another scheme that is at least intriguing. To detect the same radiation at ω_{AB}, we would create an optically thick target of atoms in state B, which, in a cooled environment, would not become superradiant, but an incoming photon at frequency ω_{AB} would trigger a superradiant avalanche.[40] Thus we would have converted one incoming photon into $\sim10^4$ outgoing photons and left 10^4 atoms in state A, which would be trivial to detect. Currently this approach has not been realized or even fully analyzed to determine its most severe limitations.

Acknowledgments

It is a pleasure to acknowledge many enjoyable and fruitful discussions during the course of this work with W. E. Cooke, K. A. Safinya, and W. Sandner.

This work was supported by the Air Force Office of Scientific Research under Contract F49620 79 C 0212.

In addition, I would like to thank F. Gounand and F. B. Dunning for critical readings of the manuscript.

References and notes

1. E. J. Beiting, G. F. Hildebrandt, F. G. Kellert, G. W. Foltz, K. A. Smith, F. B. Dunning, and R. F. Stebbings, *J. Chem. Phys. 70,* 3551 (1979).
2. T. F. Gallagher and W. E. Cooke, *Apply. Phys. Lett. 34,* 369 (1979).
3. T. F. Gallagher and W. E. Cooke, *Phys. Rev. Lett. 42,* 835 (1979).
4. M. Pimbert, *J. Phys. (Paris) 33* (1972).
5. R. Loudon, *The Quantum Theory of Light* (Oxford University Press, 1973).
6. W. E. Cooke and T. F. Gallagher, *Phys. Rev. A 21,* 588 (1980).
7. H. A. Bethe and E. A. Salpeter, *Quantum Mechanics of One- and Two-Electron Atoms* (New York: Academic Press, 1957).
8. A. Lindgard and S. A. Nielsen, *At. Data Nucl. Data Tables 19,* 534 (1977).
9. L. M. Humphrey, T. F. Gallagher, W. E. Cooke, and S. A. Edelstein, *Phys. Rev. A 18,* 1383 (1978).
10. C. H. Townes and A. L. Schawlow, *Microwave Spectroscopy* (New York: McGraw-Hill, 1955).
11. T. F. Gallagher, W. Sandner, K. A. Safinya, and W. E. Cooke, *Phys. Rev. A. 23,* 2065 (1981).
12. U. Fano and J. W. Cooper, *Rev. Mod. Phys. 40,* 441 (1968).
13. J. W. Farley and W. H. Wing, *Phys. Rev. A. 23,* 2397 (1981).
14. J. L. Hall, private communication.
15. P. R. Koch, H. Hieronymus, A. F. J. van Raan, and W. Raith, *Phys. Lett. 75A,* 273 (1980).
16. W. P. Spencer, A. G. Vaidyanathan, D. Kleppner, and T. W. Ducas, *Phys. Rev. A 24,* 2513 (1981).
17. G. F. Hildebrandt, E. J. Beiting, C. Higgs, G. J. Hatton, K. A. Smith, F. B. Dunning, and R. F. Stebbings, *Phys. Rev. A 23,* 2978 (1981).
18. M. Gross, P. Goy, C. Fabre, S. Haroche, and J. M. Raimond, *Phys. Rev. Lett. 43* 343 (1979).
19. A. G. C. Mitchell and M. W. Zemansky, *Resonance Radiation and Excited Atoms* (Cambridge University Press, 1971).
20. F. Gounand, M. Hugon, P. R. Fournier, and J. Berlande, *J. Phys. B 12,* 547 (1979).
21. W. E. Cooke and T. F. Gallagher, *Bull. Am. Phys. Soc. 23,* 1103 (1978).
22. R. H. Dicke, *Phys. Rev. 93,* 99 (1954).
23. D. R. Bates and A. Damgaard, *Phil. Trans. R. Soc. London A 242,* 101 (1949).
24. F. Gounand, *J. Phys. (Paris) 40,* 457 (1979).
25. F. Gounand, P. R. Fournier, J. Cuvellier, and J. Berlande, *Phys. Lett. 59A,* 23 (1976).
26. T. F. Gallagher and W. E. Cooke, *Phys. Rev. A 20,* 670 (1979).

27. R. F. Stebbings, C. J. Latimer, W. P. West, F. B. Dunning, and T. B. Cook, *Phys. Rev. A 12,* 1453 (1975).
28. T. F. Gallagher, W. E. Cooke, and S. A. Edelstein, *Phys. Rev. A 17,* 904 (1978).
29. P. W. Zitzewitz and N. F. Ramsey, *Phys. Rev. A 1,* 53 (1971).
30. D. Kleppner and T. W. Ducas, *Bull. Am. Phys. Soc. 21,* 600 (1976).
31. R. M. Hill and T. F. Gallagher, U. S. Patent No. 4024396 (1977).
32. W. H. Wing and W. E. Lamb, Jr., *Phys. Rev. Lett. 28,* 265 (1972).
33. T. F. Gallagher, R. M. Hill, and S. A. Edelstein, *Phys. Rev. A 13,* 1448 (1976).
34. J. A. Gelbwachs, C. F. Klein, and J. E. Wessel, *J. Quant. Electron. QE-16,* 137 (1980).
35. T. F. Gallagher, L. M. Humphrey, R. M. Hill, W. E. Cooke, and S. A. Edelstein, *Phys. Rev. A 15,* 1937 (1977).
36. T. W. Ducas and M. L. Zimmerman, *Phys. Rev. A 15,* 1523 (1977).
37. R. E. Olson, private communication.
38. H. Figger, G. Leuchs, R. Straubinger, and H. Walther, *Opt. Commun. 33,* 37 (1980).
39. T. W. Ducas, W. P. Spencer, A. G. Vaidyanathan, W. H. Hamilton, and D. Kleppner, *Appl. Phys. Lett. 35,* 382 (1979).
40. This has been independently suggested by S. Haroche, D. Kleppner, and T. F. Gallagher.

6

Theoretical approaches to low-energy collisions of Rydberg atoms with atoms and ions

A. P. HICKMAN, R. E. OLSON, AND J. PASCALE

6.1. Introduction

Considerable progress has been made in the past few years in the theoretical description of collisions involving Rydberg atoms. A characteristic feature of this work is that is has diverged from the well-established approaches for treating the interaction of atoms or molecules in the ground state or low-lying excited states. The application of these approaches in the regime of highly excited states is possible in principle but awkward in practice; hence, many new approaches have evolved that are better suited to the problem. This chapter will provide an overview of these theoretical methods and discuss their applications to some specific problems.

The contrast between the methods applicable to low-energy collisions between atoms in low-lying states and those applicable to collisions involving Rydberg atoms may be briefly described as follows. The former methods involve essentially a molecular approach, in which potential curves or surfaces for a small number of electronic states are determined as a function of the internuclear distances. The surfaces, and the coupling matrix elements between them, serve as input to a scattering calculation. However, in the case where one atom is highly excited, the most effective approaches are based on a "three-body picture," where one electron is in an orbital whose size is quite large relative to the dimensions of the core or the other atom (or molecule) involved in the collision. This electron is therefore singled out and treated as an independent particle. The ionic core and the other atom are treated as point particles. The scattering is treated by modeling the interactions among the excited electron, the ionic core, and the collision partner.

This chapter will discuss in detail the process of orbital angular momentum mixing,

$$A^{**}(nl) + B \rightarrow A^{**}(nl') + B \qquad (1)$$

where B is a rare gas and A^{**} a highly excited atom. The cross section for this process may be 10^2–10^4 Å^2 and may vary greatly with n. In terms of the three-body picture, the most important features of this system are the Coulomb interaction between A^+ and e^- and the scattering of e^- by B.

The physical picture is that the electron begins in a large, classical, ellipsoidal orbit. As the rare gas travels through the Rydberg atom, the most likely event is a collision between e^- and B that changes the shape of the electron's orbit. Most theoretical treatments have emphasized the relation between the n dependence of reaction (1) to the e^-–B scattering cross section, although Smirnov[1] and Flannery[2,3] pointed out that the A^+–B interaction may also have an important contribution.

Another process discussed herein is the scattering of Rydberg atoms and ions:

$$A^{**}(nl) + B^+ \rightarrow \begin{cases} A^{**}(n'l') + B^+ \\ A^+ + B(n''l'') \\ A^+ + B^+ + e^- \end{cases} \tag{2}$$

This system has been modeled successfully by considering the Coulomb forces among the three charged particles. Because the interaction between e^- and B^+ is stronger and longer in range than the e^-–B interaction in the preceding example, a broader distribution of final channels will be populated. This system has been studied using a Monte Carlo classical trajectory model.

A final example important to our discussion is the collision of two Rydberg atoms. Such a system may be considered an extension of the three-body model to a four-body model, where the four bodies are the two ionic cores and the two electrons. The Monte Carlo classical trajectory model also has been applied in this case to study ionization and excitation transfer.

6.2. Theory

The interaction between the perturber B and the Rydberg atom A must be defined to treat the dynamical problem of the collision. However, in the case where A is in a highly excited state ($n \gg 1$) and B a neutral, information on the scattering between a free electron and B can be used directly to calculate the collision probability (see Sect. 6.2b, under "Impulse approximation," and "Model approach").

6.2a. Definition of the interaction

Molecular potentials

The interaction between an atom A in a high Rydberg state ($n \gg 1$) with a perturber B (ion or neutral) presents all the complexity of a many-body problem. Adiabatic potential energy curves may be obtained by separating the center-of-mass motion from the relative nuclear motion, making the Born–Oppenheimer approximation, and then diagonalizing the Hamiltonian for the full system. For the Born–Oppenheimer approximation to be valid and also to obtain a meaningful definition of potential adiabatic curves, it is necessary that the relative nuclear motion be much slower than the electronic motion. This implies that n should not be too large. Ab initio potential energy curves for highly excited states are limited to relatively small molecules (see, for example, Ref. 4 for He_2^{**}).

However, most of the theoretical approaches to the collision of a high-Rydberg atom with a perturber take advantage of the fact that, on average, the Rydberg electron e^- is very distant from the core A^+ in comparison with the size of A^+. Then the interaction is reduced to a three-body interaction involving e^-, A^+, and B. For that approximation to be valid, however, the relative nuclear motion must be slow enough (e.g., thermal collisions) so that both A^+ and B may be represented by unperturbed atomic wave functions ("frozen"-core approximation). Then the problem reduces to the determination of eigenvalues of an effective Hamiltonian H for the Rydberg electron for each value of the internuclear separation $\mathbf{R}$, where

$$H(\mathbf{r},\mathbf{R}) = H_A(\mathbf{r}) + V(\mathbf{r},\mathbf{R}) \tag{3}$$

In this case, $H_A(\mathbf{r})$ is the valence-electron Hamiltonian of the atom A, $V(\mathbf{r},\mathbf{R})$ includes both the interactions between B and e^- and A^+, respectively, and a three-body interaction, and $\mathbf{r}$ and $\mathbf{R}$ are the position vectors of e^- and B, respectively, relative to A^+. The interaction $V(\mathbf{r},\mathbf{R})$, as well as $H_A(\mathbf{r})$, can be defined by model potentials or pseudopotentials.[5] When B is an ion, the interactions are relatively well known and are dominated by the Coulomb forces. When B is a neutral, both short-range and long-range interactions must be considered. Such an approach has been very useful in determining some highly excited states of alkali–rare gas atom systems,[6] although further improvements of the method have appeared necessary.[7]

Binary encounter approximation

In the case where B is a neutral, the three-body problem is often simplified further by assuming that the average distance $\langle r \rangle_{nl}$ of the Rydberg electron from A^+ is very large with respect to the range of interactions between B and e^- and between B and A^+, respectively. Then the collisions e^-–B and A^+–B are treated independently. The interaction A^+–B is generally neglected. This further approximation may not be justified for very low or very high values of n. Also, a three-body interaction is appropriate when the relative nuclear motion is large in comparison with the average velocity of the Rydberg electron. For the atomic process that has been the most often considered in recent theoretical works, the l mixing of alkali states by thermal collisions with rare gas atoms, the binary encounter approximation has been generally justified a posteriori by comparisons between experimental and theoretical results. However, Flannery[2,3] recently considered the A^+–B interaction. Therefore, the collision between the perturber B and a high-Rydberg atom A is reduced to the binary collision between B and the Rydberg electron e^-. In this way,

Fermi[8] was able to relate the pressure shift of the absorption spectrum of Rydberg series for alkali atoms perturbed by rare gas atoms to the information on the elastic scattering of an extremely low energy electron by the rare gas atom. Further studies were made by Alekseev and Sobel'man.[9] The so-called Fermi pseudopotential[10] is then expressed as

$$V(\mathbf{r}, \mathbf{R}) = 2\pi a\delta(\mathbf{r} - \mathbf{R}) \tag{4}$$

where a is the scattering length. In the case of elastic scattering of a Rydberg atom in the state nlm, the matrix element of the interaction is

$$V_{nlm,\,nlm} = 2\pi a|\psi_{nlm}(\mathbf{R})|^2 \tag{5}$$

where ψ_{nlm} is the wave function of the Rydberg electron at the position $\mathbf{R}$ with respect to A^+. For inelastic collisions, nondiagonal matrix elements are defined by[11]

$$V_{nlm,\,n'l'm'}(\mathbf{R}) = 2\pi a\psi^*_{nlm}(\mathbf{R})\psi_{n'l'm'}(\mathbf{R}) \tag{6}$$

Ivanov[12] and Omont[13] showed that the Fermi pseudopotential can be generalized in terms of the reaction matrix $R(E_{\mathbf{R}})$ for the e^--B scattering, where $E_{\mathbf{R}}$ is the kinetic energy of the electron at position $\mathbf{R}$. In the limit of very small energy $E_{\mathbf{R}}$, the result of the Fermi approximation is found.

In later works, Eq. (5) was modified to include the polarization interaction between e^- and B.[13-16] However, a more realistic interaction between e^- and B should explicitly include short- and long-range terms. The Fermi pseudopotential will be discussed later in connection with l mixing of alkali atoms by rare gas atoms.

In some cases the interaction between A^+ and B may be not negligible. For example, in the case where B is a neutral, the asymptotic form of this interaction is $U(R) = -\frac{1}{2}\alpha R^{-4}$, where α is the polarizability of the perturber. However, the assumption that this interaction should be simply added to the e^--B interaction to give the total interaction is questionable, even for large values of the quantum number n. Instead, a complete interaction, including a three-body interaction, should be used. The additional three-body term arises because the electric field that polarizes B is the vector sum of the fields caused by A^+ and e^-. The total polarization potential between A and B is thus the sum of the e^--B and A^+-B polarization interactions plus a cross term.

Coulomb forces

In the case where B is an ion or a neutral atom in a highly excited state, the simple Coulomb interaction may be used to describe the three- or four-body interaction. However, if A (or eventually B, in the case of the neutral) is excited in a relatively low Rydberg state, the polarization interactions or even the short-range interaction should not be ignored. Nevertheless, classical

Monte Carlo calculations using the Coulomb forces have been quite successful for determining the cross section for ionization of Rydberg atoms by ions or Rydberg atoms.[17-19] For inelastic transitions within the Rydberg atom that involve large energy splitting, more accurate interactions should be used because the collisions are more efficient at closer internuclear distances.

6.2b. Dynamics

Quantum-mechanical treatment of l mixing

Once the form of the interaction potential has been adopted, the quantum-mechanical theory of the dynamics may be formulated in a straightforward manner and a variety of approximations may be applied. This section will present formulae for cross sections derived from specific model interactions of the type discussed in Sect. 6.2a. We shall consider collisions of Rydberg atoms with rare gases leading to transitions of the type

$$A^{**}(nl) + B \rightarrow A^{**}(n'l') + B \tag{7}$$

and concentrate on cases where $n' = n$, but this is not essential. The methods to be discussed will include the coupled-channel approach, the Born approximation, and the impulse approximation; the relationship among these methods will also be discussed.

Coupled-channel method. The coupled-channel method of Arthurs and Dalgarno[20] was formulated by Olson[21] and Hickman[22] for collisions involving Rydberg atoms. The method assumes that a collision between a Rydberg atom and a rare gas can be described by the following Hamiltonian:

$$H = -\frac{1}{2\mu}\nabla_{\mathbf{R}}^2 + H_{\mathrm{A}}(\mathbf{r}) + V(\mathbf{r}, \mathbf{R}) \tag{8}$$

where $H_{\mathrm{A}}(\mathbf{r})$ is the Hamiltonian of the Rydberg atom, which is modeled as a one-electron atom without spin. We assume that the eigenfunctions of $H_{\mathrm{A}}(\mathbf{r})$ are hydrogenic:

$$H_{\mathrm{A}}(\mathbf{r})\psi_{nlm}(\mathbf{r}) = E_{nl}\psi_{nlm}(\mathbf{r}) \tag{9}$$

$$\psi_{nlm}(\mathbf{r}) = F_{nl}(r)\,Y_{lm}(\hat{\mathbf{r}}) \tag{10}$$

In this case, F_{nl} is a normalized radial hydrogenic wave function and Y_{lm} a spherical harmonic.

For most applications we shall assume that the structure of the core of the Rydberg atom has a negligible effect on the wave function; hence, pure hydrogenic wave functions will be used. We shall, however, take accurate values of the energy levels E_{nl}. The kinetic energy operator of the relative motion of the Rydberg atom and the rare gas projectile is $-(1/2\mu)\nabla_{\mathbf{R}}^2$, where μ is the reduced mass of the rare gas–Rydberg atom pair. For our initial discussion, we shall assume that the interaction potential is the Fermi pseudopotential given by Eq. (4). For this form of $V(\mathbf{r},\mathbf{R})$, the coupling matrix elements assume a particularly simple form.

The development of the coupled-channel equations follows the method of Arthurs and Dalgarno,[20] although we adopt a slightly different notation, as follows: $\mathbf{l}$ is the orbital angular momentum of the bound electron, with corresponding quantum numbers l and m; $\mathbf{L}$ is the angular momentum of the relative motion described by the coordinate $\mathbf{R}$, and L and M are the quantum numbers. The total angular momentum is

$$\mathbf{J} = \mathbf{L} + \mathbf{l} \tag{11}$$

Eigenfunctions of H [Eq. (8)] can be expanded in the standard way as products of radial functions of R times angular eigenfunctions of the total angular momentum $\mathbf{J}$. The set of equations that describes the coupling of levels $(n,l),(n,l+1),\ldots,(n,n-1)$ can be written as

$$\frac{\hbar^2}{2\mu}\left[-\frac{d^2}{dR^2} + \frac{L'(L'+1)}{R^2} - k_{l'l}^2\right] u_{l'L'}^{JlL}(R) = -\sum_{l''}\sum_{L''} \langle l''L''nJ|V|l'L'nJ\rangle u_{l''L''}^{JlL}(R) \tag{12}$$

where

$$k_{l'l}^2 = (2\mu/\hbar^2)(E + E_{nl} - E_{nl'}) \tag{13}$$

and

$$\langle l''L''nJ|V|l'L'nJ\rangle = \tfrac{1}{2}a[(2l''+1)(2L''+1)(2l'+1)(2L'+1)]^{1/2} \times \begin{pmatrix} J & l'' & L'' \\ 0 & 0 & 0 \end{pmatrix}\begin{pmatrix} J & l' & L' \\ 0 & 0 & 0 \end{pmatrix} F_{nl''}(R)F_{nl'}(R) \tag{14}$$

A more general form of the interaction potential than Eq. (5) can be written as an arbitrary function of R, r, and θ (the angle between $\mathbf{r}$ and $\mathbf{R}$) that is then expanded numerically in terms of Legendre polynomials:

$$V(R,r,\theta) = \sum_{\lambda} v_{\lambda}(R,r)P_{\lambda}(\cos\theta) \tag{15}$$

where

$$v_{\lambda}(R,r) = \tfrac{1}{2}(2\lambda+1)\int_0^{\pi} V(R,r,\theta)P_{\lambda}(\cos\theta)\sin\theta\, d\theta \tag{16}$$

In this case, Eq. (14) must be replaced by

$$\langle l''L''nJ|V|l'L'nJ\rangle = \sum_{\lambda=0}^{2n-2} \tfrac{1}{2}(2\lambda+1)f_{\lambda}(l''L'', l'L'; J)$$

$$\times \int_0^\infty F_{nl''}(r) v_\lambda(R,r) F_{nl'}(r) r^2 \, dr \tag{17}$$

where the f_λ coefficients may be explicitly written in terms of $3-j$ and $6-j$ coefficients.[20]

Although the preceding formulation of the l-mixing problem is expected to be quite accurate, practical difficulties will be encountered as n increases because of the large number of channels. For example, if the lowest angular momentum level considered is $l=2$, there will be $\frac{1}{2}[n(n+1)]-3$ channels. An exact solution of the coupled channels is clearly not feasible for arbitrarily high n. Fortunately, in this limit, the equations are weakly coupled, and an accurate solution may be obtained using the Born approximation or the distorted wave Born approximation. However, an intermediate regime may exist where the number of channels is too large for an exact solution and where the coupling is too strong for the application of the Born approximation.

The two-state method of Olson[21] is applicable to the strong coupling regime where the l-mixing cross section is increasing as $\sim n^4$, the geometric size of the Rydberg atom. This method, in essence, leads to a reaction radius R_c for the l-mixing process and assumes the product states are populated statistically within R_c. Attributes of the two-state method are that it is easy to apply and lengthy calculations are avoided by the inclusion of only two channels. Furthermore, this method is valid for indicating the n values at which the collision process departs from the strong coupling region, and it led to the first theoretical prediction of the maximum in the l-mixing cross sections. A deficiency of the method, however, is the application of the statistical approximation. For large n values, beyond where the cross section has maximized, such an approximation is invalid and will lead to erroneous results. In this high-n region, however, perturbation methods such as the Born approximation are valid and can be applied.

Born approximation. An approximate solution to the coupled equations presented in the preceding section may be obtained using the distorted wave Born approximation (DWBA). For this method, the total angular momentum representation is retained and an approximate solution to the coupled equations is calculated for each value of J. The total cross section is obtained by summing over all partial waves. In contrast, the total cross section is most easily obtained in the standard Born approximation by integrating over the scattering angle θ.[23,24] This section will present formulae for total cross sections in the Born approximation for a particular form of the interaction potential

$$V(\mathbf{r},\mathbf{R}) = 2\pi\tilde{a}\delta(\mathbf{r}-\mathbf{R}) - \frac{\alpha/2}{(|\mathbf{r}-\mathbf{R}|^2 + x_0^2)^2} \tag{18}$$

This potential is a useful way of modeling the interaction of an electron with a rare gas atom. The short-range behavior is controlled by the parameters $\tilde{a}$ and

x_0; the long-range behavior is determined by the polarizability α. Moreover, the results for this potential illustrate the relation between the Born approximation and the impulse approximation, which will be discussed in the next section.

The Born amplitude for excitation of the target from the state (nlm) to $(nl'm')$, coupled with the scattering of the projectile from an initial momentum $\hbar\mathbf{k}$ to final momentum $\hbar\mathbf{k}'$ may be written as a function of

$$\mathbf{Q} = \mathbf{k}' - \mathbf{k} \tag{19}$$

The amplitude is

$$f_{\mathrm{B}}(nlm \rightarrow nl'm' \,|\, \mathbf{Q}) = -(\mu/2\pi\hbar^2) \int \exp(-i\mathbf{Q} \cdot \mathbf{R}) \Psi^*_{nl'm'}(\mathbf{r}) V(\mathbf{r}, \mathbf{R}) \Psi_{nlm}(r)\, d^3r\, d^3R \tag{20}$$

The momentum change vector $\hbar\mathbf{Q}$ is related to the scattering angle Θ by

$$Q^2 = k^2 + (k')^2 - 2kk' \cos\Theta \tag{21}$$

We now substitute Eq. (18) into Eq. (20). The integral over the delta function is trivial, and the integral over the polarization term is done by changing variables to $\mathbf{r}$ and $\mathbf{x} = \mathbf{r} - \mathbf{R}$ and using the result

$$\int_0^\infty \frac{x\, dx \sin Qx}{(x^2 + x_0^2)^2} = \frac{\pi Q}{4x_0} \exp(-Qx_0) \tag{22}$$

We obtain

$$f_{\mathrm{B}}(nlm \rightarrow nl'm' \,|\, \mathbf{Q}) = \frac{\mu}{m_{\mathrm{e}}} \left(-\tilde{a} + \frac{\pi\alpha}{4x_0 a_0} \exp(-Qx_0) \right) \times \int \exp(-i\mathbf{Q} \cdot \mathbf{r}) \Psi^*_{nl'm'}(\mathbf{r}) \Psi_{nlm}(\mathbf{r})\, d^3r \tag{23}$$

The formula for the differential cross sections averaged over initial m and summed over final m' is

$$I(nl \rightarrow nl' \,|\, \Theta) = \frac{k'}{k} (2l' + 1) \sum_\lambda (2\lambda + 1) \begin{pmatrix} l & l' & \lambda \\ 0 & 0 & 0 \end{pmatrix}^2 \times \left(\frac{\mu}{m_{\mathrm{e}}} \right)^2 \left(-\tilde{a} + \frac{\pi\alpha}{4x_0 a_0} \exp(-Qx_0) \right)^2 \times \left(\int_0^\infty F_{nl'}(r) j_\lambda(Qr) F_{nl}(r) r^2\, dr \right)^2 \tag{24}$$

where j_λ is the spherical Bessel function. The total cross section is

$$\sigma(nl \rightarrow nl') = 2\pi \int_0^\pi I(nl \rightarrow nl' \,|\, \Theta) \sin\Theta\, d\Theta \tag{25}$$

Our final formulae are obtained by using Eq. (21) to transform the integral over Θ into an integral over Q. The new limits of integration are then

$$Q_{\min} = |\mathbf{k}' - \mathbf{k}| = |E_{nl'} - E_{nl}|/\hbar v_{\mathrm{i}} \tag{26}$$

$$Q_{\max} = k' + k \tag{27}$$

where v_{i} is the initial relative velocity of the Rydberg atom and its collision partner. Because of the mass difference between the electron and the collision partner, the scattering of the Rydberg atom is strongly forward peaked. Therefore, essentially all of the integral over Q comes from small values of Q, and we may replace the upper limit $Q_{\max}$ by ∞. Then

$$\sigma(nl \to nl') = \frac{2l'+1}{\hbar^2 v_{\mathrm{i}}^2} \sum_{\lambda} (2\lambda+1) \begin{pmatrix} l & l' & \lambda \\ 0 & 0 & 0 \end{pmatrix}^2 \times \int_{Q_{\min}}^{\infty} dQ\, Q \left[\left(-\tilde{a} + \frac{\pi\alpha}{4x_0 a_0} \exp(-Qx_0) \right) \times \int_0^{\infty} F_{nl'}(r) j_{\lambda}(Qr) F_{nl}(r) r^2\, dr \right]^2 \tag{28}$$

The choice of parameters $\tilde{a}$, x_0, and α for particular systems will be discussed in the next section.

Impulse approximation. When a projectile B is scattered by a Rydberg atom, it is sometimes possible to view the event as a binary interaction between A and the excited electron. We can use information about electron scattering from B as a function of momentum and then average over the momentum distribution of the excited electron to obtain an approximation to the desired scattering amplitude. This "impulse approximation" has been discussed in general terms by Newton.[25] The specific application to Rydberg atom collisions was suggested by Fermi,[8] and several recent calculations were performed by Matsuzawa.[26-29] Using the notation given in Eq. (23), we can write

$$f_{\mathrm{I}}(nlm \to nl'm' \,|\, \mathbf{Q}) = \frac{\mu}{M_{\mathrm{e}}} \int \Phi^{*}_{nl'm'}(\mathbf{k}-\mathbf{Q}) f_{\mathrm{e}}(\mathbf{k} \to \mathbf{k}-\mathbf{Q}) \Phi_{nlm}(\mathbf{k})\, d^3k \tag{29}$$

The scattering amplitude for the e–B system is given by $f_{\mathrm{e}}(\mathbf{k} \to \mathbf{k}-\mathbf{Q})$. The formula assumes that the momentum transfer $\mathbf{Q}$ in the relative motion is caused entirely by the e–B encounter. The momentum-space wave functions are assumed to be hydrogenic and are defined in the standard way:

$$\Phi_{nlm}(\mathbf{k}) = (2\pi)^{-3/2} \int d^3r \exp(i\mathbf{k}\cdot\mathbf{r}) \psi_{nlm}(\mathbf{r}) \tag{30}$$

Considerable progress in simplifying Eq. (29) can be made by adopting an approximation to f_e based on the work of O'Malley[30] and proposed by Matsuzawa.[28,29] For an electron scattering from a rare gas whose polarizability is α and whose scattering length is a, O'Malley derived the low-energy limit of f_e as follows:

$$f_e(\mathbf{k} \to \mathbf{k} - \mathbf{Q}) = -a - \frac{\pi\alpha Q}{4a_0} - \frac{4\alpha L}{3a_0} k^2 \ln ka_0 + O(k^2) \tag{31}$$

This formula includes the effect of all partial waves. It is convenient to keep only the first two terms so as to obtain a functional form of f_e that depends only on the momentum transfer

$$f_e(Q) \simeq -a - (\pi\alpha Q/4a_0) \tag{32}$$

By substituting Eq. (32) into Eq. (29) and expanding $\Phi_{nl'm'}$ and Φ_{nlm} in terms of coordinate wave functions, we obtain

$$f_I(nlm \to nl'm' \mid \mathbf{Q}) = -\frac{\mu}{m_e}\left(-a - \frac{\pi\alpha Q}{4a_0}\right) \times \int \psi^*_{nl'm'}(\mathbf{r}) \exp(i\mathbf{Q}\cdot\mathbf{r}) \psi_{nlm}(\mathbf{r})\, d^3r \tag{33}$$

Comparison of Eqs. (23) and (33) illustrates the close connection between the present impulse approximation and the Born approximation with the model potential discussed in the previous section. Let us first note that expanding the exponential

$$\exp(-Qx_0) \simeq 1 - Qx_0 + \cdots \tag{34}$$

should be justified because Q is of the order of $1/n$ and, for reasonable model potentials, $x_0 \simeq 1$. Then the two expressions for the scattering amplitude will be equal if

$$-a - \frac{\pi\alpha Q}{4a_0} = -\tilde{a} + \frac{\pi\alpha}{4x_0 a_0}(1 - Qx_0) = -\left(\tilde{a} - \frac{\pi\alpha}{4x_0 a_0}\right) - \frac{\pi\alpha Q}{4} \tag{35}$$

That is, it is necessary that

$$a = \tilde{a} - \frac{\pi\alpha}{4x_0 a_0} \tag{36}$$

The expression on the right-hand side of Eq. (36) is just the scattering length, in the Born approximation, for electron scattering by the model potential of Eq. (18).

Several points may be made about this result. First, it shows that the impulse approximation is essentially equivalent to the Born approximation with the model potential (18), giving a physical basis for this particular form of the interaction. This equivalence, in the case of the Fermi pseudopotential Eq. (4), was noted by de Prunelé.[31] Second, it suggests a systematic way of testing and improving the impulse approximation. The accuracy of a DWBA calculation using Eq. (18) may be assessed by monitoring the deviation of the S matrix from unitarity. When the DWBA S matrix is inaccurate, a better result can presumably be obtained by seeking a more accurate solution to the scattering problem but using the same model potential. Other refinements can also be included in a natural way by adding terms to the potential to describe the A^+–B interaction or the cross terms in the polarizability interaction. An improved potential of this sort would be treated by the Born approximation or the coupled-channel approach, as appropriate.

Flannery[32,33] developed a method that is a direct derivative of the impulse approximation and that essentially accounts for the interaction e^-–B by using the information concerning the elastic scattering of e^- by B. In this theory, the electron loosely bound to A^+ undergoes an elastic collision with B so that the internal energy of the Rydberg atom A is increased or decreased, leading to an excitation (or ionization) or deexcitation of the Rydberg atom A. The probabilities of transition between two levels of A are then calculated by classical mechanics, using data concerning the differential cross section for the e^-–B elastic scattering. More recently, Flannery[2,3] extended his method to include the A^+–B elastic scatterings. This theory is developed elsewhere (see Chap. 11).

Scaling formula. The preceding sections discussed the use of the coupled-channel method for small values of n and the Born and impulse approximations for large n. For many important systems, the regions of validity for these two approaches do not overlap. However, it is sometimes possible to interpolate between large and small n to obtain useful results in the intermediate regime. In this section we shall discuss an approximate scaling formula obtained by Hickman[34] that provides a rapid estimate of the l-mixing cross section for certain systems and illustrates the effects of the various physical parameters of the system.

The formula is applicable to the case where the initial level nl is separated by some ΔE from a nearly degenerate set of final levels $n, l+1; n, l+2; \ldots; n, n-1$ and where other levels are sufficiently far removed to be unimportant. Such atoms as $Na^{**}(nd)$ or $Rb^{**}(nf)$ fulfill these conditions. Then an estimate of the total l-mixing cross section is given by

$$\sigma_{l\text{mix}} = \pi n^4 a_0^2 g(\beta) f(\gamma) \tag{37}$$

where

$$\gamma = n^2 a_0 \Delta E / \hbar v \tag{38}$$

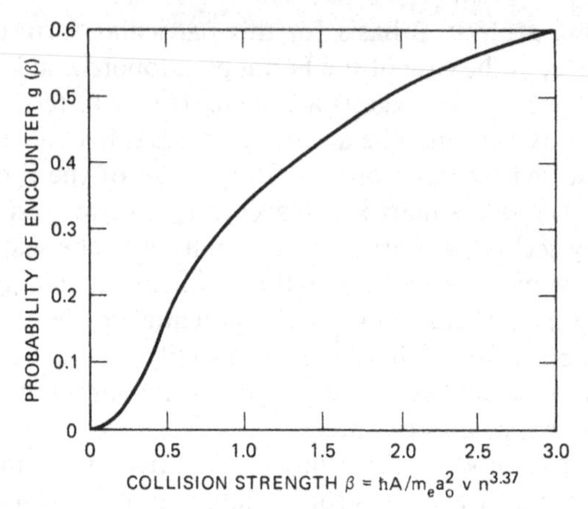

Fig. 6.1. The function $g(\beta)$, which may be interpreted as the probability of encounter between the Rydberg electron and a collision partner.

and

$$\beta = \frac{\hbar}{m_e a_0^2} \frac{1}{v n^{3.367}} \left[\frac{\sigma_e (\frac{1}{2} n^{-2})}{4\pi} \right]^{1/2} \tag{39}$$

and the functions $g(\beta)$ and $f(\gamma)$ are shown in Figs. 6.1 and 6.2, respectively. The average of the initial and final relative velocity of the Rydberg atom and its collision partner is v. This empirical formula was obtained by fitting the results of various Born approximation and coupled-channel calculations to functions of the reduced parameters β and γ; the estimated accuracy is about a factor of two.

The scaling formula is a product of three factors. The geometrical factor $\pi n^4 a_0^2$ shows that the cross section scales with the overall size of the atom. The parameter β may be interpreted as a coupling strength, and $g(\beta)$ is then the probability that the collision partner will encounter the orbiting Rydberg electron. In Eq. (39), σ_e is the electron–rare gas elastic scattering cross section at the energy $\frac{1}{2}n^{-2}$. Equation (39) shows that β increases with σ_e, which gives an effective size of the collision partner, and decreases with n because, for larger n, the electron "cloud" is more diffuse. The function $f(\gamma)$ may be interpreted as the probability that an elastic collision between the orbiting electron and the collision partner will cause a transition into a new energy level. It is interesting to note that the parameter γ may be rewritten as

$$\gamma \simeq (\Delta\delta) v_e / v \tag{40}$$

where $\Delta\delta$ is the difference in the quantum defects of initial and final states and $v_e = 1/n$ (in atomic units), the velocity corresponding to the average kinetic energy of an electron with principal quantum number n. This formula is obtained by expanding

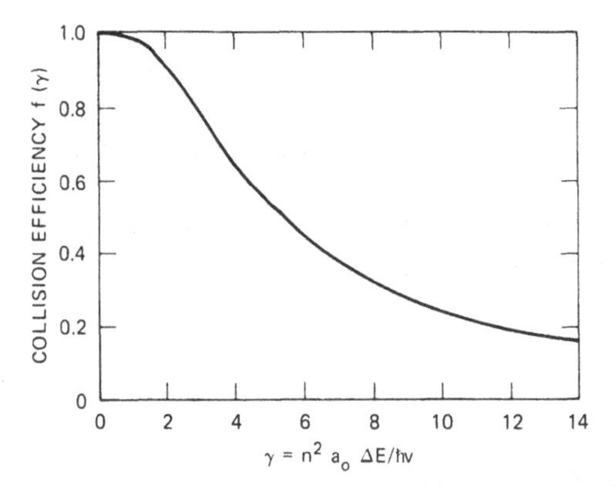

Fig. 6.2. The function $f(\gamma)$, which may be interpreted as the probability that a collision between the Rydberg electron and the collision partner will cause a change of the angular momentum level of the electron.

$$\Delta E = \frac{1}{2(n+\delta_{l'})^2} - \frac{1}{2(n+\delta_l)^2} \tag{41}$$

in the limit $\delta_l \ll n$ and substituting into Eq. (38). When γ is large, $f(\gamma) \to 0$ and hence the l mixing will be small. Eq. (40) shows that this can occur because the inelasticity is large or because v_e is large compared with v, and hence it is less likely that a collision will deflect the electron enough to change the shape of its orbit. Conversely, the l mixing will be larger when γ is small. This may occur either because the energy difference between initial and final states is small or because the electron is moving slowly relative to the collision partner and a collision tends to perturb its orbit seriously. Various calculations using this formula will be presented in Sect. 6.3a.

Semiclassical

In the semiclassical approach, we shall consider collisions of Rydberg atoms A in a state (nl) with a neutral perturber B leading to transitions of the type

$$A^{**}(nl) + B \to A^{**}(n'l') + B \tag{42}$$

Again the case where $n' = n$ will be considered in some applications.

In the semiclassical approach to the collision between two atoms, the motion of the act:ve electron is described by a wave function, whereas the relative nuclear motion is now described by a classical trajectory. We shall assume a straight-line trajectory defined by an impact parameter **b**:

$$\mathbf{R} = \mathbf{b} + \mathbf{v}t \tag{43}$$

where **v** is the relative velocity of B relative to A^+. This approximation is not essential; instead, a curved trajectory may also be used, defined by a common central potential. However, the definition of the common potential may be questionable. The assumption of a straight-line trajectory will be generally justified in most of the cases where the energy transfer involved during the collision is very small in comparison to the energy of the relative motion.

Perturbed stationary states. We assume that the interaction of A with B is described by the Hamiltonian

$$H(\mathbf{r},\mathbf{R}) = H_A(\mathbf{r}) + V(\mathbf{r},\mathbf{R}) \tag{44}$$

In the collision problem, $\mathbf{R}$ becomes a function of the time t, and the time-dependent Schrödinger equation

$$i\hbar\, \partial\psi(\mathbf{r},\mathbf{R}(t))/\partial t = H(\mathbf{r},\mathbf{R}(t))\psi(\mathbf{r},\mathbf{R}(t)) \tag{45}$$

is solved by expanding the total wave function of the system in terms of the stationary states of the isolated atom A. A system of coupled differential equations is then obtained that must be generally solved by numerical integration. We consider here the first-order perturbation theory as used by Gersten[11] and Derouard and Lombardi[35] for the l mixing of alkali atoms by the rare gas atoms. Assuming the pseudopotential of Fermi [Eq. (4)], the probability of the transition of the Rydberg electron from the state (nlm) to the state $(n'l'm')$ is then defined as

$$P_{nlm\to n'l'm'}(b,v,\Omega) = (2\pi a)^2\left|\int_{-\infty}^{+\infty}\psi_{n'l'm'}(\mathbf{R}(t))\psi_{nlm}(\mathbf{R}(t))\exp(i(E_{n'l'}-E_{nl})t)\,ds\right|^2 \tag{46}$$

where $s^2 = R^2 - b^2$.

In the case of high-Rydberg states, hydrogenic wave functions may be used [Eq. (10)]. Bates–Damgaard wave functions involving an effective quantum number may also be used, but this effect should be negligible. The average over all the orientations Ω of the impact direction leads to the transition probability from the level (nl) to the level $(n'l')$:

$$\bar{P}_{nl,n'l'}(b,v) = \frac{2\pi a^2}{2l+1}\sum_{mm'}\left|Y_{lm}\left(\frac{\pi}{2},0\right)Y_{l'm'}\left(\frac{\pi}{2},0\right)\right|^2 \times\left|\int_0^{\pi/2}F_{nl}(R)F_{nl'}(R)\exp[i(m'-m)\varphi]\,d\varphi\right|^2 \tag{47}$$

where $\tan\varphi = vt/b$.

The cross section is then

$$\sigma_{nl\to n'l'}(v) = \int_0^{\infty} 2\pi b\bar{P}_{nl\to n'l'}(b,v)\,db \tag{48}$$

and it is easy to see that[11,35]

$$\sigma_{nl\rightarrow n'l'}(v) = (a/v)^2 A_{nl\rightarrow n'l'} \tag{49}$$

where the coefficient $A_{nl\rightarrow n'l'}$ involves only the hydrogen wave functions.
The l mixing is then defined as

$$\sigma_{l\,\mathrm{mix}} = \sum_{l'\neq l} \sigma_{nl\rightarrow nl'}(v) \tag{50}$$

Derouard and Lombardi[35] calculated the l mixing of nd states of Na by He and Ne by integrating numerically the equations of the first-order perturbation theory. They showed that this approach is valid only for large values of n ($n>10$). To apply the first-order perturbation theory for smaller values of n, Gersten[11] assumed two different ranges for the values of the impact parameter in order to calculate the transition probabilities. For $0\leqslant b\leqslant b_0$, where b_0 is some critical parameter, a strong perturbation is assumed, mixing statistically the various states $l'=l$. For $b>b_0$, the perturbation is assumed to be weak and the perturbation theory is used. The results of the first-order perturbation theory are essentially the same as those of the Born approximation, using also a Fermi pseudopotential for large values of n.

Model approach. In this section we shall describe the theoretical model proposed by de Prunelé and Pascale,[36] which directly uses information concerning the scattering (elastic or inelastic) of the Rydberg electron by the perturber in place of the explicit interaction e^-–B.

In this approach it is assumed that (1) the interactions B–A^+ and B–e^- are of very short range with respect to the average distance of e^- to A^+, so that these two interactions can be treated independently (binary approximation); (2) the collision A^+–B is neglected; and (3) the interaction B–e^- is very localized. A semiclassical approach is then used to calculate the total probability for the collision (elastic or inelastic) of A with B, where e^- is described by the unperturbed wave function $\psi_{nlm}(\mathbf{r})$ and B follows a straight-line trajectory.

Assumption (3) allows us to assume that, in an impact-parameter calculation of the scattering cross section for the e^-–B scattering cross section, where e^- is considered *free*, only small values of the impact parameter ρ contribute to the cross section. Then it is assumed that the transition probability $P(\rho, v)=1$ for $\rho<a$ and $P(\rho, v)=0$ for $\rho\geqslant a$, where the radius $a=(\sigma(v)/\pi)^{1/2}$ and $\sigma(v)$ is the "true" cross section for the scattering of e^- by B.

Now e^- is considered bound to A^+. To treat the collision between the Rydberg atom A and B, the perturber, B is represented by a sphere of radius $a=(\langle\sigma_e\rangle/\pi)^{1/2}$, where $\langle\sigma_e\rangle$ is the total cross section for the scattering of a "free" electron relative to B. The quasi-free behavior of the Rydberg electron with respect to B is thus expressed. Then, the probability for the collision of the Rydberg atom with B during a very small interval of time Δt is equal to the probability of presence of the Rydberg electron e^- in the near vicinity of the sphere representing B. To define this near vicinity, de Prunelé and Pascale[36] considered the limiting case where $\langle v_e \rangle_{nl}$ (the velocity of e^- averaged over the quantum-mechanical distribution of e^- relative to A^+) is much bigger than the relative velocity v of B relative to A^+. This condition is generally well verified over a large domain of n for thermal collisions between A and B.

Because $\langle v_e \rangle_{nl} \gg v$, B can be considered at rest with respect to e^- during a very small interval of time Δt. To calculate the collision probability when B moves along a trajectory $D(\mathbf{b})$, where b is the impact parameter, a partition of $D(\mathbf{b})$ from the instant t_i to the instant t_f is assumed, each point being occupied during Δt.

The probability that no collision has occurred between B and A during the interval of time $t_f - t_i$ is first calculated for a partition of $D(\mathbf{b})$ in N equidistant points:

$$P_{N1}(b,\Omega) = \exp\left(\sum_{k=1}^{N} \ln(1 - |\psi_{nlm}(\mathbf{R}_k)|^2)\Delta\tau \right) \tag{51}$$

where $\mathbf{R}_k$ is the position at point k, the elementary volume $\Delta\tau = 4\pi a_{nl}^2$ is defined by the two concentric spheres with origin B and radii a_{nl}, and $a_{nl} + \Delta a_{nl}$. Then $a_{nl} = (\langle \sigma_e \rangle_{nl}/\pi)^{1/2}$, where $\langle \sigma_e \rangle_{nl}$ is the total cross section for the e^-–B scattering averaged over the quantum-mechanical distribution of the Rydberg electron. The element of length is taken equal to $\Delta a_{nl} = \langle v_e \rangle_{nl} \Delta\tau$. The limits $N \to \infty$ and $|t_{i,f}| \to \infty$ are then considered to define the probability that no collision has occurred between B and A from $-\infty$ to $+\infty$:

$$P_{nlm}(b,\Omega) = \exp\left(-(1/v)\langle v_e \rangle_{nl} r \langle \sigma_e \rangle_{nl} \int_{-\infty}^{+\infty} |\psi_{nlm}(\mathbf{R})|^2 \, ds \right) \tag{52}$$

with $s^2 = R^2 - b^2$.

The probability for a collision (elastic + inelastic) is therefore

$$\mathbf{P}_{nlm}(v; b, \Omega) = 1 - P_{nlm}(b, \Omega) \tag{53}$$

and

$$\mathbf{P}_{nl}(v; b, \Omega) = \frac{1}{2l+1}\left(\frac{1}{4\pi}\right) \int \mathbf{P}_{nlm}(v; b, \Omega)\, d\Omega \tag{54}$$

Finally, the angular average is replaced by upper and lower limits and the total cross sections are defined as

$$Q_{nl}^{\text{up(low)}}(v) = 2\pi \int_0^\infty db \frac{b}{A}\left[1 - \exp\left(-\frac{4}{v}\langle v_e \rangle_{nl} \langle \sigma_e \rangle_{nl}\right) \frac{A}{4\pi} \int_{-\infty}^{+\infty} ds\, F_{nl}^2 (b^2 + s^2)^{1/2}) \right] \tag{55}$$

where F_{nl} is the radial wave function and $A = 1$ for the upper limit and $A = 2l+1$ for the lower limit. For large values of n, it can be shown that

$$Q_{nl}^{\text{up(low)}}(v) \to (4/v)\langle v_e \rangle_{nl} \langle \sigma_e \rangle_{nl} \tag{56}$$

To estimate the l mixing, we make a statistical approximation, which was proved to be reasonable for large values of n:

$$\sigma_{l\,\mathrm{mix}}^{\mathrm{up(low)}}(v) = \frac{N - (2l + 1)}{N} Q_{nl}^{\mathrm{up(low)}}(v) \tag{57}$$

where N is the number of coupled states that are involved during the collision. This statistical factor goes rapidly to one with increasing values of N. For the lowest values of n, Eq. (57) should overestimate the l-mixing cross section because in that case the elastic processes are probably underestimated.

It is worthwhile to mention that, in the limiting case where $v/\langle v_e \rangle_{nl} \gg 1$ (which can be reasonably obtained for very large values of n), the model approach predicts a total cross section for the collision of A with B equal to the total cross section for the scattering of e^- by B at velocity v. But in this case, the scattering of A^+ by B should be also considered.

Classical

When the interaction of a Rydberg atom with an atom or another Rydberg atom has been modeled as a three- or four-body problem, respectively, it is feasible to treat the dynamics by classical trajectory methods. For the four-body problem or for Rydberg atom collisions with ions, the advantage is that the Coulomb potentials between each particle are explicitly known. For these cases the classical trajectory Monte Carlo (CTMC) approach is conceptually appealing and has been quite successful.[17-19] Classical electron orbits are associated with the specified initial conditions of the problem, and the phase and orientation of each electron orbit are chosen randomly for each trajectory. The classical Hamiltonian equations of motion are solved numerically, and at the termination of the trajectory, the energy and angular momentum parameters of the final electron orbit are determined and related to quantum energy levels. The method has the advantage that all possible processes are included: inelastic collisions, ionization, or electron capture by the incident particle. In the case of collisions between two Rydberg atoms, excitation transfer has also been observed in some trajectories.

The CTMC approach is successful for two main reasons. First, the interactions between the particles are relatively strong and long range, and final states over a wide energy range tend to be populated. The "graininess" due to the discrete energy levels of the states is therefore not too important. Second, the classical treatment of Coulomb scattering agrees with the exact quantum-mechanical approach. These considerations suggest that the CTMC approach is well suited to Rydberg atom collisions with ions or other Rydberg atoms but less suited to collisions with rare gases. In the latter case only a small, discrete set of final states is populated. Also, even though a model potential could be used for the electron–rare gas interaction, arbitrary assumptions must be made to compensate for the fact that the classical approach to the electron scattering by the model potential would not agree with the quantum-mechanical result.

6.3. Calculated cross sections

6.3a. Atom–Rydberg atom

l Mixing

Experimental and theoretical results for orbital angular momentum mixing have been compiled for a wide variety of systems. The most commonly measured quantity is $\sigma_{l\,\mathrm{mix}}(n)$, which is defined as

$$\sigma_{l\,\mathrm{mix}}(n) = \sum_{l'=l+1}^{n-1} \sigma(nl \rightarrow nl') \tag{58}$$

Typical initial levels are *n*d for Na and *n*f for Rb and Xe. For these cases, the inelasticity ΔE of the $nl \rightarrow nl'$ transition is very small. The cross section $\sigma_{l\,\mathrm{mix}}(n)$ exhibits a maximum as a function of *n*, but the shape and position of the peak may vary considerably. For example, Rb(*n*f) + He has a sharp peak at $n=11$, whereas Xe(*n*f) + CO_2 has a very broad maximum for $n \simeq 28$. Calculations recently performed show how such varied behavior may be related to the energy-level splittings of the Rydberg atom, the energy dependence of the electron scattering from the collision partner, and the relative velocity of the collision.

For other initial levels, ΔE may be much larger. The theoretical methods described in this chapter have not yet been applied to such systems, but many experimental studies exist (e.g., Refs. 24, 37–41).

A characteristic feature of the data is a sharp rise in the cross section as a function of *n* for small values of *n*. This increase reflects a regime of strong coupling, and the cross section rises approximately as n^4, which is proportional to the geometrical size of the atom. As *n* increases further, the cross sections level off and then begin to decrease. A greater variation in the behavior of the cross section for different collision partners is found in this region. This variation is related to the behavior of the cross section for low-energy electron scattering from the collision partner.

Because the low-energy electron-scattering cross section for scattering by helium is a weak function of energy, many theoretical calculations have been made for systems involving helium. In this case, σ_e may be approximated by a constant. Figure 6.3 compares several different calculations[11,21,22,33–35] for the system Na(*n*d) + He with the data of Gallagher et al.[42–44] All the theories reproduce the trend of the data for $n \gtrsim 10$ reasonably well. This region is characterized by weak coupling. For $n \lesssim 10$, the coupling is much stronger and the method of Derouard and Lombardi,[35] which is based on first-order time-

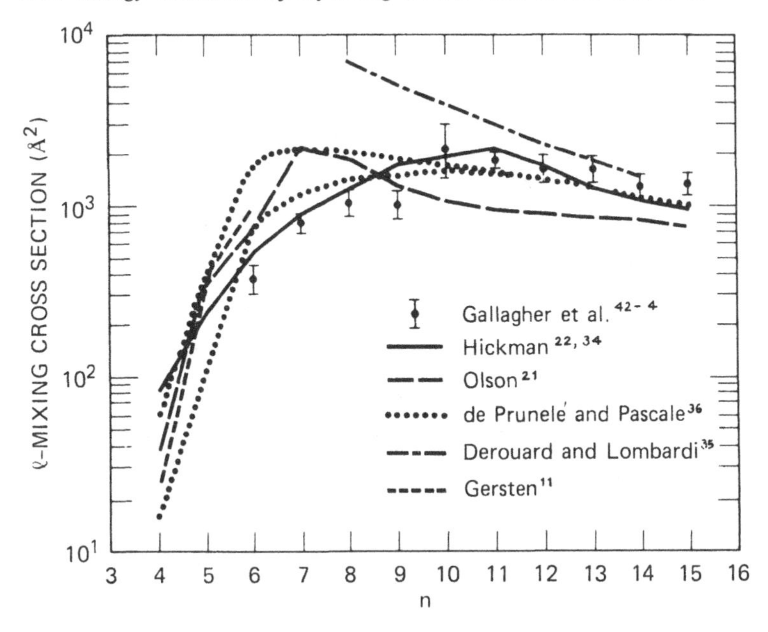

Fig. 6.3. Comparison of experimental and theoretical values of $\sigma_{l\,\mathrm{mix}}$ for Na(nd) + He with T=430 K.

dependent perturbation theory, is inapplicable. However, the approach of Gersten[11] is very similar and explicitly enforces a unitarity condition. The model approach of de Prunelé and Pascale[36] and the calculations of Olson[21] and Hickman[22, 34] are also suitable for the strong coupling region.

Omont[13] obtained an analytic result for the asymptotic form of the cross section

$$\sigma_{l\,\mathrm{mix}}(n) = 2\pi L^2/v^2n^3 \tag{59}$$

where $L = 1.19a_0$ and is the electron-helium scattering length. Derouard and Lombardi[35] modified this formula to account for the fact that l mixing to the s and p states does not occur in sodium. The modified formula is very close to the semiclassical calculations of Derouard and Lombardi.[35] Hickman[23] reported that the numerical results of the Born approximation for this system could be fit by the asymptotic form

$$\sigma_{l\,\mathrm{mix}}(n) = 4\pi AL^2/v^2n^{2.733} \tag{60}$$

where $A = 0.1788$. This result is very close to the n^{-3} behavior predicted by Omont.[13]

Other systems in which the cross section σ_e is not a strong function of energy include Rb(nf) + He and Na(nd) + N_2. Theoretical and experimental

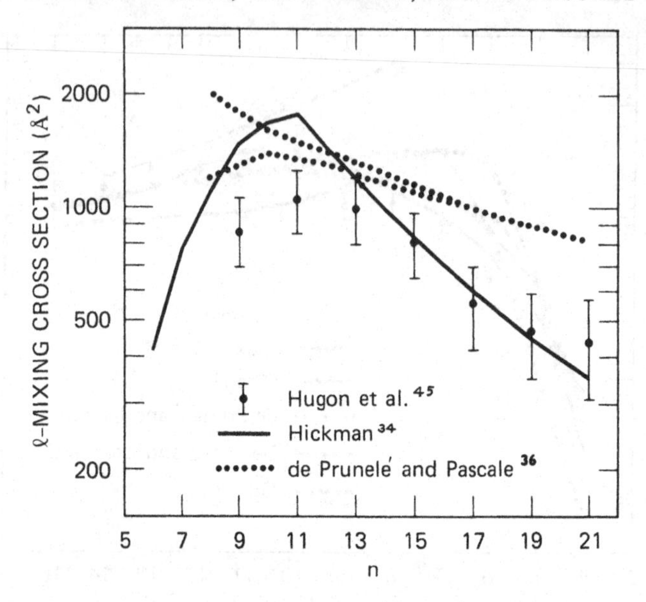

Fig. 6.4. Comparison of experimental and theoretical values of $\sigma_{l\,\mathrm{mix}}$ for Rb(nf)+He with $T = 520$ K.

results for these systems are shown in Figs. 6.4 and 6.5.[34, 36, 43–45] Reasonable accuracy is obtained even when the collision partner is a small molecule such as N_2. This result suggests that the internal structure of the molecule plays only a minor role. It is important to remember, however, that N_2 has no dipole moment. Cross sections for collisions of Rydberg atoms with polar molecules were calculated by Latimer[46] and Matsuzawa;[27, 29] Smith et al.[47] performed experimental studies.

The energy dependence of the cross section σ_e is treated differently in different theories, but the result is qualitatively the same in each case. The simplest approach is the scaling formula of Eq. (37). For a particular n, the effective electron-scattering cross section is σ_e evaluated at the average kinetic energy of the electron in its orbit. As n increases, the effective cross section will correspond to σ_e evaluated at successively lower energies. Another approach is to average the σ_e over the momentum wave function of the initial state.[36] This approach is probably more realistic.

An illustrative example is provided by collisions of Rydberg atoms with argon. The electron–argon elastic cross section shows a Ramsauer minimum at $E = 0.37$ eV. This energy corresponds approximately to $n = 8$. Hence, for larger n, the cross section evaluated for the average kinetic energy of the electron whose principal quantum number is n, $\sigma_e(\frac{1}{2}n^{-2})$, is a monotonically increasing function of n. The average of σ_e over the momentum wave function for the level n shows qualitatively the same behavior. In other words, as the

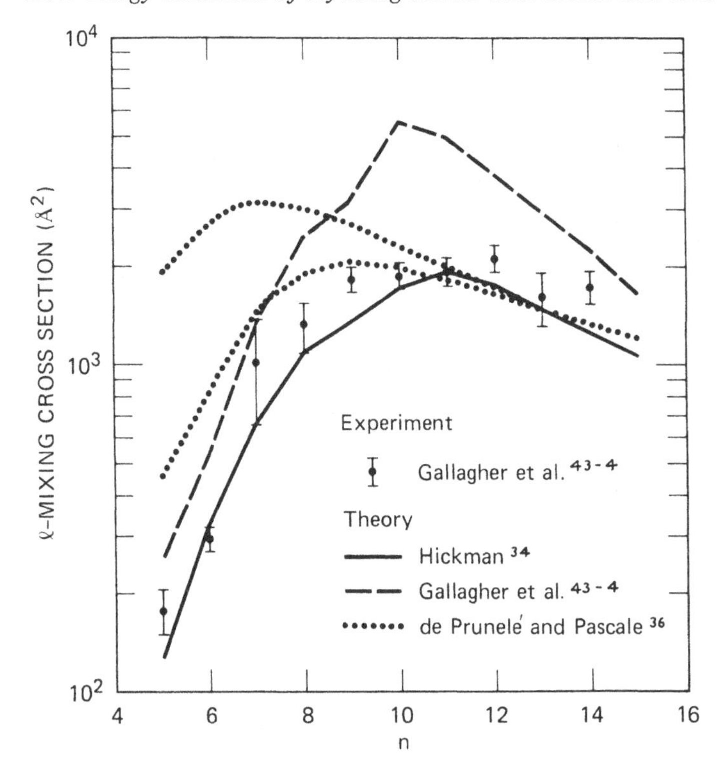

Fig. 6.5. Comparison of experimental and theoretical values of $\sigma_{l\,\mathrm{mix}}$ for Na(nd) + N_2 with $T = 430$ K.

quantum number n increases, the effective size of the perturber argon increases. This effect modifies the n dependence $\sigma_{l\,\mathrm{mix}}$ and causes $\sigma_{l\,\mathrm{mix}}$ to decrease more slowly with n, as shown by the results in Fig. 6.6 for Rb(nf) + Ar.[34, 36, 45]

The magnitude of σ_e determines the region of n where the transition from strong to weak coupling occurs. For example, σ_e for Xe is much larger than for He. As shown in Fig. 6.7,[34, 36, 45] $\sigma_{l\,\mathrm{mix}}(n)$ for Rb(nf) + Xe increases until $n \simeq 20$; the corresponding point for Na(nd) + He is $n \simeq 10$. The maximum $\sigma_{l\,\mathrm{mix}}(n)$ will occur near the smallest value of n for which a perturbation theory is valid. This value can be estimated using relations given by Omont[13] or in terms of the functions of reduced parameters given by Eqs. (37)–(39).

The cross sections for l mixing of Xe(nf) by CO_2 shown in Fig. 6.8 are also quite large because of the large e–CO_2 scattering cross section.[34, 48] In both this and the preceding case, an additional factor that tends to make $\sigma_{l\,\mathrm{mix}}$ larger is the low collision velocities for the relatively heavy particles. Physically, it is reasonable that the cross section should increase because a slower projectile

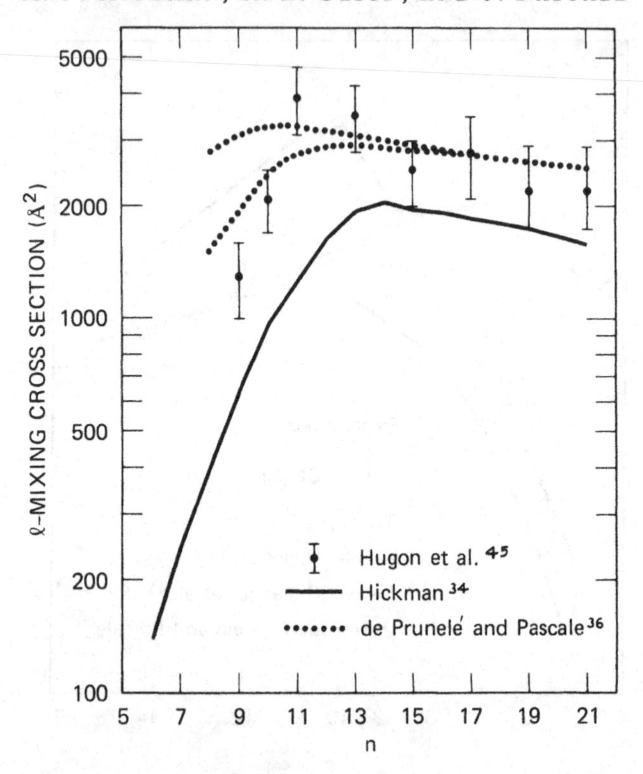

Fig. 6.6. Comparison of experimental and theoretical values of $\sigma_{l\,\mathrm{mix}}$ for Rb(nf) + Ar with $T = 520$ K.

will spend more time in the region of the Rydberg atom. Hickman[34] discussed this effect in terms of the scaling rule, Eq. (37), and showed that it was asymptotically valid. The result of Omont,[13] Eq. (59), showed the same effect.

The different behaviors of the cross sections with n for various perturbers were also explained in the model approach of de Prunelé and Pascale[36] by the differences in the averaged cross sections $\langle\sigma_e\rangle_{nl}$. The model approach finds that, for large values of n, the l-mixing cross sections should be proportional to a^2/nv, in contrast to the impulse and the Born approximations for a Fermi pseudopotential, which predict a behavior of a^2/n^3v^2. Note that a formula similar to Eq. (56) was also derived by Alekseev and Sobelman,[9] Omont,[13] and Matsuzawa[27-28] and may also be derived from the binary encounter theory of Flannery.[36]

In considering the different asymptotic scaling laws predicted by different theories, we must remember that the total cross section may include processes such as ionization or $n \to n'$ transitions not included in $\sigma_{l\,\mathrm{mix}}$. Further experiments are desirable to check the behavior of the cross sections for large values of n.

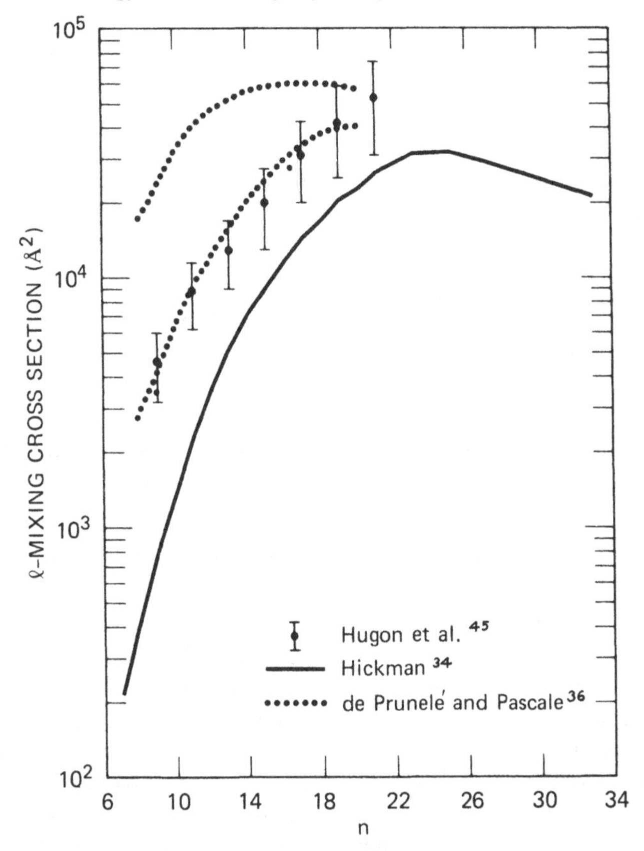

Fig. 6.7. Comparison of experimental and theoretical values of $\sigma_{l\,\mathrm{mix}}$ for Rb(nf) + Xe with $T = 520$ K.

It should be noted that the preceding discussion is based on the assumption that the e^-–B interaction is the most important mechanism for the l-mixing process.

Associative ionization

An interesting process that can arise in a thermal-energy collision of a high-lying Rydberg atom with its parent ground-state atom is molecular ionization. The associative ionization reaction

$$A^{**} + A \rightarrow A_2^+ + e^- \tag{61}$$

has been termed in the literature the Hornbeck–Molnar process.[49] Another molecular ionization reaction,

$$A^{**} + A \rightarrow A^+ + A + e^- \tag{62}$$

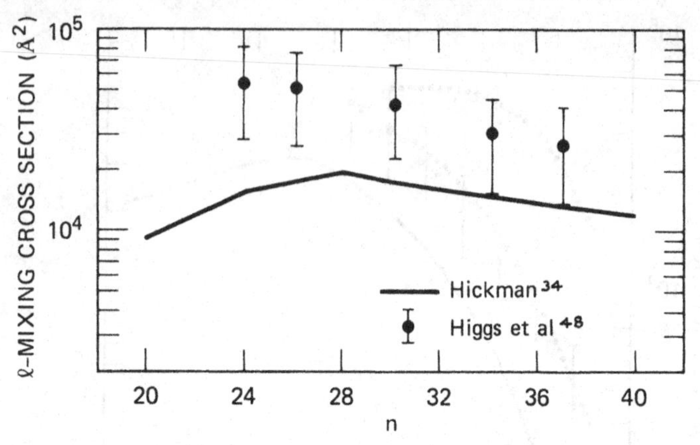

Fig. 6.8. Comparison of experimental and theoretical values of $\sigma_{l\,\mathrm{mix}}$ for Xe(nf)+CO_2 with $T = 300$ K.

is possible if the sum of the excitation energy of A** and the translational energy exceeds the ionization potential of atom A.

A simplified understanding of the collision mechanism for reactions (61) and (62) can be undertaken with the help of Fig. 6.9. Because the reaction is symmetric, several gerade and ungerade molecular potentials emanate from the separated atom limits. One or more may be attractive and be molecular Rydberg states of the A_2^+ ion. However, several will be constant or replusive and lead to curve crossings into the continuum for autoionization.

Related collision work indicates that there is almost unit probability of ionization for the collision trajectories that lead into the continuum. The difficulty is then to estimate the number of accessible molecular states that cross into the continuum. One attempt at this problem was made by Liu and Olson[50] in their analysis of ionization observed in Ca**+Ca collisions.[51] Molecular configurations of the excited states were deduced and, because of the energetics, only s-wave electrons were allowed to be ejected.

Cross sections were then estimated using

$$\sigma_{\mathrm{ion}} = \pi P R_*^2 \tag{63}$$

where P is the probability that the particles cross into the continuum and eject an s-wave electron and the crossing distance is denoted by R_*. For high-lying Rydberg atoms, their ionization potential is essentially hydrogenic and equal to $0.5/n^2$ in atomic units. Because the long-range form of the A_2^+ interaction potential is determined by the point-charge-induced dipole interaction, an estimate of R_* using Eq. (63) can be obtained by solving

$$\alpha_{\mathrm{d}}/2R_*^4 = 0.5/n^2 \tag{64}$$

where α_{d} is the dipole polarizability of atom A. Combining Eqs. (63) and (64) gives the following formula for the ionization cross section:

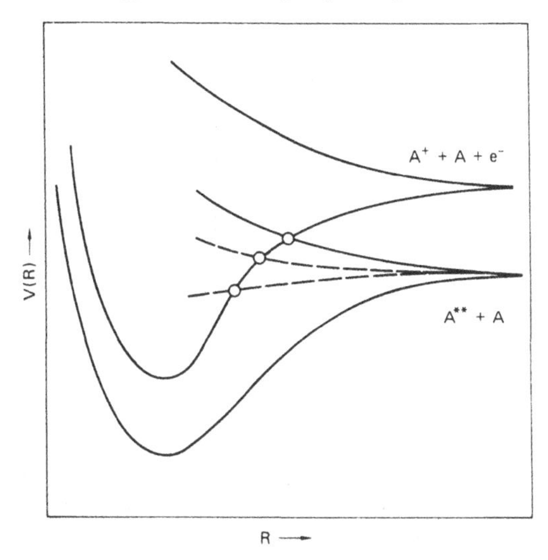

Fig. 6.9. Schematic of the potential energy curves followed during the collision of a high-lying Rydberg atom with its parent ground-state atom to form $A_2^+ + e^-$ or $A^+ + A + e^-$.

$$\sigma_{ion} = \pi P n \alpha_d^{1/2} \qquad (65)$$

Hence, σ_{ion} is expected to increase linearly with n and depend on the square root of the dipole polarizability of ground-state A.

Because the ionization reaction is determined by a relatively short-range interaction (when compared with the dimensions of the Rydberg atom), the cross sections are not exceedingly large. For a Rydberg atom in the $n=20$ level, representative cross sections for reactions (61) and (62) are on the order of 10^{-15} to 10^{-14} cm^2, leading to reaction rates in the 10^{-10}–10^{-9}-cm^3 s^{-1} region.

An interesting application of the molecular ionization reaction is to determine a limit on the dissociation energy of A_2^+. Experimentally, molecular ions will be observed only when the ionization potential of the Rydberg atom minus its translational energy is less than or equal to the D_e of the molecular ion. Indeed, Solarz et al.[52] used this method to determine D_e for Ca_2^+, Sr_2^+, and Ba_2^+, which are experimentally observed to lie between ~0.7 and 1.1 eV.

Negative ion formation

For the collision of a Rydberg atom with a ground-state atom that can support an extra electron to form a stable negative ion, it is possible to form a positive–negative ion pair via the reaction

$$A^{**} + B \rightarrow A^+ + B^- \qquad (66)$$

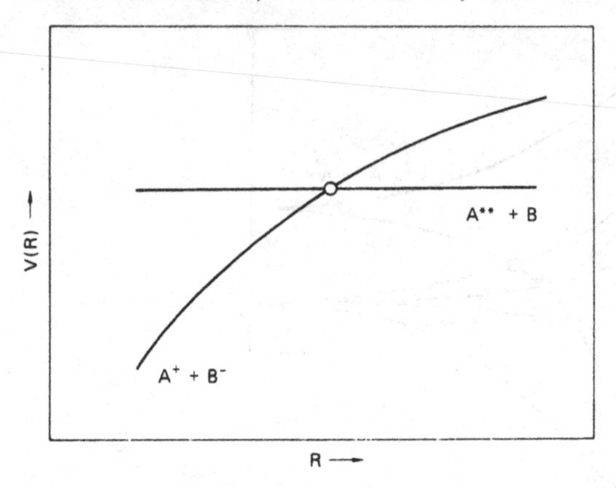

Fig. 6.10. Schematic of potential energy curves displaying the interaction between the covalent potential of a Rydberg atom and a ground-state atom and the ion-pair Coulomb potential leading to negative and positive ions.

The reaction is endothermic (see Fig. 6.10) and thus requires a collision energy of at least 1 eV for the process to proceed.

Considerable experimental and theoretical work has been carried out for the reverse reaction to (66): ion–ion mutual neutralization.[53] From these studies, it is clear that to expect an appreciable cross section for ion-pair formation the Rydberg levels must lie approximately 0.5–2.0 eV below the asymptotic limit of the ion-pair Coulomb curve. Because experiments using accelerated beams of Rydberg atoms will be extremely difficult, it may be years before negative ion formation is observed for collisions of Rydberg atoms on ground-state atoms.

Although ion-pair formation may not be readily observable, the effects of the attractive Coulomb curve will give rise to enhanced mixing between low-lying Rydberg levels. Such effects will be especially noticeable for collisions of Rydberg atoms with ground-state alkali atoms. Because the alkalis have electron affinities of ~0.5 to 0.6 eV, collisions between them and a Rydberg atom in a $n \leq 6$ level should give an observable increase in the "l-mixing" and "n-mixing" cross sections. This effect has been seen by Gallagher et al.[43-44] in collisions of Na Rydberg atoms with N_2.

6.3b. Ion–Rydberg atom

For the ion–Rydberg atom collisions, we mainly consider the intermediate velocity regime where the collision velocity is approximately equal to the orbital velocity of the Rydberg atom's outer electron ($v_e = 1$ a.u./n = 2.19×10^8 cm s$^{-1}/n$). Percival and Richards[54] presented an excellent review of this topic, which emphasizes the use of classical treatments in determining the collisional cross sections; in addition, Banks et al.[55] published cross sections for the $H^+ + H^{**}(n)$ reaction. We shall continue along these lines with a short review of recent numerical calculations and related experimental observations.

Our emphasis will be on the use of the classical trajectory Monte Carlo (CTMC) method to determine the cross sections for the electron capture reaction

$$A^{+q} + B^{**}(nl) \rightarrow A^{+q-1}(n'l') + B^+ \tag{67}$$

and the ionization reaction

$$A^{+q} + B^{**} \rightarrow A^{+q} + B^+ + e^- \tag{68}$$

In reactions (67) and (68), A^{+q} may be a single- or multicharged positive ion in charge state q. The CTMC method will be quite reliable for these processes because the collisions take place well in the regime where the classical correspondence principle is valid.

Electron capture and direct ionization

The CTMC method is based on solving Hamilton's equations of motion for a three-body, three-dimensional system,[56] which includes the incident ion, the target nucleus, and an electron initially bound to the target nucleus.[57] Six random numbers are used to generate the distribution of initial conditions. For each distribution, the classical trajectories of the nuclei are calculated from a large internuclear separation to the distance of the closest approach and out again to a large internuclear separation. The Coulomb forces among all three bodies are included in the calculation. If, at the end of an individual trajectory, the electron is still bound to the target nucleus, it is cataloged as no reaction. However, if the electron is found to be bound to the projectile ion, the reaction is cataloged as electron capture, and if the electron is bound to neither nucleus, it is cataloged as ionization. The cross sections for the various processes are then directly compared with the ratio of successful tries for that process to the number of trajectories calculated.

The CTMC method includes the Coulomb forces among all three bodies, the electron and two point charges for the nuclei, for which the angular scattering between two point charges is the same in both classical and quantum-mechanical frameworks. A very important consideration is the classical description of the reactant Rydberg atom's electron position and momentum distributions. However, Abrines and Percival[57] showed that it is possible to use Kepler's equation of planetary motion to represent hydrogenic atoms with a randomly determined set of initial conditions that are constrained to yield the binding energy of the atom; a microcanonical set of classical descriptions of the hydrogenic particle yields the same momentum distribution for the electron equally populated in n^2 of its corresponding (l, m) states as is found

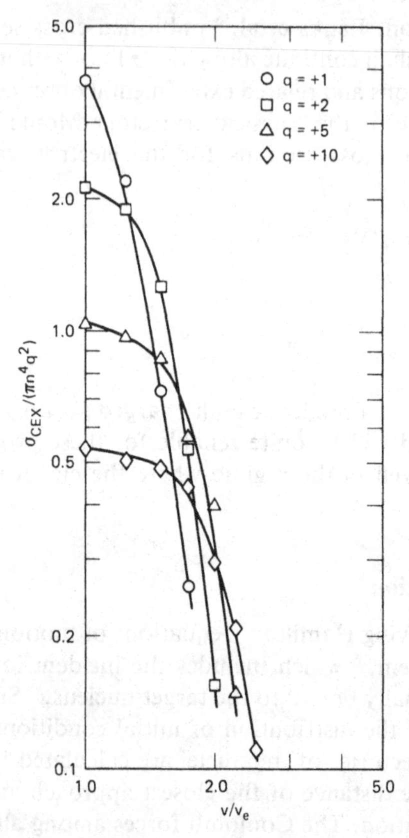

Fig. 6.11. Scaled electron capture cross sections for collisions of incident ions in charge states $q = +1, +2, +5$, and $+10$ with hydrogenic atoms in arbitrary principal quantum level n. The heavy particle relative velocity has been scaled by the orbital velocity of the hydrogenic electron v_e $(= 1/n$ a.u.). Statistical errors on the displayed points are less than ±10%.

quantum mechanically. Hence, because of these two important comparisons between classical and quantum-mechanical descriptions, it is expected that the CTMC method will yield a very realistic description of ion–Rydberg atom scattering. Supporting this conclusion is the fact that the CTMC method has been shown to yield accurate cross sections for the multiply charged ion, ground-state atomic hydrogen systems.[58] However, deficiencies of the CTMC method are that it does not account for quantum tunneling or interference effects and it is difficult to apply to a process that is a small component of the sum of the many scattering events.

From the classical correspondence principle, it can be predicted that the scattering cross sections for reactions (67) and (68) at a given charge state q will scale with the geometric size of the Rydberg atom, $\pi n^4 a_0^2$. Likewise, the heavy particle collision velocity will scale with the orbital velocity v_e of the

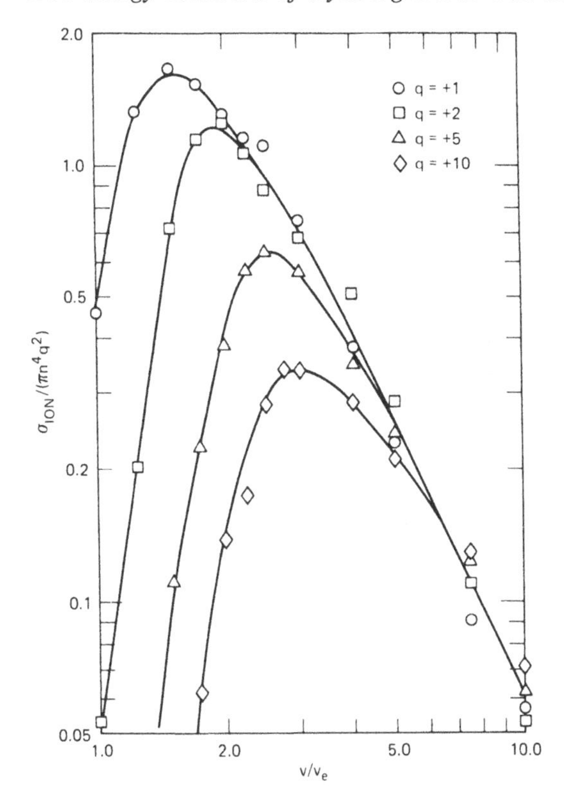

Fig. 6.12. Ionization cross sections with the same notation as in Fig. 6.11.

Rydberg electron. CTMC calculations were performed for Rydberg atoms in various n levels where the n^2 sublevels of the (l, n) states were equally populated.[18, 19]

The results of these cross section calculations for electron capture [Eq. (67)], ionization [Eq. (68)], and the sum of these two cross sections are shown in Figs. 6.11, 6.12, and 6.13, respectively. As expected, electron capture dominates at low v/v_e and then decreases precipitously for $v/v_e \gtrsim 1.5$. However, as the electron capture cross sections decrease, the ionization rises rapidly to a maximum at $v/v_e \simeq 2.0$, so that the overall electron removal cross section is smooth and monotonic.

Several scaling dependences arise from the calculated cross sections. At high velocities, $v/v_e \gtrsim 5$, the ionization cross sections approach the functional form

$$\sigma_{\text{ion}} = 6\pi n^2 q^2 a_0^2 / v^2 \tag{69}$$

where the heavy particle velocity is expressed in atomic units (1 a.u. of velocity $\equiv 2.19 \times 10^8$ cm s^{-1}). Because these are classical results, tunneling is

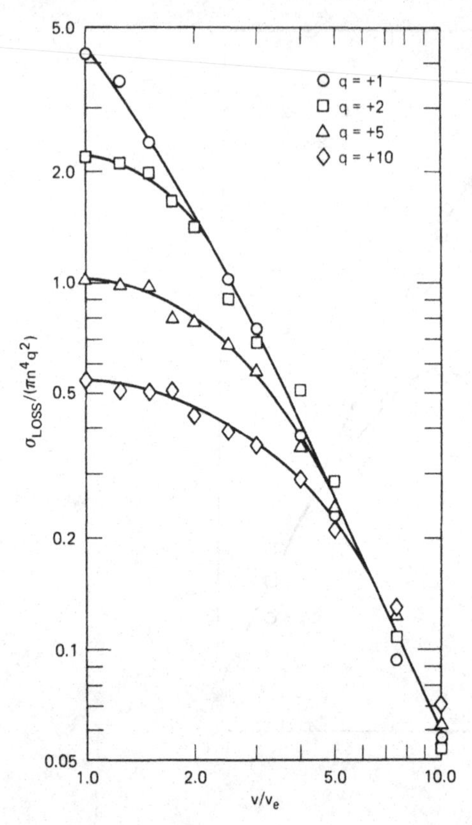

Fig. 6.13. Sum of the electron capture and ionization cross sections with the same notation as in Fig. 6.11.

neglected so that at high v/v_e the cross sections are underestimated; at $v/v_e \simeq 10$, Eq. (69) leads to values approximately 30% below the first Born results.[59]

At the lowest velocities, $v/v_e \simeq 1$, the electron capture cross sections approach the limit

$$\sigma_{CEX} = 5.5\pi n^4 q a_0^2 \tag{70}$$

The cross sections are then proportional to the geometric size of the Rydberg atom and are linearly proportional to the charge state of the incident ion and independent of the collision velocity. The functional form presented in Eq. (70) can be rationalized using the simple classical arguments of Bohr and Lindhard.[60] At low velocities, the collision time becomes comparable to the time for the target electron to orbit about the nucleus and the cross section is determined from the internuclear separation, where the force on the electron from the incident projectile A^{+q} is equal to that from the singly charged B nucleus. It is then assumed the electron will be removed from the target atom for all impact parameters smaller than this separation. Within this simple approximation, it can be shown that $\sigma_{CEX} \simeq 3\pi n^4 q a_0^2$, which has a functional form identical to Eq. (70).

It is possible to compare the sum of the electron capture and ionization cross sections presented in Fig. 6.13 with the experimental data of Koch and Bayfield[61] for the $H^+ + H$ ($44 \leqslant n \leqslant 50$) collision system. In the velocity range $1.35 \leqslant v/v_e \leqslant 3.31$, the observed cross sections had magnitudes of $\sim 4 \times 10^{-9}$ cm^2 to $\sim 1 \times 10^{-9}$ cm^2. Using Fig. 6.13 and a nominal n value of 47, we find the experimental values are almost exactly a factor of 3.65 larger than our theoretical calculations throughout the velocity regime. The quoted uncertainty in the absolute values of the experimental cross sections is a factor of two, with almost negligible error in the relative cross sections. Hence, we would conclude that the theoretical calculations are in excellent agreement with the velocity dependence of the observed cross sections, but there still is a discrepancy in the absolute magnitudes of the cross sections.

Another test of the theoretical cross sections can be made using the data of Burniaux et al.,[62] who investigated the electron capture reaction

$$He^{2+} + H^{**}(n) \rightarrow He^+ + H^+ \tag{71}$$

for n ranging from 8 to 24 and relative energies from 0.25 to 478 eV. Their data confirm the rapid decrease in the electron capture cross sections for $v/v_e \gtrsim 1.5$ (see Fig. 6.11). Because these authors measured absolute cross sections, it is also possible to compare the experimental results with the low-velocity theoretical predictions given by Eq. (70). The agreement is satisfactory and within the experimental error limits of $\pm 23\%$.

Recently, Kim and Meyer[63] measured the cross sections for electron removal from hydrogenic Rydberg atoms by collisions with N^{3+}:

$$N^{3+} + H^{**}(n) \rightarrow \begin{cases} N^{3+} + H^+ + e^- \\ N^{2+} + H^+ \end{cases} \tag{72}$$

The experiments were conducted at a velocity of 2.8×10^8 cm s^{-1} for n values ranging from 9 to 24. For these collisions, v/v_e varies from 11.5 to 30.7, so that it is accurate to predict that the electron removal process will be dominated by ionization. Contrary to the n^2 scaling dependence predicted by classical [Eq. (69)] and the first Born results, these authors found that the cross sections scale as $n^{3.12}$.

Such results are rather surprising and prompt a closer look at the experimental setup. Experimentally, it is necessary to deflect the H^+ formed via reaction (72) into an analyzer for counting. However, the deflector voltages used, 1–1.5 kV cm^{-1}, are such that they will also Stark-ionize any excited H^{**} formed during the collision via

$$N^{3+} + H^{**}(n) \rightarrow N^{3+} + H^{**}(n') \tag{73}$$

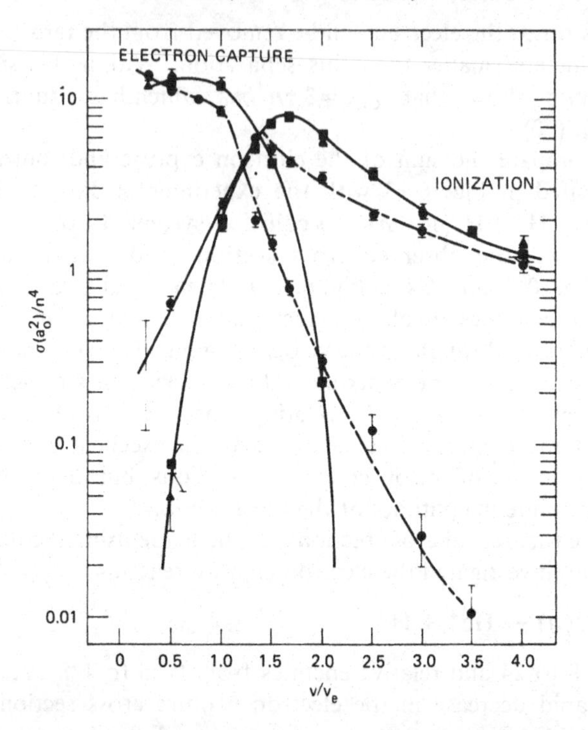

Fig. 6.14. CTMC ionization and electron capture cross sections, with statistical standard deviations, for specific initial l vs. reduced velocity. Circles: $2/15 \leqslant l_c/n \leqslant 3/15$; squares: $14/15 \leqslant l_c \leqslant 1$. (Data from MacKellar and Becker[59])

with final n' values greater than 28 (1.5 kV cm^{-1}). Thus, the true cross sections measured by Kim and Meyer were the sum of ionization and simple excitation to $n' \gtrsim 27$. Correspondingly, as the initial n state approaches 27 (as in the $n=24$ measurements), the observations primarily measure the excitation process, which has a geometric cross section $\sim n^4$, rather than the much smaller ionization cross section that scales as n^2.

We used the CTMC method to try to reproduce the Kim and Meyer results and found that the $n=9$ values were primarily due to the ionization process.[64] Moreover, the importance of the excitation process increased dramatically with n, until, for $n=24$, the excitation cross sections for $n'>27$ were more than an order of magnitude larger than the ionization cross sections. Including both reactions (72) and (73) to simulate the experiments, the classical method reproduced the $\sim n^3$ scaling observed by Kim and Meyer.

Throughout this section, we have made comparisons with theoretical calculations performed for a given n level where the (l, m) sublevels were distributed statistically among their n^2 possible values. However, it is also possible to "bin" the l initial values in CTMC calculations by making use of the relationship to the eccentricity of the electron's orbit. Such calculations were performed by MacKellar and Becker[59] for the scattering of singly charged particles from hydrogenic Rydberg atoms in the $n=15$ and $l=2$ and $l=14$ states. The results of the calculations are shown in Fig. 6.14. The circular electron orbits, $l=14$, and the highly eccentric, almost straight-line orbitals, $l=2$, give rise to very different behaviors, especially in the reduced velocity regime $v/v_e \lesssim 2$. At high velocities, the ionization cross sections for the two highly different l values merge, as seen in Fig. 6.14.

Final state distributions

There is considerable interest, both theoretical and experimental, in investigating the collisions of ions and Rydberg atoms in sufficient detail to determine product state distributions. In many instances, it is accurate to use classical methods to interpret and predict the scattering data.[65] Correspondingly, CTMC calculations are being used to determine quantitatively the cross sections and final state distributions.

Because of the small change in electronic energy that is required, "l-changing" and "n-changing" collision processes

$$A^{+q} + B^{**}(nl) \rightarrow A^{+q} + B^{**}(n'l') \tag{74}$$

give rise to extremely large cross sections. In fact, measurements by MacAdam et al.[66] show that the cross section for changing the electronic level of a Na(nd) atom ($20 \leqslant n \leqslant 34$) by a collision with He^+ at 450 to 600 eV is on the order of 10^{-8} cm^2. The cross sections are larger than geometric and are found to scale as $\sim n^5$.

Theoretical calculations have been directed to the "n-changing" reaction. It is possible to "quantize" the classical calculations by determining the electronic energy after the collision and using the hydrogenic relationship

$$E = (-q^2/2n_c^2) \tag{75}$$

The noninteger classical principal quantum number n_c is then defined[59] as belonging to electronic level n if it lies between the values

$$[(n-1)(n-\tfrac{1}{2})n]^{1/3} \leqslant n_c \leqslant [n(n+\tfrac{1}{2})(n+1)]^{1/3} \tag{76}$$

For large n, this interval is approximately $n-\frac{1}{2} \leqslant n_c \leqslant n+\frac{1}{2}$. Similarly, the classical orbital angular momentum quantum number l_c is "quantized" if it lies within the limits $l \leqslant l_c \leqslant l+1$.

Cross sections for the "n-changing" process have been calculated for the collision of an ion in charge state +1 with a Rydberg atom in state $n=10$. The cross sections are shown in Fig. 6.15 for a velocity of $v=2v_e$ and compared with the analytical result of Lodge et al.[67] As expected, the cross sections are large and exceed the geometric value $\pi n^4 a_0^2 = 8.8 \times 10^{-13}$ cm^2 for small changes in n. The CTMC results are generally within 20% of the analytical

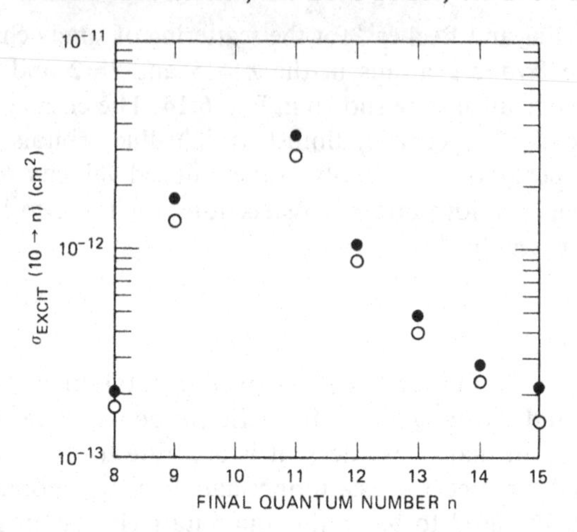

Fig. 6.15. "n-changing" cross sections for the collision of a $q=+1$ ion with a Rydberg atom in the $n=10$ state at $v=2v_e$. Open circles: the CTMC results; solid circles: computed from the analytical form given by Lodge et al.[67]

result of Lodge et al.[67] As predicted by Percival and Richards,[54] at higher velocities, the CTMC method leads to an underestimate of the "n-changing" cross sections because a large contribution arises from the classically inaccessible region.

Another interesting class of reactions involving product state distributions is the investigation of the electron capture process

$$A^{+q} + B(nl) \rightarrow A^{+q-1}(n'l') + B^{+} \quad (77)$$

for which CTMC calculations have been performed for the investigation of the product n levels populated after electron capture for collisions of singly and multiply charged ions with Rydberg atoms.[19] A representative sample of the type of information gained is displayed in Fig. 6.16.

Because these investigations were conducted for various charge states of the incident ion and for numerous initial Rydberg levels of the target atom, it is possible to determine some of the systematics of the electron capture–collision process. From the analysis of the product distributions, it is apparent that the electron on the Rydberg atom attempts to preserve two important physical quantities – its orbital size and initial electronic energy. If we denote n_i, n_f and z_i, z_f as the initial and final principal quantum numbers and nuclear charge states, respectively, of the Rydberg electron before and after collision, then to preserve orbital size, the following relationship holds:

$$n_f = (z_f/z_i)^{1/2} n_i \quad (78)$$

Likewise, to preserve electronic energy, we obtain

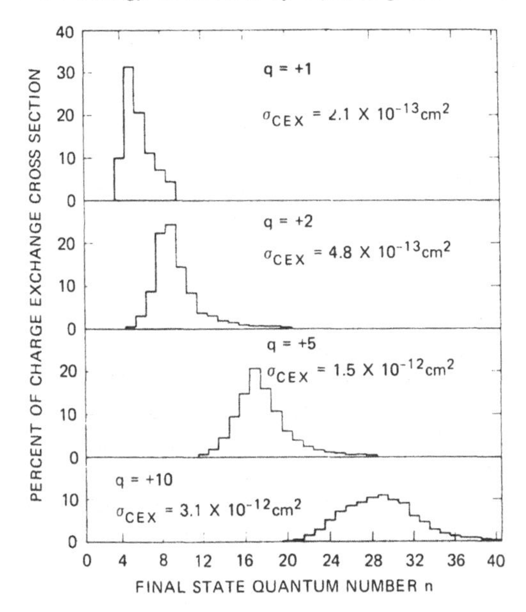

Fig. 6.16. Percentage of the charge-exchange cross section residing in principal quantum level n after collisions of $q=+1, +2, +5$, and $+10$ ions with a hydrogenic atom originally in the $n=5$ state. The collision velocity is $v=v_c$.

$$n_f = (z_f/z_i)n_i \tag{79}$$

For singly charged incident ions, both of these quantities can be satisfied simultaneously with the maximum in the electronic product distribution equal to the original level. However, for a multiply charged incident ion, it is impossible to satisfy both Eqs. (78) and (79). Nevertheless, the calculations of Olson[19] indicate that the maximum in the distributions corresponds to the compromise of both initial orbital size and energy, with the n_f maximum equal to the product of the square roots of the right-hand side of Eqs. (78) and (79); then

$$n_f = (z_f/z_i)^{3/4}n_i \tag{80}$$

Equation (80) has not been experimentally confirmed. However, an indirect test is available from the experiments of Burniaux et al.[61] Their experimental setup was such that in the electron-capture process

$$He^{2+} + H^{**}(n) \rightarrow He^{+}(n') + H^{+} \tag{81}$$

if the product Rydberg ions were formed in $n'>40$, they would be Stark ionized and not detected as electron-capture products. Measurable electron-capture cross sections were observed for initial $n \leqslant 24$, but for $n \geqslant 25$, the cross sections could not be detected. Such observations are consistent with the

CTMC results and Eq. (80), both of which predict that the maximum in the product n distribution will exceed $n'=40$ for initial n states, $n>25$.

Theoretical work must still be performed to determine both the n and l distributions after electron capture. The calculations are possible, but they are extremely time-consuming because of the difficulty in obtaining good statistics in a CTMC calculation when there are a large number of product channels.

6.3c. Rydberg atom–Rydberg atom

Interactions between two Rydberg atoms take place high in the continuum for autoionization where an almost infinite number of channels are available for ionization and excitation transfer. This complexity makes it difficult to use a quantum-mechanical method to describe the scattering such as is used in Penning ionization problems, but such interactions can readily be subjected to a classical analysis. Because Flannery, in Chapt. 11, discusses the applicability of semiquantal and quantum methods to Rydberg atom–Rydberg atom scattering, we shall limit ourselves to descriptions of the four-body CTMC method and results.

Ionization

In this section, we will be specifically concerned with collision processes that lead to positive ion formation

$$A^{**}(n) + A^{**}(n) \rightarrow A^{+} + \cdots \tag{82}$$

when both Rydberg atoms are in the same principal quantum level n. We expect that the cross sections will depend only slightly on the orbital angular momentum quantum number l when $l \ll n$.

Application of the three-dimensional, four-body CTMC method requires for each trajectory the solution of a set of 18 coupled first-order differential equations, which are Hamilton's equations of motion (for details, see Ref. 68). The CTMC method includes the Coulomb forces among all four bodies – the two electrons and two point charges for the nuclei – for which the angular scattering between two point charges is the same in both classical and quantum-mechanical frameworks.

Several authors[17-19,69,70] used the CTMC method to obtain results for reaction (82) for velocities $v \gtrsim v_e$. Because no data are available for Rydberg atom–Rydberg atom scattering, comparisons were made to the nonclassical H(1s)+H(1s) system for which data exist.[71,72] The tests were successful in reproducing the experimental data and showing the applicability of this classical method.

Of interest then is whether or not the CTMC method can provide a good estimate of the cross section for reaction (82) when the collision velocity is low and approaches that of thermal energy. The classical method should be especially applicable to Rydberg atom collisions because the product channels are

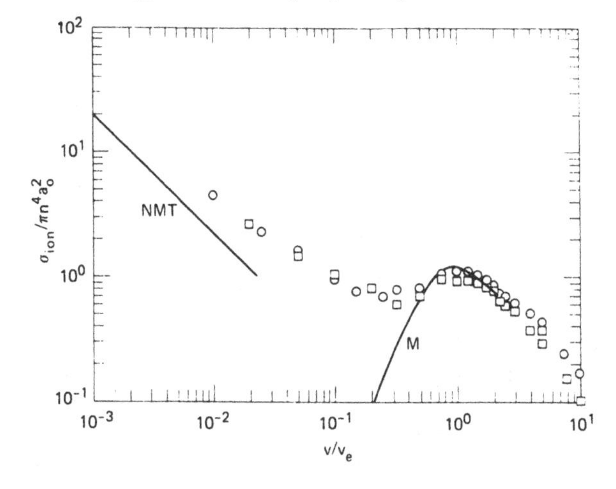

Fig. 6.17. Classical trajectory Monte Carlo Rydberg atom–Rydberg atom ionization cross sections for $n=10$ (squares) and $n=20$ (circles), scaled as $\sigma_{ion}/\pi n^4 a_0^2$ and v/v_e. The H(1s) + H(1s) high-velocity measurements of McClure[72] and the He* ($n=2$) + He* ($n=2$) of Neynaber et al.[73] are depicted by the lines, M and NMT, respectively.

almost continuous and molecular effects are of relatively short range. Calculations along these lines were performed by Olson[18] and are shown in Fig. 6.17.

Analysis of the trajectories leading to positive ion formation indicates that at high velocities, $v > v_e$, the mechanism is predominantly impact ionization (i.e., close collisions of the electrons and nuclei) with double ionization (both electrons ejected) being an important component at the highest velocities. At intermediate velocities, $v \simeq v_e$, the impact-ionization component is still important, but classical exchange or capture to form a transient negative ion must also be taken into account. The classical exchange maximizes when the collision velocity is comparable to the orbital velocities of the electrons, $v \simeq v_e$, and gives rise to the slight maximum in the cross section. At low velocities, $v \ll v_e$, which reach into the thermal-energy regime, another collision mechanism dominates. This mechanism is electronic deexcitation of one Rydberg atom with the simultaneous transfer of the electronic energy to the ionization of the electron on the second Rydberg atom. In essence, the heavy nuclei are spectators to the ionization process. This long-range electron–electron interaction does not change the momenta of the heavy nuclei and results in "n-changing" cross sections for changing n to $n' \leqslant n/\sqrt{2}$ being very nearly equal to the ionization cross section. Clearly, the low-velocity classical cross sections presented are expected to be valid only when there is almost a continuum of electronic states available in the $n/\sqrt{2}$ region.

Figure 6.17 compares the CTMC cross section results with experimentally derived values. The H(1s) + H(1s) ionization cross sections of McClure[72] are shown at high velocities. Good agreement for $v \gtrsim 0.5 v_e$ is obtained because the ionization mechanisms (impact and exchange) are well described by a classical model. However, at lower velocities, the CTMC method greatly overestimates the cross section. The reason for the difference is that the H(1s)+(H(1s) system is extremely nonclassical and the electron–electron simultaneous de-excitation and ionization mechanism predicted by the classical model is not allowed because both electrons are already in their ground quantum states.

At low velocities, the calculations are compared with the He* $(n=2)+$ He* $(n=2)$ results of Neynaber et al.[73] Although this system is not in a high-Rydberg state, it does possess some of their characteristics; that is, a lower quantum level is available for the electron–electron ionization mechanism. To plot the He* data in Fig. 6.17, a hydrogenic model was used and scaled to the He* ionization energy to obtain an effective quantum number $n=(2I)^{1/2}=1.71$; the ionization potential I chosen for the He* system corresponds to the 87% He($2\,^3$S) and 13% He($2\,^1$S) beam composition reported by Neynaber et al.[73] The experimental data are depicted by line NMT and display almost the same velocity dependence as that predicted by the classical model. As expected, the magnitude of the experimental cross sections is less than the CTMC results because a true continuum of product electronic levels is not available for the electron–electron ionization mechanisms.

An interesting comparison, moreover, is in the apparent agreement between theory and experiment in the low-velocity $\sim v^{-0.65}$ dependence on the cross sections. Naively, we would expect the classical cross sections to be proportional to the time of the collision and behave as v^{-1}. However, the calculated velocity dependence is less, indicating that the particles probably avoid the strong electron–electron interactions necessary for ionization by inducing almost adiabatic changes at large internuclear separations during each collision event. This latter process gives rise to large "n-changing" cross sections for small changes in n.

It is also of interest to note that according to the Langevin criterion,[74] a van der Waals interaction will yield an $E^{-1/3}$ or $v^{-2/3}$ cross section dependence for thermal-energy collisions, almost identical to what is obtained in the exact classical calculations, which assume only Coulomb interactions.

Thus, it appears that an initial step has been made toward understanding Rydberg atom–Rydberg atom interactions. The ionization cross sections have been calculated over almost six orders of magnitude in collision energy and yield values that are accurate to better than a factor of two. Hence, the four-body CTMC method appears to be especially well adapted to this problem and gives a physical insight into the collision mechanisms.

Excitation transfer

Excitation transfer in Rydberg atom–Rydberg atom collisions is expected to be a major process with extremely large cross sections.[75] Unfortunately, at low velocities, there are no data or calculations with which to make quantitative predictions.

Only preliminary values are available from CTMC calculations for two like Rydberg atoms; these values place the "n-changing" cross sections approximately an order of magnitude larger than the ionization values at thermal energies. This leads to "n-changing" values over two orders of magnitude larger than geometric. Because "l-changing" collisions require an even softer interaction, these cross sections can easily reach magnitudes that are three to four orders of magnitude larger than geometric. Definitely more research into these collision problems is warranted.

6.4. Concluding remarks

Studies on a variety of systems involving Rydberg atoms and neutral targets have provided analyses of the relationship between the scattering and the energy-level splittings of the Rydberg atom, the energy dependence of the electron scattering from the collision partner, and the relative velocity of the collision. At the present level of understanding, several theoretical approaches provide satisfactory explanations of some of the available data. However, the theories differ in their predictions of the n and v dependence of the cross sections for n beyond the range of the present measurements. Refined experiments will be necessary for further progress.

For collisions involving Rydberg atoms and ions, the classical trajectory Monte Carlo approach has been shown to be quantitatively reliable and also of tremendous value for providing physical insight into the mechanisms involved.

Acknowledgments

This work was supported by the Office of Naval Research under Contract N00014-79-C-0182. The chapter was prepared while one of us (J. Pascale) was a visiting scientist at SRI International, on leave of absence for one year from C.E.N. Saclay. J. Pascale thanks the Molecular Physics Laboratory at SRI International for its kind hospitality and is very grateful to Dr. E. de Prunelé for fruitful discussions.

References and notes

1. V. A. Smirnov, *Opt. Spektrosk. 37,* 407 (1974) [*Opt. Spectrosc. 37,* 231 (1974)].
2. M. R. Flannery, *J. Phys. B 13,* L657 (1980).
3. M. R. Flannery, *Phys. Rev. 22,* 2408 (1980).
4. J. S. Cohen, *Phys. Rev. A 13,* 86 (1976).
5. J. N. Bardsley, *Case Stud. At. Phys. 4,* 299 (1974).
6. J. Pascale and J. Vandeplanque, *J. Chem. Phys. 68,* 2278 (1974).
7. M. Philippe, F. Masnou-Seeuws, and P. Valiron, *J. Phys. B 12,* 2493 (1979).

8. E. Fermi, *Nuovo Cimento 11,* 157 (1934).
9. V. A. Alekseev and I. I. Sobel'man, *Sov. Phys. JETP 22,* 882 (1966).
10. G. Breit, *Phys. Rev. 71,* 215 (1947).
11. J. I. Gersten, *Phys. Rev. A 14,* 1354 (1976).
12. G. K. Ivanov, *Opt. Spektrosk. 40,* 965 (1976) [*Opt. Spectrosc. 40,* 554 (1976)].
13. A. Omont, *J. Phys. (Paris) 38,* 1343 (1977).
14. B. M. Smirnov, *Sov. Phys. JETP 24,* 314 (1967).
15. E. Roueff, *Astron. Astrophys. 7,* 4 (1970).
16. R. K. Janev, *J. Phys. B 4,* 215 (1971).
17. R. E. Olson, *J. Phys. B 12,* L109 (1979).
18. R. E. Olson, *Phys. Rev. Lett. 43,* 126 (1980).
19. R. E. Olson, *J. Phys. B 13,* 483 (1980).
20. A. M. Arthurs and A. Dalgarno, *Proc. R. Soc. London 256,* 540 (1960).
21. R. E. Olson, *Phys. Rev. A 15,* 631 (1977).
22. A. P. Hickman, *Phys. Rev. A 18,* 1339 (1978).
23. A. P. Hickman, *Phys. Rev. A 19,* 994 (1979).
24. M. Hugon, F. Gounand, P. R. Fournier, and J. Berlande, *J. Phys. B 13,* 1585 (1980).
25. R. G. Newton, *Scattering Theory of Waves and Particles* (New York: McGraw-Hill, 1966), pp. 587–91.
26. M. Matsuzawa, *J. Chem. Phys. 55,* 2685 (1971); *Erratum 58,* 2674 (1973).
27. M. Matsuzawa, *Phys. Rev. A 18,* 1396 (1978).
28. M. Matsuzawa, *J. Phys. B 12,* 3743 (1979).
29. M. Matsuzawa, *Phys. Rev. A 20,* 860 (1979).
30. T. F. O'Malley, *Phys. Rev. 130,* 1020 (1963).
31. E. de Prunelé, Unpublished thesis, University of Paris, 1979.
32. M. R. Flannery, *Ann. Phys. (N.Y.) 61,* 465 (1970).
33. M. R. Flannery, *Ann. Phys. (N.Y.) 79,* 480 (1973).
34. A. P. Hickman, *Phys. Rev. A 23,* 87 (1981).
35. J. Derouard and M. Lombardi, *J. Phys. B 11,* 3875 (1978).
36. E. de Prunelé and J. Pascale, *J. Phys. B 12,* 2511 (1979).
37. R. S. Freund, T. A. Miller, B. R. Zegarski, B. Jost, M. Lombardi, and A. Dorelon, *Chem. Phys. Lett. 51,* 18 (1977).;
38. F. Gounand, P. R. Fournier, and J. Berlande, *Phys. Rev. A 15,* 2212 (1977).
39. T. F. Gallagher, W. E. Cooke, and S. A. Edelstein, *Phys. Rev. A 17,* 125 (1978).
40. T. F. Gallagher, W. E. Cooke, and S. A. Edelstein, *Phys. Rev. A 17,* 904 (1978).
41. T. F. Gallagher, and W. E. Cooke, *Phys. Rev. A 19,* 2161 (1979).
42. T. F. Gallagher, S. A. Edelstein, and R. M. Hill, *Phys. Rev. Lett. 35,* 644 (1975).
43. T. F. Gallagher, S. A. Edelstein, and R. M. Hill, *Phys. Rev. A 15,* 1945 (1977).
44. T. F. Gallagher, R. E. Olson, W. E. Cooke, S. A. Edelstein, and R. M. Hill, *Phys. Rev. A 16,* 441 (1977).
45. M. Hugon, R. Gounand, P. R. Fournier, and J. Berlande, *J. Phys. B 12,* 2707 (1979).
46. C. J. Latimer, *J. Phys. B 10,* 1889 (1977).
47. K. A. Smith, F. G. Kellert, R. D. Rundel, F. B. Dunning, and R. F. Stebbings, *Phys. Rev. Lett. 40,* 1362 (1978); F. G. Kellert, K. A. Smith, R. D. Rundel, F. B. Dunning, and R. F. Stebbings, *J. Chem. Phys. 72,* 3179 (1980).
48. C. Higgs, K. A. Smith, F. B. Dunning, and R. F. Stebbings, *J. Chem. Phys. 75,* 745 (1981).
49. J. A. Hornbeck and J. P. Molnar, *Phys. Rev. 84,* 621 (1951).

50. B. Liu and R. E. Olson, *Phys. Rev. A 18,* 2498 (1978).
51. J. A. Armstrong, P. Esherick, and J. J. Wynne, *Phys. Rev. A 15,* 180 (1977).
52. R. W. Solarz, E. F. Worden, and J. A. Paisner, *Opt. Eng. 19,* 85 (1980).
53. J. T. Moseley, R. E. Olson, and J. R. Peterson, *Case Stud. At. Phys. 5,* 1 (1975).
54. I. C. Percival and D. Richards, *Adv. At. Mol. Phys. 11,* 1 (1975).
55. D. Banks, K. S. Barnes, and J. McB. Wilson, *J. Phys. B 9,* L414 (1976).
56. M. Karplus, R. N. Porter, and R. D. Sharma, *J. Chem. Phys. 43,* 3259 (1965).
57. R. Abrines and I. C. Percival, *Proc. Phys. Soc. (London) 88,* 861 (1966).
58. R. E. Olson and A. Salop, *Phys. Rev. A 16,* 531 (1977).
59. A. D. MacKellar and R. L. Becker, private communication.
60. N. Bohr and J. Lindhard, *Dan. Mat. Fys. Medd. 28,* 1 (1954).
61. P. M. Koch and J. E. Bayfield, *Phys. Rev. Lett. 34,* 448 (1975).
62. M. Burniaux, F. Brouillard, A. Jognaux, T. A. Govers, and S. Szucs, *J. Phys. B 10,* 2421 (1977).
63. J. H. Kim and F. W. Meyer, *Phys. Rev. Lett. 44,* 1047 (1980).
64. R. E. Olson, *Phys. Rev. A 23,* 3338 (1981).
65. I. C. Percival, in *Electronic and Atomic Collisions,* ed. G. Watel (Amsterdam: North-Holland Publ., 1978), pp. 569–78.
66. K. B. MacAdam, D. A. Crosby, and R. Rolfes, *Phys. Rev. Lett. 44,* 980 (1980); *Phys. Rev. A 24,* 1286 (1981).
67. J. G. Lodge, I. C. Percival, and D. Richards, *J. Phys. B 9,* 329 (1976).
68. D. L. Bunker, in *Methods in Computational Physics,* eds. B. Alder, S. Fernbach, and M. Rotenberg (New York: Academic Press, 1971), pp. 287–325.
69. R. L. Becker and A. D. MacKellar, *J. Phys. B 12,* L345 (1979).
70. M. A. Prasad and K. Unnikrishnan, *Phys. Rev. A 22,* 514 (1980).
71. A. B. Wittkower, G. Levy, and H. B. Gilbody, *Proc. Phys. Soc. (London) 91,* 306 (1967).
72. G. W. McClure, *Phys. Rev. 166,* 22 (1968).
73. R. H. Neynaber, G. D. Magnuson, and S. Y. Tang, *J. Chem. Phys. 68,* 5112 (1978).
74. G. Gioumousis and D. P. Stevenson, *J. Chem. Phys. 29,* 294 (1958).
75. K. A. Safinya, J. F. Delpech, F. Gounand, W. Sandner, and T. F. Gallagher, *Phys. Rev. Letts. 47,* 405 (1981); *Phys. Rev. A 25,* 1905 (1982).

7

Experimental studies of the interaction of Rydberg atoms with atomic species at thermal energies

F. GOUNAND AND J. BERLANDE

7.1. Introduction

Because of their large size and the weak binding of the outer electron, we may assume that highly excited (Rydberg) atoms are very sensitive to any kind of external perturbation, in particular, that perturbation resulting from the interaction with another atomic partner. Accordingly, this sensitivity should increase when the outer electron, referred to by its principal quantum number n, is increasingly excited, i.e., when n increases. It is such properties that have instigated much research by atomic physicists.

The large size of the Rydberg atoms suggests that the outer electron and the ionic core, far away from each other on the average, behave as separate scatterers during an atomic collision, i.e., to describe the dynamics of the collision, we simply add the contribution of the electron–perturber interaction to that of the ionic-core–perturber interaction.[1] This would make theoretical studies of collisional processes much simpler for Rydberg states than for low-lying excited states. We may even go a step further and speculate whether only one of these interactions is predominant in any collisional process. This is one of the crucial questions that experimentalists (and theoreticians) have tried to answer in the past few years. Conversely, information on electron–atom and ion–atom scattering at low energies might be obtained from studies of Rydberg atom–atom collisions that would be impossible to obtain by other techniques.

More practical reasons justify the interest of the scientific community in Rydberg studies. Because, under many circumstances, Rydberg states are a necessary step between the ionization continuum and the ground or low excited states of an atom, the Rydberg states and specially those of high degeneracy (i.e., those of high angular momentum l) play an important role in astrophysical and laboratory plasmas when collisional–radiative recombination processes govern the macroscopic properties of the medium. For isotope separation by lasers, for example, collisions may affect the overall efficiency of the technique.

Experimental studies of Rydberg atom–atom interactions started in 1934 when Amaldi and Segré[2] measured the shift of spectral lines associated with transitions between the ground state and highly excited *n*P states of alkali atoms. Fermi, in a famous paper,[1] analysed the basic physical phenomena responsible for this effect and laid the foundation for the theoretical treatment of Rydberg atom–atom interactions.

Recently, because of the development of new experimental techniques, there has been a strong revival of interest in Rydberg studies from both the experimentalists and the theoreticians. Many publications have appeared during the past 5 years and certainly many will appear in the upcoming years.

This chapter is intended as a review of the present status of the field of experimental studies of Rydberg atom–atom interactions at thermal energies. Basically we shall include two types of collisional processes:

1. Those that lead to loss from a given Rydberg state
2. Those that affect only the spectroscopic characteristics of the spectral lines emitted by these levels (i.e., their width and their position)

It is traditional to consider the first type of process as belonging to collision physics and the second to spectroscopy. However, for highly excited states, the border between these two fields is not as clear as for low-lying excited states.

In Sect. 7.2, we shall review briefly the experimental techniques that are most commonly used for these studies. Because they are well known to the reader, being used in other experimental fields, attention will be paid primarily to the points that are of interest for the theme of this chapter. The basic limitations of these techniques will also be discussed.

In Sect. 7.3, we shall present and discuss the recent results, with a brief summary of previous works when it is needed for clarity. Section 7.4 will include both prospective and concluding remarks.

Among previous review articles on collisional properties of Rydberg atoms, we would like to mention those of Stebbings,[3,4] Edelstein and Gallagher,[5] and Gallagher.[6] Mention should also be made of an article by Omont,[7] which, although dealing primarily with the theoretical aspects of Rydberg atom collisions, includes a large discussion of experimental results; we shall make use of this article in Sect. 7.3. The reader will notice that most of the experiments reported here were performed on alkali Rydberg states, perturbed either by a rare gas atom or an alkali atom of the same species; this is not surprising in view of the techniques (see Sect. 7.2) that are used for these studies.

7.2. Survey of experimental techniques

Although most of the experimental work reported here is highly dependent on optical techniques and on the most recent advances in laser technology, we shall treat separately those methods used for collisional studies and those used for line-profile studies because they have different requirements concerning resolution and sensitivity.

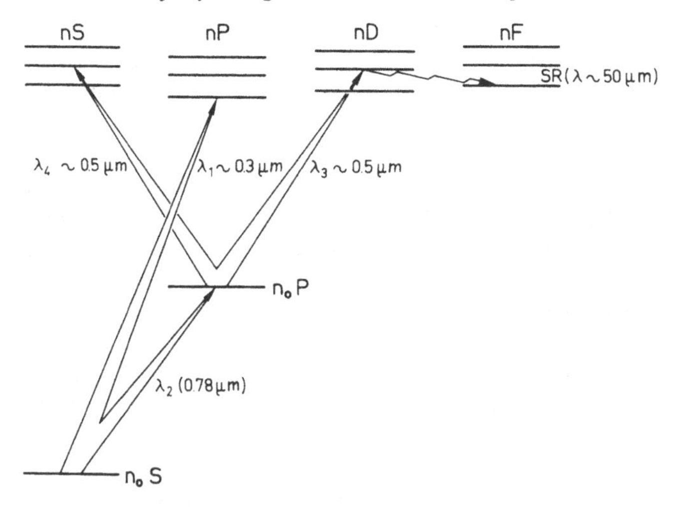

Fig. 7.1. Excitation scheme for alkali Rydberg states (the figure refers to rubidium). SR denotes superradiant emission.

7.2a. Experimental study of deexcitation processes

The cross sections for deexcitation of Rydberg levels, as a consequence of collisions with neutral atoms, are generally extracted from the evolution of the population of one or more Rydberg levels following the pulsed light excitation (or perturbation) of one of them.

Excitation of Rydberg states

Selective excitation of Rydberg levels is most commonly achieved by pulsed dye lasers (generally pumped by N_2 lasers) because of their high power, their flexibility, and their particular suitability for time-resolved detection.[8] The spectral width of the laser source lies usually in the range of tenths of an angstrom, allowing selective excitation of well-defined Rydberg levels and it can even be reduced when necessary. Using various dyes, continuous tuning of the wavelength can be achieved from the UV (when frequency-doubling devices are required) to the near-IR region, with typical peak powers of 0.1 to 50 kW, depending on the wavelength and the power of the pump laser.

For alkali atoms, the pulsed dye lasers allow the (one- or two-step) excitation of highly excited states according to Fig. 7.1. The nP states can be pumped directly from the n_0S ground state (the corresponding λ_1 wavelength lies in the UV region).[9] Because of the low power usually available after frequency doubling as well as the low oscillator strength values for the n_0S–nP transitions, only low populations ($\sim 10^{-3}$–10^{-4} of the ground-state population) are usually achieved. The nS and nD states can be populated using a two-step scheme,[10,11] the intermediate level being one of the two fine-structure, resonant n_0P levels (for this step, alkali resonance lamps as well as dye lasers have been used, owing to the low power required to achieve a high population of the resonance level). The wavelength (λ_3 or λ_4) for the second step lies in the visible, thus allowing significant population of S or D Rydberg levels ($\sim 10^{-2}$ of the ground-state population). By using selection rules for electric dipole transitions or by improving the wavelength resolution, population of individual Rydberg fine-structure levels can also be easily achieved.[11] Note that the polarization properties of the laser light can be used in order to produce some coherence in the excited states if necessary (for example, for depolarization studies).[12]

Recently, a method was demonstrated that very efficiently populates the alkali F states using a three-step scheme (see Fig. 7.1).[13-15] A high-D-state population was first achieved using the previously described method with an intense source of light for λ_3. Then, a strong superradiant emission ($\lambda \sim 50$ μm) transferred most of the D-state population to the neighboring F state that is energetically located just below the D level, in a time (~ 10 ns) that is much shorter than expected (~ 1 μs) from normal radiative decay. This method can be viewed as a radiative cascading transfer process for which only one way (between closely spaced levels) would be efficient. This is because superradiance for long-wavelength transitions is easy to initiate. Although the principle of the method is quite simple, great care has to be taken, especially in the adjustment of the D-state population, in order to avoid parasitic effects.[14]

Similar techniques may be used for other atomic species (alkaline earths[16] or uranium,[17] for example). For the case of the rare gases, optical excitation from the ground state is difficult (though possible using, for example, synchrotron radiation). A stationary discharge[18] or an afterglow[19] first provides a sufficient population of metastable atoms from which selective excitation (or perturbation) is again achieved using pulsed laser light.

Detection methods

Generally what is observed is the time evolution of the fluorescence (Fig. 7.2) from either the initially pumped level only (studies of total depopulation) or from both this level and the neighboring levels (studies of transfer between well-defined levels).[20]

In the former case, the equation that governs the time evolution of the population of the pumped level can be written (to first order) as

$$dN_i/dt = -AN_i \tag{1}$$

where A accounts for both radiative and collisional deexcitation. Solving Eq. (1) leads to the basic equation, for the time constant τ, of the decay rate of the population N_i:

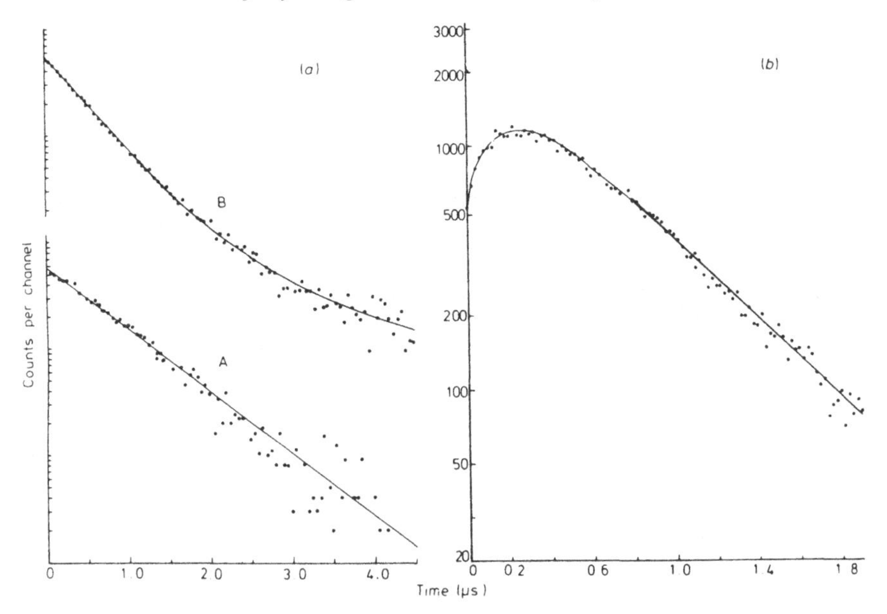

Fig. 7.2. Log-linear plots of fluorescence intensity (counts per channel) against time. (*a*) Direct fluorescence from cesium $13D_{3/2}$ level for $N_{Cs}=5.2\times10^{10}$ cm^{-3} (curve A) and $N_{Cs}=9.5\times10^{12}$ cm^{-3} (curve B). Curve B shows the effect of repopulation from neighboring states (see text); curve A exhibits a pure monoexponential decay. (*b*) Sensitized fluorescence from 11F level of cesium after excitation at $t=0$ to 15S level. (From Pendrill[39])

$$1/\tau = (1/\tau_{\text{rad}}) + K \tag{2}$$

where τ is the (measured) effective lifetime, τ_{rad} the radiative lifetime (that can be, for Rydberg states, strongly affected by blackbody effects[21]), and K the total depopulation (quenching) rate from collisions. As usual in cell experiments, K can be written as

$$K = NQv \tag{3}$$

if only collisions with neutral atoms are taken into account, where N is the density of the neutral perturbers, Q the quenching cross section, and v the Maxwellian-averaged velocity of the colliding partners. Additional terms of similar form have to be added in Eq. (3) if collisions with electrons and other charged particles are present.

Equations (2) and (3) clearly show that the Q value can be extracted from measurements of the variation of the effective lifetime τ as a function of the perturbing gas density N.

The latter case (simultaneous observation of the fluorescence originating from both the pumped level and its neighbors) can be treated in a similar,

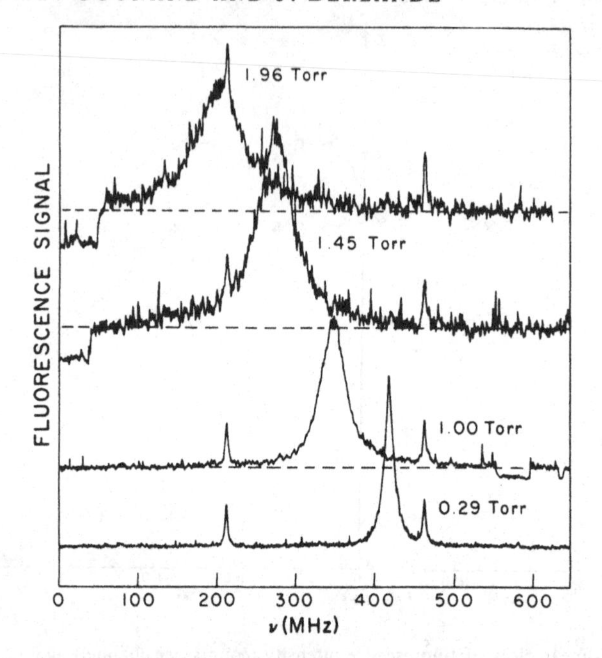

Fig. 7.3. Recorder trace of $5S_{1/2}$–$24S_{1/2}$ signal in rubidium at various argon pressures as a function of laser detuning. The sharp spikes are superimposed signals from a 250-MHz reference cavity. (From Brillet and Gallagher[29])

though more complicated, way. For a given level j, the time evolution of the N_j population is governed by the following equation:

$$dN_j/dt = -AN_j + \sum_k B_k N_k \tag{4}$$

where the second term on the right-hand side of Eq. (4) accounts for the population transfer from neighboring k states. Thus the time evolution of the N_j state population is deduced from the solution of a set of linear differential equations of the same type as Eq. (4). For a given level, the population decay exhibits a multiexponential behavior (see Fig. 7.2*b*) from which the transition rates (and hence the cross sections) can be extracted from observations performed at various perturbing gas pressures.

Note that the reliability of the data analysis obviously depends on the validity of the basic equations [Eqs. (1) or (4)] (see Sect. 7.2c). The time scale of the observed decays generally lies in the 0.1–10 μs range. A good time resolution as well as a high signal-to-noise ratio (especially in the case of a multiexponential decay) are needed. Moreover, small signals are generally observed thus requiring the use of photon-counting or appropriate analog techniques.

When cw laser excitation is used,[22] the results are extracted from time-integrated fluorescence measurements. This technique is less suitable for Rydberg states than for low-lying states owing to the low powers available from cw lasers.

Fluorescence methods obviously suffer from serious limitations, mainly because of their overall poor sensitivity. The use of selective field ionization techniques,[23] that have already been extensively used for spectroscopic studies, should, in that respect, be of great help for future experimental studies of deexcitation processes.[24]

The study of ions produced during Rydberg atom–atom collisions can be achieved by the classical methods of charged-particle detection.[25] Mass selection of the ions is most useful.[26] The data analysis is of the same type as those already described for neutral excited species.

Finally, mention should be made of a special technique that has been cleverly used for studying the deexcitation of Rydberg helium levels.[18] Although selective, it does not require optical excitation. Using the anticrossing properties of Zeeman atomic levels, a magnetic field alters the initial population equilibrium of two excited levels of helium in a dc discharge. Thus any modification of the population of other atomic levels (observed by the classical fluorescence technique) at the anticrossing point arises from radiative or collisional transfer from the initially perturbed levels. This method does not seem to be of general use for studying Rydberg atom–atom interactions because of its particular requirements.

7.2b. Experimental study of broadening and shift of spectral lines

Until recently, studies of broadening and shift of spectral lines have been made by using classical absorption techniques.[27] Usually the light source is a high-pressure xenon lamp. A high-dispersion spectrograph is used for the detection of the absorption profiles. The main feature of the method is that these profiles include the apparatus function that has to be carefully determined to obtain reliable results. The experiments can only be performed at high pressures ($\sim$1 atm) of perturbing gas in order to overcome the Doppler broadening.

Two novel techniques that do not suffer from this strong limitation have been used recently. The first technique is based on the simultaneous absorption by an atom of two photons of equal energy and opposite momentum (thus initial and final states have to be of the same parity).[28] The resulting absorption profile is determined only by the homogeneous broadening (i.e., natural and collisional). The elimination of Doppler broadening allows experiments to be performed at much lower pressures (in the millitorr range) than previously, the resolution being that of the excitation source. Continuous-wave lasers easily provide the required narrow linewidth ($\lesssim$10 MHz) and stable mode operation. The laser is frequency-swept (see Fig. 7.3) through the profile (the width of which is typically a few hundred megahertz). The absorption of light is detected either by a classical fluorescence technique[29] (this is only possible for n values lower than about 35 because of the low efficiency of the two-photon excitation) or by means of a thermionic detector,[30] i.e., a diode running in the space-charge-limited mode.[31] In this latter case, excited states produced by the absorption of light create some ions by collisional processes: these ions spend a long time in the space-charge region around the cathode leading to an efficient reduction of the negative potential barrier (under appropriate conditions, 10^6 electrons are obtained for every ion). This very efficient detection method requires a sufficiently high alkali pressure ($\sim 10^{-2}$ Torr) as well as the absence of any stray electric field for proper operation.

The second technique, used for line profile studies, is a trilevel echo method.[32] This method refers to a broad class of echo methods.[33] Its detailed description lies outside the scope of this chapter. Note only that this technique needs a well-defined sequence of three dye laser pulses of appropriate characteristics to allow the emission of an echo pulse. The broadening rate can be directly extracted from the observation of the variation of the echo intensity as a function of the perturbing gas pressure. The method does not require narrow linewidth lasers and, as in the case of two-photon absorption, is intrinsically free from Doppler effects.

7.2c. Limitations of experimental techniques and sources of error

It seems appropriate here to review briefly the limitations of the techniques previously described and the sources of uncertainty that might affect the results. We shall distinguish between those that affect the reliability of the data and those connected with the data analysis. Although the former have to be carefully analyzed, the latter are of much more concern for our purpose because, in many cases, they strongly limit the application of the already described experimental methods to the study of Rydberg atom–atom interactions.

In all cases, measurements (decay time constant, line broadening, etc.) are performed as a function of the perturbing gas pressure. Thus a precise measure of the pressure is needed. These determinations are more difficult to achieve in the case of alkali atoms than for rare gases.[34] Other sources of error, whose importance depends on the particular experiment, have to be considered also. We only list some of them: residual vacuum, gas purity, pulse pile-up[35] (in case of photon-counting measurements), thermal escape,[36] polarization effects,[37] and measurements of the decay time constants.[20,38]

Some features of the Rydberg states lead to specific difficulties in data analysis in the deexcitation experiments. First, following the collision of a given Rydberg atom with a neutral perturber, many exit channels are, in principle, accessible even at thermal energies. This point is clearly of major importance in assessing the validity of any model used for data analysis (see Sect. 7.2a, under "Detection methods"). For example, the use of a one-level model [see Eq. (1)] for the determination of the quenching rate implies that repopulation from other nearby states (that can be populated by various radiative or collisional processes) is negligible. This has to be carefully checked to determine the appropriate experimental conditions for which a pure exponential decay is observed.[9] Similarly, in the case of excitation transfer measurements, it must be determined if all the relevant levels have been included in the set of differential equations [see Eq. (4)] used for data analysis.[39] Note that use of a multilevel model requires highly reliable data.

Second, Rydberg states are more sensitive than low-lying excited states to superradiance,[13] blackbody effects,[21] and collisions with charged particles.[40] Superradiance can lead to a very rapid redistribution of populations between nearby levels, especially if high densities of excited species are present. Blackbody effects can also give rise to a noticeable redistribution of population, especially if high-n states are involved. Finally, cross sections for charged-particle–Rydberg-atom collisions are very large. In particular, the influence on the measured processes of collisions with electrons (created by photoionization and collisional processes in cell experiments or present in a dc discharge or in an afterglow) has to be evaluated.

These remarks explain why most of the deexcitation experiments have been performed for n values lower than about 30. For the reasons previously mentioned, a cell-type experiment in a neutral vapor is the easiest way to perform Rydberg-atom–atom studies at moderately high n values. Generally, precise experimental checks are possible for assessing the validity of the model used to analyze the data. Experiments performed in a dc discharge or in an afterglow are more difficult to interpret. Even low electron densities can affect the measurements. One can account for electron collisions, but in that case a complicated model that takes into account numerous levels has to be used. The validity of such a model is difficult to assess. Clear evidence for that point is demonstrated by the significant scatter between the results published by two groups, concerning the same process.[18] Finally, to study collisional ionization processes, an elaborate model is needed for data analysis, their corresponding rates being generally lower than those of other inelastic processes (see Sect. 7.3).

The two novel techniques just presented open an interesting and promising way for investigation of properties of high-n states. Although some problems should arise for high-n values because of the overlap of line components, the measurements seem mainly limited by the detection efficiency. In that respect, spectroscopic studies have already demonstrated that two-photon absorption measurements can be performed up to very high n values.[41]

7.3. Results and discussion

In the first part of this section, we shall present and discuss experimental data concerning two types of collisional processes involving highly excited states: (1) those that lead to a broadening and a shift of atomic spectral lines and

(2) those that are responsible for deexcitation (i.e., quenching, excitation transfer, ionization) of a given Rydberg level. In the final part of this section, we shall attempt to discuss those features common to the data presented.

Because a good understanding of broadening and shift studies (especially the most recent ones) is only achieved when inelastic collision processes are taken into account, we shall present first data concerning deexcitation processes, for which a wide range of results is available. Some guidelines are necessary for presentation. It has been recognized for a long time in atomic collision physics that the asymptotic energy defect ΔE between the entrance and exit channels of a reaction process is a most important parameter in the collision dynamics.[42] Generally speaking, the greater the ΔE, the lower the cross section at a given energy. It is interesting to note that the results first obtained on collisions involving Rydberg states, i.e., Na (nD)–rare gas collisions[10] on one hand, Rb (nP)–rare gas collisions[9] on the other, exhibit quite different orders of magnitude for the quenching cross section: "large" for Na ($\sim 10^{-13}$ cm^2), "small" for Rb ($\sim 10^{-15}$ cm^2). This appeared at first surprising: remember that large cross sections were expected for all Rydberg states. It has been clearly shown in more recent studies to be due to the large difference between the energy defects involved in the two processes.[43] All the results now available show the major importance of the magnitude of the inelasticity of the process considered on the cross section values. For this reason, we shall divide, somewhat arbitrarily, the review of the deexcitation processes into two parts, one devoted to quasi-elastic processes (such as angular momentum mixing, fine-structure transitions, isotopic excitation transfer) for which low energy defects are involved, and the other devoted to inelastic processes (such as transfer of excitation to a bound atomic level and collisional ionization). For quasi-elastic processes, the energy defect ΔE has to be on the order of 1 cm^{-1}, and for inelastic processes, it is much larger (10–100 cm^{-1}, for example).

7.3a. Collisional processes

Quasi-elastic processes

We shall review in this section three processes: angular momentum mixing, fine-structure transitions, and isotopic excitation energy transfer. At the end of this section, we shall review briefly results concerning collisional disalignment.

Angular momentum mixing process. The study of this process has played an important role in the development of collisional studies involving Rydberg states. Historically, it was the first process reported in the literature.[10] Moreover, the experimental results have stimulated the interest of theoreticians. In fact, most of the theoretical studies of collisions involving Rydberg states have been devoted to this reaction, which can be written as

$$A(n, l) + B \rightarrow A(n, l' > l) + B - \Delta E$$

where $A(n, l)$ is an alkali atom in a quasi-hydrogenic state (n, l), i.e., the corresponding quantum defect δ is very small, $A(n, l' > l)$ indicates that the exit channels are the neighboring states with the same n value and higher orbital quantum numbers, B is the perturbing atom, and ΔE the asymptotic energy defect between the (n, l) and $(n, l+1)$ levels [the $(n, l' > l+1)$ levels are very close to the $(n, l+1)$ states, owing to the energy ordering of the alkali atoms]. The quasi-elastic nature of the reaction is obvious from the fact that all the involved states are quasi-hydrogenic.

At present, results are available for nD sodium states[10,44] ($6 \leqslant n \leqslant 15$; ΔE decreases from 21 cm^{-1} for $n=6$ to about 0.9 cm^{-1} for $n=15$), nF sodium states[45] ($13 \leqslant n \leqslant 15$; ΔE being on the order of 0.08 cm^{-1}), both perturbed by He, Ne, and Ar, and also for nF rubidium levels[46] ($9 \leqslant n \leqslant 21$; ΔE ranges from 3 cm^{-1} for $n=9$ to 0.2 cm^{-1} for $n=21$) in collision with He, Ar, Xe or ground-state Rb atoms. Finally, measurements of nF cesium states ($7 \leqslant n \leqslant 11$; ΔE ranges from 9.3 to 1.6 cm^{-1}) in collision with ground-state Cs atoms have also been reported.[15] For D states, classical two-step pumping is used while nF states are produced either by radio-frequency mixing[45] or superradiant population transfer[46] from previously excited D states. In all experiments, cross sections are derived from the variation of the measured effective lifetimes as a function of the pressure of the perturbing gas (see Sect. 7.2). Measurements have to be performed at low pressures because the l-mixing cross sections are so large that, at moderate pressures, the initial D state population is transferred rapidly to the long-lived higher-l states: all that can be seen then is the average lifetime of all the l states. The n variation of this pressure independent lifetime ($\sim n^{4.5}$) provides clear evidence of the nature of the final states. Moreover, Gallagher et al.[45] showed, using a field-ionization technique, that all the $l' > l$ states were noticeably populated after the collision, i.e., no particular Δl selection rule holds for the process.

Results obtained for Rb (nF) states are shown in Fig. 7.4. These particular results will be used to discuss the angular momentum mixing process. Other results lead to similar conclusions. The cross section starts to increase for low-n values before reaching a maximum after which a more-or-less pronounced decrease is observed. The position of the maximum shifts toward higher-n values when the mass and the polarizability of the perturber increase (the maximum seems just reached for Xe and probably occurs at high-n values for Rb). For a given n value, the magnitude of the cross section strongly depends on the nature of the perturbing atom. Continuous increase from He to Rb is observed. This observation is clearly related to the well-known behavior of the e^--atom elastic scattering cross sections at low energies[47] that increase in the following order: Ne, He, Ar, Kr, Xe, ground-state alkali. A simple picture[44] that takes into account only the Rydberg-electron–perturber interaction explains the observed n behavior of the angular momentum mixing cross sections; for low-n values, the passage of the perturber through the small elec-

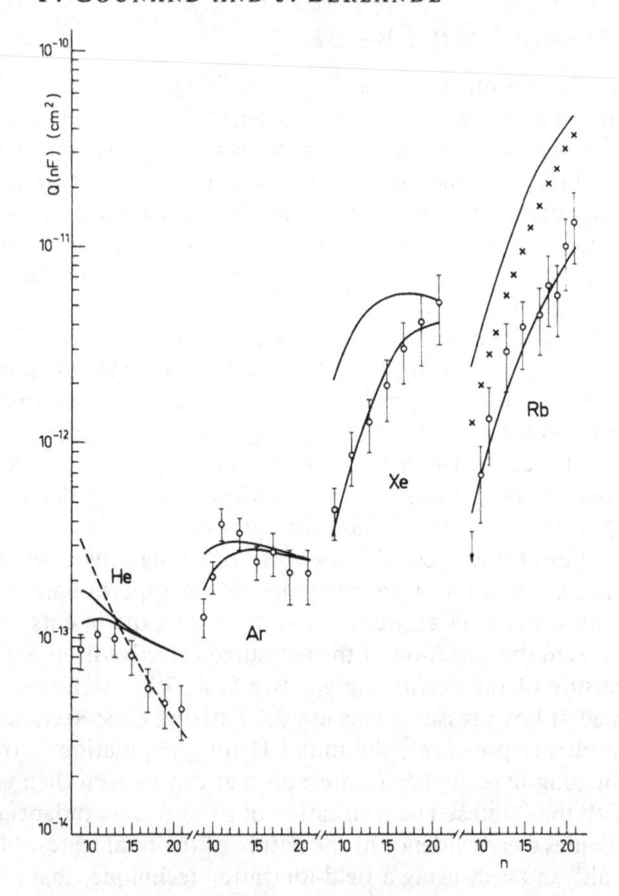

Fig. 7.4. Angular momentum mixing cross sections of Rb (nF) levels for various perturbing atoms. Experimental results (circles) are from Ref. 46. Solid lines indicate lower and upper limits for the de Prunelé-Pascale model.[48] The broken dotted line corresponds to the results of a calculation using the first Born approximation.[58] The values of the geometrical cross section σ_G (crosses) are also included. See text.

tronic cloud induces the transition with high probability due to the short-range e^--perturber interaction, but the growing extension of the electronic cloud as n increases decreases the probability of inducing the transition and leads to the decrease of the cross sections for high-n values. The stronger the e^--perturber interaction is, the higher are the n values for which the decrease occurs.

Numerous calculations taking into account only the e^--perturber interaction are available for detailed comparison with the experimental data (see, for example, Chap. 6). Some results are included in Fig. 7.4. If one excepts the binary encounter theory,[49] which always underestimates the cross section, the other approaches give satisfactory agreement with experiment in the helium case,[50-52] for which the use of a Fermi-type potential is clearly valid (for $n \gtrsim 10$). The theoretical results confirm that no Δl selection rule occurs: partial cross sections for $(n, l) \rightarrow (n, l')$ transitions decrease only slightly for increasing l' values.[52] An n^{-4} dependence of these partial cross sections for high-n values leads to an n^{-3} variation for the l-mixing process (because of the increasing number of partial cross sections that contribute to the process) in reasonable agreement with experimental findings.[53] Moreover, the fact that the partial cross sections are almost independent of ΔE (for sufficiently small ΔE values) leads to similar cross sections for the angular mixing of both Na (nD) and Na (nF) states (although the corresponding ΔE values differ by about one order of magnitude) in agreement with experimental observations[45] when properly interpreted.[48]

When going to heavier perturbers, the comparison between theory and experiment becomes more difficult because of the necessary inclusion in the theoretical treatment of correction terms to the Fermi potential, the diffusion length approximation being no longer valid[7] (except when one uses the de Prunelé–Pascale model,[48] which only needs the experimental data for elastic e^--perturber scattering at low energies and provides satisfactory agreement with experiment for all the studied perturbers, including ground-state Rb atoms; see Fig. 7.4). However, computations have been performed for heavier rare gases and provide satisfactory agreement with experiment for Ne, Ar, and Xe.[53,54] In this latter case, the comparison is only possible for the highest-n values investigated.

In the case of alkali–alkali collisions, no theoretical calculations are available (except those of de Prunelé–Pascale). The measured cross sections increase roughly as the geometrical cross section of the Rydberg atom σ_G (i.e., as n^4),[55] their values being about three times smaller than σ_G for Rb and one order of magnitude larger for Cs. These unexpectedly large values[15] for Cs [even larger than the depolarizing cross sections of Cs (nD) states[12]] have not yet been explained, although no doubt exists as to the nature of the measured process.

To summarize, the angular momentum mixing process has clearly been demonstrated to be due to the Rydberg-electron–perturber interaction.[56]

Fine-structure transitions. Only the D states of heavy alkalis have been investigated. The studied process is the following:

$$A(nD_{3/2}) + B \rightarrow A(nD_{5/2}) + B - \Delta E$$

where $A(nD_{3/2})$ and $A(nD_{5/2})$ represent the alkali atom in an $nD_{3/2}$ and $nD_{5/2}$ state, respectively, B the perturbing atom (alkali atom A in its ground state or helium), and ΔE corresponds to the asymptotic fine-structure interval. The nD ($6 \leqslant n \leqslant 10$) levels of cesium[22] have been studied by time-integrated fluorescence techniques using cw laser excitation. Only one laser is used, its frequency being tuned either to the $6P_{1/2} \rightarrow nD_{3/2}$ or $6P_{3/2} \rightarrow nD_{3/2}$ transi-

Table 7.1. Comparison between experimental Q_{exp} and theoretical Q_{th} cross section values (in square angstroms) for fine-structure ($nD_{3/2} \rightarrow nD_{5/2}$) transitions of rubidium $nD_{3/2}$ states induced by collision with helium atoms

Level	Q_{exp} (Ref. 58)	Q_{th} (Ref. 58)	Q_{th} (Ref. 7)
$9D_{3/2}$	510 ± 100	980	520
$10D_{3/2}$	455 ± 155	580	380
$11D_{3/2}$	220 ± 60	380	300

tion. The population of the $6P_{1/2,3/2}$ level is achieved mainly by photodissociation of excited Cs_2 molecules. The derivation of the cross sections requires the knowledge of nD lifetimes as well as of the branching ratios of the observed transitions (this can lead to some systematic errors in the quoted values). The Cs (nD)[11] ($8 \leqslant n \leqslant 14$) and Rb ($nD$)[57,58] ($9 \leqslant n \leqslant 13$) levels have been investigated by classical time-resolved techniques using a two-step excitation scheme (see Fig. 7.1). In fact, only total depopulation cross sections are obtained for the cesium $nD_{3/2}$ states, but, as demonstrated by a careful examination of the fluorescence decay curves, the main part of the quenching is due to $nD_{3/2} \rightarrow nD_{5/2}$ fine-structure transitions.

Let us first examine the results obtained for Cs (nD) states (where ΔE ranges from 43 cm^{-1} for $n=6$ to about 1.3 cm^{-1} for $n=14$) in collisions with ground-state Cs atoms. The measured cross sections increase roughly as n^{*4}, their magnitude being on the order of the geometrical cross section σ_G. Quite similar observations were reported for the total depopulation of the rudibium $nD_{3/2}$ states ($9 \leqslant n \leqslant 13$; the fine-structure interval ΔE ranging from 0.7 to 0.15 cm^{-1}) because of collisions with ground-state Rb atoms. Moreover, the measurements of the 13D quenching cross sections without allowance for fine structure was observed to be only 20% of the $13D_{3/2}$ depopulation cross section, clearly indicating that the quenching of the $nD_{3/2}$ states is mainly due to fine structure transitions.

All the experimental data concerning fine-structure mixing in collision of alkali Rydberg atoms with ground-state alkali atoms show large cross sections, where the interaction that is responsible for the process is between the outer electron and the ground-state alkali atom. Because this interaction is strong, the cross sections (only observed for $n \leqslant 14$) roughly reflect the geometrical size of the atom, according to the simple picture mentioned for the case of angular momentum mixing. Deviation from this behavior (though not yet observed) is only expected for very high n values. As for the case of angular momentum mixing, no theoretical data are available for comparison with experiment.

The only measurements available for fine-structure mixing by rare gases are those of Ref. 58. The cross sections (reported in Table 7.1) exhibit an n dependence opposite to that observed in the case of alkali–alkali collisions, their values being about two orders of magnitude smaller. Similar n-decreasing behavior was observed for the angular momentum mixing process induced by helium. In the helium case (the only one that has been experimentally studied), theoretical calculations that take into account only the e^-–helium interaction can be carried out. They predict an n^{*-4} dependence of the cross sections, in satisfactory agreement with the measured values. For example, the first Born approximation[58] as well as the estimates given by Omont[7] show reasonable agreement with the reported data (see Table 7.1).

Isotopic excitation exchange energy transfer. Recently,[59] the rate constant for isotopic excitation transfer between ^{3}He and ^{4}He was reported for the following reaction:

$$^3\text{He}\,(n = 9) + {}^4\text{He}\,(n = 1) \rightarrow {}^4\text{He}\,(n = 9) + {}^3\text{He}\,(n = 1) + \Delta E$$

where ΔE is on the order of 1.5 cm^{-1}. Selective laser perturbation in a room-temperature helium afterglow is used. A four-level model (including the ^{3}He (n=9), ^{4}He (n=9), ^{3}He (n=8), and ^{4}He (n=8) states, which are efficiently populated by superradiance under the experimental conditions) allows the derivation of the rate constant from time-resolved fluorescence measurements. The measured rate, $(5.7 \pm 1.0) \times 10^{-10}\ \text{cm}^3\ \text{s}^{-1}$, is equal to that calculated by the authors for the pure charge exchange process between $^3\text{He}^+$ and ^{4}He ($5.3 \times 10^{-10}\ \text{cm}^3\ \text{s}^{-1}$). The authors conclude that the dominant interaction for this particular reaction is that between the ionic core and the neutral perturber.

In concluding this section, we should like to mention the measurements of the disalignment cross sections of Cs ($n\text{D}_{3/2}$) levels (n=12, 13) induced by collision with ground-state Cs atoms.[12] To our knowledge, it is the only collision experiment that deals with a purely elastic process. The damping of hyperfine quantum beats is observed as a function of Cs density. The disalignment cross sections are then extracted by means of tensorial algebraic analysis. These cross sections, respectively, $(4.1 \pm 0.8) \times 10^4$ Å^2 and $(5.3 \pm 0.9) \times 10^4$ Å^2 for n=12 and 13, are observed to be roughly three times larger than the previously measured depopulation cross sections (see previous discussion) that are on the order of the geometrical cross sections σ_G. No theoretical computation is, at present, available for this process, but the magnitude of the cross sections clearly indicates, as is the case for most of the other processes previously reported in this section, that the process is governed by the outer-electron–perturber interaction.

Inelastic processes

Collisional deexcitation. We shall include in this section the measurements of the total depopulation (quenching) cross sections of a given Rydberg state. Although no identification of the final products was made, it

Table 7.2. Experimental cross section values (in square angstroms) for total depopulation of nonhydrogenic alkali states induced by collisions with various perturbers; effective quantum number n^* and quantum defect δ values are indicated

						Perturber			
Element	Ref.	Level	δ	n^*	Alkali	He	Ne	Ar	Xe
Na	60	7S	1.3	5.7		3.8 ± 0.6		0.48 ± 0.15	46 ± 7
		8S	1.3	6.7		5.5 ± 0.9		5.3 ± 0.8	32 ± 3
		9S	1.3	7.7		7.3 ± 1.0		18.9 ± 3.0	29 ± 6
		10S	1.3	8.7		14.4 ± 3.0		18.1 ± 3.0	56 ± 9
		11S	1.3	9.7		21.0 ± 5			97 ± 20
Rb	57, 58	12S	3.1	8.9	$(1.4 \pm 0.4) \times 10^4$	110 ± 20			
		14S	3.1	10.9	$(2.1 \pm 0.5) \times 10^4$	205 ± 50			
		16S	3.1	12.9	$(3.9 \pm 1.0) \times 10^4$	145 ± 40			$(2.6 \pm 0.5) \times 10^4$
		18S	3.1	14.9	$(7.0 \pm 1.4) \times 10^4$	325 ± 65			
Rb	9	12P	2.65	9.35	$(2.3 \pm 0.8) \times 10^3$	38 ± 7	4.9 ± 0.9	15 ± 3	
		14P	2.65	11.35	$(7.1 \pm 2.5) \times 10^3$	56 ± 10	12.0 ± 2.4	25 ± 5	
		17P	2.65	14.35	$(1.0 \pm 0.35) \times 10^4$	58 ± 11	11.0 ± 2.1	28 ± 5	
		22P	2.65	19.35	$(1.6 \pm 0.6) \times 10^4$	60 ± 12	13.0 ± 2.5	30 ± 6	
Rb	57, 58	9D	1.35	7.65		11 ± 2			
		10D	1.35	8.65		15 ± 2			
		11D	1.35	9.65		29 ± 5			
		13D	1.35	11.65	$(7 \pm 4) \times 10^3$				
		15D	1.35	13.65		130 ± 40			
Cs	11	10S	4.05	5.95	$(0.7 \pm 0.5) \times 10^4$				
		11S	4.05	6.95	$(0.4 \pm 0.5) \times 10^4$				
		12S	4.05	7.95	$(1.0 \pm 0.6) \times 10^4$				
		13S	4.05	8.95	$(1.9 \pm 0.7) \times 10^4$				
		14S	4.05	9.95	$(3 \pm 1) \times 10^4$				
Cs	22	6D	2.5	4.5	$(1.1 \pm 0.6) \times 10^2$				
		7D	2.5	5.5	$(5.3 \pm 1.5) \times 10^2$				
		8D	2.5	6.5	$(3.6 \pm 1.5) \times 10^2$				
		9D	2.5	7.5	$(6.3 \pm 2.3) \times 10^2$				
		10D	2.5	8.5	$(1.55 \pm 0.4) \times 10^3$				

can be easily shown that these depopulations lead to a transfer of excitation to neighboring states of the Rydberg atom. The typical energy defects involved lie in the 10^1–10^2-cm^{-1} range. Most of the experiments were performed using time-resolved fluorescence measurements.

We shall begin this section with a presentation of the results concerning alkali Rydberg states, which have been the most widely studied and therefore provide a good basis for comparison with theoretical approaches. Table 7.2 summarizes the data. We discuss alkali–alkali and alkali–noble gas collisions separately because the theoretical data are available only in the latter case, for reasons already mentioned.

Let us first consider the results concerning alkali–alkali collisions. They display two features. First, the cross sections are large (10^{-13}–10^{-12} cm^2) and always increase with n^*, as was the case for quasi-elastic processes. For this reason, the main interaction leading to depopulation may be considered to be the strong outer electron–alkali interaction that occurs for large internuclear distances. Ionic core–perturber interactions would lead to much smaller, only slightly n dependent, cross sections.

The second feature is that for similar n^* values the cross sections are observed to be generally larger when δ is close to an integer; in this case, quasi-hydrogenic levels of high multiplicity are energetically close to the considered Rydberg level. It seems therefore that there is an effect of the energy defect ΔE on the magnitude of the cross sections observed.

When considering the rare gases as perturbers, the situation is quite different (if the low-lying Na (nS) levels are omitted). The cross sections have low values (~10 Å^2) and show no sharp increase with n. They seem to increase slightly before reaching an almost constant value for $n^*>10$. A clear influence of the energy defect on the magnitude of the cross sections is observed. For example, the cross sections observed for the Rb (nS) levels[58] (for which $\delta=3.1$), which are energetically close to the neighboring $(n-3)$ F, G, H,... states, are much larger than those measured for the more isolated Rb (nP) levels[9] (for which $\delta=2.65$). Finally, the values of the cross sections concerning levels of similar quantum defects are roughly independent of the nature of the alkali atom, for the same perturber (21 Å^2 for Na (11S) level and 29 Å^2 for Rb (11D) level in collision with He).

One may wonder whether the e^--perturber interaction previously associated with large cross sections (>10^3 Å^2) is still responsible for the inelastic processes under study. An attempt to answer this crucial question was made by Gallagher and Cooke.[60] They suggest that, instead of the e^--perturber interaction, the ionic core–perturber interaction is responsible for the inelastic processes. Two reasons were given. First, the poor overlap between the electronic wave functions corresponding to the entrance and exit channels should lead to very small coupling elements in calculations using methods like the first Born approximation or impulse approach. Therefore, they would give small cross section values. Second, an interaction occurring at large impact parameters, such as in the case of the e^--perturber interactions, would imply

Table 7.3. Comparison between experimental Q_{exp} and theoretical Q_{th} cross section values (in square angstroms) for total depopulation of some nonhydrogenic alkali levels induced by collision with helium atoms

Element	Level	Q_{exp}	Ref.	Q_{th}	Ref.
Na	10S	14.4 ± 3.0	60	16	58
	11S	21 ± 5	60	26	58
Rb	12S	110 ± 20	58	289	58
	14S	205 ± 50	58	378	58
	16S	145 ± 40	58	391	58
	18S	325 ± 65	58	349	58
	12P	38 ± 7	9	23	62
	14P	56 ± 10	9	41	62
	17P	62 ± 11	9	63	62
	22P	60 ± 12	9	82	62

some marked n-dependence of the cross sections. They proposed instead a simple model based on the short-range alkali ion–rare gas interaction (for which potentials are available[61]). This model considers the formation of a transient dipole (or multipole) moment that splits the electronic energy levels. In that case, diabatic transitions between the levels can occur. A crude estimate of the cross sections for such a process gives an n-independent cross section of about 7, 19, and 48 Å^2 for Na–He, Na–Ar, and Na–Xe collisions, respectively. For the highest Na (nS) states investigated, reasonable agreement with experiment is thus obtained, in view of the crudeness of the estimate.

On the other hand, Hugon et al.[58] gave various arguments in favor of the electron–perturber interaction being responsible for the inelastic processes. The regular evolution of the cross sections as a function of the energy defect ΔE when going from quasi-elastic to inelastic processes suggests that no drastic change in the nature of the interaction responsible for the process occurs. Moreover, the cross sections obtained for the Rb (nS) levels in collisions with He and Xe are much larger than those calculated using the model of Gallagher and Cooke (about 20 and 80 Å^2, respectively). Hugon et al. calculated the depopulation cross sections of nonhydrogenic levels induced by collisions with helium atoms, taking into account only the outer e^-–helium interaction. The method used the first Born approximation with a Fermi-type potential.[51] Table 7.3 shows the results, which are observed to be in satisfactory agreement with the experimental data. Note that the calculations show the smooth n dependence as well as the influence of the energy defect (marked n dependence is only expected for small ΔE values, i.e., $\Delta E \lesssim 1$ cm^{-1}). Finally, the calculations indicate that the neighboring high angular momentum states of high-multiplicity are those that contribute most significantly to the quenching process.[62] Thus the obtained cross sections for depopulation arise from the addition of small contributions from numerous partial cross sections.

Table 7.4. Cross section values (in square angstroms) for some inelastic processes involving helium Rydberg levels in collisions with ground-state He atoms

n	$n\,^3S \rightarrow n\,^3D$ Transfer[18]	Quenching of triplet states[19]	Quenching of $n\,^3S$ levels[63]
6	4.3±0.8	—	22.4±3.5
7	11.6±2.1	—	29.2±7.3
8	16.0±2.8	14.9	35.6±5.5
9	28.9±5.2	13.8	38.6±6.3
10	—	11.2±3.0	47.7±8.6
12	—	11.5	—
14	—	15.5	—
17	—	21.9	—

It is difficult to define precisely the range of ΔE for which the theoretical treatment of Hugon et al. is strictly valid. However, consideration of only the e^--perturber interaction leads to a satisfactory agreement for both quasi-elastic and inelastic processes induced by collisions with helium. The case of the other rare gases requires more detailed calculations, including corrections to the Fermi potential, before any conclusion can be made.

To conclude, it is our opinion that the main interaction responsible for the depopulation due to collisions with rare gas atoms of the Rydberg nonhydrogenic levels considered here is the e^--perturber interaction. Some contribution from the ionic core-perturber interaction is probable, especially when small cross sections (~ 10 Å^2) are observed.

The experiments concerning collisional processes involving Rydberg states of other atomic species provide information on inelastic processes that do not modify the basic ideas previously reported. Many experiments were done on Rydberg helium levels perturbed by ground-state helium. Some typical results are given in Table 7.4.

Cross sections for angular momentum transfer between $n\,^{1,3}S$ and $n\,^{1,3}D$ states of helium ($5<n<9$) perturbed by ground-state helium atoms have been reported.[18] The corresponding energy defects are of the order of 100 cm^{-1} or larger. Results were obtained using an anticrossing spectroscopic method. Although the derivation of the cross sections can be affected by some systematic errors, as noted by the authors, the data clearly show the features already noted in the case of alkali Rydberg states: the cross sections are small and only slightly n dependent. No comparison with theoretical results was given.

Total depopulation rates due to collisions with ground-state helium of states of different angular momentum but the same quantum number n ($8<n<17$) were measured in a helium afterglow.[19] The results are extracted from a complex model (that includes collisions with both electrons and ground-state helium atoms). The authors assumed that under their experimental conditions there was a very fast mixing (with a time constant of ~ 1 ns) of the various l

levels corresponding to the same n state. This seems somewhat contradictory with the mixing time (~50 ns) deduced from the experimental cross sections for He–He angular momentum transfer (~50 Å^2, see Table 7.4).[18] It was observed that the cross sections were on the order of 10 Å^2 and weakly n dependent. Comparison was made by the authors only with theoretical calculations using the binary encounter approach,[49] which was proved to provide unsatisfactory results in case of alkali atoms.[9,48] In view of the disagreement observed between calculated and measured values, the authors discarded the e^--perturber interaction as the main interaction responsible for the process.

Using a nonselective approach, Hitachi et al. recently reported quenching cross sections for various $n\,^1S$ and $n\,^3S$ helium levels in collisions with ground-state helium (see Table 7.4) as well as with the other noble gases.[63] Results concerning the total depopulation of two Rydberg levels of uranium perturbed by helium and neon were published.[17] The reported cross sections are ~10 Å^2 for ΔE on the order of 29 cm^{-1}.

Collisional ionization. When a Rydberg atom A** collides with a ground-state atom of the same chemical nature, there exists some probability of ionization according to

$$A^{**} + A \rightarrow A_2^+ + e^- \qquad (5)$$

$$\rightarrow A^+ + A + e^- \qquad (6)$$

when energy requirements are met and

$$A^{**} + A \rightarrow A^- + A^+ \qquad (7)$$

for alkali atoms.[64] Homonuclear associative ionization of Rydberg alkali atoms [reaction (5)] has been known to occur for some 50 years.[65] However, reliable quantitative studies of collisional ionization [processes (5)–(7)] are very scarce because experimental data are difficult to interpret (see Sect. 7.2).

Lee and Mahan[66] measured the ratio of the concentrations [A^+] and [A_2^+] resulting from collisional ionization of various nP states ($8<n<16$) of Rb and Cs produced by photoexcitation of a neutral vapor. This ratio, as expected, increases with increasing values of n; it starts from zero for $n \simeq 8$–10, where only A_2^+ ions are formed (the dissociation energy of A_2^+ is ~0.5–0.7 eV) and reaches 0.33 for $n=12$. Under the conditions of the experiment (alkali pressure $\simeq 0.1$ Torr), it is easily shown using the results of Ref. 9 for collisional transfers and those of Ref. 67 for radiative constants that the results of Lee and Mahan (for $n>10$) refer to a group of levels (the selectivity of the light excitation is lost) rather than to a single level.

Measurements of the rate of associative ionization [process (5)] for various nS, nP, and nD levels ($8<n<12$) of cesium have been made during the 1970s by two groups of Russian physicists.[68] A cesium vapor is illuminated by a cesium lamp (dc discharge) to produce nP states; a weak discharge in the cesium vapor allows, using the resonant 6P level as a step, excitation of the nS and nD states. Identification of ionic products is not done in most cases. The results of these works are often contradictory.

Recently, using optical excitation of rubidium vapor, Klucharev et al.[25] measured the ionization rate coefficients for various nP levels ($7<n<14$) at 520 K. There was no mass analysis of the ionic products. Except for $n=7,8$, the ionization cross sections were in the 10^{-14} cm^2 range; they showed a slight maximum at $n=11$ and then a decrease at higher-n values. However, one may wonder whether some systematic error is present at high-n values because the data analysis made use of a very simple balance equation for the population of the considered nP level. If the extrapolation procedure were done with data obtained at $N_0>10^{14}$ cm^{-3} (N_0 being the ground-state atom density), it can be easily shown from Ref. 9 that, for $n=12$, 30% of the ions observed come from neighboring levels. The ionization cross sections of Klucharev et al. compare well with the recent theoretical results of Mikajlov and Janev.[25]

Another difficulty in studying ionization processes involving Rydberg states was pointed out by Worden et al.[26] They looked at Sr^+ and Sr_2^+ signals when they photoexcited the high members of the 5s ns and 5s nd series in strontium vapor. They found strong evidence that electrons, initially produced by photoionization, play an important role in ionizing high-Rydberg states through an avalanche mechanism. This strongly affects the chemistry of these states at high excitation density and high number density.

7.3b. Broadening and shift of alkali spectral lines

Measurements of the shift and broadening of optical lines have provided an efficient tool for investigating the interactions of highly excited Rydberg atoms with atomic species (only the upper level of the studied transition is affected by the addition of a foreign gas). Numerous experimental studies concerning the principal series n_0S–nP of alkali atoms have been performed, especially by Russian physicists, using classical absorption techniques. Recently, the development of tunable lasers has initiated a strong revival of the field. It seems appropriate to present, before any review of the experimental data, a brief theoretical survey of the field in order to facilitate the interpretation of the experimental data. (In this chapter, we shall only discuss the wide set of data available on alkali Rydberg states, although some fragmentary data exist on other atomic species; see Ref. 72. Note, however, that these data provide support for the basic interpretation given here.)

Shift and broadening will be noted by Δ and γ, respectively, where γ refers to the FWHM of the line profile. They can be expressed, as usual, in term of collision rates, i.e.,

$$\Delta = K^{\Delta}N \tag{8}$$

$$\gamma/2 = K^{\gamma}N = \sigma^{\gamma}vN \tag{9}$$

where K^{γ} and K^{Δ} are the collision rates for broadening and shift, respectively,

N the density of perturbing particles, σ^{γ} the broadening cross section, and v the averaged velocity of the alkali-perturber pair.

Atomic units will be used throughout this section, unless otherwise specified, $e=m=\hbar=a_0=1$. Some units of interest are: velocity $v=2.19\times10^8$ cm s^{-1}, cross section $\sigma=0.28$ Å^2, polarizability $\alpha=0.148$ Å^3, and rate constant $K=6.13\times10^{-9}$ rad cm^3 s$^{-1}=9.78\times10^{-16}$ MHz cm$^3=3.26\times10^{-20}$ cm^{-1}/cm^{-3}. For the scales on the figures, we generally use the (various) units given by the authors.

In 1934, Fermi[1] calculated the shift of the highest members of the principal series of the alkali atoms perturbed by foreign gases, assuming that the motion of the Rydberg electron was that of a slow, free electron through the gas of foreign perturbers. For sufficiently high pressures of perturbing atoms (~1 atm), the orbit can enclose a great number of foreign atoms (~10^4). According to this picture, Fermi noted that the shift of the line, when a perturbing gas was added, was due to two effects: (1) the Rydberg electron is scattered by the neutral perturbers within its orbit and (2) the ionic core of the alkali atom polarizes these neutral perturbers, thus leading to an energy lowering of the system. Accordingly, Fermi calculated the contribution of these two effects (scattering and polarization effects will be hereafter noted by the subscripts sc and pol, respectively) to be

$$\Delta_{\mathrm{sc}} = \pm(\pi\sigma_0)^{1/2}N \tag{10}$$

$$\Delta_{\mathrm{pol}} \simeq -10\alpha N^{4/3} \tag{11}$$

where α is the polarizability of the perturber and σ_0 the zero-energy cross section for elastic scattering of a free electron by the perturbing atom. If we assume the diffusion length approximation to be valid, $\sigma_0=4\pi L^2$ (L being the scattering length of the perturber),[69] thus leading to $\Delta_{\mathrm{sc}}=2\pi LN$. Finally, the total shift was assumed to be the sum of the two contributions

$$\Delta = \Delta_{\mathrm{sc}} + \Delta_{\mathrm{pol}} \tag{12}$$

It is easy to show that, under usual experimental conditions ($N<10^{20}$ cm^{-3}), the main contribution to Δ is that of the scattering effect. Note that Eq. (10) only holds for vanishing energy of the outer electron, i.e., for very high n values.[70] In that case, the shift rate K^{Δ}, derived from Eqs. (8) and (10), is expected to have a constant value that depends only on the nature of the perturbing atom. This asymptotic value provides a measurement of σ_0, the value of which may be difficult to obtain by other experimental methods. Note, finally, that Fermi did not derive the expression for the broadening γ.

Alekseev and Sobel'man[71] extended somewhat the formulation of Fermi. They expressed the shift and the width of spectral lines due to scattering effects in terms of the phase shifts δ_l for the e^--perturber elastic scattering:

$$\Delta_{\mathrm{sc}} = N\int\left[(\pi/q)\sum_l(2l+1)\sin 2\delta_l\right]W(q)\,dq \tag{13}$$

$$\gamma_{sc} = N\int\left[(4\pi/q)\sum_l (2l+1)\sin^2\delta_l\right]W(q)\,dq$$
$$= N\int q\sigma(q)W(q)\,dq \tag{14}$$

where $W(q)$ and $\sigma(q)$ are, respectively, the momentum distribution of the Rydberg electron and the cross section for the elastic scattering of a free electron with momentum q by the perturber. Equation (13) reduces to Eq. (10) if only s scattering is taken into account. These authors also derived the following expressions for both γ and Δ due to polarization effects:

$$\gamma_{pol} = 7.18(\alpha^2 v)^{1/3}N \tag{15}$$

$$\Delta_{pol} = -6.22(\alpha^2 v)^{1/3}N \tag{16}$$

All the formulas (13)–(16) were obtained within the framework of the impact approximation of line-shape theory.[72] (This explains why Eq. (16) differs from Eq. (11), which was obtained using the statistical approximation.) The velocity v_e of the electron being always much greater than v (for $n<100$), the validity of the impact approximation requires the following conditions:

$$(\rho_{pol})^3 N = \left(\frac{\pi}{4}\frac{\alpha}{v}\right)N \ll 1; \qquad \rho_{pol} \ll n^2 \tag{17}$$

where ρ_{pol} is the Weisskopf radius for the polarization interaction.

Recently, Omont[7] showed that Eq. (14), which reduces to

$$\sigma^{\gamma} v = \sigma_0 \langle v_e \rangle \tag{18}$$

only holds when inelastic collisions are dominant in the scattering process, contrary to the assertion of Alekseev and Sobel'man. This case corresponds to high-n values ($n>40$). Omont derived an estimate of γ_{sc}:

$$\gamma_{sc} \simeq (8L^2/n)N \tag{19}$$

Moreover, for intermediate-n values, Omont provided estimates of the broadening rate that can be summarized as follows:

$$\tfrac{1}{2}\gamma = \tfrac{1}{2}\gamma_{sc} + \tfrac{1}{2}\gamma_{pol} = vN(\sigma^{\gamma}_{sc} + \sigma^{\gamma}_{pol}) = (K^{\gamma}_{sc} + K^{\gamma}_{pol})N \tag{20}$$

with

$$\sigma^{\gamma}_{pol} = \pi(\alpha/v)^{2/3} \tag{21}$$

$$\sigma^{\gamma}_{sc} = \begin{cases} 2\pi\rho_{sc}(2n^2 - \rho_{pol}) & \text{if} \quad \rho_{sc} < \rho_{pol} \\ \pi(4\rho_{sc}n^2 - \rho^2_{sc} - \rho^2_{pol}) & \text{if} \quad \rho_{sc} \geqslant \rho_{pol} \end{cases} \tag{22}$$

where ρ_{sc} and ρ_{pol} are the following Weisskopf radii:

$$\rho_{pol} \simeq (\alpha/v)^{1/3}, \qquad \rho_{sc} = 0.14(L/n^3 v)^2 \tag{23}$$

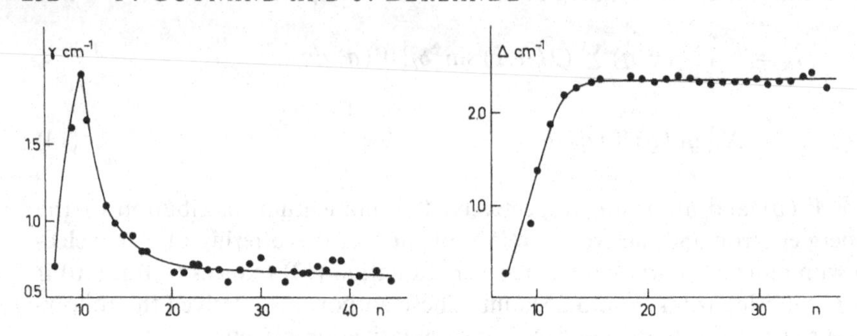

Fig. 7.5. The n dependence of the width γ and the shift Δ of cesium 6S–nP lines due to perturbation by argon atoms ($N_{Ar} = 6.7 \times 10^{19}$ cm^{-3}). (From Mazing and Vrublеskaya[74])

In these formulae, the elastic scattering is assumed to be dominant; the polarization contribution is about 15% smaller than the one calculated by Akelseev and Sobel'man [Eq. (15)].

There is no fundamental reason according to the previous theoretical survey for dividing, as will be done, the presentation and the discussion of the experimental results into two parts: one devoted to the case where the perturber is a rare gas atom and the other concerned with the case where the perturber is a ground-state alkali atom. However, we shall retain this presentation for two reasons: (1) the Δ and γ values corresponding to the two cases as well as their n dependence were experimentally found to be quite different and (2) comparison with theoretical results is relatively easy when the perturber is a rare gas atom and more difficult when the perturber is a ground-state alkali.

We shall first recall for clarity the most important experimental works published before 1977 as well as the status of their interpretation, which was already discussed by Omont. We shall then present the results obtained during the period 1977–80 by using laser techniques. In fact, these experiments, performed on S and D Rydberg states (whereas previous studies were only concerned with P levels) at low perturbing gas pressure for which the impact approximation is valid, have provided significant new insights.

Collisions with noble gas atoms

Numerous results were obtained before 1977 using classical absorption methods.[2,71,73,74] All these measurements concern the n_0S–nP series of the alkali atoms. Owing to the large instrumental width of conventional spectroscopy (~1 cm^{-1}), the measurements were taken at high pressures (>1 atm) in order to overcome the Doppler broadening. Under these conditions, as pointed out by Alekseev and Sobel'man,[71] the basic assumptions of the impact treatment were only marginally satisfied. On the other hand, the works of

Mazing et al.[27,74] and of Serapinas[75] were performed under experimental conditions that more readily permit comparison with theory.

The following three points should be noted. First, all the quoted works show the same n behavior for Δ and γ. Starting from low-n values, the shift Δ is always increasing before reaching an asymptotic value (for $n \simeq 20$), whereas the broadening γ first increases up to $n \simeq 10$ and then decreases to its asymptotic value. If we limit our discussion to Rydberg lines ($n > 10$), two ranges have to be considered: an intermediate range ($n \simeq 10$–20), where, roughly speaking, Δ and γ show opposite n dependence, and a high-n range, where they both have reached their asymptotic values. Figure 7.5 shows a typical example.[74] In fact, this high-n range can exhibit a slight n dependence before the asymptotic limit is obtained (for very high n values, say $n > 40$).

Second, concerning the shift Δ, the rapid increase at low-n values occurs in an n range where the impact approximation is clearly not valid. Thus a more complex treatment is needed. On the other hand, the asymptotic limit generally agrees well with the theoretical predictions. For example, for the case shown in Fig. 7.5 (Cs perturbed by Ar), Eqs. (10) and (16) give, after conversion to wavenumber units, an asymptotic limit for Δ of about 2.7 cm^{-1}, which is in reasonable agreement with the experimental finding (~ 2.3 cm^{-1}). The asymptotic shift is dominated by the scattering effect with only a small amount due to polarization (~17% in our example). Thus the observed asymptotic value provides a good estimate for σ_0. Moreover, as theoretically predicted, the asymptotic shift does not depend on the nature of the considered alkali, and its sign, according to Eq. (10), indicates if the perturber presents a Ramsauer effect for free e^- elastic scattering (in that case a red shift is observed). Finally, the intermediate-n range for which slight n dependence occurs is also observed to be in good agreement with the theoretical predictions, taking into account corrections to the diffusion-length approximation.[70]

Third, neglecting the low-n region for which the broadening γ exhibits a sharp increase, we shall discuss the observed decrease of γ for intermediate-n values and the asymptotic value of γ. The latter is well predicted by Eq. (15). The polarization effect is dominant in this asymptotic region, the scattering contribution of the inelastic collisions [Eq. (19)] being small (except for heavy rare gases for which it can be of the same order of magnitude owing to their high-σ_0 values). For our example (Cs–Ar), the total broadening is calculated, using Eqs. (15) and (19), to be about 0.65 cm^{-1}, the scattering contribution being 20%. The intermediate-n region was extensively discussed by Omont.[7] Good agreement between theory [Eqs. (20)–(23)] and experiment was observed.

Recently two experiments, using the novel techniques described in Sect. 7.2, were performed. Kachru et al.[76] made extensive measurements of noble-gas-induced broadening of transitions to Rydberg nS and nD sodium states by using a trilevel echo technique. Their broadening cross sections [cf. Eq. (9)],

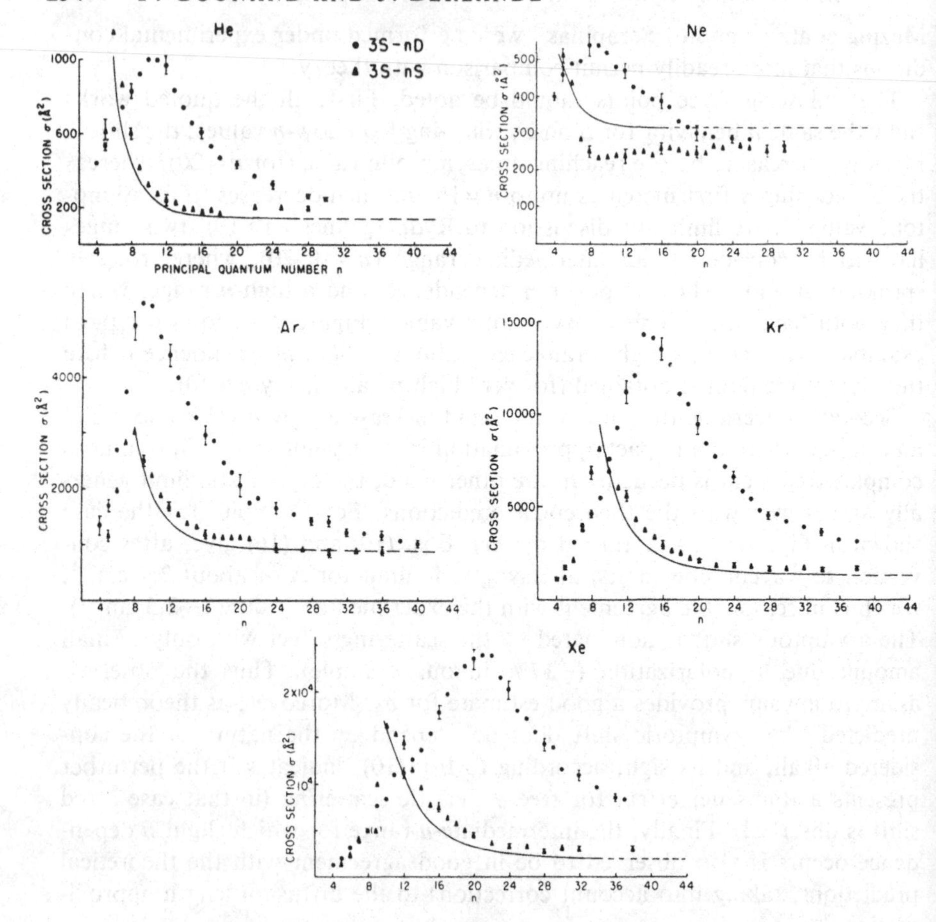

Fig. 7.6. Broadening cross sections (in square angstroms; recall that 1 a.u. = 0.28 Å²) for *n*S and *n*D sodium states perturbed by rare gas atoms. The broken line represents the theoretical asymptotic limit [cf. Eq. (15)] and the solid line the results obtained by using Eqs. (20)–(23) in the intermediate-*n* range. The circles are for 3S–*n*D and the triangles for 3S–*n*S lines. (From Kachru et al.[76])

obtained for both intermediate- and high-*n* values, are reported in Fig. 7.6. Consider first the data concerning the *n*S levels. Overall good agreement is observed between theory and experiment (except for Ne). However, the experimental results are (except for Ne) always slightly larger than theoretically predicted, indicating probably a small influence of inelastic collisions (even for intermediate-*n* values), which is consistent with the small inelastic cross sections for depopulation of Na (*n*S) states.[60] (This is due, as pointed out in Sect. 7.3a, to the nonhydrogenic nature of these states.) Note that Eq. (19) cannot provide a reasonable estimate of the inelastic part of the broadening because it only holds for quasi-hydrogenic (or closely spaced) levels. Table 7.5

Table 7.5. Comparison between experimental K^γ(S) and K^γ(D) and theoretical K^γ_{tot} broadening rates of S and D states of sodium perturbed by rare gas atoms in the high-n region ($n \simeq 40$); the contributions of scattering K^γ_{sc} and polarization K^γ_{pol} effects are also reported; all values are in atomic units

	Experiment (Ref. 76)		Theory		
Perturber	K^γ(S)	K^γ(D)	K^γ_{sc}	K^γ_{pol}	K^γ_{tot}
He	0.51	0.64	0.14	0.40	0.54
Ne	0.36	0.39	0.01	0.51	0.52
Ar	1.11	1.28	0.29	1.26	1.55
Kr	1.86	3.46	1.37	1.60	2.97
Xe	3.41	8.95	4.29	2.21	6.50

gives the experimental K^γ values (in atomic units) as well as our theoretical estimates [Eqs. (19) and (15)] for the highest-n value. We can see that K^γ_{pol} agrees within 20% (except for Ne and Xe) with the experimental K^γ(S) values. However, as previously noted, Eq. (19) cannot reliably provide the contribution of the scattering because the addition of K^γ_{sc} to K^γ_{pol} leads to K^γ_{tot} values which are much larger than the measured K^γ(S) ones.

The broadening of the nD levels exhibits a quite different behavior (see Fig. 7.6). This is clearly due to a large contribution from inelastic scattering and is quite consistent with the large depopulation cross sections obtained by Gallagher et al.[44] for these hydrogenic states (see Sect. 7.3a). If one simply adds to the experimental results of Kachru et al.[76] for nS levels (for n ranging from 8 to 15) in collision with He, Ne, and Ar the corresponding inelastic contributions, as given in Ref. 44 for D states, reasonable agreement is obtained with the measurements of Kachru et al. for D states, indicating good self-consistency of their data. Note that for the highest-n values K^γ (D) (see Table 7.5) shows better agreement with K^γ_{tot} than do the K^γ (S) values. This clearly indicates that, as expected owing to the hydrogenic nature of the Na (nD) states, Eq. (19) provides more reasonable estimates of the inelastic broadening rates in the case of the D levels that it does for the nonhydrogenic nS states. Finally, we must mention that the somewhat anomalous behavior of the neon case has, to our knowledge, not yet been explained.

Brillet et al.[29] recently measured the broadening and shift of two-photon Rydberg 5s–ns lines in rubidium perturbed by foreign gases ($15 \leqslant n \leqslant 35$). They used the Doppler-free two-photon absorption method with classical fluorescence detection (see Sect. 7.2). The shift was observed to be constant in the n=15–30 range (except, perhaps, for a small increase in the neon case). Moreover, the measured values agreed well with our theoretical predictions

Table 7.6. Comparison between experimental K^{Δ}_{exp} and theoretical K^{Δ}_{th} asymptotic shift rate values for rubidium S states perturbed by rare gas atoms; the contribution of polarization K^{Δ}_{pol} and scattering K^{Δ}_{sc} effects are also reported, all values in atomic units; negative values refer to red shifts

Perturber	K^{Δ}_{exp} [29]	K^{Δ}_{th}	K^{Δ}_{pol}	K^{Δ}_{sc}
He	7.82	6.78	−0.70	7.48
Ne	0.66	0.67	−0.84	1.51
Ar	−11.9	−12.7	−2.00	−10.7
Kr	−23.8	−25.7	−2.46	−23.2
Xe	−47.7	−44.1	−3.32	−49.8

from Eqs. (10) and (16) as shown in Table 7.6. The polarization contribution is only noticeable in the neon case, as expected from the low-σ_0 value that leads to a small contribution to the scattering effect. The broadening rates, shown in Fig. 7.7, clearly exhibit the influence of the inelastic collisions. This is consistent with the fact that, owing to their quantum defect ($\delta = 3.1$), the depopulation cross sections of Rb (nS) levels have nonnegligible values, as measured by Hugon et al.[58] (see Sect. 7.3a).

To summarize, the recent line-profile studies have shown the influence of the inelastic collisions on the broadening rates, especially for the case of hydrogenic states. Previous studies dealing with the nonhydrogenic nP states could not reveal such an effect. An interesting connection between collisional and line-shape studies thus appears. For example, the broadening cross sections measured by Kachru et al.[76] for Na (nD) states perturbed by Kr and Xe allow, after subtraction of the polarization contribution, determination of the corresponding depopulation cross sections of the Na (nD) states for $n > 18$ with reasonable accuracy. Following the same procedure, we can show that the depopulation cross sections for Rydberg alkali states in collisions with heavier rare gases decrease, with increasing n values, as expected but not yet observed by classical collisional techniques.

Collisions with ground-state alkali atoms

Measurements of self-broadening and shift of the principal series of K, Rb, and Cs were performed prior to 1976.[27,75,77] The results were discussed by Omont.[7] Their main features are the following. Broadening and shift have nearly constant values for $n > 30$. For $n \lesssim 30$, monotonic n behavior (for K and Rb) as well as oscillatory n behavior (for Cs) were observed for both γ and Δ. The measured asymptotic K^{γ} and K^{Δ} values were found to be much larger than those corresponding to an interaction with a rare gas atom. For example, in Cs, $K^{\gamma} = 170$ a.u. and $K^{\Delta} \simeq 190$ a.u. compared with $K^{\gamma} \simeq 1.6$ a.u. and $K^{\Delta} \simeq$

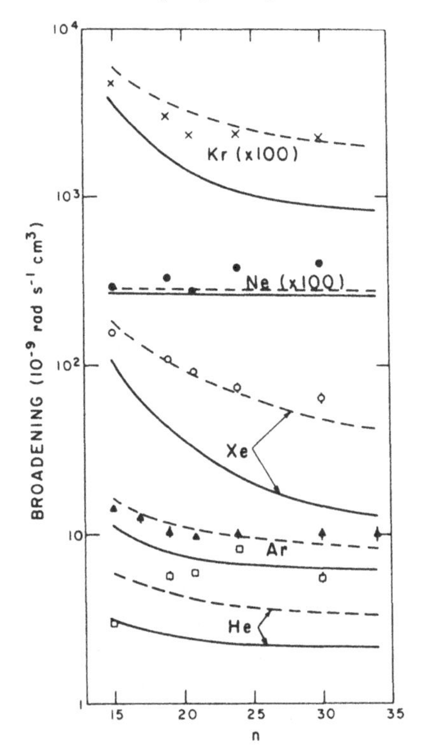

Fig. 7.7. Broadening rates of 5S-nS transitions in rubidium perturbed by rare gas atoms.[29] The solid lines are the theoretical values computed from Eqs. (20)-(23); the broken lines correspond to the sum of those values and the inelastic collision contribution [Eq. (19)].

5 a.u. for Cs (nP) states perturbed by Ar. Because the calculated K^{γ}_{pol} and K^{Δ}_{pol} [respectively, 11 and 19 a.u. according to Eqs. (15) and (16)] are much smaller than the observed K^{γ} and K^{Δ} values, the scattering contribution is dominant for both the shift and the broadening. The inelastic contribution is probably small (on the order of the polarization contribution, at most) if one refers to the alkali-alkali quenching rates obtained by Gounand et al.[9] for Rb (nP) levels. It seems difficult to evaluate the elastic scattering part because the basic assumptions of Eqs. (10) and (22) are no longer valid (the scattering length approximation has to be modified). No precise experimental data concerning the e^--alkali elastic scattering are available,[47] which leads to very imprecise knowledge of the corresponding phase shifts that, according to Eqs. (13) and (14), govern the scattering contribution. Presnyakov[78] attempted to evaluate the scattering contribution but the method used seems questionable.[7] Therefore, comparison between theory and experiment is difficult.

Recently, measurements of self-broadening of D lines in cesium,[30] as well as of self-broadening and shift of S and D lines in rubidium[79, 80] were performed

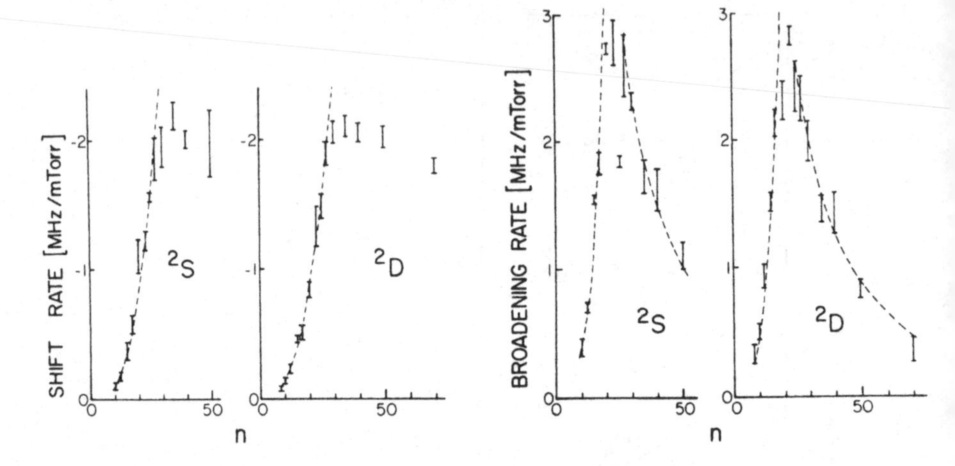

Fig. 7.8. Observed shift and broadening rates for Rydberg nS and nD two-photon lines of Rb, induced by ground-state Rb atoms. (From Stoicheff and Weinberger[80])

using Doppler-free two-photon excitation and thermionic detection (see Sect. 7.2). The broadening rate for Cs (nD) states[30] increases monotonically up to $n \simeq 28$ and then decreases slightly. For $n \simeq 40$, K^{γ} is about 85 a.u. (the polarization contribution is calculated to be on the order of 10 a.u. and the inelastic scattering contribution is probably on the same order of magnitude). These observations are compared with the results of Mazing and Serapinas,[27] who found a K^{γ} value on the order of 170 a.u. for P states. In fact, no major difference is expected between D and P states for the same alkali atom in view of Eq. (14). The authors point out the difficulties of classical absorption methods and suspect the oscillations observed by Russian physicists in the n dependence of the broadening of the nP states to be the result of the strong influence of satellites leading to non-Lorentzian profiles. Note, however, that Kaulakis et al.[81] recently explained the oscillations observed in Ref. 27 in terms of a new resonance scattering process.

References 79 and 80 report measurements of both self-broadening and shift of S and D two-photon lines in rubidium. Results of Ref. 80 are shown in Fig. 7.8. Consider first the shift: it was observed to increase up to $n \simeq 30$ before reaching a nearly constant value. The results of the two groups are in very good agreement, leading (for $n \gtrsim 30$) to $K^{\Delta} \simeq 104$ and 110 a.u. for S and D lines, respectively. The polarization contribution is expected to be about 17 a.u. for both cases. These experimentally obtained K^{Δ} values are to be compared with the ones previously published for nP states (120 a.u. for K and 190 a.u. for Cs). An agreement seems, therefore, to exist between the results concerning different alkali and spectral series for the shift induced by a perturber of the same chemical nature. No direct comparison between theory and experiment is possible.

Concerning the broadening, it increases up to $n \simeq 20$ and then decreases continuously to $n \simeq 70$. Agreement is observed between the two sets of results for the maximum values of K^γ (~ 80 a.u. for both S and D lines); the values obtained by Stoicheff and Weinberger[80] are slightly smaller than those measured by Weber and Niemax[79] (for $n=60$, $K^\gamma \simeq 19$ a.u. and 27 ± 7 a.u. respectively for D states).[82] It should be noted that, contrary to the nP levels of K and Cs, the polarization contribution (~ 10 a.u.) accounts for a significant amount in the observed K^γ values, especially for D states. A significant contribution from inelastic collisions is also probable in view of the values of the total depopulation rates reported by Hugon et al.[58] for Rb ($n\mathrm{D}_{3/2}$) and Rb (nS) levels in collisions with ground-state Rb atoms.[83] Thus the observed decrease of K^γ in the $n=30$–50 range is due to the decrease of both the elastic and the inelastic scattering contributions. These contributions are difficult to evaluate theoretically, contrary to the case of rare gas perturbers.

Pronounced periodic oscillations as a function of n were observed (for $10 \lesssim n \lesssim 30$) by Stoicheff and Weinberger[80] for the broadening of the S levels; for the D states, oscillations, if present, are much less pronounced (these observations are not inconsistent with the data of Ref. 79, in view of their large scatter). The authors tentatively attribute these oscillations to an interaction between the valence electron and a quasi-molecular ion formed during the collision between a ground-state Rb atom with the ionic Rb^+ core. Recently, oscillations in the n dependence of both linewidths and frequency shifts were observed for S and D series of potassium.[84] Similar oscillations in the n dependence of the linewidths for the D series of cesium were observed by Niemax and Weber.[85] These authors suggest that the oscillations are associated with different ion production rates for different atomic states. Note that Kaulakis et al.[81] and Presnyakov[86] explained the oscillations observed for both the linewidth and the shift of various alkali Rydberg lines by a resonance-scattering mechanism.

Clearly, more experimental as well as theoretical work is needed for a better understanding of the broadening and shift of Rydberg alkali lines due to perturbation by alkali ground-state atoms. It will require, in particular, a satisfactory evaluation of the scattering contribution, which is, as previously mentioned, very difficult to obtain.

7.3c. Summarizing discussion

From the information provided in this chapter, we conclude that there exist clearly some features common to the available experimental data on Rydberg atom–atom interactions at thermal energies.

The Rydberg levels exhibit a relative stability to perturbations due to collisions with atomic species. Although large depopulation cross sections are generally observed, their n dependence shows a saturation or even a decrease in the high-n range (for inelastic processes, this saturation is not observed for all cases but should probably occur at sufficiently high-n values). Similar observations are also made for the broadening and shift of spectral lines. This behavior is undoubtedly remarkable.

The parameter ΔE has an indisputable influence on the cross section values for depopulation processes as well as, in many cases, on the broadening and shift rates (see, for example, the broadening of alkali hydrogenic states by rare gases).

Many processes reviewed here are predominantly due to the interaction of the perturber with the Rydberg outer electron. This interaction was shown to prevail for quasi-elastic processes and for many inelastic processes. This is also true for line-broadening processes (see, for example, the broadening of hydrogenic states by rare gases and the broadening and shift in the case of alkali–alkali interactions).

In this chapter we have shown a clear connection between collisional de-excitation and line-shape studies. This might be usefully considered in the future.

7.4. Concluding remarks

It is clear from the presentation in this chapter that some lines of research deserve more attention in the future.

The e^--perturber interaction is, in many cases considered here, the dominant interaction during Rydberg atom–atom collisions, i.e., it was shown to be responsible for many of the elastic and inelastic processes observed. It is necessary, however, to investigate more completely the relative influence of different interactions for the other collisional processes. In particular, depopulation experiments involving nonhydrogenic states in collision with various perturbers would more precisely define the relative contribution of the ionic core–perturber interaction. Similarly, studies of both depopulation processes and line profiles involving high-n states can be expected to provide significant insight to that problem (let us recall that in the high-n range only the broadening by rare gas atoms has already been clearly shown to display a major influence of the ionic core–perturber interaction). Many of these future experiments will require more sensitivity of the data acquisition system and therefore refinements of the available methods as well as a search for new techniques are necessary.

Clearly, a large theoretical effort is still required in order to provide a precise comparison of the existing data with theoretical results; this is of particular concern for alkali–alkali collisions. Such an effort would certainly provide invaluable help for guiding and interpreting future experimental work.

No mention was made in this chapter of the energy (or temperature) dependence of the cross sections, which may provide some checks of the theoretical computations. Experiments on interatomic excitation transfer are of great practical interest. Finally, there is a lack of data concerning collisional ionization.

At the end of this chapter, it is worth mentioning that Fermi was the first to consider that, during the collision of a Rydberg atom with a neutral perturber, the ionic core of the Rydberg atom and the outer electron might act as separate scatters for the perturber. This is certainly the most interesting feature of the Rydberg atom–atom interaction, as evidenced by this chapter. In that respect, these collisions occupy an almost unique situation in collision physics.

Acknowledgments

The authors would like to thank M. Hugon for many helpful comments as well as for a careful reading of the manuscript. They also acknowledge the helpful discussions with T. F. Gallagher.

Addendum

Since initial preparation of this manuscript further papers dealing with collisional processes involving Rydberg atoms have been published, confirming the vitality of the field. Some new lines of research have emerged; we would like to mention them briefly here.

One new and promising line of research appears to be the study of collisions involving Rydberg atoms in the presence of a dc electric field.[87,88] Two effects are expected: a change in the cross section value and a modification in the distribution of the end products. Both effects have been observed but no detailed calculations are available to account for these measurements quantitatively. Another new line of research is the study of Rydberg atom–Rydberg atom collisions. They have been studied under both resonant and nonresonant conditions. In the first case,[89,90] a small dc electric field was used in order to obtain exact energy balance between the entrance and exit channels. Huge cross sections ($\sim 10^9$ Å^2) were reported, in good agreement with theoretical estimates. In the second case,[91] the study of Rydberg spectral lines at high densities of Rydberg atoms led to interesting observations. This emphasizes the clear connection between the line shape and collisional studies dealing with Rydberg atoms. A model has recently been proposed[92] that allows one to connect the wide set of data already available for the alkali Rydberg states and lines perturbed by rare gases; it is worthwhile to mention here that measurements[93,94] at high-n values ($n \simeq 40$) performed by using a field-ionization technique, have confirmed the dominant contribution of the outer electron-perturber interaction to the quenching process.

These examples show that all the aspects of Rydberg atom–atom collisions have certainly not yet been fully explored. Moreover they confirm and justify, if necessary, the strong interest of the scientific community in this exciting field of research.

References and notes

1. E. Fermi, *Nuovo Cimento 11,* 157 (1934).
2. E. Amaldi and E. Segré, *Nature 133,* 141 (1934); *Nuovo Cimento 11,* 145 (1934).
3. R. F. Stebbings, in Proceedings, *X ICPEAC,* ed. G. Watel (Amsterdam: North-Holland Publ., 1978), p. 547.
4. R. F. Stebbings, *Advances in Atomic and Molecular Physics,* eds. D. R. Bates and B. Bederson (New York: Academic Press, 1979), vol. 15, p. 77.
5. S. A. Edelstein and T. F. Gallagher, in *Advances in Atomic and Molecular Physics,* eds. D. R. Bates and B. Bederson (New York: Academic Press, 1978), vol. 14, p. 365.
6. T. F. Gallagher, in Proceedings, *XI ICPEAC,* eds. N. Oda and K. Takayanagi (Amsterdam: North-Holland Publ., 1980), p. 473.
7. A. Omont, *J. Phys. (Paris) 38,* 1343 (1977).
8. F. P. Schäfer, ed., *Dye Lasers* (New York: Springer-Verlag, 1973).
9. F. Gounand, P. R. Fournier, and J. Berlande, *Phys. Rev. A 15,* 2212 (1977).
10. T. F. Gallagher, S. A. Edelstein, and R. M. Hill, *Phys. Rev. Lett. 35,* 644 (1975).
11. J. S. Deech, R. Luypaert, L. R. Pendrill, and G. W. Series, *J. Phys. B 10,* L137 (1977).
12. L. R. Pendrill and G. W. Series, *J. Phys. B 11,* 4049 (1978).
13. F. Gounand, M. Hugon, P. R. Fournier, and J. Berlande, *J. Phys. B 12,* 547 (1979).
14. M. Hugon, F. Gounand, and P. R. Fournier, *J. Phys. B 11,* L605 (1978).
15. A similar method was developed by J. Marek and M. Ryschka [*Phys. Lett. 74 A,* 51 (1979)]. In this work, superradiant transfer occurs at the second step of the pumping rather than at the third step.
16. W. E. Cooke, T. F. Gallagher, S. A. Edelstein, and R. M. Hill, *Phys. Rev. Lett. 40,* 178 (1978).
17. H. L. Chen and C. Borzilieri, *J. Chem. Phys. 72,* 858 (1980).
18. R. S. Freund, T. A. Miller, B. R. Zegarski, R. Jost, M. Lombardi, and A. Dorelon, *Chem. Phys. Lett. 51,* 18 (1977).
19. F. Devos, J. Boulmer, and J. F. Delpech, *J. Phys. (Paris) 40,* 215 (1979).
20. A. Corney, *Adv. Electron. Electron Phys. 29,* 119 (1970).
21. T. F. Gallagher and W. E. Cooke, *Phys. Rev. Lett. 42,* 835 (1979).
22. A. C. Tam, T. Yabuzaki, S. M. Curry, M. Hou, and W. Happer, *Phys. Rev. A 17,* 1862 (1978).
23. T. F. Gallagher, L. M. Humphrey, R. M. Hill, and S. A. Edelstein, *Phys. Rev. Lett 37,* 1465 (1976); see also J. L. Vialle and H. T. Duong, *J. Phys. B 12,* 1407 (1979).
24. Some qualitative collisional results have been obtained using this techniques [T. F. Gallagher, W. E. Cooke, and S. A. Edelstein, *Phys. Rev. A 17,* 904 (1978)], which has already been widely used for Rydberg atom–molecule collisions.
25. A. N. Klucharev, A. V. Lazarenko, and V. Vrijnovic, *J. Phys. B 13,* 1143 (1980). For a theoretical interpretation of these results, see A. A. Mihajlov and R. K. Janev, *J. Phys. B 14,* 1639 (1981).
26. E. F. Worden, J. A. Paisner, and J. G. Conway, *Optics Lett. 3,* 156 (1978).
27. M. A. Mazing and P. D. Serapinas, *Sov. Phys. JETP 33,* 294 (1971).
28. L. S. Vasilenko, V. P. Chevotayev, and A. V. Shiskaev, *Sov. Phys. JETP Lett. 12,* 113 (1970); B. Cagnac, G. Grynberg, and F. Biraben, *Phys. Rev. Lett. 32,* 643 (1974); see also N. Bloembergen and M. D. Levenson, in *High Resolution Laser Spectroscopy,* ed. K. Shimoda (New York: Springer-Verlag, 1976), p. 315.

29. W. L. Brillet and A. Gallagher, *Phys. Rev. A 22,* 1012 (1980).
30. K. H. Weber and K. Niemax, *Opt. Commun. 28,* 317 (1979); see also B. P. Stoicheff and E. Weinberger, *Phys. Rev. Lett. 44,* 733 (1980).
31. D. Popescu, I. Popescu, and J. Richter, *Z. Phys. 226,* 160 (1969).
32. A. Flusberg, R. Kachru, T. Mossberg, and S. R. Hartmann, *Phys. Rev. A 19,* 1607 (1979).
33. R. Kachru, T. W. Mossberg, E. Whittaker, and S. R. Hartmann, *Opt. Commun. 31,* 223 (1979).
34. A. Gallagher and E. L. Lewis, *J. Opt. Soc. Am. 63,* 864 (1973).
35. V. I. Goldanskii, A. V. Kutsenko, and M. I. Podgoretskii, *Counting Statistics of Nuclear Particles* (Delhi: Hindustan Publ., 1962).
36. J. L. Curtis and P. Erman, *J. Opt. Soc. Am. 67,* 1218 (1977).
37. J. S. Deech, R. Luypaert, and G. W. Series, *J. Phys. B 8,* 1406 (1975).
38. R. D. Dyson and I. Isenberg, *Biochemistry 10,* 3233 (1971).
39. L. R. Pendrill, *J. Phys. B 10,* L469 (1977).
40. I. C. Percival, *Nucl. Fusion 6,* 182 (1966).
41. B. P. Stoicheff and E. Weinberger, *Can. J. Phys. 57,* 2143 (1979).
42. See, for example, A. Gallagher, *Phys. Rev. 157,* 68 (1967).
43. The concept of energy defect ΔE is clear only if the nature of the exit channel is known. For a quenching process, ΔE has to be understood as the typical energy defect between the level studied and its neighbors.
44. T. F. Gallagher, S. A. Edelstein, and R. M. Hill, *Phys. Rev. A 15,* 1945 (1977).
45. T. F. Gallagher, W. E. Cooke, and S. A. Edelstein, *Phys. Rev. A 17,* 904 (1978).
46. M. Hugon, F. Gounand, P. R. Fournier, and J. Berlande, *J. Phys. B 12,* 2707 (1979).
47. H. S. W. Massey, E. H. S. Burhop, and H. B. Gilbody, *Electronic and Ionic Impact Phenomena* (Oxford University Press, 1969), vol. 1, chap. 6.
48. E. de Prunelé and J. Pascale, *J. Phys. B 12,* 2511 (1979).
49. M. R. Flannery, *Ann. Phys. (NY) 61,* 465 (1970).
50. R. E. Olson, *Phys. Rev. A 15,* 631 (1977).
51. A. P. Hickman, *Phys. Rev. A 18,* 1339 (1978).
52. J. Derouard and M. Lombardi, *J. Phys. B 11,* 3875 (1978).
53. M. Matsusawa, *J. Phys. B 12,* 3743 (1979).
54. A. P. Hickman, *Phys. Rev. A 19,* 394 (1979).
55. $\sigma_G = \pi\langle r^2\rangle = (\pi/2)n^{*2}[5n^{*2}+1-3l(l+1)]a_0^2$, where a_0 is the Bohr radius and n^* the effective quantum number, i.e., $n^* = n - \delta$.
56. Angular mixing has also been studied in the case of Rydberg helium levels,[18] but, because of the large energy defect involved (>100 cm^{-1}), we shall review these data in Sect. 7.3a, under "Inelastic process."
57. M. Hugon, F. Gounand, and P. R. Fournier, *J. Phys. B 13,* L109 (1980).
58. M. Hugon, F. Gounand, P. R. Fournier, and J. Berlande, *J. Phys. B 13,* 1585 (1980).
59. J. Boulmer, G. Baran, F. Devos, and J.-F. Delpech, *Phys. Rev. Lett. 44,* 1122 (1980).
60. T. F. Gallagher and W. E. Cooke, *Phys. Rev. A 19,* 2161 (1979).
61. Y. S. Kim and R. G. Gordon, *J. Chem. Phys. 61,* 1 (1974).
62. F. Gounand, unpublished thesis, University Paris-Sud, Orsay, 1980.
63. A. Hitachi, C. Davies, T. A. King, S. Kubota, and T. Doke, *Phys. Rev. A 22,* 856 (1980).
64. Although ionization is possible from a collision with an atom B of different chemi-

cal nature (if energy requirements are met), no experiment has been reported concerning this process.
65. F. Mohler and C. Boeckner, *J. Res. Natl. Bur. Stand.* *5,* 51 (1930); *5,* 399 (1930).
66. Y. T. Lee and B. H. Mahan, *J. Chem. Phys.* *42,* 2893 (1965).
67. F. Gounand, *J. Phys.* (*Paris*) *40,* 457 (1979); unpublished Commissariat à l'energie atomique reports, 1979.
68. B. V. Dobrolezh, A. N. Klucharev, and V. Y. Sepman, *Opt. Spektrosk.* *38,* 1090 (1975) [*Opt. Spectrosc.* (*USSR*) *38,* 630 (1975)]. Y. P. Korchevoi, V. I. Lukashenko, and J. N. Khil'ko, *Zh. Tekh. Fiz.* *46,* 2302 (1976) [*Sov. Phys. Tech. Phys.* *21,* 1356 (1976)]; and references therein.
69. For these calculations, the L value reported by A. Omont [*J. Phys.* (*Paris*) *38,* 1343 (1977)] for rare gas atoms was used.
70. O. B. Firsov {*Zh. Eksp. Teor. Fiz.* *21,* 627 and 634 (1951)} and G. K. Ivanov {*Opt. Spektrosk.* *40,* 965 (1976) [*Opt. Spectrosc.* (*USSR*) *40,* 554 (1976)]} extensively discussed the validity of Fermi's treatment.
71. V. A. Alekseev and I. I. Sobel'man, *Zh. Eksp. Teor. Fiz.* *49,* 1274 (1965) [*Sov. Phys. JETP 22*, 882 (1966)].
72. We only give the impact theory results because they are sufficient for the discussion of most of the experimental results.
73. S. Y. Ch'en and M. Takeo, *Rev. Mod. Phys 29,* 20 (1957) and references therein. For recent articles on alkaline earth atoms, see, for example, J. R. Rubbmark, S. A. Borgstjöm, and K. Bochasten, *J. Phys. B 10,* 421 (1977) or K. Veda, T. Fujimoto, and K. Fukuda, *J. Phys. Soc. Jap.* *49,* 1147 (1980).
74. M. A. Mazing and N. A. Vrubleskaya, *Zh. Eksp. Teor. Fiz.* *50,* 343 (1966) [*Sov. Phys. JETP 23,* 228 (1966)].
75. P. D. Serapinas, in *Physics of Atomic Collisions* (New York: Consultants Bureau, 1971), vol. 51, p. 81.
76. R. Kachru, T. W. Mossberg, and S. R. Hartmann, *Phys. Rev. A 21,* 1124 (1980); see also A. Flusberg, R. Kachru, T. Mossberg, and S. R. Hartmann, *Phys. Rev. A 19,* 1607 (1979).
77. K. Wendt and H. J. Kusch, *Z. Phys. A 275,* 111 (1975).
78. L. P. Presnyakov, *Phys. Rev. A 2,* 1720 (1970).
79. K. H. Weber and K. Niemax, *Opt. Commun.* *31,* 52 (1979).
80. B. P. Stoicheff and E. Weinberger, *Phys. Rev. Lett.* *44,* 733 (1980).
81. B. P. Kaulakis, L. P. Presnyakov, and P. D. Serapinas, *Zh. Eksp. Teor. Fiz.* *30,* 60 (1979) [*Sov. Phys. JETP Lett.* *30,* 53 (1979)].
82. These K^{γ} values have the same magnitude as those obtained for the Rb (nP) states[75] ($K^{\gamma}=42$ a.u. for $n=32$).
83. It is, in fact, the broadening of the fine-structure levels that is measured by high-resolution, two-photon, Doppler-free absorption.
84. B. P. Stoicheff, D. C. Thompson, and E. Weinberger, in *Proceedings of the Fifth International Conference on Spectral Line Shapes,* eds. J. Seidel and B. Wende (Berlin: de Gruyter, 1981), p. 1071.
85. K. Niemax and K. H. Weber, in *Program and Abstracts of the Fifth International Conference on Spectral Line Shapes,* eds. J. Seidel and B. Wende (Berlin: de Gruyter, 1980), p. 115.
86. L. P. Presnyakov, private communication.
87. M. Slusher, C. Higgs, K. A. Smith, F. B. Dunning, and R. F. Stebbings, in Proceedings, *XII ICPEAC* Gatlinburg, (1981), p. 1111; *Phys. Rev. A 26,* 1350 (1982).

88. R. Kachru, T. F. Gallagher, F. Gounand, K. A. Safinya, and W. Sandner, *Phys. Rev. A*. (1982), submitted.
89. K. A. Safinya, J. F. Delpech, F. Gounand, W. Sandner, and T. F. Gallagher, *Phys. Rev. Lett. 47,* 405 (1981).
90. T. F. Gallagher, K. A. Safinya, F. Gounand, J. F. Delpech, W. Sandner, and R. Kachru, *Phys. Rev. A 25,* 1905 (1982).
91. J. M. Raimond, G. Vitrant, and S. Haroche, *J. Phys. B 14,* L655 (1981).
92. F. Gounand, J. Szudy, M. Hugon, B. Sayer, and P. R. Fournier, *Phys. Rev. A 26,* 831 (1982).
93. M. Hugon, P. R. Fournier, and E. de Prunelé, *J. Phys. B 14,* 4041 (1981).
94. M. Hugon, B. Sayer, P. R. Fournier, and F. Gounand, *J. Phys. B* (1982), submitted.

8

Theoretical studies of collisions of Rydberg atoms with molecules

MICHIO MATSUZAWA

8.1. Introduction

Recently, atoms in highly excited Rydberg states (which we shall refer to for simplicity as high-Rydberg atoms) have attracted the interest of many laboratories because of their many possible applications, i.e., isotope separation,[1] nuclear fusion,[2] and radioastronomy.[3] Thus collisional properties of high-Rydberg atoms have been the target of active theoretical and experimental research. The behavior of these atoms is quite different from that of atoms in the ground or lower excited states when they interact with neutral species because of the characteristic properties of high-Rydberg atoms. In Table 8.1, we briefly summarize a few of the properties of a high-Rydberg atom that are essential to our theoretical considerations of Rydberg–neutral collisions. Numerical values are shown for the ground state, i.e., principal quantum number $n=1$, and for a high-Rydberg state with $n=30$, typical of those currently being studied in the laboratory.

Usually we can employ the Born–Oppenheimer separation for the slow atom–atom (molecule) collision. Thus adiabatic potential energy curves give an adequate basis for the description of slow atomic collision processes. The applicability of such potential energy curves comes from the fact that the electrons in the ground state or in lower excited states move very fast compared with the relative motion and can adjust themselves adiabatically to it. In the high-Rydberg states, the excited Rydberg electron moves rather slowly along the large orbit. For example, the root-mean-square velocity v_n of the electron in a high-Rydberg state with principal quantum number $n \simeq 30$ is 7.3×10^6 cm s^{-1}, which is still fast compared with those of heavy particles at thermal energies. However, the condition for the validity of the Born–Oppenheimer separation becomes less satisfied as n increases. Therefore, we must resort to a completely different way of describing high-Rydberg–neutral collisions.

In Table 8.1, one immediately sees that the radius of the high-Rydberg orbit is usually large compared with an effective range of interaction between a charged particle and a neutral species. Thus, when a high-Rydberg atom A^{**} interacts with a neutral species B, the Rydberg electron e_R and the atomic ion

Table 8.1. Some properties of high-Rydberg atoms essential to the understanding of Rydberg–neutral collisions

Property	n-Dependence	Numerical values $n=1$	$n=30$
Mean radius r_n	$n^2 a_0$	5.3×10^{-9} cm $(=a_0)$	4.8×10^{-6} cm
Root-mean-square velocity of the Rydberg electron v_n	v_0/n	2.2×10^{8} cm s^{-1} $(=v_0)$	7.3×10^{6} cm s^{-1}
Period of the electronic motion τ_n	$n^3\tau_1$	1.5×10^{-16} s $(=\tau_1)$	4.1×10^{-12} s
Binding energy E_n	R/n^2	13.6 eV $(=R)$	15 meV
Level spacing between adjacent levels ΔE_n	$2R/n^3$	~10 eV	1 meV

core A^+ behave as independent scatterers. Hence it is quite unlikely that two types of interaction, i.e., those between e_R–B and A^+–B play decisive roles simultaneously in the Rydberg–neutral collision processes with high n under usual conditions. This essential feature of the Rydberg–neutral interaction was first recognized by Fermi,[4] who used it to explain experimental data on pressure shifts of high-Rydberg series perturbed by a neutral species. If one observes the absorption lines of the high-Rydberg series of an alkali atom in 1-atm pressure of a rare gas, one still sees very sharp absorption lines even for very high principal quantum numbers, say $n=20$–30, as first observed by Amaldi and Segrè.[5] In this case, although the levels of the high-Rydberg series shift considerably, i.e., by several wave numbers, they broaden relatively little. As shown in Table 8.1, a high-Rydberg atom with $n=30$ has a large radius and contains about 10^4 atoms within its orbit if it is placed in a rare gas at atmospheric pressure at 0°C. Thus the Rydberg electron interacts with a large number of rare gas atoms impusively while it moves slowly along its large orbit. However, this excited atom is still stable. This is quite surprising because its small binding energy is known to make the high-Rydberg atom very fragile (see Table 8.1).

Fermi[4] ascribed this phenomenon to the following two effects (1) elastic scattering of the quasi-free Rydberg electron e_R by rare gas atoms B and (2) polarization of the rare gas atoms B within the Rydberg orbit by the electric field caused by the atomic ion core A^+. In this phenomenon, the e_R–B and A^+–B interactions coexist, but they are statistically independent of each other. Further in the scattering effect, Fermi proposed a model in which the Rydberg electron behaved as if it were "free" and slow. (Hereafter, for simplicity, we shall refer to this model as the "free" electron model.) His model was successfully applied to analysis of experimental data on the pressure shift. (For more details on this subject, the reader may refer to review articles.[6-10]) Later, Russian groups undertook refined treatments of this phenomenon, both theoretically[11,12] and experimentally.[13,14]

In the beginning of the 1970s several authors, i.e., Flannery,[15,16] Matsuzawa,[17-20] and Fowler and Preist,[21,22] independently pointed out that collision processes between high-Rydberg atoms and neutral species

$$A^{**}(\alpha) + B(\beta) \rightarrow A^{**}(\alpha') + B(\beta') \quad (1)$$

can be largely understood in terms of information on elastic and inelastic scattering of slow electrons by B, i.e.,

$$e + B(\beta) \rightarrow e + B(\beta') \quad (2)$$

where the sets of quantum numbers $\alpha(\alpha')$ and $\beta(\beta')$ denote the states of the high-Rydberg atoms and the neutral species, respectively. Here we assume that B is in the ground state or a lower excited state. Therefore we can use the "free" electron model and assume that the e_R–B interaction plays a decisive role in the collision process (1).

For fast collisions, Butler and May[23] predicted, based on the "free" electron model, that the cross sections σ_i for the collisional ionization of $H^{**}(n)$ by H(1s) at velocity V should be approximately equal to the total scattering cross section σ_e^t for free electrons colliding with H(1s) at the same velocity V.

These considerations lead us to the conclusion that the neutral species B does not interact with the high-Rydberg atom A^{**} as a whole. Therefore the state-to-state cross section for the Rydberg–neutral collision should be smaller than the geometrical cross section of the high-Rydberg atom and does not increase as the latter increases.[24] [More precisely the Rydberg–neutral cross sections should be smaller than the geometrical cross section multiplied by the statistical weight g_f of the final channel if $g_f \gg 1$.] On the other hand, if the Rydberg atom A^{**} interacts with a charged particle B^+, any interaction between two of the particles A^+, e_R, and B^+ is of Coulombic character. Hence the charged particles B^+ can interact with A^{**} as a whole and the cross sections for the A^{**}–B^+ collisions will increase proportionally to the size of the geometrical cross section [see the review articles on this subject by Percival and Richards[25] and Takayanagi[26]]. Thus we see that there is sharp contrast between the n dependence of the cross sections for Rydberg–neutral collisions and that for Rydberg–charged-particle collisions.

In this chapter, we wish to review the theoretical studies of collisions of high-Rydberg atoms with molecules. In Sect. 8.2, we shall discuss theoretical treatments based on the impulse approximation and the "free" electron model, putting emphasis on the understanding of the essential physics of the collision process (1). To make our arguments complete for collision process (1), we shall also briefly review some theoretical models other than the "free" electron one. In Sects. 8.3 and 8.4, details of the theoretical results for slow and fast collisions with molecules will be given and compared with selected experimental data. In Sect. 8.5, we shall discuss the extension of the theoretical treatments given in Sect. 8.2 to collisions involving high-Rydberg ions. In the last section, we shall present concluding remarks and discuss the future problems in this field.

8.2. Theoretical treatments

Let us consider a polar molecule as the neutral collision partner B. Specifically, we take as collision partners H**(30) and a polar molecule with dipole moment $D \simeq 1ea_0$, where e is the electronic charge. Several authors calculated the cross sections for the rotational excitation of a polar molecule by slow electron impact. For example, the cross section for $J=0 \rightarrow 1$ excitation of HCl is theoretically estimated to be about 10^{-13} cm^2 at thermal energies (see, for instance, Itikawa and Takayanagi;[27] also recent review articles by Itikawa[28] and Lane[29]). The cross section for rotational excitation (or deexcitation) is largest for electron–molecule scattering. On the other hand, the cross sections for the $J=0 \rightarrow 1$ rotational excitation of HCl by slow proton impact, i.e., $E \lesssim 0.1$ eV, are estimated to be at most of the order of 10^{-13} cm^2.[30] Using these data and the relation $\sigma \simeq \pi r^2$, we have

$$r(e\text{-HCl})/r_{30} \simeq 0.04 \quad \text{and} \quad r(\text{H}^+\text{-HCl})/r_{30} \lesssim 0.04$$

where $r(e\text{-HCl})$ and $r(\text{H}^+\text{-HCl})$ are the effective ranges of the e-HCl and H^+-HCl interactions, respectively. For rotational excitation of a polar molecule by electron impact, the cross section is inversely proportional to E_e, where E_e is the incident energy of the electron. This means that $r(e\text{-HCl})$ increases proportionally to n. However, the condition $r(e\text{-HCl})/r_n \ll 1$ becomes better satisfied for higher n because $r_n \simeq n^2 a_0$. If we take CN with a smaller rotational constant as the polar molecule, these conditions become less well satisfied. (Hereafter we shall use the symbol e_R if we wish to stress that it is a high-Rydberg electron.) Therefore, for the following theoretical consideration, it is convenient to categorize the Rydberg–neutral collisions into two ranges of n: (A) high n and (B) intermediate n depending on whether the conditions

$$r(e_R\text{-B}) \ll r_n \quad (\simeq n^2 a_0) \tag{3}$$

$$r(\text{A}^+\text{-B}) \ll r_n \quad (\simeq n^2 a_0) \tag{4}$$

are satisfied. For collisions with the polar molecules with $D \lesssim 1ea_0$, we can safely conclude that conditions (3) and (4) are satisfied for high n, say, $n \simeq 30$. The boundary values of n_b that separate these regions depend critically on the nature of the interaction between the charged particle and the neutral one. If B is a rare gas atom instead of the polar molecule, we see that n_b reduces to a

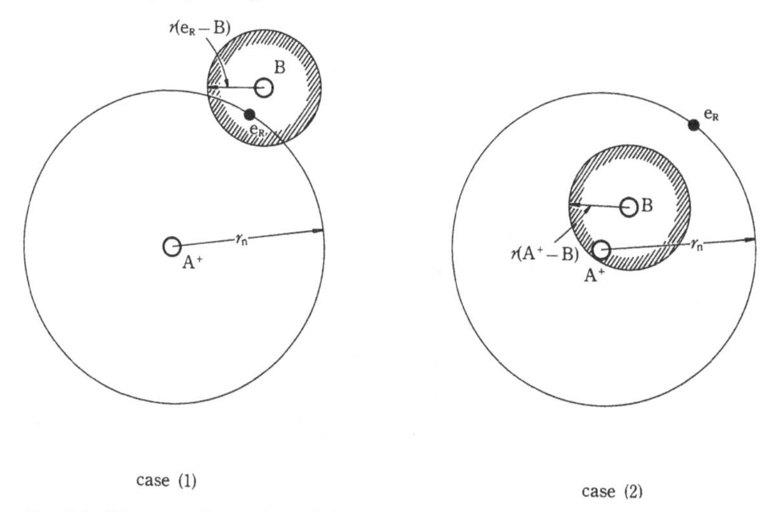

Fig. 8.1. Schematic illustration of the two extreme cases for high-Rydberg-neutral collisions: case (1), where the e_R–B interaction plays a decisive role; and case (2), where the A^+–B interaction is important.

lower value. In other words, for fixed n, conditions (3) and (4) are better satisfied for collisions with rare gas atoms than with polar molecules. In the intermediate-n range, we must take into account the fact that both the e–B and A^+–B interactions play important roles simultaneously. This requires an approach different from that of Fermi.

8.2a. Collisions involving Rydberg atoms with high n: validity of the impulse approximation and the "free" electron model

For the present case, two extreme classes of collisions can be identified: (1) the case where the e–B interaction plays a decisive role and (2) the case where the A^+–B interaction is important (Fig. 8.1). Theoretically speaking, we can treat them independently, though both cases may coexist experimentally. For example, the measured pressure shift of the high-Rydberg series can be given as a simple sum of these two contributions, i.e., the scattering effect and the polarization effect. However, because of conditions (3) and (4), we see immediately that collision process (1) accompanied with a change of quantum numbers denoting the state of e_R results mainly from the e_R–B interaction. On the other hand, process (1) involving changes of the internal state of A^+ occurs chiefly because of the A^+–B interaction. In both cases, changes of the internal state of B may also take place simultaneously. In collision experiments, we can, in principle, determine the quantum numbers. Hence, we can identify both cases experimentally for high n, even though the scattering cross sections arising from both types of interactions are of the same order of magnitude.

The case where the interaction between the Rydberg electron and the neutral species plays a decisive role

Table 8.1 shows that the binding energy of the Rydberg electron is very small and the velocity of the electron in its orbit is roughly equal to thermal velocities. Furthermore, the Coulomb field caused by A^+ can be considered to be almost constant where the e_R-B interaction has values appreciably different from zero. This suggests that the effect of the existence of A^+ can be taken into account as the momentum distribution of a quasi-free and slow electron. In the high-Rydberg state, the period τ_n of the electronic motion is usually long compared with the collision time τ_c between e_R and B. These features come from the semiclassical nature of the Coulomb field, particularly at its peripheral part (see Table 8.1). Taking the collision pair H**(30)-HCl as a specific example, we have as the velocities of the electron v_1 and v_{30} and of the relative motion V at room temperature,

$$v_1 \simeq 2.2 \times 10^8 \text{ cm s}^{-1} \qquad v_{30} \simeq 7 \times 10^6 \text{ cm s}^{-1}$$

and

$$V \simeq 2 \times 10^5 \text{ cm s}^{-1}$$

Thus we obtain $v_1/V \simeq 10^3$ and $v_{30}/V \simeq 40$. This comparison shows how the condition of Born–Oppenheimer separation becomes less satisfied as n increases, even at thermal energies. Further, we can estimate τ_c using $r(e_R\text{-HCl}) \simeq 2 \times 10^{-7}$ cm as

$$\tau_c = r(e_R\text{-HCl})/v_{30} \simeq 3 \times 10^{-14} \text{ s}$$

Table 8.1 gives $\tau_{30} = 4.1 \times 10^{-12}$ s, so

$$\tau_c/\tau_{30} \simeq 0.01$$

Therefore, we have, for high n,

$$\tau_c/\tau_n \ll 1 \tag{5}$$

This means that for case (1), e_R collides with B impulsively while A** is interacting with B at thermal velocities. Thus the impulse approximation is applicable to the present case for which $V \ll v_n$.[11,17,24,31–33]

It should be noted that conditions (3)–(5) are also satisfied for fast collisions of high-Rydberg atoms with neutral species, i.e., $V \gg v_n$.[34,35] This can be seen as follows: For the fast collisions, the orbital velocity v_n of the Rydberg electron can be neglected compared with the relative velocity V. Therefore we can put approximately

$$v_{eB} \simeq V, \qquad v_{A^+B} \simeq V$$

where v_{eB} and v_{A^+B} are the velocities of e_R and A^+ relative to B. For high velocities, the e_R–B and A^+–B collision cross sections usually become smaller while r_n is constant. Thus conditions (3) and (4) for fast collisions become better satisfied than for slow collisions. Further, the collision time τ_c also becomes small because $\tau_c \simeq r(e_R\text{-B})/v_{eB}$ and $v_{eB} \simeq V$. Hence we see that condition (5) is valid except for $v_n \simeq V$. This is rather a natural consequence because the impulse approximation was originally designed to describe fast collisions.[31,32,36]

Thus far we have mainly discussed collisions with polar molecules. The charge–dipole interaction is the longest-range interaction between a charged particle and a neutral one. Thus this is the least-favorable case for the Fermi-type approach. Hence, if the validity of this approach is proved for the present case, we see that conditions (3)–(5) are better satisfied for all other types of interaction between the charged particle and the neutral one. (For a more quantitative discussion, see Appendix.) Thus, for collisions of Rydberg atoms with high n, we can categorize the high-Rydberg–neutral collisions as slow, intermediate, and fast, depending on whether $V \ll v_n$, $V \simeq v_n$, and $V \gg v_n$, respectively. It should be noted that a thermal collision is not always equivalent to a slow collision. If we limit our discussion to high-Rydberg atoms with $n \lesssim 1000$, a thermal collision corresponds to a slow one. Later, however, we shall distinguish between the terminologies "slow" and "thermal" to make our argument unambiguous.

For collision process (1) with the momenta of relative motion $\hbar\mathbf{K}$ and $\hbar\mathbf{K}'$ before and after collision, the impulse approximation[11,17,31–33] yields a general relation between the scattering amplitude $f(\alpha, \beta, \mathbf{K} \to \alpha', \beta', \mathbf{K}')$ for process (1) and $f_e(\beta, \boldsymbol{\kappa} \to \beta', \boldsymbol{\kappa}')$ for process (2), i.e.,

$$f(\alpha, \beta, \mathbf{K} \to \alpha', \beta', \mathbf{K}') = \frac{\mu}{\mu_{eB}} \int G^*(\alpha' \mid \mathbf{q}') G(\alpha \mid \mathbf{q}) f_e(\beta, \boldsymbol{\kappa} \to \beta', \boldsymbol{\kappa}')\, d\mathbf{q} \qquad (6)$$

$$\mathbf{q} + [M_a/(M_a + m_e)]\mathbf{K} = \mathbf{q}' + [M_a/(M_a + m_e)]\mathbf{K}' \qquad (7)$$

Here $G(\alpha \mid \mathbf{q})$ and $G(\alpha' \mid \mathbf{q}')$ are the momentum-space wave functions of the Rydberg electron, $\mathbf{q}$ and $\mathbf{q}'$ the wave vectors relative to A^+, μ the reduced mass of the A^{**}–B system, μ_{eB} the reduced mass of the e–B system, and $\boldsymbol{\kappa}$ and $\boldsymbol{\kappa}'$ the wave vectors relative to B, i.e.,

$$\boldsymbol{\kappa} = \frac{M_B}{M_B + m_e}\mathbf{q} - \frac{m_e(M_a + m_e + M_B)}{(M_a + m_e)(M_B + m_e)}\mathbf{K} \qquad (8)$$

and

$$\boldsymbol{\kappa}' = \frac{M_B}{M_B + m_e}\mathbf{q}' - \frac{m_e(M_a + m_e + M_B)}{(M_a + m_e)(M_B + m_e)}\mathbf{K}' \qquad (9)$$

where M_a, M_B, and m_e are the masses of A^+, B, and e, respectively. Expression (6) is derived from the plane-wave approximation for the motion of the e–B system relative to A^+. Because of the smallness of the electron mass, the relative motion between A^+ and $(B + e)$ is essentially undistorted.

Expression (8) can be rewritten as $\boldsymbol{\kappa}=\mu_{eB}\mathbf{v}_{eB}/\hbar$ where

$$\mathbf{v}_{eB} = [M_a/(M_a + m_e)]\mathbf{v} - \mathbf{V} \tag{10}$$

Hereafter we shall suppress the suffix n on the orbital velocity v_n, except for the case where we need to specify it explicitly.

For the slow collisions, i.e., $V \ll v$, the second term of $\mathbf{v}_{eB}$ can be neglected. Thus, after making the mass-disparity approximation, i.e., $M_a, M_B \gg m_e$, we obtain

$$\boldsymbol{\kappa} \simeq \mathbf{q} = m_e\mathbf{v}/\hbar$$

Hence expressions (6) and (7) can be rewritten as[17]

$$f(\alpha, \beta, \mathbf{K} \to \alpha', \beta', \mathbf{K}') = \frac{\mu}{m_e} \int G^*(\alpha' \mid \mathbf{q}')G(\alpha \mid \mathbf{q})f_e(\beta, \mathbf{q} \to \beta', \mathbf{q}')\, d\mathbf{q} \tag{11}$$

where $\mathbf{q}+\mathbf{K}=\mathbf{q}'+\mathbf{K}'$.

The forward scattering amplitude of expression (11) for elastic collisions was derived by Alekseev and Sobel'man[11] and was successfully applied to the analysis of the pressure shift of high-Rydberg series perturbed by foreign gas atoms.

The relation is simpler if f_e depends on the momentum transfer only. In this case, relation (11) reduces to

$$f(\alpha, \beta, \mathbf{K} \to \alpha', \beta', \mathbf{K}') = (\mu/m_e)\langle\alpha'|\exp(i\mathbf{Q}\mathbf{r})|\alpha\rangle f_e(\beta \to \beta', Q) \tag{12}$$

where $\hbar\mathbf{Q}$ ($\mathbf{Q}=\mathbf{K}-\mathbf{K}'=\mathbf{q}'-\mathbf{q}$) is the momentum transfer and $\langle\alpha'|\exp(i\mathbf{Q}\mathbf{r})|\alpha\rangle$ is a transition form factor of the high-Rydberg atom. This simplification is quite useful for practical applications.

For the intermediate velocity range, i.e., $V \simeq v$, we may have $v_{eB} \simeq 0$. Thus the collision time τ_c may become long and comparable with τ_n because τ_c is approximately equal to $r(e_R\text{-B})/v_{eB}$, which invalidates condition (5). This means that the A**-B interaction becomes stronger than in the cases of slow and fast collisions and may lead to the breakdown of the perturbative approach, i.e., the adoption of the plane wave for the relative motion. The collisions of A**(n) with B for $V \simeq v$ will involve many coupled states and is very difficult to model theoretically.

For fast collisions, i.e., $V \gg v$, condition (5) becomes again satisfied. Therefore, expressions (6) and (7) are applicable to this case. Because $V \gg v$, the first term of $\mathbf{v}_{eB}$ [see Eq. (10)] can be neglected. Thus we have

$$\mathbf{v}_{eB} \simeq -\mathbf{V}$$

Then, ignoring the dependence of $\boldsymbol{\kappa}(\boldsymbol{\kappa}')$ on $\mathbf{q}$, expression (6) can be reduced to a simplified form quite similar to expression (12), i.e.,[34]

$$f(\alpha, \beta, \mathbf{K} \to \alpha', \beta', \mathbf{K}') = (\mu/m_e)\langle\alpha'|\exp(i\mathbf{Q}\mathbf{r})|\alpha\rangle f_e(\beta, \boldsymbol{\kappa} \to \beta', \boldsymbol{\kappa}') \tag{13}$$

where

$$\mathbf{Q} = \mathbf{K} - \mathbf{K}' = \boldsymbol{\kappa}' - \boldsymbol{\kappa} \tag{14}$$

Comparison of this expression with expressions (11) and (12) shows that, for the fast collisions, the binding effect of the Rydberg electron can be always considered as the transition-form factor. Again under the mass-disparity assumption, we have

$$\boldsymbol{\kappa} = -m_e \mathbf{V}/\hbar \tag{15}$$

From expressions (12) and (13), we see that the transition form factor of the Rydberg atom is a key quantity for the understanding of the Rydberg–neutral collisions.

Quite independently of Fermi's work, Bates and Khare[37] investigated electron–ion recombination in a dense neutral gas, i.e.,

$$\mathrm{A}^+ + e + \mathrm{B} \rightarrow \mathrm{A}^{**}(n) + \mathrm{B} \tag{16}$$

based on a model essentially equivalent to the "free" electron model, where B is an atom. Later, Flannery[15,16] developed this treatment into the semiquantal theory that is suitable for the description of slow collisions between A** and B. The key idea of this theory is that the rate constant for process (1) is considered to be the scattering cross section for process (2) multiplied by the velocity of the electron relative to B and averaged over the scattering direction based on classical mechanics and over the normalized distribution with respect to the velocities of the Rydberg electron. Thus this theory does not give the relation between the scattering amplitudes but that between the differential cross sections for processes (1) and (2). A detailed account of the semiquantal theory is presented in Chap. 11 by Flannery. Therefore we cite only the final expression given by this theory:

$$\sigma(\epsilon, \Delta_\mathrm{B}, V)\, d\epsilon = \frac{d\epsilon}{V^2} \int_{v_{\min}}^{\infty} \frac{\mathfrak{F}(v_e)}{v_e}\, dv_e \int_{v_{eB}^-}^{v_{eB}^+} \frac{v_{eB}\, v'_{eB}\, dv_{eB}}{\gamma(v_e, V)} \times \int_{\psi^-}^{\psi^+} \frac{\sigma_e(v_{eB}, v'_{eB}, \psi)\, d(\cos\psi)}{[(\cos\psi^+ - \cos\psi)(\cos\psi - \cos\psi^-)]^{1/2}} \tag{17}$$

Here ϵ is the internal energy of the Rydberg atom and Δ_B the change of the internal energy of the neutral species B (for an increase in the internal energies of A** and B, $\Delta\epsilon$, $\Delta_\mathrm{B} > 0$), v_e the electron velocity relative to the center of mass of the (A^+–e) system, $\mathfrak{F}(v_e)$ the velocity distribution of the Rydberg electron, ψ the scattering angle for collision process (2), σ_e the differential cross section for the e–B scattering (2),

$$\gamma^2 = \frac{\mu_{eB}^2 v_{eB}^2}{(1+c)^2} [(1 + c)(v_e^2 + cV^2) - cv_{eB}^2]$$

and

$$c = \frac{M_a M_B}{m_e(m_e + M_a + M_B)}$$

Further, v_{min}, $v_{eB}^{\pm}$, and $\psi^{\pm}$ are given by kinematics. For later discussions, we note that if the scattering amplitude f_e for process (2) depends only on the momentum transfer P $(=\hbar Q)$, expression (17) is reduced to a simpler form:

$$\sigma(\epsilon, \Delta_B, V)\, d\epsilon = \frac{\pi\, d\epsilon}{\mu_{eB}^2 V^2} \int_{v_{min}}^{\infty} \frac{\mathcal{F}(v_e)}{v_e}\, dv_e \int_{P-}^{P+} |f_e(\beta \to \beta', Q)|^2\, dP \qquad (18)$$

where the upper and lower limits of the momentum transfer $P^{\pm}$ are determined by kinematics [see Eq. (54) in Ref. 16].

Fowler and Preist[21] also proposed a model essentially equivalent to the "free" electron model to explain the observed ionization rates of alkali metals in hydrocarbon flames (see also Ref. 22). Based on the Chew–Low approximation,[38] they showed that the cross section for process (1) can be written in terms of the squared scattering amplitude $|f_e|^2$ and the momentum-space wave function of the Rydberg electron. The small mass ratio m_e/μ simplifies this relation into their final expression for the collisional ionization of the high-Rydberg atoms by collision with the molecules, i.e.,

$$\sigma(K, E_n, E_\beta, E_{\beta'}) = (\mu/m_e)^{1/2}(\bar{q}/K)[(q/q')\sigma_e(q)]\bar{q} \qquad (19)$$

where $\sigma_e(q)$ is the cross section for process (2) when the Rydberg electron has momentum $\hbar q$ corresponding to an energy $\epsilon' = E_\beta - E_{\beta'} - E_n$, symbols E_β and $E_{\beta'}$ denote the internal energies of the molecule B, E_n is the binding energy of the Rydberg electron, and $\bar{q}$ an average of the wave vector of the Rydberg electron. In expression (19), the quantity $[(q/q')\sigma_e(q)]$ is the squared scattering amplitude $|f_e|^2$, but it is nonphysical because ϵ' is below the threshold for excitation. They argued that this quantity can be obtained by analytical continuation of $|f_e|^2$ into nonphysical values of q and q'. This expression has been applied to the collisional ionization of high-Rydberg atoms by polar and quadrupolar molecules.

Later, Smirnov[39] also investigated the collisional ionization of high-Rydberg atoms by neutral species based on the classical impact-parameter method. Assuming a straight line trajectory for B, he obtained the following expression for the ionization probability W_i of the Rydberg atom:

$$W_i = \int \frac{dz}{V} |\psi(\mathbf{R})|^2 |\mathbf{V} - \mathbf{v}| \int_{\Delta \geqslant E_n} d\sigma$$

where $d\sigma$ is the differential cross section for process (2), Δ the energy exchange between e_R and B, $|\psi(\mathbf{R})|^2$ the electron density at a given point on the trajectory, $\mathbf{V}$ the relative velocity, $\mathbf{v}$ the electron velocity, and the straight line trajectory is denoted by the z axis. The integration of the differential cross section must be carried out over the scattering angles under the condition $\Delta \geqslant E_n$.

Finally, after the integration over impact parameter **b**, the collisional ionization cross section is written as

$$\sigma_i = \int W_i \, d\mathbf{b} = \int d\mathbf{b} \, dz \, |\psi(\mathbf{R})|^2 \frac{|\mathbf{V}-\mathbf{v}|}{V} \int_{\Delta \geqslant E_n} d\sigma = \left\langle \frac{|\mathbf{V}-\mathbf{v}|}{V} \int_{\Delta \geqslant E_n} d\sigma \right\rangle \tag{20}$$

In his final expression, the bracket means averaging over the electron distribution in the Rydberg atom.

Since 1975, there has been a surge of experimental activity in the study of collisions involving high-Rydberg atoms because of the availability of tunable dye lasers.[40-42] This has stimulated theoretical activities, which are based on the "free" electron model and which have been mainly applied to thermal collisions of high-Rydberg atoms with rare gas atoms.[43-51]

There are some differences among the previously mentioned theoretical approaches based on the "free" electron model. For example, expression (6) relates the scattering amplitude f for process (1) to that f_e for process (2), whereas all other expressions (17), (19), and (20) give relations between the cross sections for process (1) and those for process (2). In the case where information on the phase of the scattering amplitude is needed, for example, in a phenomenon such as the pressure shift, relation (6) must be used. In cases where only the relation between the cross sections for two processes is needed, expressions (17), (19), and (20) can be used. They are often cast into the form in which the rate constant for process (1) is written as that for process (2) averaged over the momentum distribution of the Rydberg electron.

Matsuzawa[24,51] discussed the mutual relationships among the previously mentioned theoretical approaches. If we take the imaginary part of expression (6) for the forward-scattering amplitude for elastic scattering, i.e., $K=K'$, $\alpha=\alpha'$, and $\beta=\beta'$, we have

$$\operatorname{Im} f(\alpha, \beta, \mathbf{K} \rightarrow \alpha, \beta, \mathbf{K}) = (\mu/\mu_{eB}) \int |G(\alpha \mid \mathbf{q})|^2 \operatorname{Im} f_e(\beta, \boldsymbol{\kappa} \rightarrow \beta, \boldsymbol{\kappa}) \, d\mathbf{q} \tag{21}$$

The optical theorem gives, for the total scattering cross section $\sigma_e^t(\kappa)$ for process (2),

$$\operatorname{Im} f_e(\beta, \boldsymbol{\kappa} \rightarrow \beta, \boldsymbol{\kappa}) = (\kappa/4\pi)\sigma_e^t(\kappa) \tag{22}$$

because no artificial assumption has been made concerning the e–B interaction. For expression (6), we can no longer apply the optical theorem for process (1) in an exact sense because the "free" electron model has been employed to derive expression (6). However, because of conditions (3) and (4), it is quite unlikely that both the e_R–B and A^+–B interactions play significant roles simultaneously. Therefore one may interpret the left-hand side of Eq. (21) as being related to that part of the total cross section, i.e., $\sigma_{e\text{-B}}^t(K)$, for processes arising only from the e–B interaction, i.e.,

$$\mathrm{Im} f(\alpha, \beta, \mathbf{K} \to \alpha, \beta, \mathbf{K}) = (K/4\pi)\sigma^{t}_{e\text{-}B}(K) \tag{23}$$

Substitution of Eqs. (22) and (23) into (21) yields

$$V\sigma^{t}_{e\text{-}B}(K) = \int v_{eB}\sigma^{t}_{e}(\kappa)\,|G(\alpha\,|\,\mathbf{q})|^2\,d\mathbf{q} \tag{24}$$

For slow collisions, i.e., $V \ll v$, this expression leads to a simpler relation,

$$V\sigma^{t}_{e\text{-}B}(K) = \int v\sigma^{t}_{e}(q)\,|G(\alpha\,|\,\mathbf{q})|^2\,d\mathbf{q} \tag{25}$$

because $v_{eB} \simeq v$, and $\boldsymbol{\kappa} \simeq \mathbf{q}$. This was first derived by Alekseev and Sobel'man[11] and was applied to the pressure broadening of high-Rydberg series perturbed by rare gas atoms. This was also applied to collisions with molecules, which will be discussed in more detail in the following sections.

For fast collisions, i.e., $V \gg v$, expression (24) reduces to another simple relation,

$$\sigma^{t}_{e\text{-}B}(V) = \sigma^{t}_{e}(V) \tag{26}$$

because $v_{eB} \simeq V$ and $\kappa \simeq m_e V/\hbar$. This expression is closely related to that predicted by Butler and May.[23]

We immediately see that expression (24) has a strong similarity to expression (20), which was given by Smirnov,[39] under the assumption that the particular transition is specified for process (1) only from the energetic point of view, in spite of the different averaging procedures over the electronic motion, i.e., in momentum space for expression (24), while in configuration space for expression (20).

Judging from the key idea of semiquantal theory, we see that expression (25), under the previously mentioned assumption, has a strong similarity to expressions given by Flannery [see Eq. (40) of Ref. 16]. Actually, he recently showed that the semiquantal expression can be derived from the impulse approximation.[35]

Recently, Nakamura et al.[52] also discussed the mutual relationship between the impulse approximation and the semiquantal theory for a particular case where f_e depends on momentum transfer only. They showed by changing integral ordering that the impulse approximation[31] with the binary encounter approximation for the transition-form factor coincided with the semiquantal expression (18) with the mass-disparity approximation, i.e., m_e/M_a, $m_e/M_B \simeq 0$, and with the electron-velocity distribution determined by the initial Rydberg state.

Furthermore, it was shown[51] that the perturbative approach combined with the Fermi pseudopotential[43,45–47] was essentially equivalent to expression (11) under the scattering-length approximation for f_e.

The optical theorem comes from probability conservation and gives an upper limit on the cross section if it is based on the exact scattering amplitude. However, expression (23) is derived from the approximate impulse-scattering

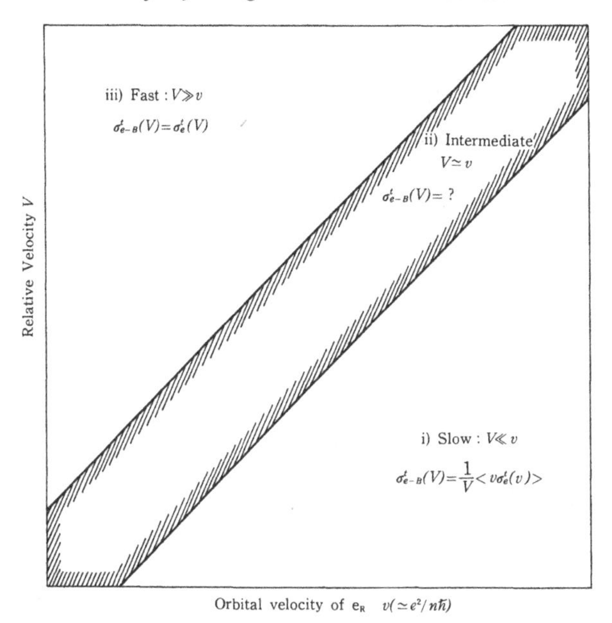

Fig. 8.2. Relation between the total cross sections for the high-Rydberg-neutral collision (1) and electron-molecule scattering, process (2). The bracket denotes the average over the momentum distribution of the initial Rydberg state. Concerning the quantitative validity of these relations, see the discussions given for the specific examples in Sects. 8.3a and 8.4.

amplitude (6). [To make our argument unambiguous, we shall hereafter refer to Eq. (23) as an approximate "optical" theorem.] Thus we must question its validity until we verify that expression (6) does really satisfy probability conservation. Hence, we cannot claim a priori, without any test of its internal consistency, that the derivatives of the approximate "optical" theorem (23), i.e., Eqs. (24)–(26), give an upper limit on the cross section for the process arising from the e–B interaction. This is contrary to the claim recently made by Flannery.[35] Despite this limit to their validity, expressions (25) and (26) are quite attractive because they are easily interpreted based on the "free" electron model and give quite simple relations between the total cross sections for two-collision processes, i.e., process (1) and (2) as shown in Fig. 8.2. At least, we may use these relations as rough estimates for the cross sections for high-Rydberg-neutral collisions. For slow collisions, i.e., $V \ll v$, we have relation (25), where the momentum distribution of the Rydberg electron must be taken into account explicitly. On the other hand, for fast collisions, i.e., $V \gg v$, we obtain a simple relation, Eq. (26). In the former case, we can view the Rydberg atom as a slow and free electron "beam" that has a definite momentum distribution determined by the quantum state. In the latter case, we can also view the Rydberg atom as an electron "beam" with the velocity corresponding to that of the relative motion, though the orbital motion of the Rydberg electron tends to reduce the monochromaticity of this electron "beam." The validity of these relations will be discussed in more detail for the specific examples investigated later.

The case where the interaction between the atomic ion core and the neutral species is important

For this case, the Rydberg electron e_R behaves like a spectator while the atomic ion core A^+ is interacting with the neutral species B (or BC) [see Fig. 8.1, case (2)]. Thus we can expect strong correlation between process (1) and collision processes between A^+ and the neutral species. The latter can be charge transfer, i.e.,

$$A^+ + B \to A + B^+ \tag{27}$$

or an ion–molecule reaction such as

$$A^+ + BC \to AB^+ + C \tag{28}$$

Corresponding to process (27) or (28), we may specify process (1) as

$$\begin{aligned} A^{**}(n) + B &\to A + B^{**}(n') \\ &\to A + (B^+ + e) \end{aligned}$$

or

$$\begin{aligned} A^{**}(n) + BC &\to AB^{**}(n') + C \\ &\to (AB^+ + e) + C \end{aligned}$$

respectively. Further, we can also expect a weak n dependence of the cross section for process (1) at fixed velocities, except for $V \simeq v_n$ because e_R is not directly involved in the collision dynamics. Recently there has appeared some experimental evidence for this type of correlation.[53-55]

To our knowledge, no extensive theoretical study has been made to pursue this correlation quantitatively. Quite recently, Flannery[35] proposed a mechanism for l changing in thermal collisions of high-Rydberg atoms with rare gas atoms based on the noninertiality of A^+ during the A^+–B interaction. However, this was found to give no appreciable contribution to l changing in thermal collisions with neutrals for high n.[56] This is consistent with an intuitive physical picture based on conditions (3) and (4).

8.2b. Collisions of Rydberg atoms with intermediate n

As n decreases below n_b, conditions (3) and (4) become less well satisfied. Therefore we must consider both the e_R–B and the A^+–B interactions simultaneously. Hence no direct correlation to any other simpler process can be

Table 8.2. Summary of the theoretical considerations for Rydberg atom–neutral collisions

Range of n	Type of interactions	Remarks (including velocity range)
A. High n Case 1 and case 2 can coexist, but they are statistically independent of each other.	1. The e_R–B interaction plays a decisive role.	*a*. Slow, $V \ll v_n$. The "free" electron model and the impulse approximation are valid. *b*. Intermediate, $V \simeq v_n$. The impulse approximation breaks down because of $\tau_c \simeq \tau_n$. A close-coupling-type calculation is needed. *c*. Fast, $V \gg v_n$. The "free" electron model and the impulse approximation are valid.
	2. The A^+–B interaction is important.	Strong correlation to A^+–B collisions is anticipated. A weak n dependence of the cross section at fixed velocities is predicted because the Rydberg electron behaves as a spectator.
B. Intermediate n	Both the e_R–B and A^+–B interactions play significant roles simultaneously or successively.	No direct correlation to any other simpler process can be expected.

anticipated. Quite recently, for slow collisions of A** and B, Janev and Mihajlov[57] proposed a model in which transitions of the Rydberg electron take place as a result of its resonant coupling with transitions in the quasi-molecular subsystem AB^+. They applied the model only to Rydberg atom–atom collisions, though it can be considered to be potentially applicable to collisions with molecules.

To explain their experimental data on collisional depopulation of Na**(ns) by rare gases, Gallagher and Cooke[58] also proposed a qualitative model in which the n state of the Rydberg electron is changed by the transition-dipole moment induced in the rare gas atom during the Na^+–rare gas collision. It should be noted that the cross sections estimated using such models decrease rapidly as n increases. This is a natural consequence of conditions (3) and (4).

Before closing this section, we summarize in Table 8.2 the theoretical understanding obtained here for Rydberg atom–neutral collisions, which may be useful for the overall understanding of Rydberg–neutral collisions. In principle, the theoretical treatments given here can be applied to collisions with atoms.[24,51] However, two other chapters are devoted to this subject. Therefore, in the following sections, we shall concentrate on theoretical studies of collisions with molecules and compare them with experimental data. Further emphasis will be put on the case A–1, i.e., the "free" electron model combined with the impulse approximation, because the theoretical studies on collisions with molecules have been undertaken mainly with this approximation.

8.3. Slow collisions

In slow electron–molecule scattering (2), there are some types of collision processes, i.e., elastic scattering, rotational excitation or deexcitation, and vibrational excitation or deexcitation, that may cause state changing or collisional ionization of high-Rydberg atoms in collision process (1). In other words, the molecule has internal degrees of freedom from which the Rydberg electron can gain (or release) energy. As will be seen later, the cross sections for process (1) increase dramatically as the resonance condition becomes satisfied, i.e., the energy defect for collision process (1) becomes small. Because of the internal degrees of freedom, resonant conditions are more likely to be satisfied in collisions with molecules than in those with atoms. In particular, a rare gas atom has no internal degrees of freedom that the Rydberg electron can excite or deexcite. Therefore we can expect that the discreteness of the Rydberg atom energy levels can affect collision dynamics more directly in collisions with rare gas atoms than in those with molecules because the rare gas atom has no ability to compensate for energy differences of the Rydberg atom between initial and final channels.

8.3a. Collisions with polar molecules

Hotop and Niehaus[59] measured the cross sections for collisional ionization of high-Rydberg atoms by polar and quadrupolar molecules. The cross sections for the polar molecules were measured to be of the order of $\sim 10^{-12}$–10^{-13} cm^2 at room temperature and are large compared with cross sections for collisions of atoms in the ground or lower excited states with neutral molecules. At room temperature, the molecules are usually in rotationally excited states. Therefore, to explain these data, Matsuzawa[17] proposed the mechanism of energy transfer from molecular rotation to electronic motion of the Rydberg electron in the high-Rydberg state based on the "free" electron model for which the resonance condition is likely to be satisfied because of the rotational level spacings being comparable to the binding energy of the Rydberg electron. It is well known that rotationally inelastic scattering dominates in thermal-electron–polar-molecule collisions.[28,60] Therefore, the proposed mechanism is considered to be dominant for process (1).

Hence we may consider collisional ionization as a special case of (1), i.e.,

$$A^{**}(n,l) + B(J,\gamma) \rightarrow (A^+ + e) + B(J',\gamma') \tag{29}$$

and rotational deexcitation by slow electron impact as a special case of (2), i.e.,

$$e + B(J,\gamma) \rightarrow e + B(J',\gamma') \tag{30}$$

where J and J' $(J>J')$ are the rotational quantum numbers and γ and γ' denote those quantum numbers other than J that specify the rotational states of the molecule. Here l is the angular momentum of the initial high-Rydberg state. It is assumed that the averaging procedure was carried out over the magnetic quantum number. The continuum of the final Rydberg state may be denoted by the set of quantum numbers including the energy of the ionized electron ϵ.

Takayanagi[61] showed that the Born approximation combined with a point-multipole interaction can adequately describe rotationally inelastic scattering of slow electrons by polar molecules with dipole moments $D \lesssim 1 ea_0$.

In all the theoretical studies of collisional ionization,[16,17,20-22] Born amplitudes for the rotational deexcitation were employed as input data to relations, i.e., (12), (19), and (20), between processes (1) and (2). The calculated cross sections were of the order of $\sim 10^{-12}$–10^{-13} cm^2 and were in reasonable agreement with the experimental data of Hotop and Niehaus,[59] though the calculated values depended fairly sensitively on the magnitudes of the dipole moments.

Now we shall give a detailed description of the theoretical studies and make more quantitative comparison between theory and experiment. For rotational deexcitation of polar molecules by slow-electron impact, we have as the squared Born scattering amplitude for the dipole-allowed deexcitation of the molecule,

$$|f_e(J \to J-1, Q)|^2 = \frac{4}{3} a_0^2 \left(\frac{D}{ea_0}\right)^2 \frac{J}{2J+1} \frac{1}{(Qa_0)^2} \tag{31}$$

for a linear molecule,

$$|f_e(J, K_r \to J, K_r - 1, Q)|^2 = \frac{4}{3} a_0^2 \left(\frac{D}{ea_0}\right)^2 \frac{J^2 - K_r{}^2}{J(2J+1)} \frac{1}{(Qa_0)^2} \tag{32}$$

for a symmetric-top molecule, and

$$|f_e(J, \tau \to J', \tau', Q)|^2 = \frac{4}{3} a_0^2 \left(\frac{D}{ea_0}\right)^2 \frac{S(J, \tau \to J', \tau')}{2J+1} \frac{1}{(Qa_0)^2} \tag{33}$$

for an asymmetric-top molecule.[61-64] Here K_r is the rotational angular momentum about the body-fixed symmetric axis ($K_r = -J, -J+1, \ldots, J$) for the symmetric-top molecule. The quantity $S(J, \tau \to J', \tau')$ is the line strength of the asymmetric top where τ designates each of $(2J+1)$ sublevels for a given J ($\tau = -J, -J+1, \ldots, J$).

Substituting these expressions into Eq. (12) and neglecting the energy exchange between the internal degrees of freedom and the relative motion, Matsuzawa[20] derived the following expressions for the collisional ionization cross sections arising from dipole-allowed deexcitation, i.e., process (29),

$$\sigma_i^d(nl, J \to J-1) = \frac{16\pi a_0^2}{3} \frac{e^2}{\hbar V} \left(\frac{D}{ea_0}\right)^2 \frac{J}{2J+1} I_d(nl, \epsilon_0) \tag{34}$$

for the linear molecule,

$$\sigma_i^d(nl, J, K_r \to J-1, K_r) = \frac{16\pi a_0^2}{3} \frac{e^2}{\hbar V} \left(\frac{D}{ea_0}\right)^2 \frac{J^2 - K_r^2}{J(2J+1)} I_d(nl, \epsilon_0) \quad (35)$$

for the symmetric-top molecule, and

$$\sigma_i^d(nl, J, \tau \to J', \tau') = \frac{16\pi a_0^2}{3} \frac{e^2}{\hbar V} \left(\frac{D}{ea_0}\right)^2 \frac{S(J, \tau \to J', \tau')}{2J+1} I_d(nl, \epsilon_0) \quad (36)$$

for the asymmetric-top molecule with $n \gtrsim n_0$ $\{=(R/\Delta E_r)^{1/2}\}$, where

$$I_d(nl, \epsilon_0) = \int_0^\infty \left.\frac{dF_{nl}(Q, \epsilon)}{d(\epsilon/2R)}\right|_{\epsilon=\epsilon_0} d(Qa_0)$$

Here $\epsilon_0 = \Delta E_r - E_{nl}$, $dF_{nl}(Q,\epsilon)/d(\epsilon/2R)$ is the form-factor density of the high-Rydberg atom for the transition from the (n, l) state to the continuum, R (=13.6 eV) the Rydberg constant, ϵ the energy of the ionized electron, E_{nl} the binding energy of the Rydberg electron in the (n, l) state, and ΔE_r the rotational energy released in the rotational deexcitation given by

$$\Delta E_r = 2BJ \quad (37)$$

for the dipole-allowed deexcitation $J \to J-1$ of the linear molecule and for the $(J, K_r) \to (J-1, K_r)$ transition in a symmetric-top molecule, where B is the rotational constant compatible with the international nomenclature.[65] For collisional ionization, we can use the binary encounter theory (BET) for evaluation of the form factor because the BET form factor was shown to be a good approximation.[66] Thus the quantity $I_d(nl)$ is easily calculated, and shows the quite weak l dependence of the cross section. Preist's results for linear molecules were found to yield values slightly larger than expression (34).[22]

Figure 8.3 shows a comparison of the theoretical results[20] on collisional ionization with the absolute measurements of σ_i^d by Hotop and Niehaus[59] and the relative measurements by Chupka (quoted by Matsuzawa).[20] In the experiment of Hotop and Niehaus, the high-Rydberg rare gas atoms were produced by electron impact. Thus the theoretical cross sections at thermal energies were averaged over the assumed distribution (i.e., $1/n^3$) over n. Here the l distribution was not taken into account because of the weak l dependence of the theoretical cross sections. The assumption for the n distribution was based on the theoretical calculation of the excitation cross section[67] and later experimentally confirmed.[68] Furthermore, the cross sections were also averaged over the thermal distribution of the rotational states. Thus we have, at room temperature,[20]

$$\sigma_i^d = 4.7 \times \mu^{1/2} D^2 \langle I_d \rangle \times 10^{-13} \text{ cm}^2 \quad (38)$$

where μ is given in atomic mass units, D in atomic units, and $\langle I_d \rangle$ denotes the quantities $\{J/(2J+1)\} I_d(n, \epsilon_0)$, $\{(J^2 - K_r^2)/J(2J+1)\} I_d(n, \epsilon_0)$, and

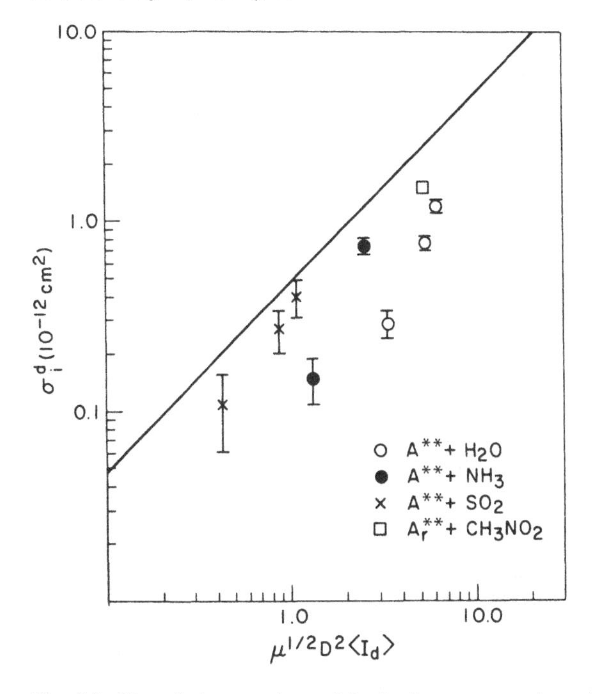

Fig. 8.3. Plot of the experimental ionization cross section σ_i^d vs. $\mu^{1/2}D^2\langle I_d \rangle$ (μ in atomic mass units, D in atomic units). The data on CH_3NO_2 by Chupka are normalized to the absolute value of Hotop and Niehaus.[59] The straight line shows the theoretical values given by Eq. (38). As for the data of Hotop and Niehaus, A** stands for the high-Rydberg atoms He**, Ne**, and Ar** as $\mu^{1/2}D^2$ increases. There are no experimental data for the Ne–NH_3 system. (From Matsuzawa[20])

$\{S(J, \tau \rightarrow J', \tau')/(2J+1)\}I_d(n, \epsilon_0)$ averaged over the assumed n distribution and over the thermal distribution of the rotational states. In Fig. 8.3, the theoretical result, i.e., Eq. (38), is denoted by a straight line. This figure indicates that the dependence of the experimental cross section on the dipole moments is well explained by the "free" electron model combined with the mechanism of energy transfer from molecular rotation to the Rydberg electron.

Shibata et al.[69] investigated isotope effects in the collisional ionization of high-Rydberg hydrogen atoms by H_2O and found good agreement with the velocity dependence of expressions (34)–(36).

Based on the previously mentioned mechanism, theory predicts that the experimental collisional ionization cross sections averaged over the rotational states of linear polar molecules with large rotational constants should show steplike structures as a function of n.[70,71] This comes from the fact that a polar molecule with rotational quantum number J can only ionize high-Rydberg atoms with binding energies $E_n \leqslant 2BJ$ [see Eq. (37)] because energy exchange is not likely to occur between the internal degrees of freedom and

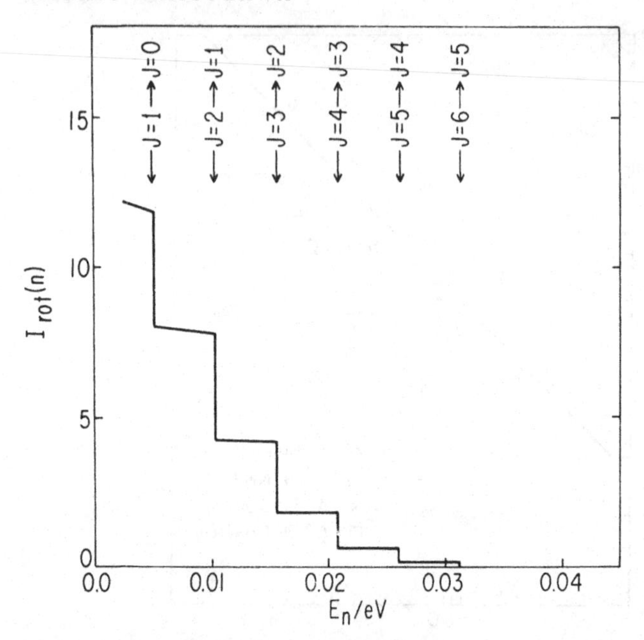

Fig. 8.4. Steplike structure in the n dependence of the theoretical collisional-ionization cross section for the H** + HF system. The symbol $I_{\rm rot}$ denotes the quantity $\{J/(2J+1)\}I_{\rm d}(n, \epsilon_0)$ averaged over the rotational distribution. The arrows denote the position of $E_n = 2BJ$ $(J=1, 2, 3, \ldots)$, where $2B = 0.005\,19$ eV. The rotational constant B of the HF molecule is taken from Herzberg.[65] (From Matsuzawa and Chupka[71])

the relative motion at thermal energies. Therefore the steplike structure should appear at every binding energy that satisfies the relation $E_n = 2BJ$ $(J=1, 2, \ldots)$ as shown for the H** + HF system in Fig. 8.4.

Chupka[71,72] experimentally detected this steplike structure in the Kr**(np) + HF and Kr**(np) + HCl systems. By mass spectroscopy, he compared SF_6^- production in Kr–SF_6 mixtures both with and without an admixture of HF or HCl. The difference in the SF_6^- ion production rate was found to be proportional to the rate of direct ionization of the Kr**(np) by collision with HCl or HF. Figure 8.5 shows the influence of HF on SF_6^- production. The steplike structure occurs at photon wave lengths corresponding to the rotational states that begin to contribute to the collisional ionization, i.e., $E_n = 2BJ$, though the sharpness of the theoretically predicted step tends to be dulled by the fact that a small amount of energy is actually exchanged between rotational and translational motion (see Fig. 8.4). This is the first clear-cut experimental evidence to support the mechanism of energy transfer from molecular rotation to electronic motion.

If we assume that the particular transition for process (1) can be specified only from an energetic point of view and replace $\sigma_{\rm e}^{\rm t}(q)$ by the cross section $\sigma_{\rm e}(\beta \to \beta', q)$ for the particular process, we have, from Eq. (25),

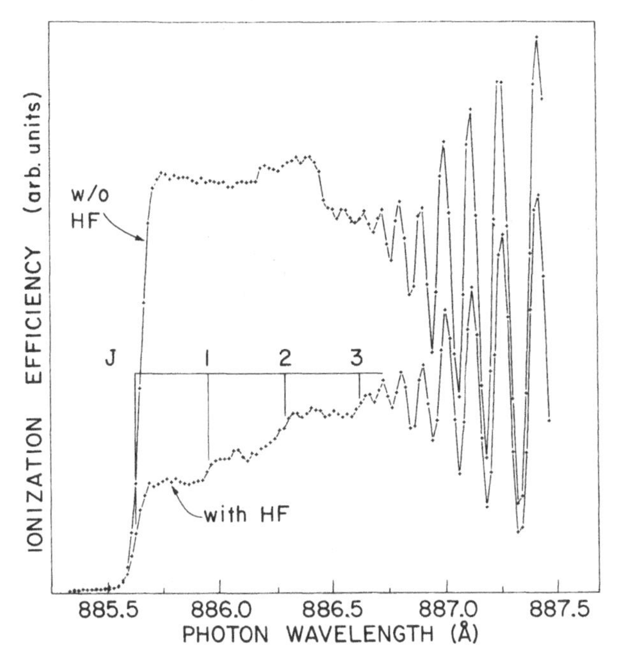

Fig. 8.5. SF_6^- ion signal observed in Kr–SF_6 mixtures with and without HF as a function of wavelength. The symbol J denotes the rotational quantum number of the initial molecular state. (From Matsuzawa and Chupka[71])

$$\sigma(nl,\beta,K \to n',\beta',K') = (1/V)\int_{q_{\min}}^{\infty} v\sigma_e(\beta \to \beta',q)\,|g_{nl}(q)|^2 q^2\,dq \tag{39}$$

where $g_{nl}(q)$ is the radial wave function of the Rydberg electron in momentum space, $q_{\min}$ is determined by energy conservation for process (2), i.e.,

$$q_{\min} = \begin{cases} [(2m/\hbar^2)(E_{\beta'} - E_\beta)]^{1/2} & \text{for} \quad E_\beta < E_{\beta'} \\ 0 & \text{for} \quad E_\beta \geqslant E_{\beta'} \end{cases}$$

and Eq. (39) can be interpreted similarly to Eq. (25). Further, we can immediately see that expression (39) recovers the expression employed by Latimer:[73]

$$\sigma(nl,\beta,K \to n',\beta',K') = \frac{\langle v\rangle}{V}\int_0^{\infty} \sigma_e(\beta \to \beta',q)\,|g_{nl}(q)|^2 q^2\,dq \tag{40}$$

if we make use of the approximation

$$\langle v\sigma_e(\beta \to \beta',q)\rangle \simeq \langle v\rangle\langle\sigma_e(\beta \to \beta',q)\rangle \tag{41}$$

where $\langle v\rangle$ is the average velocity of the Rydberg electron. At thermal energies, we have $v \gg V$. Therefore, $\langle v\rangle$ is equal to the average velocity of the electron in process (2). However, there is some ambiguity of definition of $\langle v\rangle$.

If we replace $\sigma_e(\beta \to \beta')$ by $\langle v\rangle_{\text{ion}}\,\sigma_e(\beta \to \beta')/v_f$, expression (39) recovers the expression given for collisional ionization by Fowler and Preist,[21] where v_f

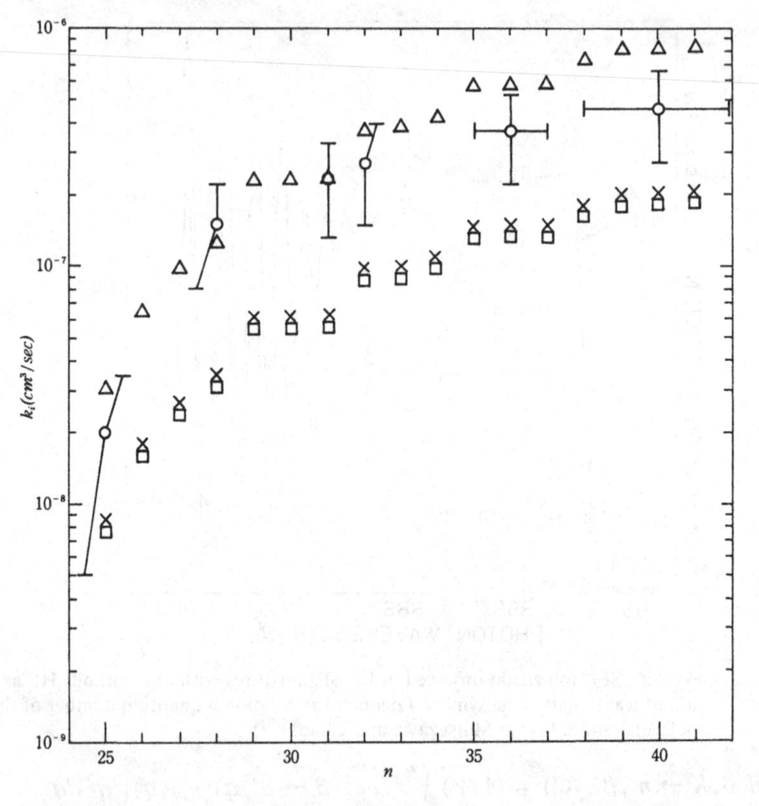

Fig. 8.6. Comparison of theoretical and experimental rate constants for collisional ionization of Xe**(nf) atoms by ammonia molecules: circles, Kellert et al.;[76] squares, expression (35); crosses, expression (39); triangles, expression (40).

is the electron velocity of the final channel in rotational deexcitation (30) and $\langle v \rangle_{ion}$ the average velocity of the electron ejected in collisional ionization (29). Using expression (40), Latimer[73] calculated the cross section $\sigma_i^d(nf)$ for collisional ionization of Xe**(nf) by polar molecules based on the mechanism of energy transfer from molecular rotation to electronic motion. However, he did not consider the effect of the inversion splitting of the NH_3 molecules. After discussing mutual relationships among various expressions given by the different authors, Matsuzawa[24] calculated the collisional ionization cross sections $\sigma_i^d(nf)$ for NH_3 using expressions (35) and the derivative of expression (25), i.e., (39) and (40), taking into account inversion splitting of the NH_3 molecule. In Eq. (40), he put $\langle v \rangle = v_0/\nu_{nl}$ $(\nu_{nl}=n-\delta_{nl})$ as Latimer did. Rundel[42] also calculated the same cross section using the semiquantal theory of Flannery, i.e., expression (18). (The calculated cross sections given in this reference should be divided by π.[74]) Figure 8.6 shows the results of these theoretical calculations for collisional ionization by NH_3 together with experimental data.[75,76]

At first sight, expression (40) employed by Latimer[73] seems to be in quantitative agreement with experimental data. Approximation (41) may be justified under the condition that the cross section $\sigma_e(\beta \rightarrow \beta', q)$ be a slowly varying function of electron velocity. Unfortunately this condition is not satisfied because $\sigma_e(\beta \rightarrow \beta', q)$ has a sharp peak for the rotational deexcitation of the polar molecule by electron impact. Actually the calculated rate constants based on expression (39) are smaller than those of expression (40) and almost agree with those obtained by the impulse approximation, i.e., expression (35). As Matsuzawa[24] discussed, according to the argument by Nakamura et al.[52] (see also the relevant discussion given in Sect. 8.2a), expression (18) should agree with expression (35), i.e., the derivative of expression (12) for the collisional ionization. Thus Rundel's results (not shown in Fig. 8.6) almost coincide with expression (35) except for a slight difference.[42] This difference comes from the adoption of a different data set on the energy gap of the inversion splitting and from the small energy exchange between the relative motion and the internal degrees of freedom. Figure 8.6 shows that, except for the results calculated by expression (40), the theoretical rate constants, which are mutually consistent with one another, explain well the experimental n dependence of the collisional ionization but are systematically smaller than the absolute experimental cross section by a factor of three or four.

For collisional ionization with polar molecules, expression (39), the derivative of the approximate "optical" theorem, yields almost the same values as Eqs. (34)–(36), i.e., the cross sections directly calculated from the scattering amplitude (11) or (12). Therefore, for this case, expression (39) may be used to evaluate the cross sections instead of Eqs. (34)–(36). On the other hand, for collisions with rare gas atoms, Gounand et al.[77] calculated expression (25) for l-mixing collisions of Na**(nd) with He. The calculated results based on expression (25) were found to be smaller than the theoretical estimates calculated directly from the scattering amplitude (11) or (12) by a factor of three or four for $n=15$. This seems to show an internal inconsistency within the theoretical model. In this case, the latter cross sections are found to be consistent with experimental findings.[78,79] However, this apparent internal inconsistency can be explained as follows.[51] The derivation of expressions (25), (39), and (40) depends critically on the approximate "optical" theorem (23). It is well known that the first Born approximation does not satisfy the "optical" theorem because of its perturbative approach. Similar arguments can be applied to expressions (25), (39), and (40). This arises because the perturbative approach for the relative motion does not adequately describe probability conservation for the range of n where experimental data are available at present. This comes most likely from the short range of the e_R–B interaction for collisions with the rare gas atoms. Thus theoretical cross sections directly evaluated from scattering amplitudes (6) are more reliable than those calculated from expressions based on the approximate "optical" theorem (23) if they are not consistent with one another. Further, the validity of expression (25) and its analogs needs to be quantitatively tested for high-Rydberg–polar-molecule collisions. (Regarding expression (26), see discussion in Sect. 8.4.)

As far as the n dependence of the cross sections $\sigma_{\mathrm{i}}^{\mathrm{d}}$ (equivalently the rate constants) is concerned, the experimental data (see Fig. 8.6) are well explained by the "free" electron model combined with the mechanism of the energy transfer from the molecular rotation to the Rydberg electron. It should be noted that the increase of $\sigma_{\mathrm{i}}^{\mathrm{d}}$ with n does not come from the increase of the geometric cross sections as already discussed in the Introduction but from the increase in the numbers of rotational states that can contribute to the ionization of the Rydberg atom as the condition $\Delta E_{\mathrm{r}} \geqslant E_n$ is satisfied. Kellert et al.[76] could not detect the steplike structure for the collisional ionization by NH_3, probably because the inversion splitting tends to smear the steplike structures together with the effects of energy exchange between the internal degrees of freedom and the relative motion. There is still some systematic discrepancy in the absolute magnitudes of the cross sections, say, a factor of three or four between theory and experiment for the collisional ionization of $\mathrm{Xe}^{**}(n\mathrm{f})$ with NH_3.[24,42,76] The reason for this quantitative discrepancy is not yet understood.

Now it would be relevant to make a brief remark on electron–ion recombination in a dense molecular gas. This process can be considered as the inverse of collisional ionization because the electron is quite likely to be recombined into high-Rydberg states. Bates et al.[80] studied this subject theoretically, based on a model essentially equivalent to the "free" electron model, and considered inelastic collisions of the molecules including rotational and vibrational transitions. Clear-cut information on collisional ionization can thus help achieve a better quantitative understanding of recombination processes.

Within the framework of the "free" electron model, energy transfer from molecular rotation to the Rydberg electron can also cause state-changing collisions, i.e., n-changing collisions. These occur in high-Rydberg–polar-molecule collisions if the energy released in the rotational deexcitation is smaller than the ionization potential of the Rydberg atom or if the high-Rydberg atom is deexcited by energy transfer to rotational motion of the molecules. Again, based on this mechanism, Matsuzawa[81,82] investigated the state-changing collisions of high-Rydberg atoms with various types of polar molecules. Expression (12) yields, for the state-changing cross sections,

$$\sigma(n,l,J,\gamma \rightarrow n',l',J',\gamma') = 2\pi(e^2/\hbar V)^2 \int_{|K-K'|}^{K+K'} F_{nl,n'l'}(Q)\,|f_{\mathrm{e}}(J,\gamma \rightarrow J',\gamma',Q)|^2 (Qa_0)\, d(Qa_0) \tag{42}$$

where $F_{nl,n'l'}(Q)$ is the squared form factor, i.e.,

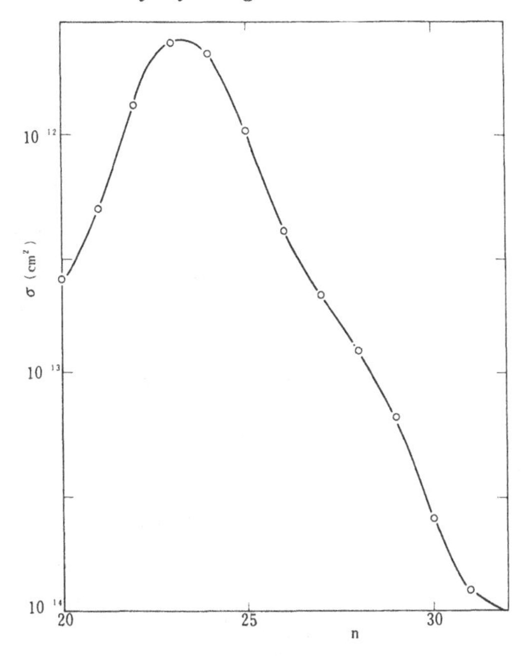

Fig. 8.7. Theoretical l-changing cross sections for $Xe^{**}(nf) + H_2O(6_{-5}) \rightarrow Xe^{**}(ng) + H_2O(5_{-1})$, $V = 6.30 \times 10^4$ cm s^{-1}, $\Delta E_r = 0.742$ cm^{-1}, and $(D/ea_0)^2 S/(2J+1) = 2.244 \times 10^{-2}$. (From Matsuzawa[81])

$$F_{nl,n'l'}(Q) = (2l+1)^{-1} \sum_{m=-l}^{l} \sum_{m'=-l}^{l} |\langle n'l'm'|\exp(i\mathbf{Q}\mathbf{r})|nlm\rangle|^2$$

If we substitute one of expressions (31)–(33) into Eq. (42), depending on the type of molecule, we can easily evaluate the desired cross sections.

For l-changing collisions, the resonant condition suggests that this process is likely to occur if asymmetric-top molecules are chosen as collision partners. Figure 8.7 indicates typical theoretical results[81] for l changing in collisions of high-Rydberg $Xe^{**}(nf)$ atoms with $H_2O(J_\tau)$ molecules, i.e.,

$$Xe^{**}(nf) + H_2O(6_{-5}) \rightarrow Xe^{**}(ng) + H_2O(5_{-1}) \tag{43}$$

We see that the n dependence of the cross section shows a sharp peak as the resonance condition is satisfied and that the maximum cross sections are $\sim 10^{-11}$–10^{-12} cm^2.

As to n-changing collisions, it is convenient to classify them into two classes: a small n change and a large n change. The resonance condition suggests that the former processes are likely to occur for the asymmetric-top molecules and the latter for the linear and symmetric-top molecules with large rotational constants.

Table 8.3. Theoretical cross sections for $Xe^{**}(nf)+H_2O(J_\tau) \rightarrow Xe^{**}(n+1, f)+H_2O(J'_{\tau'})$

	Transition $J_\tau \rightarrow J'_{\tau'}$			
	$5_{-4} \rightarrow 4_0$		$4_{-3} \rightarrow 3_1$	
n	ΔE (cm^{-1})	σ (cm^2)	ΔE (cm^{-1})	σ (cm^2)
25	—	—	0.632	3.44×10^{-14}
26	1.014	2.9×10^{-15}	−0.822	1.14×10^{-14}
27	−0.236	1.67×10^{-13}	—	—
28	−1.316	4.3×10^{-16}	—	—

For n-changing collisions with small Δn, Table 8.3 shows typical theoretical results for the n-changing cross sections for the process[82]

$$Xe^{**}(nf) + H_2O(J_\tau) \rightarrow Xe^{**}(n+1, f) + H_2O(J'_{\tau'}) \tag{44}$$

Again, we see that the cross sections decrease dramatically as the energy defect ΔE increases. This was pointed out for $H^{**}+H_2$ collision by Flannery.[16] The cross sections for processes (43) and (44) are small compared with the geometrical cross sections $\pi\langle r^2\rangle_{25f} \simeq 10^{-10}$ cm^2, as theoretically predicted. There are some processes for which the cross sections become comparable with the geometrical cross sections,[81] which can be understood if the statistical weight of the final state g_f ($\gg 1$) is taken into account. At present, there are no experimental data to directly compare with the theoretical results for l-changing collisions and for n-changing collisions with small Δn.

For large Δn, Matsuzawa[82] investigated the following processes theoretically:

$$Xe^{**}(nf) + NH_3(J, K_r) \rightarrow Xe^{**}(n'l') + NH_3(J-1, K_r) \tag{45}$$

and

$$Xe^{**}(nf) + HCl(J) \rightarrow Xe^{**}(n'l') + HCl(J-1) \tag{46}$$

using expression (42) combined with expressions (34) and (35). He evaluated the cross sections using the binary encounter theory (BET) form factor, which was also shown to be a good approximation for a high-Rydberg atom and large Δn, similar to the case for the discrete-continuum transition.[66]

Figures 8.8 and 8.9 show the velocity and n' dependence of the theoretical n-changing cross section $\sigma(27f, 7 \rightarrow n', 6)$ for process (45), where the cross sections were averaged over the thermal distribution [i.e., at temperature T, where $V = (8kT/\pi\mu)^{1/2}$] with respect to the rotational quantum number K_r and summed over the final angular momentum states. These figures indicate that, because of the resonance condition and the dipole-selection rule, the $Xe^{**}(nf)$ are selectively excited to groups of discrete, more highly excited states. Further, we see from these figures that a fairly large number of Ryd-

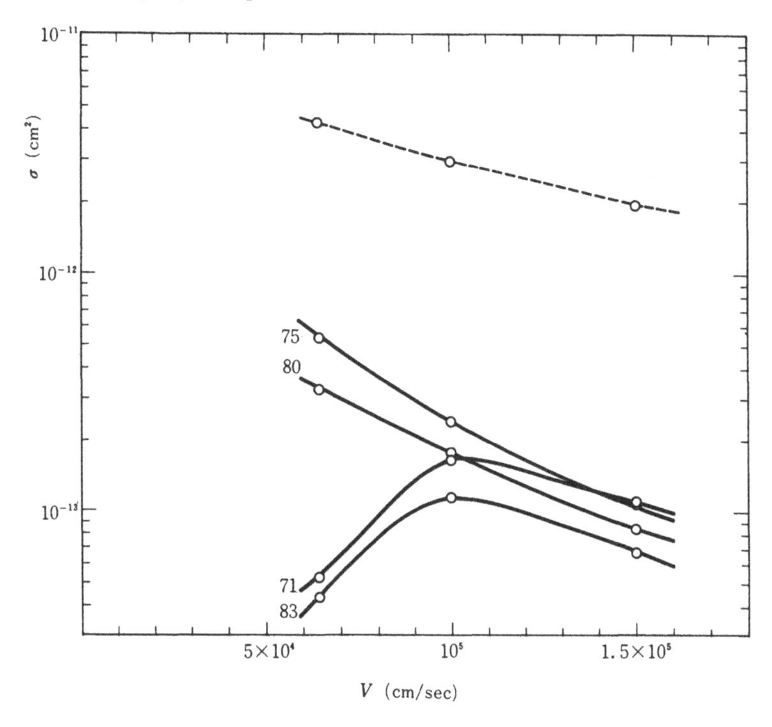

Fig. 8.8. Velocity dependence of the theoretical cross sections for $Xe^{**}(27f)+NH_3(7) \rightarrow Xe^{**}(n')+NH_3(6)$. The solid lines denote the cross sections; numbers attached to each line specify the principal quantum number n' of the final state; broken line shows the theoretical cross section summed over n'. The molecular data used are $D=1.47$ D (Ref. 83) and $B=9.444$ cm^{-1} (Ref. 84). (From Matsuzawa[82])

berg states can be excited simultaneously. This is possible because even if internal energy resonance is not exactly satisfied, the high level density at large values of n' requires that only a small amount of energy be transferred between translation and internal motions. It should be noted that this energy exchange is very small because of the small energy spacing at high n' ($\simeq 75$) and was previously neglected in collisional ionization.[20]

Further, deexcitation of the high-Rydberg atom (i.e., rotational excitation of the polar molecule) in processes such as (45) and (46) is quite unlikely except for an accidentally resonant case. This is because of the low density of high-Rydberg states at low n'. This inevitably leads to large $|\Delta E|$ for the state-changing collision, which causes a drastic decrease in the cross sections. In Fig. 8.8, the dashed line indicates the theoretical cross sections $\sigma(nl, \sum n')$ summed over n'. This theoretical cross section shows that the velocity dependence is almost inversely proportional to V. Thus $k(nl, \sum n')$ $[=V\sigma(nl, \sum n')]$ is nearly constant and the values of $k(27\text{f}, \sum n')$ for processes (45) and (46) are 2.77×10^{-7} cm^3 s^{-1} and 2.26×10^{-7} cm^3 s^{-1} respectively, at 300 K. The

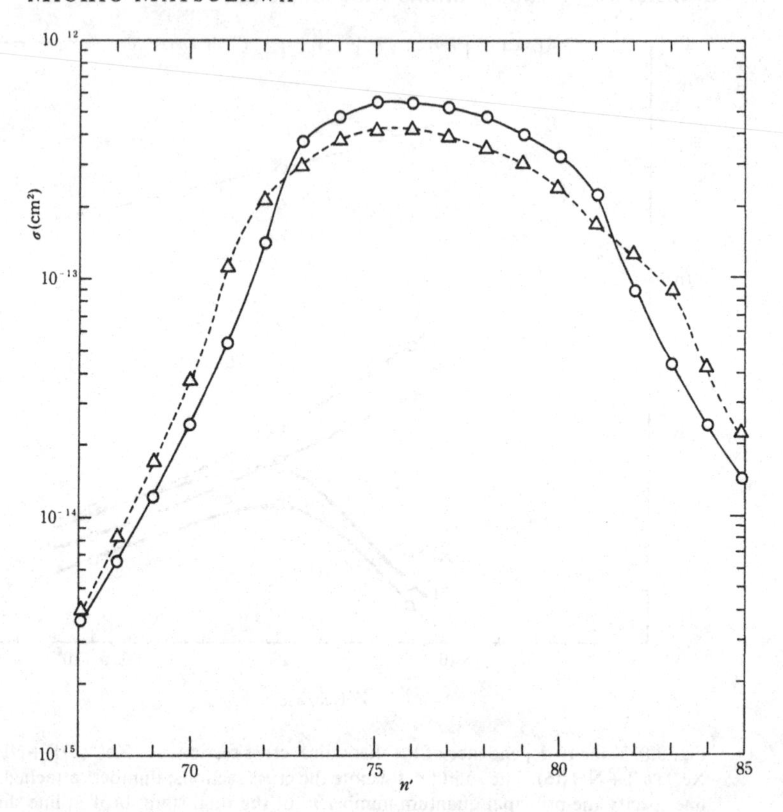

Fig. 8.9. The n' dependence of the theoretical cross sections for $Xe^{**}(27f) + NH_3(7) \rightarrow Xe^{**}(n') + NH_3(6)$, with $T = 300\,K$ and $V = 6.47 \times 10^4$ cm s^{-1}. The broken line shows the theoretical cross sections taking into account inversion splitting; see also the discussion in the text.

wave function of the high-Rydberg electron resembles that of the slow, ionized electron in the Coulomb field. Therefore, theoretical treatment of state-changing collisions to higher n' is quite similar to that of collisional ionization if the condition $\Delta n \gg 1$ is satisfied. Thus, in this case, we also obtain the V^{-1} velocity dependence of the cross sections $\sigma(nl, \sum n')$ [cf. Eqs. (34)-(36)] though individual cross sections for $nl \rightarrow n'$ transitions do not show this behavior, as indicated in Fig. 8.8. The n dependence of the theoretical rate constant $k(nl, \sum n')$ is also found to be weak. For example, $k(nf, \sum n')$ for

$$Xe^{**}(nf) + NH_3(5) \rightarrow Xe^{**}(\textstyle\sum n') + NH_3(4)$$

is evaluated to be 3.08×10^{-7} cm^3 s^{-1} for $n = 27$ and 3.44×10^{-7} cm^3 s^{-1} for $n = 33$ at 300 K. These theoretical results are found to be in qualitative agreement with experimental findings.[75]

Recently Rundel[85] also calculated the state-changing cross sections for pro-

Table 8.4. Comparison of theory with experiment for the rate constants for $Xe^{**}(31f) + NH_3(J) \rightarrow Xe^{**}(\Sigma n') + NH_3(J-1)$

	k (10^{-7} cm^3 s^{-1})[a]	
$(J \rightarrow J-1)$	Calculated[b]	Measured[c]
$(1 \rightarrow 0)$	0.01	—
$(2 \rightarrow 1)$	0.32	—
$(3 \rightarrow 2)$	0.53	—
$(4 \rightarrow 3)$	0.48	—
$(5 \rightarrow 4)$	0.44	—
$(5 \rightarrow 4) + (4 \rightarrow 3) + (3 \rightarrow 2)$	1.45	4

[a]The value given as the rate constant for each transition is the product of the usual rate constant and the population of the thermal distribution over the rotational states.
[b]Data from Rundel;[74,85] the values given in this reference should be divided by π.
[c]Data from Kellert et al.[76]

cesses (45) using expression (18) with the Born amplitude (32) based on the semiquantal theory of Flannery,[16] taking into account the inversion splitting of the ammonia molecules. Table 8.4 shows detailed comparisons between theory[84] and experiment.[76] For the state-changing collisions with large Δn, it should be noted that expression (18) with the electron velocity distribution of the initial Rydberg state is equivalent to expression (42), i.e., the derivative of expression (11) with the BET transition form factor, as was previously discussed for collisional ionization. This does not apply to state-changing collisions with small Δn because the BET transition form factor is no longer a good approximation for this case. Thus, again, we see that, quantitatively speaking, there is a systematic discrepancy between theory and experiment, i.e., a factor of three for the state-changing collisions. The reason for this discrepancy is not yet understood. Further theoretical and experimental studies are needed to clarify this situation. Rundel[85] also noted the effect of inversion splitting, i.e., the double-peaked structure in the n' dependence of the cross section. However, this seems to come from the adoption of cruder data on the level spacing of the inversion splitting. Actually, Fig. 8.9 shows that the K_r dependence of this level spacing[86] tends to smear this structure and that the n' distribution broadens a little.

So far we have restricted ourselves to the mechanism of energy transfer from molecular rotation to electronic motion. In principle, within the framework of the "free" electron model, we can consider another possibility, i.e., elastic scattering, for process (2), particularly for the l-changing collisions of $Xe^{**}(nf)$ with NH_3 experimentally investigated by Kellert et al.[76] However, we do not as yet have a well-founded understanding of elastic scattering of extremely slow electrons by polar molecules, which prevents us from attacking this problem. Gallagher et al.[87] carried out theoretical studies on the angular momentum mixing of high-Rydberg Na** atoms with CO molecules based on the treatment by Olson.[44] This is based on a mechanism in which *e*-CO elastic scattering causes the angular momentum change of Na** as in Na**-rare gas collisions. Their results seem to be in agreement with experimental data,[87] though no attention was paid to the particular properties of the polar molecule. This probably comes from the small dipole moment of CO, i.e., $D = 0.113ea_0$. Further theoretical study is needed to understand l-changing collisions of high-Rydberg atoms with polar molecules with $D \simeq 1ea_0$.

Before ending this subsection, it would be useful to make some comments on the remarkable stability of high-Rydberg atoms in pressure shift measurements of high-Rydberg series of alkali metals perturbed by the rare gas atoms in contrast with their fragility in collisions with the polar molecules. The rare gas atom has no internal degrees of freedom that the Rydberg electron can excite or deexcite. Thus the electronic motion of the Rydberg atom can only couple with the relative motion, which is quite unlikely to occur at thermal energies. Further, the high-Rydberg alkali atom is optically excited from the ground state to high-np states, which are usually energetically well separated from other levels. This situation leads to the stability of high-Rydberg (np) alkali atoms because even the l-changing process is unlikely to take place for well-isolated np states. This stability may break down for Rydberg atoms with higher l because the l changing is more likely to occur, as is seen for the l mixing of Na**(nd) states by collisions with the rare gas atoms,[78] owing to the smaller energy gap between the nd state and the same n manifold with higher $l \geqslant 3$.

8.3b. Collisions with quadrupolar molecules

For collisional ionization and state-changing collisions with quadrupolar molecules, we can again consider the mechanism for the energy transfer from molecular rotation to the electronic motion of the Rydberg electron within the framework of the "free" electron model.[16,17,21] It is again well known that the Born approximation can describe rotational transitions of quadrupolar molecules induced by slow electron impact, particularly near threshold energies.[60]

The squared Born amplitude for the rotational deexcitation of a linear molecule by electron impact is written as[88]

$$|f_e(J \to J-2, Q)|^2 = \frac{2}{15} a_0^2 \left(\frac{Q_p}{ea_0^2} \right)^2 \frac{J(J-1)}{(2J+1)(2J-1)} \tag{47}$$

where Q_p is the quadrupolar moment of the molecule. For collisional ionization of high-Rydberg atoms by quadrupolar molecules, i.e.,

$$A^{**}(nl) + B(J) \to (A^+ + e) + B(J-2) \tag{48}$$

procedures similar to those for polar molecules yield,[17,20] as the cross sections for process (48),

$$\sigma_i^q(nl, J \to J-2) = \frac{8\pi a_0^2}{15}\frac{e^2}{\hbar V}\left(\frac{Q_p}{ea_0^2}\right)^2 \frac{J(J-1)}{(2J+1)(2J-1)} I_q(n, \epsilon_0) \tag{49}$$

for $n \gtrsim n_0$ $[=(R/\Delta E_r)^{1/2}]$, where

$$I_q(n, \epsilon) = \int_0^{\infty} \frac{dF_{nl}(Q, \epsilon)}{d(\epsilon/2R)} (Qa_0)^2 d(Qa_0)$$

The averaging procedure is similar to that for polar molecules and gives the collisional ionization cross section for a quadrupolar molecule σ_i^q at thermal velocities as[20]

$$\sigma_i^q = 4.7 \times \mu^{1/2} Q_p^2 \langle I_q \rangle \times 10^{-14} \text{ cm}^2 \tag{50}$$

where μ is given in atomic mass units and Q_p in atomic units. The symbol $\langle I_q \rangle$ denotes $\{J(J-1)/(2J+1)(2J-1)\} I_q(n, \epsilon_0)$ averaged over the assumed n distribution and over the thermal distribution of the rotational states.

For the quadrupolar molecules, Hotop and Niehaus[59] were not able to detect collisional ionization. However, Chupka (quoted by Matsuzawa[20]) obtained the relative ratio

$$\{\sigma_i^q(N_2)/\sigma_i^d(H_2O)\}_{exp} \simeq 10^{-3}$$

Theoretical estimates of this ratio based on expressions (38) and (50) yield

$$\{\sigma_i^q(N_2)/\sigma_i^d(H_2O)\}_{theor} \simeq 10^{-4}$$

Though this value is smaller than the experimental one by one order of magnitude, we see that the "free" electron model can explain the experimental trends.

Using the semiquantal theory, i.e., expression (17), Flannery[16] investigated n-changing collisions of high-Rydberg H** atoms with H_2 molecules, based on energy transfer from electronic to rotational motion:

$$H^{**}(n) + H_2(J) \to H^{**}(n') + H_2(J') \tag{51}$$

which is of astrophysical interest. Within the framework of the "free" electron model, he calculated the cross sections for small Δn for near-resonant cases such as

$$H^{**}(9) + H_2(J=0) \to H^{**}(8) + H_2(J'=2) \qquad \Delta E = 5 \times 10^{-4} \text{ eV} \tag{52}$$

$$H^{**}(12) + H_2(J=1) \to H^{**}(9) + H_2(J'=3) \qquad \Delta E = -5 \times 10^{-5} \text{ eV} \tag{53}$$

$$H^{**}(7) + H_2(J=2) \to H^{**}(6) + H_2(J'=4) \qquad \Delta E = -2.6 \times 10^{-3} \text{ eV} \tag{54}$$

and

$$H^{**}(9) + H_2(J=0) \to H^{**}(8) + H_2(J'=0) \qquad \Delta E = 4.45 \times 10^{-2} \text{ eV} \tag{55}$$

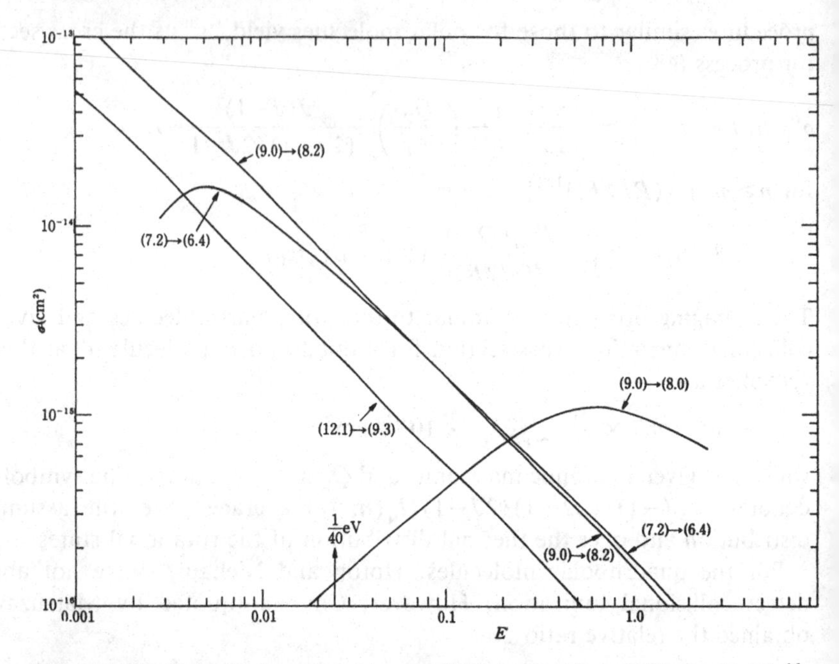

Fig. 8.10. The theoretical cross sections for $H^{**}(n)+H_2(J) \to H^{**}(n')+H_2(J)$.[16] The numbers in parentheses indicate the particular $(n, J) \to (n', J')$ transition.

In this calculation, he adopted the rotational excitation cross sections given by the elaborate close-coupling calculations of Henry and Lane[89] but assumed that the angular distribution was isotropic. Figure 8.10 indicates his results for processes (52)-(55). The cross sections for these processes are of the order of 10^{-14}-10^{-15} cm^2 at thermal energies, say, $E \simeq 1/40$ eV, and are smaller than those for polar molecules by $\sim 10^{-2}$-10^{-3}. The cross section for (55) is smaller than that for process (52) at thermal energies though the e-H_2 elastic scattering cross section is at least an order of magnitude greater than the rotational excitation cross section. This arises from the larger energy defect $|\Delta E|$ for this process.

Nishina[90] investigated state-changing collisions of high-Rydberg $Xe^{**}(nf)$ atoms with H_2 and D_2 and large Δn based on procedures similar to those used for polar molecules.[81,82] Figure 8.11 shows the n' dependence of the theoretical cross sections for the process

$$Xe^{**}(18f) + D_2(J=3) \to Xe^{**}(n') + D_2(J=1)$$

Comparison of this figure with Fig. 8.9 shows that the calculated cross sections are smaller than those for process (45) with $n=27$, $J=7$ by 10^4. He found that the polarization effect in electron–molecule scattering can be neglected. (This effect has been taken into account according to the procedure by Dalgarno and Moffet.[91]) To our knowledge, there have been no experimental data

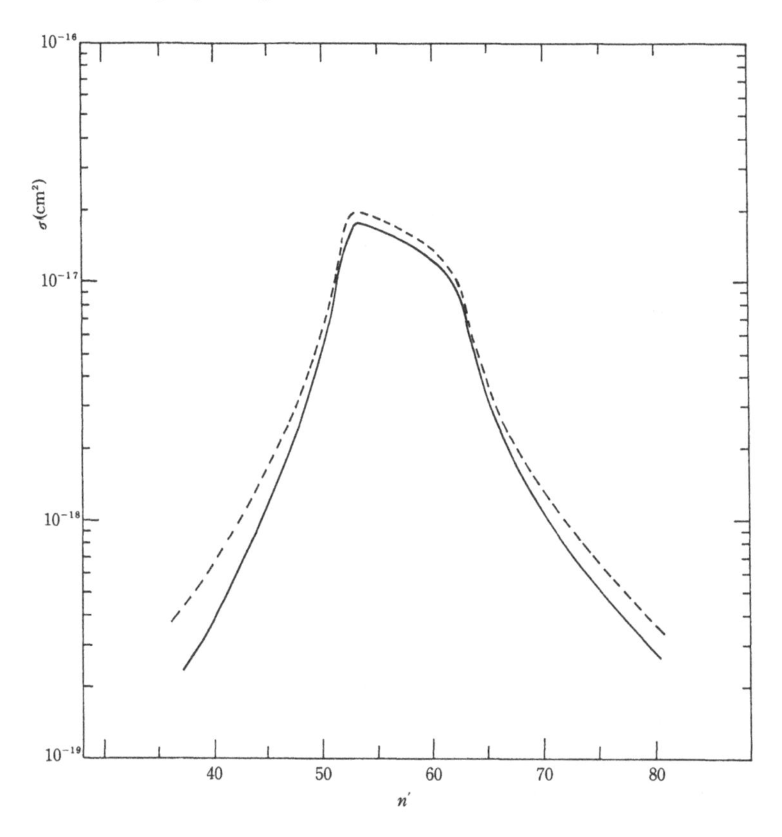

Fig. 8.11. The n' dependence of the theoretical cross sections for $Xe^{**}(18f)+D_2(J=3) \rightarrow Xe^{**}(n')+D_2(J=1)$, with $T=300$ K and $V=1.28\times10^5$ cm s^{-1}. The broken line denotes results with the polarization effect included.

on state-changing collisions with quadrupolar molecules for comparison with the theoretical results.

In general, the cross sections for collisional ionization and state changing in collisions with the quadrupolar molecules are smaller than those for polar molecules. This comes about because the effective range of the charge–quadrupole interaction is short compared with that of the charge–dipole interaction.

In electron–quadrupolar-molecule scattering, elastic scattering dominates over rotational excitation as far as some simple molecules such as H_2 and N_2 are concerned. This is in remarkable contrast with electron–polar-molecule scattering and means that for collisions with quadrupolar molecules elastic scattering in process (2) is relatively more likely to cause l-changing collisions in process (1) with a small energy defect such as

$$Na^{**}(nd) + N_2 \rightarrow Na^{**}(n, \Sigma l') + N_2 \qquad (5 \lesssim n \lesssim 15) \tag{56}$$

than for collisions with polar molecules. Actually, the maximum cross section is measured to be about 2000 Å^2 for the l-changing process (56) at thermal energies.[87] This value is larger than that for the n-changing collision cross section for processes (52) and so forth. Several authors[43–48, 50, 51] have developed theoretical treatments for l-mixing (changing) collisions of high-Rydberg atoms with rare gas atoms, a detailed account of which is given in another chapter. Under the assumption that the internal degrees of freedom of the molecule do not play a significant role, these theoretical treatments are potentially applicable to l-changing collisions with quadrupolar molecules, as suggested by the reasonable agreement between the experimental data on process (56) and the theoretical estimates.[87] In the previously mentioned theoretical treatments for intermediate n ($\lesssim 10$), the A^+–B interaction may interfere with the e_R–B interaction, as shown in the work based on the resonant energy conversion mechanism,[57] and it may lead to modification of the theoretical results obtained. This situation should be taken into account more seriously for state-changing collisions such as processes (52)–(55) with intermediate n because the cross sections obtained are comparable with those based on the resonant conversion mechanism.

8.3c. Collisions with electron-attaching molecules

The "free" electron model also enables us to relate electron-transfer processes from high-Rydberg atoms A** to electron-attaching molecules B at thermal energies, i.e.,

$$A^{**}(\alpha) + B(\beta) \rightarrow A^+ + B^-(\gamma) \tag{57}$$

to thermal electron attachment (EA) to B:

$$e + B(\beta) \rightarrow B^-(\gamma) \tag{58}$$

where γ describes the vibrational states of the negative ion. A similar argument can be applied to relate the following two processes, i.e., electron transfer with simultaneous molecular dissociation (ETMD) at thermal energies

$$A^{**}(\alpha) + BC(\beta) \rightarrow A^+ + BC^-(\gamma) \rightarrow A^+ + B + C^- \tag{59}$$

and thermal electron dissociative attachment (EDA) to the molecule BC,

$$e + BC(\beta) \rightarrow BC^-(\gamma) \rightarrow B + C^- \tag{60}$$

For this case, the quantum number γ also describes dissociative motion between B and C^- together with the internal state of the negative ion BC^-. In processes (57) and (59), the Rydberg electron becomes bound after collision. Therefore, expression (11) should be modified. Assuming that the negative ions produced in processes (57) and (59) are the same species as those produced in processes (58) and (60) respectively, and that a plane wave can be employed for describing the relative motion, Matsuzawa[18,19] obtained the following expressions for the rate constants k_{ET} $(=V\sigma_{ET})$ and k_{ETMD} $(=V\sigma_{ETMD})$

$$k_{ET}(nl,\beta) = \int v\sigma_a(q)\,|g_{nl}(q)|^2 q^2\,dq \tag{61}$$

and

$$k_{ETMD}(nl,\beta) = \int v\sigma_{da}(q)\,|g_{nl}(q)|^2 q^2\,dq \tag{62}$$

where σ_a and σ_{da} are the cross sections for EA and EDA, respectively. These expressions are derived without aid of the optical theorem. However, they are similar to expression (25) and are easy to interpret based on the "free" electron model. Namely, the rate constants k_{ET} and k_{ETMD} are, respectively, equal to the rate constants for EA and EDA over the momentum distribution of the Rydberg electron in the initial state. Later, Smirnov[39] obtained an expression similar to Eq. (61) by substituting σ_a into expression (20). However, there seems to be some ambiguity in the averaging procedure because he stated that the averaging procedure should be performed over a velocity distribution of the Rydberg electron different from that given in expression (29). We find, from expressions (61) and (62), that k_{ET} and k_{ETMD} become independent of n if we assume that s wave capture occurs in EA and EDA.[18,19]

Some researchers[92-94] have experimentally investigated processes (57) and (59) for collisions of $Xe^{**}(nf)$ $(25 \leqslant n \leqslant 40)$ with various electron-attaching molecules. West et al.[92] measured the absolute rate constants $k_{ET}(n)$ for the process

$$Xe^{**}(nf) + SF_6 \rightarrow Xe^+ + SF_6^- \tag{63}$$

at thermal velocities. They found that $k_{ET}(n)$ for (63) was essentially independent of n for $25 \leqslant n \leqslant 40$ and obtained $k_{ET} \simeq 4 \times 10^{-7}$ cm^3 s^{-1}. These values were found to be consistent with the data on k_a, the rate constant for EA. Foltz et al.[93] and Hildebrandt et al.[94] measured the absolute rate constants k_{ETMD} for various electron-attaching molecules and compared them with the rate constants for EDA. Both rate constants were found to be in close agreement with one another, though they were averaged over different velocity distributions. Figure 8.12 shows a typical example:

$$Xe^{**}(nf) + CCl_4 \rightarrow Xe^+ + CCl_3 + Cl^- \tag{64}$$

for $25 \leqslant n \leqslant 40$. As to the more detailed comparison beteen k_{ET} (k_{ETMD}) and k_a (k_{da}), the reader may refer to Chap. 9 by Dunning and Stebbings and also to the review article by Compton.[98] The close agreement between k_a (k_{da}) and k_{ET} (k_{ETMD}) seems to lend support to the "free" electron model.

However, some qualifications are appropriate on the theoretical and experimental sides. In the theoretical treatments, the Coulomb attraction in the final channels of processes (57) and (59) has not been properly taken into

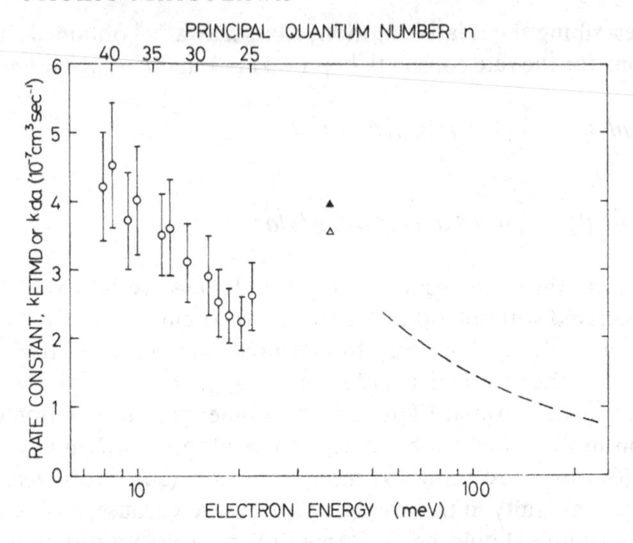

Fig. 8.12. Comparison of the rate constant k_{ETMD} for (64) with k_{da} for thermal dissociative attachment to CCl_4: for k_{ETMD}, circle, Foltz et al.;[93] for k_{da}, filled triangle, Warman and Sauer[95] microwave; broken line, Christodoulides and Christophorou[96] swarm; open triangle, Davies et al.[97] swarm. (From Matsuzawa[24])

account, though it may not be neglected, particularly at thermal energies.[99] Recent experimental studies have suggested that processes (57) and (59) are well separated into two independent steps.[100, 101] The first step is thought to be a simple attachment of the quasi-free Rydberg electron to the electron-attaching molecule at a large separation between A^+ and B (or BC), roughly equal to the radius of the Rydberg orbit. In the second step, the product negative ion B^- (or C^-) interacts with the atomic ion core A^+ because of the Coulomb attraction. In process (57), B^- may be stabilized by its interaction with A^+. In process (59), the kinetic energy released from the molecular dissociation may be enough to throw the product ion fragment C^- from the field of the atomic ion core A^+. Therefore the condition of the adoption of the plane wave for the relative motion may be better justified for this case. Astruc et al.[101] concluded that the "free" electron model satisfactorily described the first step, i.e., the formation of the negative ion B^- (or BC^-) but that the existence of the atomic ion core A^+ could not be ignored in thermal collisions. Thus the electron transfer process, particularly at thermal energies, may constitute an exceptional case where the atomic ion core A^+ plays a significant role even under condition A.1.*a* listed in Table 8.2 because of the Coulomb attraction in the final channel after rearrangement. In this context, the validity of the "free" electron model depends critically on the separability between these two steps. Further experimental and theoretical studies are needed to clarify this situation.

8.4. Fast collisions

From an intuitive discussion, Butler and May[23] predicted, based on the "free" electron model, that ionizing collisions of H**(n) with H(1s) can be identified with the total cross section for scattering of an electron, moving with the same speed as H**(n), by H(1s). Smirnov[39] investigated the ionization of high-Rydberg atoms by collisions with neutral species. He obtained, from expression (20) under the condition $V \gg v$,

$$\sigma_{\mathrm{i}} = \left\langle \int_{\Delta E > E_n} d\sigma \right\rangle \tag{65}$$

Starting from this expression, he showed that collisional ionization σ_{i} of A** by collision with an atom or molecule B is equal to the elastic or inelastic scattering of the electrons by B. De Prunelé and Pascale[50] also pointed out that the total cross section for fast collisions of A** with rare gas atoms is equal to the total cross section for free-electron scattering by rare gas atoms.

Quite recently, Matsuzawa[34] showed that the "free" electron model coupled with the impulse approximation, i.e., expression (6), can give a comprehensive theoretical description of fast collisions of A**(n) with B as well as of slow collisions. Adoption of the approximate "optical" theorem, i.e., Eq. (23), leads to relations (24) and (26) between the total cross sections for processes (1) and (2) from Eq. (6). In particular, Eq. (26) is a more general derivation of the relation for fast-Rydberg–rare-gas collisions pointed out by de Prunelé and Pascale.[50] We are led from Eq. (13), i.e., the derivative of Eq. (6), to the following asymptotic expression for the elastic scattering:[34]

$$\sigma_{\mathrm{el}}(nl, \beta, V) \simeq \frac{2\pi C_l}{n^4}\left(\frac{e^2}{\hbar V}\right)^2 |f_{\mathrm{e}}(\beta, \boldsymbol{\kappa} \to \beta, \boldsymbol{\kappa})|^2 \tag{66}$$

where C_l ($\simeq 1$) is determined by the form factor of the high-Rydberg atom. Further, Eq. (13) yields an expression for the ionization of A**(n), i.e.,

$$\sigma_{\mathrm{i}}(n, \beta, K \to \beta') = 2\pi a_0^2\left(\frac{e^2}{\hbar V}\right)^2 \int_{K_{\min}}^{K_{\max}} |f_{\mathrm{e}}(\beta, \boldsymbol{\kappa} \to \beta, \boldsymbol{\kappa}')|^2 \mathfrak{I}_n(Q)Q\,dQ \tag{67}$$

where

$$\mathfrak{I}_n(Q) = \int_0^{\epsilon(Q)} |\langle \epsilon|\exp(i\mathbf{Qr})|n\rangle|^2\, d\epsilon \tag{68}$$

Here, $\langle \epsilon|\exp(i\mathbf{Qr})|n\rangle$ is the transition form factor of the Rydberg atom for the discrete–continuum transition and $K_{\min}$, $K_{\max}$, and $\epsilon(Q)$ are determined by energy conservation and other kinematic relations. The profile function $\mathfrak{I}_n(Q)$ determines how the e–B differential cross sections should be modified in the profile of the Q dependence for process (1). Adoption of the binary encounter theory (BET) form factor, which is justified for the high-Rydberg state,[66] leads to the following Q dependence of the profile function for the high-n limit:

$$\mathcal{I}_n(Q) = \begin{cases} 1 & \text{for} \quad \kappa_{\min} < Q < \kappa_{\max} \\ 0 & \text{otherwise} \end{cases} \tag{69}$$

where $\kappa_{\min}$ and $\kappa_{\max}$ are determined by energy conservation for process (2). Thus for the high-n limit, expression (67) can be reduced to an integral cross section for the e–B collision (2) if we take into account relation (15), i.e., $\boldsymbol{\kappa} = -m_e\mathbf{V}/\hbar$. In other words, the ionization cross section σ_i for process (1) converges to the integrated cross section for process (2) in the high-n limit, i.e.,

$$\lim_{n\to\infty} \sigma_i(n, \beta, V \to \beta') = \sigma_e(\beta, V \to \beta') \tag{70}$$

This also leads to the relation

$$\lim_{n\to\infty} \sigma_i(n, \beta, V) = \sigma_e^t(\beta, V) \tag{71}$$

where $\sigma_i(n, \beta, V) = \sum_{\beta'} \sigma_i(n, \beta, V \to \beta')$ and

$$\sigma_e^t(\beta, V) = \sum_{\beta'} \sigma_e(\beta, V \to \beta')$$

Equation (71) recovers the prediction intuitively given by Butler and May.[23] Expressions (66) and (67) seem to be consistent with expressions (24) and (26), i.e., the derivatives of the approximate "optical" theorem, in contrast with the case of thermal l-changing collisions with the rare gas atoms.[77] Thus for the present case, we may use expression (26) as the upper limit to the cross sections arising solely from the e–B interaction (see Fig. 8.2).

Quite recently, with the previously mentioned prediction in mind, Koch[102] experimentally measured the total ionization cross section σ_i for

$$D^{**}(35 \leqslant n \leqslant 50) + N_2 \to (D^+ + e) + N_2(\Sigma\beta') \tag{72}$$

and the total destruction cross sections σ_d for

$$D^{**}(n \simeq 46) + N_2 \to D^{**}(n \lesssim 28 \text{ and } n \gtrsim 61) + N_2(\Sigma\beta') \tag{73}$$

In this experiment, the energy range of the high-Rydberg D atom ($E_D = 6$–13 keV) was chosen to correspond to the electron energy range $1.5 < E_e < 3.5$ eV, where free e–N_2 scattering has a large σ_e^t dominated by a series of $N_2^-\,(^2\Pi_g)$ compound-state-induced resonances. The observed resonant structure in and the absolute magnitudes of σ_i and σ_d confirmed that the Rydberg electron bound to the fast atom scattered quasi-freely for $D^{**}(n) + N_2$ collisions.[102] These experimental results are explained more quantitatively by Eqs. (26), (66), (67), and (70). The cross section σ_d is larger than σ_i, which is thought to result from the nearly n independent contribution (see Table 8.2) from the D^+–N_2 collision, i.e., the charge-transfer process (with cross section σ_{10}),

$$D^+ + N_2 \to D + N_2^+$$

Comparison of Eq. (26) with Eq. (66) shows that the contribution of state-changing collisions with small Δn to σ_d is negligibly small in the high-n limit.[34]

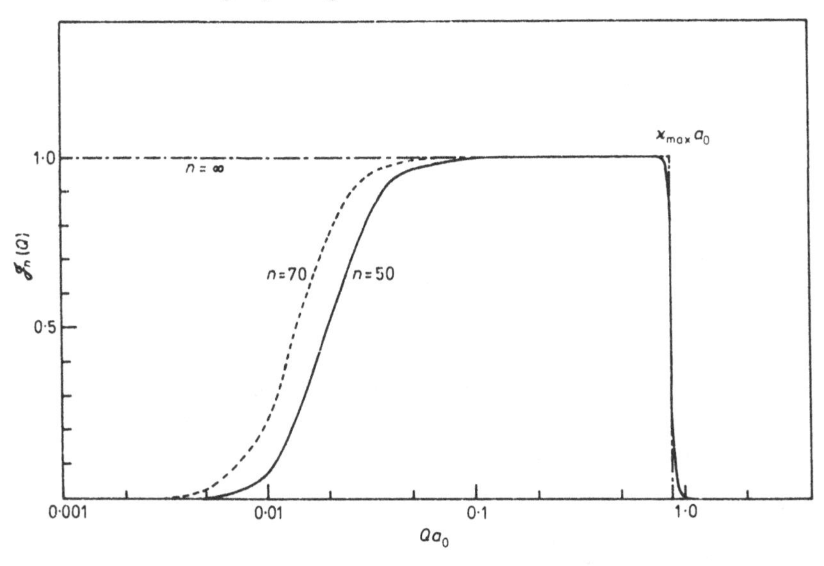

Fig. 8.13. The profile function $\mathcal{J}_n(Q)$ for $D^{**}(n)+N_2$ for $E_D = 10$ keV. The figure shows the case where $\beta=\beta'$ (i.e., $\Delta=R/n^2$). Thus $K_{min}>\kappa_{min}=0$. Around $Qa_0\simeq 1$, only $\mathcal{J}_{50}(Q)$ is shown because the broken curve cannot be distinguished from the full line. (From Matsuzawa[34])

Thus we have $\sigma_d \simeq \sigma^t_{e\text{-}B} + \sigma_{10} \simeq \sigma_i + \sigma_{10}$, which is consistent with experimental findings for fast collisions.[102] In this case, expression (26), based on the approximate "optical" theorem (23), explains the experimental findings. This probably comes from the fact that the perturbative approach for the relative motion is better justified for fast collisions (see Fig. 8.2). Further, Koch[102] observed the weak n dependence of σ_d. This can be explained by the weak n dependence of the profile function,[34] which is shown in Fig. 8.13. Thus we see that the "free" electron model combined with the impulse approximation can also describe fast collisions of high-Rydberg atoms with neutral species. In other words, we can view the high-Rydberg atom as a "beam" of free electrons. However, this "beam" has the energy width arising from the momentum distribution of the Rydberg electron.[102] In principle, expression (6) enables us to investigate in more detail how the monochromaticity of this "beam" is reduced by the momentum distribution of the Rydberg electron for fast collisions.

8.5. Extension to collisions of ions in high-Rydberg states

Recently, Matsuzawa[34, 103, 104] showed that the "free" electron model can be extended to collisions involving hydrogenlike ions $A^{(Z-1)+}$ in high-Rydberg states with $n/Z \gg 1$, i.e.,

$$A^{**(Z-1)+}(\alpha) + B(\beta) \rightarrow A^{**(Z-1)+}(\alpha') + B(\beta') \tag{74}$$

Table 8.5. Classification of hydrogenlike ions based on the binding energy of the Rydberg electron

	Strength of the binding	
n_r	$Z = 1$(H)	$Z \gg 1$
$n_r \gg 1$	Loose	Loose
$n_r \simeq 1$	Normal	Normal
$n_r \ll 1$	—	Tight

For the hydrogenlike ions $A^{**(Z-1)+}$ in high-Rydberg states, we have the well-known scaling laws for the energy levels E_n^Z and other physical quantities, i.e.,

$$E_n^Z = Z^2 E_n = -Z^2 R/n^2$$

$$v_n^Z = Z v_n = Z e^2/n\hbar$$

$$r_n^Z = r_n/Z \simeq n^2 a_0/Z$$

$$\tau_n^Z = \frac{\tau_n}{Z^2} \simeq \frac{2\pi n^3}{Z^2} \frac{\hbar^3}{m_e e^4}$$

Here v_n is the root-mean-square velocity of the Rydberg electron and the period τ_n of its electronic motion is given by $2\pi r_n/v_n$. The subscript and the superscript Z denote the atomic number of the ion. If we define a "reduced" quantum number $n_r = n/Z$, we may rewrite the scaling laws as

$$E_n^Z = E_{n_r} \tag{75}$$

$$v_n^Z = v_{n_r} \tag{76}$$

$$r_n^Z = Z r_{n_r} \tag{77}$$

$$\tau_n^Z = Z \tau_{n_r} \tag{78}$$

Here n_r is reduced to the usual principal quantum number of the hydrogen atom if $Z=1$. Particularly for large Z, we can categorize high-Rydberg ions into three groups according to the binding energy of the electron, i.e., $n_r \gg 1$, $n_r \simeq 1$, and $n_r \ll 1$ (see Table 8.5). We usually consider that ions with $n \gg 1$ are in high-Rydberg states. However, it is more reasonable to classify high-Rydberg ions with $n \gg 1$ in two groups, $n_r \gg 1$ and $n_r \simeq 1$. In the latter case, the binding energy of the Rydberg electron is still of a "normal" order of magnitude and comparable with that of a valence electron in the ground state or in lower excited states. Therefore, we must treat the Rydberg electron as one of the valence electrons of the colliding system when the high-Rydberg ion interacts with the neutral species. In charge-transfer processes involving highly stripped ions in high-Rydberg states such as O^{7+} $(n=8)$, the Rydberg electron e_R has been treated in such a manner.[105] In the former, the Rydberg electron is loosely bound and energetically well separated from all other electrons of the colliding system when the Rydberg ion interacts with the neutral species. According to Eqs. (77) and (78), for the collisions of the high-Rydberg ions with $n_r \gg 1$, we can easily see that conditions (3)-(5) are also satisfied because $Z>1$, i.e., $r(e_R\text{-B}) \ll Zr_{n_r}$, $r(A^{Z+}\text{-B}) \ll Zr_{n_r}$, $\tau_c \ll Z\tau_{n_r}$. Thus we can employ the "free" electron model to study collisions of high-Rydberg ions with $n_r \gg 1$ with neutral species, i.e., process (74) if the e_R–B interaction plays a decisive role in the collision.

Thus we obtain the following expression for the scattering amplitude for process (74) instead of expression (6) for the high-Rydberg atom:

$$f_Z(\alpha, \beta, \mathbf{K} \to \alpha', \beta', \mathbf{K}') = (\mu/\mu_{eB}) \int G_Z^*(\alpha' \,|\, \mathbf{q}') G_Z(\alpha \,|\, \mathbf{q}) f_e(\beta, \boldsymbol{\kappa} \to \beta', \boldsymbol{\kappa}') \, d\mathbf{q} \tag{79}$$

where $G_Z(\alpha\,|\,\mathbf{q})$ and $G_Z(\alpha'\,|\,\mathbf{q}')$ are the momentum space wave functions of the Rydberg electron in the Coulomb field caused by the A^{Z+} ion. This expression can be physically understood from the fact that, for fixed n_r, the e_R–B system feels the same Coulomb attraction from A^{Z+} as that produced by A^+. Therefore the same argument as that for the neutral high-Rydberg atom may be applied to the present case if the collisions take place exclusively at large impact parameters such as $b \simeq Zr_{n_r}$.

Under the conditions stated, the approximate "optical" theorem, i.e., Eq. (23) is also applicable to expression (79), i.e.,

$$\operatorname{Im} f_Z(\alpha, \beta, \mathbf{K} \to \alpha, \beta, \mathbf{K}) = (K/4\pi)\sigma_{\text{e-B}}^{Z,\text{t}}(\alpha, \beta, K) \tag{80}$$

where $\sigma_{\text{e-B}}^{Z,\text{t}}$ is the total cross section for process (74) solely from the e–B interaction. This yields the relation

$$V\sigma_{\text{e-B}}^{Z,\text{t}}(\alpha, \beta, K) = \int v_{eB}\, \sigma_e^{\text{t}}(\kappa) \,|G_Z(\alpha\,|\,\mathbf{q})|^2 \, d\mathbf{q} \tag{81}$$

which leads to the following expressions:

$$V\sigma_{\text{e-B}}^{Z,\text{t}}(\alpha, \beta, K) = \int v\sigma_e^{\text{t}}(q) \,|G_Z(\alpha\,|\,\mathbf{q})|^2 \, d\mathbf{q} \tag{82}$$

for slow collisions and

$$\sigma_{\text{e-B}}^{Z,\text{t}}(\alpha, \beta, V) = \sigma_e^{t}(\beta, V) \tag{83}$$

for fast collisions. These expressions, based on the approximate "optical" theorem, need to be tested as in the Rydberg atom case.

Furthermore, for slow collisions where f_e depends only on the momentum transfer $\hbar Q$, we are led from Eq. (6) to the following expression instead of expression (12):

$$f_Z(\alpha, \beta, \mathbf{K} \to \alpha', \beta', \mathbf{K}') = (\mu/m_e)\langle \alpha' | \exp(i\mathbf{Q}\mathbf{r}) | \alpha \rangle_Z f_e(\beta \to \beta', Q) \tag{84}$$

For fast collisions, we obtain, instead of expression (13),

$$f_Z(\alpha, \beta, \mathbf{K} \to \alpha', \beta', \mathbf{K}') = (\mu/m_e)\langle\alpha'|\exp(i\mathbf{Qr})|\alpha\rangle_Z f_e(\beta, \boldsymbol{\kappa} \to \beta', \boldsymbol{\kappa}') \tag{85}$$

Here $\langle\alpha'|\exp(i\mathbf{Qr})|\alpha\rangle_Z$ is a transition form factor of the high-Rydberg ion and other quantities are the same as those previously defined. Therefore the scaling properties of the transition form factor yield relations between the cross sections or the rate constants for processes (1) and (74), i.e., some scaling laws.

For collisional ionization of high-Rydberg ions by polar molecules, procedures similar to those in Sect. 8.3a yield the following expression for the cross sections:

$$\sigma_i^Z(n, J \to J-1) = \frac{16\pi a_0^2}{3} \frac{e^2}{\hbar V} \left(\frac{D}{ea_0}\right)^2 \frac{J}{2J+1} I_d^Z(n, \epsilon_0) \tag{86}$$

The scaling law of the BET form factor, i.e.,

$$dF_n^Z(Q, \epsilon)/d(\epsilon/2R) = dF_{n_r}(Q, \epsilon)/d(\epsilon/2R) \tag{87}$$

gives a simple scaling law for the rate constant k_i^d $(= V\sigma_i^d)$, i.e.,

$$k_i^{d,Z}(n, J \to J-1) = k_i^d(n_r, J \to J-1) \tag{88}$$

Here the scaling property for the binding energy, i.e., Eq. (75), was used in the energy conservation law for process (74). Physically this means that the collisional ionization rates should be equal if the orbital velocity of the Rydberg electron is the same [see Eq. (76)]. The same scaling law is also applicable to quadrupolar molecules.

For the fast collisions, we are led from expression (85) to some scaling laws. For example, for elastic scattering, the correspondence principle (see Percival and Richards[25]) gives the following:

$$|\langle\alpha|\exp(i\mathbf{Qr})|\alpha\rangle_Z|^2 = |\langle\alpha|\exp(i\mathbf{Qr}/Z)|\alpha\rangle|^2 = \mathfrak{F}(n^2 Q a_0/Z)$$

which yields a simple scaling law for elastic scattering, i.e.,

$$\sigma_{el}^Z(n, \beta, V) \simeq \sigma_{el}(n_r, \beta, V)/Z^2 \tag{89}$$

For fast collisional ionization, we also have only to note the scaling property (87), which yields

$$\sigma_i^Z(n, \beta, V \to \beta') = \sigma_i(n_r, \beta, V \to \beta') \tag{90}$$

where V is also equal to the electron velocity relative to B.

To our knowledge, there have been no experimental studies of high-Rydberg ions in well-defined states. It thus remains to test the scaling properties, i.e., (88)–(90), experimentally.

8.6. Concluding remarks

In the previous sections, we saw that the Fermi-type approach was quite powerful for understanding the essential physics of collision process (1). Particularly for case A.1 in Table 8.2, where the *e*–B interaction plays a decisive role, the "free" electron model can reasonably explain various aspects of the experimental findings on high-Rydberg–neutral collisions (including collisions with rare gas atoms not discussed here), though there are still some quantitative discrepancies. To our knowledge, there have been no theoretical studies corresponding to the intermediate velocity range, i.e., case A.1.*b* in Table 8.2, where a close-coupling-type or similarly elaborate calculation is needed. For case A.2 in Table 8.2, where the A^+–B interaction plays a decisive role, there have been no extensive theoretical studies that pursue quantitatively the correlation between process (1) and process (27) or (28), though some experimental studies have appeared. Further, as to the intermediate-*n* range, i.e., case B, where the *e*–B and A^+–B interactions play roles simultaneously and may interfere with one another, there has been only one theoretical approach and it has not been applied to collisions with molecules.

As this brief survey indicates, there are many problems still remaining to be solved before we have a full understanding of Rydberg–neutral collisions or equivalent interactions between high-Rydberg atoms and neutral species.

If we restrict ourselves to the "free" electron model, we believe that it is quite important to understand the reason for the quantitative discrepancy between theory and experiment. This is particularly true because the "free" electron model for slow collisions suggests the possibility of using high-Rydberg atoms to probe extremely low-energy electron–atom (molecule) scattering provided this model is quantitatively correct. Actually, pressure-shift measurements have already been used to determine zero-energy scattering lengths for rare gas atoms.[9] Recently, it was shown theoretically that the pressure-shift measurements enable the detection of resonances in the low-energy electron–atom (molecule) collisions.[106-109] Similarly the quasi-elastic collisions of high-Rydberg atoms with rare gas atoms also yield information on low-energy electron–rare gas collisions.[110]

Appendix: Estimates of the fractional error of the impulse approximation in case A.1 in Table 8.2

There are two classes of interactions *U* for which the impulse approximation is valid, i.e., weak binding and semiclassical.[32,33] The fractional error arising from the impulse approximation may be estimated from Eq. (24) of the paper by Chew and Goldberger.[33] For high-Rydberg–neutral collisions, we have[56]

$$\delta \simeq \langle \Delta U \rangle \frac{f}{\lambda} \frac{1}{\hbar} \left(\frac{\hbar}{\epsilon} + \tau \right) \ll 1 \tag{91}$$

Here $\langle \Delta U \rangle$ is the average variation of the Coulomb interaction *U* over the range of the *e*–B interaction, *f* the scattering amplitude of the *e*–B collision, and τ the delay time of the *e*–B collision. It should be noted that δ is proportional to $\langle \Delta U \rangle$.

For slow collisions, i.e., $V \ll v_n$, ϵ is the kinetic energy of the Rydberg electron and ƛ the reduced de Broglie wavelength. For fast collisions, i.e., $V \gg v_n$, ϵ is the energy of the relative motion and ƛ its reduced de Broglie wavelength. If there is a resonance in the e–B collision, τ may be interpreted as the resonance lifetime τ_r ($\simeq \hbar/\Gamma$). Condition (5) may be replaced by condition (91).

For the Coulomb field, we have

$$\langle \Delta U \rangle \simeq r(e\text{–B})(e^2/n^4 a_0^2)$$

where we have put $r \simeq n^2 a_0$. For nonresonant cases, we can put $\tau = 0$. Therefore, for slow collisions, we have

$$\delta \simeq \frac{2}{n^3}\left(\frac{r(e\text{–B})}{a_0}\right)\left(\frac{f}{a_0}\right) \ll 1 \tag{92}$$

For thermal collisions with polar molecules such as $H^{**}(30) + HCl$, we may take $f \simeq r(e\text{–HCl}) \simeq 2 \times 10^{-7}$ cm and obtain $\delta \simeq 0.1$. This also applies to thermal l-changing collisions with HCl. For thermal collisions with Xe atoms, we have $r(e\text{–B}) \simeq f \simeq a$ ($\simeq 4 \times 10^{-8}$ cm), where a is the zero-energy scattering length. Equation (92) yields $\delta \simeq 0.004$. Hence we see that the semiclassical nature of the Coulomb field guarantees the validity of the impulse approximation at thermal energies.

For nonresonant fast collisions, condition (91) is rewritten as

$$\delta \simeq \frac{2}{n^3}\left(\frac{v_n}{V}\right)\left(\frac{f}{a_0}\right)\left(\frac{r(e\text{–B})}{a_0}\right) \ll 1 \tag{93}$$

We see that condition (93) is well satisfied for $V \gg v_n$ and $n \gg 1$. Thus nonresonant fast collisions of high-Rydberg atoms with neutrals is the most favorable case for the impulse approximation because both the weak binding and the semiclassical conditions for U are simultaneously met. If there is a resonance in the e–B collision, we have

$$\delta \simeq \frac{r(e\text{–B})e^2}{n^4 a_0^2}\,\frac{f}{ƛ}\,\frac{\tau_r}{\hbar} \ll 1 \tag{94}$$

because $\tau_r \gg \hbar/\epsilon$. For fast collisions such as $D^{**}(50) + N_2$, i.e., Eqs. (72) and (73), we may take $\epsilon \simeq 10$ keV, $\Gamma \simeq 1$ eV ($\tau_r \simeq \hbar/\Gamma$), and $\sigma_e^! \simeq 4\pi f^2 \simeq 3.5 \times 10^{-15}$ cm^2, where $r(e\text{–B}) \simeq f$, at the resonance energy. Then Eq. (94) yields $\delta \simeq 0.07$.

Acknowledgments

I wish to take this opportunity to thank Dr. Mitio Inokuti of the Argonne National Laboratory, Dr. K. Katsuura of Kuraray Co. Ltd., and Professor Tsutomu Watanabe of the University of Tokyo for their valuable discussions and useful comments in the early stages when I initiated theoretical studies in the field of the Rydberg-neutral collisions. The writing of the preliminary version of this chapter was begun during my visiting fellowship at the Centre for Chemical Physics (CCP), University of Western Ontario, and then discontinued because of the rapid progress in this field around 1976.

I am also indebted to the encouragement of Professor J. Wm. McGowan, University of Western Ontario, chairman of CCP at that time. The final version was completed at the University of Electro-Communications (UEC). This work was supported in part by a Grant-in-Aid for Scientific Research from the Ministry of Education, Science and Culture of Japan. I am also indebted to Ms. J. Tsukimura of UEC for typing the manuscript.

References and notes

1. V. S. Letkhov and C. B. Moore, *Sov. J. Quantum Electron. 6,* 129 (1976).
2. A. C. Rivière and D. R. Sweetman, in *Atomic Collision Process,* ed. M. Rc. McDowell (Amsterdam: North-Holland Publ., 1964), p. 734.
3. A. K. Dupree and L. Goldberg, *Annu. Rev. Astron. Astrophys. 8,* 231 (1970).
4. E. Fermi, *Nuovo Cimento 11,* 157 (1934).
5. E. Amaldi and E. Segrè, *Nuovo Cimento 11,* 145 (1934).
6. H. Margenau and W. W. Watson, *Rev. Mod. Phys. 8,* 22 (1936).
7. S. Y. Chen and M. Takeo, *Rev. Mod. Phys. 29,* 20 (1957).
8. R. G. Breen, in *Handbuch der Physik,* ed. S. Flügge (New York: Springer-Verlag, 1964), vol. 27, p. 72.
9. H. S. W. Massey and E. H. S. Burhop, *Electronic and Ionic Impact Phenomena,* 2nd ed. (Oxford University Press, 1969), vol. 1, pp. 414–18, 544, 556.
10. B. Bederson, *Comments At. Mol. Phys. 2,* 116 (1970).
11. B. A. Alekseev and I. I. Sobel'man, *Zh. Eksp. Teor. Fiz. 49,* 1274 (1965) [*Sov. Phys. JETP 22,* 882 (1966)].
12. L. P. Presnyakov, *Phys. Rev. A 2,* 1720 (1970).
13. M. A. Mazing and N. A. Vrublebskaya, *Zh. Eksp. Teor. Fiz. 50,* 343 (1965) [*Sov. Phys. JETP 23,* 228 (1966)].
14. M. A. Mazing and P. D. Serapinas, *Zh. Eksp. Teor. Fiz. 60,* 541 (1971) [*Sov. Phys. JETP 32,* 294 (1971)].
15. M. R. Flannery, *Ann. Phys. (N. Y.) 61,* 465 (1970).
16. M. R. Flannery, *Ann. Phys. (N. Y.) 79,* 480 (1973).
17. M. Matsuzawa, *J. Chem. Phys. 55,* 2685 (1971); errata *58,* 2674 (1973).
18. M. Matsuzawa, *J. Phys. Soc. Japan 32,* 1088 (1972).
19. M. Matsuzawa, *J. Phys. Soc. Japan 33,* 1108 (1972).
20. M. Matsuzawa, *J. Electron. Spectrosc. Relat. Phenom. 4,* 1 (1974).
21. G. N. Fowler and T. W. Preist, *J. Chem. Phys. 56,* 1601 (1972).
22. T. W. Preist, *J. Chem. Soc. Farad. Trans. I, 68,* 661 (1972).
23. S. T. Butler and R. A. May, *Phys. Rev. 137A,* 10 (1965).
24. M. Matsuzawa, in Invited Papers and Progress Reports, *XI ICPEAC,* eds. N. Oda and K. Takayanagi (Amsterdam: North-Holland Publ., 1980), p. 493; and references therein.
25. I. C. Percival and D. Richards, *Advances in Atomic and Molecular Physics,* eds. D. R. Bates and B. Bederson, (New York: Academic Press, 1975), vol. 11, p. 1.
26. K. Takayanagi, *Comments At. Mol. Phys. 6,* 177 (1977).
27. Y. Itikawa and K. Takayanagi, *J. Phys. Soc. Japan 26,* 1254 (1969).
28. Y. Itikawa, *Phys. Rep. 46,* 117 (1978).
29. N. F. Lane, *Rev. Mod. Phys. 52,* 29 (1980).
30. K. Takayanagi, *J. Phys. Soc. Japan 45,* 976 (1978).
31. G. F. Chew, *Phys. Rev. 80,* 196 (1950).
32. G. F. Chew and G. C. Wick, *Phys. Rev. 85,* 636 (1952).

33. G. F. Chew and M. L. Goldberger, *Phys. Rev. 87,* 778 (1952).
34. M. Matsuzawa, *J. Phys. B 13,* 3201 (1980).
35. M. R. Flannery, *Phys. Rev. A 22,* 2408 (1980).
36. N. F. Mott and H. S. W. Massey, *Theory of Atomic Collisions,* 3rd ed. (Oxford University Press, 1965), p. 334.
37. D. R. Bates and S. P. Khare, *Proc. Phys. Soc. London 85,* 231 (1965).
38. G. F. Chew and F. E. Low, *Phys. Rev. 113,* 1640 (1959).
39. B. M. Smirnov, in *The Physics of Electronic and Atomic Collisions,* eds. J. S. Risley and R. Geballe (University of Washington Press, Seattle, 1976), p. 701.
40. R. F. Stebbings, in *Electronic and Atomic Collisions,* ed. G. Watel (Amsterdam: North-Holland Publ., 1978), p. 549.
41. T. F. Gallagher, in Invited Papers and Progress Reports, *XI ICPEAC,* eds. N. Oda and K. Takayanagi (Amsterdam: North-Holland Publ., 1980), p. 473.
42. R. D. Rundel, in Invited Papers and Progress Reports, *XI ICPEAC,* eds. N. Oda and K. Takayanagi (Amsterdam: North-Holland Publ., 1980), p. 481.
43. J. I. Gersten, *Phys. Rev. A 14,* 1354 (1976).
44. R. E. Olson, *Phys. Rev. A 15,* 631 (1977).
45. A. Omont, *J. Phys. (Paris) 38,* 1343 (1977).
46. J. Derouard and M. Lombardi, *J. Phys. B 11,* 3875 (1978).
47. A. P. Hickman, *Phys. Rev. A 18,* 1339 (1978).
48. A. P. Hickman, *Phys. Rev. A 19,* 994 (1979).
49. T. Shirai, H. Nakamura, K. Iguchi, and Y. Nakai, *J. Phys. B 11,* 1039 (1978).
50. E. de Prunelé and J. Pascale, *J. Phys. B 12,* 2511 (1979).
51. M. Matsuzawa, *J. Phys. B 12,* 3743 (1979).
52. H. Nakamura, T. Shirai, and Y. Nakai, *Phys. Rev. A 17,* 1892 (1978).
53. C. A. Kocher and A. J. Smith, *Phys. Rev. Lett. 39,* 1516 (1978).
54. H. Hiraishi, T. Kondow, T. Fukuyama, and K. Kuchitsu, *Chem. Phys. Lett. 66,* 9 (1979).
55. J. Boulmer, G. Brau, F. Devos, and J.-F. Delpech, *Phys. Rev. Lett. 44,* 1125 (1980).
56. M. Matsuzawa, *J. Phys. B 14,* L553 (1981).
57. R. K. Janev and A. A. Mihajlov, *Phys. Rev. A 20,* 1890 (1979); *A 21,* 819 (1980).
58. T. F. Gallagher and W. E. Cooke, *Phys. Rev. A 19,* 2161 (1979).
59. H. Hotop and A. Niehaus, *J. Chem. Phys. 47,* 2506 (1967).
60. K. Takayanagi and Y. Itikawa, *Advances in Atomic and Molecular Physics,* eds. D. R. Bates and B. Bederson (New York: Academic Press, 1970), vol. 6, p. 105.
61. K. Takayanagi, *J. Phys. Soc. Japan 21,* 507 (1966).
62. O. H. Crawford, *J. Chem. Phys. 47,* 1100 (1967).
63. Y. Itikawa, *J. Phys. Soc. Japan 30,* 835 (1971).
64. Y. Itikawa, *J. Phys. Soc. Japan 32,* 217 (1972).
65. G. Herzberg, *Molecular Spectra and Molecular Structure,* Vol I: *Spectra of Diatomic Molecules,* 2nd ed. (New York: Van Nostrand, 1950), p. 536.
66. M. Matsuzawa, *Phys. Rev. A 9,* 241 (1974).
67. G. M. Cunningham and C. C. Lin, *J. Chem. Phys. 41,* 3268 (1964).
68. J. A. Schiavone, D. E. Donohue, D. R. Herrick, and R. S. Freund, *Phys. Rev. A 16,* 48 (1978).
69. T. Shibata, T. Fukuyama, and K. Kuchitsu, *Chem. Lett. 1974,* 75 (1974).
70. M. Matsuzawa, *Bull. Am. Phys. Soc. 19,* 69 (1974).
71. M. Matsuzawa and W. A. Chupka, *Chem. Phys. Lett. 50,* 373 (1977).
72. W. A. Chupka, *Bull. Am. Phys. Soc. 19,* 70 (1974).

73. C. J. Latimer, *J. Phys. B 10,* 1889 (1977).
74. R. D. Rundel, private communication.
75. K. A. Smith, F. G. Kellert, R. D. Rundel, F. B. Dunning, and R. F. Stebbings, *Phys. Rev. Lett. 40,* 1362 (1978).
76. F. G. Kellert, K. A. Smith, R. D. Rundel, F. B. Dunning, and R. F. Stebbings, *J. Chem. Phys. 72,* 3179 (1980).
77. F. Gounand, P. R. Fournier, and J. Berlande, *Phys. Rev. A 15,* 2212 (1977).
78. T. F. Gallagher, S. A. Edelstein, and R. M. Hill, *Phys. Rev. Lett. 35,* 644 (1975).
79. T. F. Gallagher, S. A. Edelstein, and R. M. Hill, *Phys. Rev. A 15,* 1945 (1977).
80. D. R. Bates, V. Malaviya, and N. A. Young, *Proc. R. Soc. London, Ser. A 320,* 437 (1971).
81. M. Matsuzawa, *Phys. Rev. A 18,* 1396 (1978).
82. M. Matsuzawa, *Phys. Rev. A 20,* 860 (1979).
83. R. D. Nelson, D. R. Lide, and A. A. Maryott, *Selective Values of Electronic Dipole Moments for Molecules in the Gas Phase,* Standard Data Series, National Bureau of Standards, Publ. No. 10 (Washington D.C.: U.S. GPO, 1967).
84. G. Herzberg, *Molecular Spectra and Molecular Structure,* Vol. III: *Electronic Spectra and Electronic Structure of Polyatomic Molecules* (New York: Van Nostrand, 1966).
85. R. D. Rundel, in *Electronic and Atomic Collisions, Abstracts of Papers,* eds. N. Oda and K. Takayanagi (Amsterdam: North-Holland Publ., 1980), p. 936.
86. M. S. Cord, J. D. Peterson, M. S. Loiko, and R. H. Haas, *Microwave Spectral Tables,* Vol. IV: *Polyatomic Molecules without Internal Rotation,* National Bureau of Standards Monograph 70, (Washington, D.C.: U.S. GPO, 1968), p. 369.
87. T. F. Gallagher, R. E. Olson, W. E. Cooke, S. A. Edelstein, and R. M. Hill, *Phys. Rev. A 16,* 441 (1977).
88. E. Gerjuoy and S. Stein, *Phys. Rev. 97,* 1671 (1955).
89. R. J. W. Henry and N. F. Lane, *Phys. Rev. 183,* 221 (1969).
90. S. Nishina, Master's Thesis (in Japanese) submitted to the Univ. of Electro. Comm. (1980).
91. A. Dalgarno and R. J. Moffet, *Natl. Acad. Sci. India A 33,* 51 (1963).
92. W. P. West, G. W. Foltz, F. B. Dunning, C. J. Latimer, and R. F. Stebbings, *Phys. Rev. Lett. 36,* 854 (1976).
93. G. W. Foltz, C. J. Latimer, G. F. Hildebrandt, F. G. Kellert, K. A. Smith, W. P. West, F. B. Dunning, and R. F. Stebbings, *J. Chem. Phys. 67,* 1352 (1977).
94. G. F. Hildebrandt, F. G. Kellert, F. B. Dunning, K. A. Smith, and R. F. Stebbings, *J. Chem. Phys. 68,* 1349 (1978).
95. J. M. Warman and M. C. Sauer, Jr., *J. Chem. Phys. 52,* 6428 (1970).
96. A. A. Christodoulides and L. G. Christophorou, *J. Chem. Phys. 54,* 4691 (1971).
97. F. J. Davies, R. N. Compton, and D. R. Nelson, *J. Chem. Phys. 59,* 2324 (1973).
98. R. N. Compton, in Invited Papers and Progress Reports, *CI ICPEAC,* eds. N. Oda and K. Takayanagi (Amsterdam: North-Holland Publ., 1980), p. 251.
99. M. Matsuzawa, *J. Phys. B 8,* 2114 (1975); corrigendum, *9,* 2559 (1976).
100. C. E. Klots, *J. Chem. Phys. 66,* 5240 (1977).
101. J. P. Astruc, R. Barbe, and J. P. Schermann, *J. Phys. B 12,* L377 (1979).
102. P. M. Koch, *Phys. Rev. Lett. 43,* 432 (1979).
103. M. Matsuzawa, Abstracts, *X ICPEAC,* ed. G. Watel (Amsterdam: North-Holland Publ., 1978), p. 168.
104. M. Matsuzawa, *Contrib. Res. Group At. Mol.* (*Japan*) *14,* 231 (1978).

105. R. E. Olson, in Invited Papers and Progress Reports, *XI ICPEAC,* eds. N. Oda and K. Takayanagi (Amsterdam: North-Holland Publ., 1980), p. 391; and references therein.
106. M. Matsuzawa, *J. Phys. B 8,* L382 (1975).
107. M. Matsuzawa, *J. Phys. B 10,* 1543 (1977).
108. B. P. Kaulakis, L. P. Presnyakov, and P. P. Serapinas, *Pis'ma Zh. Eksp. Teor. Fiz. 30,* 60 (1979) [*Sov. Phys. JETP Lett. 30,* 53 (1980)].
109. B. P. Kaulakis, *Zh. Opt. Spectrosk. 48,* 1047 (1980).
110. K. Sasano and M. Matsuzawa, *J. Phys. B 14,* L91 (1981).

9

Experimental studies of thermal-energy collisions of Rydberg atoms with molecules

F. B. DUNNING AND R. F. STEBBINGS

9.1. Introduction

The study of collisions of Rydberg atoms[1] with molecules provides unusual opportunities and challenges for the experimenter. Rydberg atoms have such short lifetimes that collision experiments must typically be completed within a few microseconds following Rydberg excitation. As measurement techniques become more sophisticated, it is evident that many effects not normally of consequence in studies of atoms in ground or low-lying excited states become important. For example, weak electric fields can have pronounced effects on Rydberg excitation and collision mechanisms. Rydberg atoms are only weakly bound and hence are extremely fragile. Thermal-energy collisions can lead to rapid mixing among Rydberg levels, with rate constants as large as $10^{-6}\ \mathrm{cm^3\ s^{-1}}$, or can result in ionization. Thus, even if excitation results in the production of atoms in a single Rydberg state, collisional mixing can rapidly lead to the evolution of a complex, time-dependent Rydberg population distribution. Interactions with background 300-K thermal radiation must also be considered.

Thermal-energy collisions involving an atom in a high-Rydberg state, with principal quantum number n and angular momentum quantum number l, and a molecular target can result in reactions such as

$$\mathrm{A}(nl) + \mathrm{BC} \rightarrow \mathrm{A} + \mathrm{BC}^+ + e \qquad \text{(Penning ionization)} \qquad (1)$$

$$\rightarrow \mathrm{ABC}^+ + e$$
$$\rightarrow \mathrm{AB}^+ + \mathrm{C} + e \qquad \text{(associative ionization)} \qquad (2)$$

$$\rightarrow \mathrm{A} + \mathrm{B}^+ + \mathrm{C} + e \qquad \text{(dissociative ionization)} \qquad (3)$$

which make use of the internal energy of the atom. Similar reactions have been studied using metastable atoms. However, for Rydberg atoms, a variety of additional collision processes are possible:

$$\mathrm{A}(nl) + \mathrm{BC} \rightarrow \mathrm{A}(nl') + \mathrm{BC} \qquad (l\ \text{changing}) \qquad (4)$$

$$\rightarrow \mathrm{A}(n'l') + \mathrm{BC} \qquad (n\ \text{changing}) \qquad (5)$$

$$\rightarrow \mathrm{A}^+ + \mathrm{BC} + e \qquad \text{(collisional ionization)} \qquad (6)$$

$$\rightarrow \mathrm{A}^+ + \mathrm{BC}^- \qquad \text{(electron transfer)} \qquad (7)$$

$$\rightarrow \mathrm{A}^+ + \mathrm{B}^- + \mathrm{C} \qquad \text{(dissociative electron transfer)} \qquad (8)$$

Of particular interest are processes (4) and (5), which can lead to the formation of a large number of different final Rydberg states. Frequently several of the preceding reactions proceed concurrently, each contributing to the overall collisional depopulation of the initial Rydberg state. The ultimate challenge to the experimenter is to identify the reaction processes occurring and to determine the cross sections and/or rate constants for each.

Experiments undertaken in a number of laboratories have provided data concerning many of these reactions. Until quite recently, many such measurements utilized Rydberg atoms in a broad and indeterminate range of quantum states that were produced by electron impact excitation. In consequence, many of the details of the collision process under investigation were obscured. Nonetheless, these pioneering experiments provided considerable insight into the nature of Rydberg collisions and paved the way for later studies.

Ionization effects ascribed to collisions involving atoms in an indeterminate mixture of highly excited states were reported by Čermák and Herman.[2] Associative ionization was noted by Kupriyanov[3] in collisions of highly excited helium and argon atoms with H_2. Hotop and Niehaus[4] observed the Penning ionization products H_2^+ and HD^+ in collisions of Ar** with H_2 and HD, respectively. Collisional ionization was investigated by Hotop and Niehaus[5] and Shibata et al.[6] These data[5] indicated that for the molecules H_2O, NH_3, SO_2, and C_2H_5OH the collisional ionization cross sections were large, $\sim 10^{-13}$–10^{-12} cm^2. Electron transfer to SF_6 was also observed,[5] the cross section for this process being $\sim 10^{-12}$ cm^2. Other workers detected negative ion formation through electron transfer to a variety of other molecules.[7–9] The first study involving atoms excited to a single, well-defined Rydberg state was reported by Chupka.[10,11]

Most theoretical investigations of Rydberg collision processes make use of a simple model, proposed by Fermi,[12] which assumes that the separation between the Rydberg electron and its ionic core is so large that they both do not interact simultaneously with a target particle. The collision is then analyzed in terms of a binary encounter between the Rydberg electron and the target particle or between the ionic core and the target. In thermal collisions with molecular targets, it is common to treat the collision as an interaction between the Rydberg electron and the target molecule in which the Rydberg electron behaves as if it were free except that it has a momentum distribution characterized by its quantum state.[13–16] However, Flannery[17] recently pointed out that for quasielastic (e.g., l-changing) collisions at thermal energies effects due to the core may not always be negligible.

9.2. Experimental methods

In an ideal experiment, atoms in a single Rydberg state would be allowed to interact with a target whose quantum state was also well defined. External radiation and fields would be reduced to unimportant levels. The various charged and excited reaction products would be identified and the corresponding absolute rate constants determined, together with their dependence on such parameters as the quantum states of the Rydberg and target particles. In this section, progress toward this goal will be discussed.

9.2a. Production of high-Rydberg atoms

Rydberg states may be populated by, for example, electron-impact excitation,[18] electron capture,[19] and photoexcitation. Only photoexcitation, however, affords the resolution necessary to permit excitation of a single Rydberg state, and most of the work to be described in this chapter makes use of Rydberg atoms created by laser-induced photoexcitation.

It is important to recognize that the character of the photoexcited states will depend on the electric field conditions in the excitation region. In a field-free environment, l is a good quantum number and, neglecting nuclear spin, photoexcitation results in the production of a Rydberg state $|nljm_j\rangle$. Because of the la Porte selection rule $\Delta l = \pm 1$, only states of low l may be populated. In the presence of an electric field, however, every Stark state can be described by a superposition of zero-field $|nljm_j\rangle$ states and thus all low-$|m_l|$ Stark states can be excited optically.[20] At low field values, the Stark splitting is small and typical laser linewidths are such that several Stark states may be populated simultaneously. Thus the creation of atoms in a single Rydberg level $|nljm_j\rangle$ requires excitation in a field-free environment using a narrow-linewidth optical source.

9.2b. Detection of Rydberg atoms

Rydberg atoms may be detected by observing the radiation they emit in spontaneous decay,[21] by rf multiphoton ionization,[22] and by photoexcitation to autoionizing states.[23] More commonly, however, they are detected by field ionization, a technique that has found wide application in the study of Rydberg collisions because it permits the absolute detection of Rydberg atoms, affords the possibility of state analysis, and enables the study of long-lived high-n states.

Field ionization and the Stark effect

The Schrödinger equation describing a single electron in a Coulomb potential is separable in both spherical and parabolic coordinates.[24] Because it remains separable in parabolic coordinates following the application of an external electric field, it is convenient to discuss the Stark effect with reference to parabolic states. In zero field, each state can be labeled by the parabolic quantum numbers n, n_1, n_2, $|m_l|$ related via

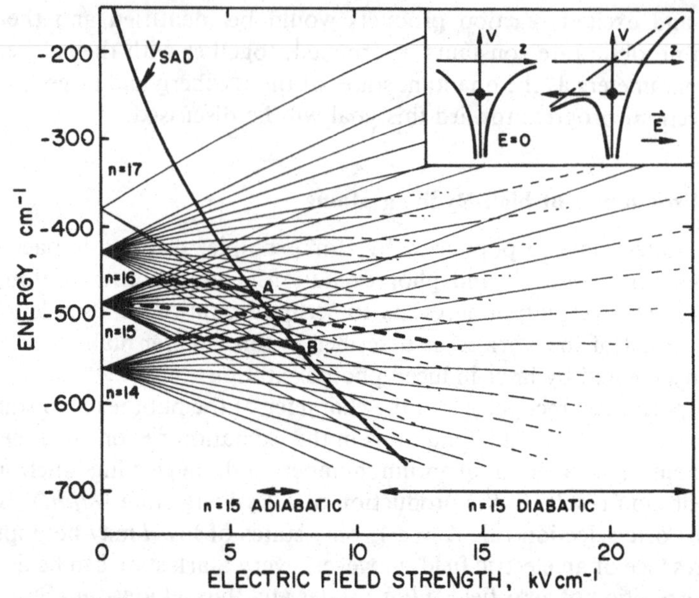

Fig. 9.1. $m_l=0$ Stark states in the vicinity of $n=15$. The high-field end of each state is dashed to indicate the range of fields over which the ionization rate increases from 10^7 to 10^{11} s^{-1}. The heavy solid lines give examples of adiabatic passage to ionization; the heavy dashed line diabatic passage. Inset shows the potential energy of the Rydberg electron in (*left*) the absence and (*right*) the presence of an external field directed along the z axis. The height of the resultant saddle point as a function of applied field is shown by the line labeled V_{SAD}.

$$n = n_1 + n_2 + |m_l| + 1 \tag{9}$$

The eigenfunctions[24] are asymmetrical with respect to the plane $z=0$. For $n_1 > n_2$, the electron probability density is greatest for positive values of z, and the opposite is true for $n_1 < n_2$.

For an external electric field F applied in the $+z$ direction, the potential takes the form

$$V = -(1/r) + Fz \quad \text{(a.u.)} \tag{10}$$

The presence of the electric field breaks some of the energy degeneracy (n^2 neglecting spin) and results in the Stark effect. The resultant energy levels, calculated by use of fourth-order perturbation theory,[25,26] are shown in Fig. 9.1 as a function of applied field for $m_l=0$ states in the vicinity of $n=15$. States with $n_1 > n_2$ have a dipole moment aligned antiparallel to the applied field and increase in energy with applied field. States with $n_1 < n_2$ have a dipole moment parallel to the applied field and decrease in energy.

Consideration of the inset in Fig. 9.1 shows that the presence of an electric field results in a saddle point on the $-z$ axis. Classically, field ionization of a

Rydberg state occurs when it lies above the saddle point, the height of which is given by[27]

$$V_{SAD} = -2F^{1/2} \quad \text{(a.u.)} \tag{11}$$

The line labeled V_{SAD} in Fig. 9.1 shows this height as a function of applied field and separates the region to the right where ionization is classically possible from that to the left where it is not.

In quantum-mechanical treatments of field ionization, the electron may escape from the atom by tunneling (quantum-mechanical tunneling, QMT), and the Stark states shown in Fig. 9.1 exhibit rapidly increasing ionization rates in the dashed regions.[28] Thus in the quantum picture, the majority of Stark states are relatively stable in wide regions where ionization is classically possible. This is readily understood because the high-lying Stark states for which $n_1 > n_2$ have wave functions with very little electron probability density in the vicinity of the saddle point, where electron escape may most readily occur, and thus ionization is inhibited. However, the lowest members of each Stark manifold have probability densities that are maximum in the vicinity of the saddle point and, hence, attain large ionization rates at critical fields close to those predicted by saddle-point theory.

Passage to ionization

Most Rydberg collision experiments are carried out in zero electric field, and it is necessary to understand how the atoms respond as the applied electric field is increased from zero to the value at which ionization occurs.

The ionization of sodium nd $^2D_{3/2,5/2}$ states will be discussed as a specific example, although similar considerations generally apply. In zero field, the states are described, neglecting nuclear spin, using the quantum numbers n, l, j, and m_j. At "intermediate" field strengths, where the quadratic Stark energy shift is greater than the fine-structure splitting, the atoms can be adequately described by the quantum numbers n, l, m_l, and m_s. At "high" fields, an essentially linear Stark effect is observed,[25] until states from adjacent Stark manifolds overlap, and the states are best described by use of the parabolic quantum numbers. It should be noted that at all field strengths the component of the total angular momentum along the field axis is conserved, i.e., $m_j = m_l + m_s$. Because m_l remains a good quantum number during passage from intermediate to high fields, the transition from states characterized by $|m_j|$ to those characterized by $|m_l|$ occurs between zero and intermediate fields.

Passage from zero to intermediate fields can be discussed by reference to the correlation diagrams shown in Fig. 9.2. If the electric field is increased sufficiently slowly, there will be a gradual adiabatic evolution of each initial zero-field state into a single corresponding intermediate-field state resulting in

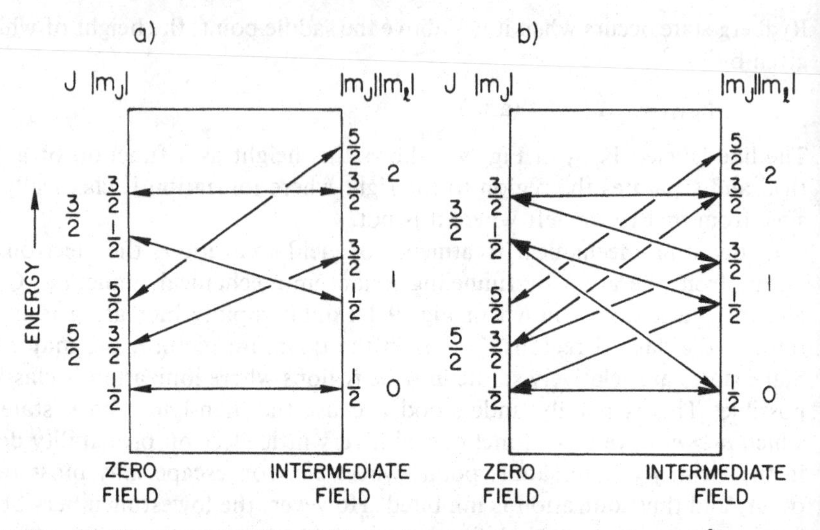

Fig. 9.2. Zero-to-intermediate-field correlation diagrams for sodium *nd* $^2D_{3/2,5/2}$. States of the same J and $|m_l|$ have been artificially separated for clarity. Adiabatic passage from low to intermediate fields results in the correlation diagram shown in (*a*); diabatic passage that in (*b*).

the correlation diagram shown in Fig. 9.2*a*. If, however, the electric field is applied sufficiently rapidly, each zero-field state will be projected onto the intermediate-field states from which it is composed, yielding the correlation diagram shown in Fig. 9.2*b*. As is evident from Fig. 9.2, the contribution from each of the initial $J=\frac{3}{2},\frac{5}{2}$ states to the final $|m_l|=0,1,2$ populations is strongly dependent on whether the passage to intermediate fields is adiabatic or diabatic. For $n\simeq 30$, electric field strengths of a few volts per centimeter constitute intermediate fields.

The energies of the $|m_l|=0$ and 1 Stark states of sodium in the vicinity of $n=15$ are shown in Fig. 9.3 as a function of applied field.[25] At electric field strengths below those at which states from adjacent Stark manifolds overlap, each sodium state exhibits an essentially linear Stark effect. At higher field strengths, avoided crossings occur because of coupling, between states of the same $|m_l|$, that results from the presence of non-Coulombic terms in the Hamiltonian.[25,29] These avoided crossings are more pronounced for the $m_l=0$ than for the $|m_l|=1$ states. Also shown are the corresponding hydrogenic Stark manifolds for which no such interaction exists and adjacent manifolds cross.

The similarities between sodium and hydrogen Stark structure evident in Fig. 9.3 enable the response of sodium Rydberg atoms to an increasing electric field to be discussed using energy-level diagrams and ionization rates calculated by use of relatively simple hydrogenic Stark theory,[26,28] but taking into account the presence of, and possible effects because of, avoided crossings.

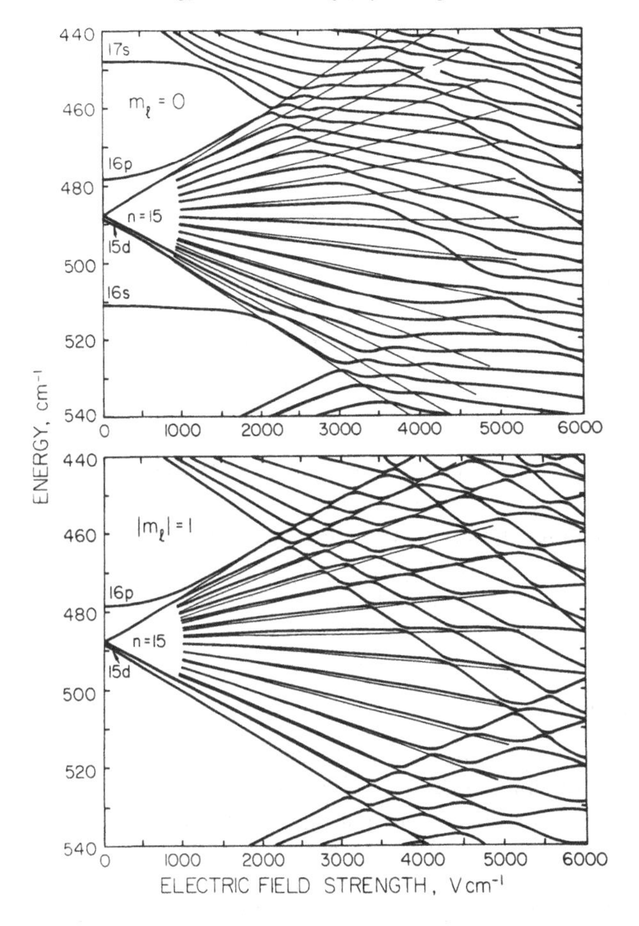

Fig. 9.3. Stark structure of sodium $|m_l| = 0, 1$ states in the vicinity of $n = 15$.[25] Hydrogenic $|m_l| = 0, 1$ Stark states are indicated by the thin lines.

To follow a given state to ionization, it must first be identified with the appropriate hydrogenic Stark state. It is apparent from Fig. 9.3 (*top*), for example, that ionization of the $m_l = 0$ state resulting from sodium 15d excitation is to be discussed by initially considering the second lowest member of the $n = 15$, $m_l = 0$ hydrogenic manifold, i.e., the $|15, 1, 13, 0\rangle$ Stark state.

The behavior of an atom as the field is increased through the region where adjacent Stark manifolds overlap depends on the strength of their mutual interactions.[30-40] For a state that is not strongly coupled to other states, the minimum energy separation at avoided crossings is small. Thus, as the electric field is increased, there is a high probability that avoided crossings will be traversed diabatically. The atom then follows a path to ionization like that shown by the heavy dashed line in Fig. 9.1 and ionization occurs at the quantum-mechanical limit. Diabatic ionization of states in the $n=15$ manifold thus occurs over the range of field strengths indicated in Fig. 9.1. On the other hand, if the atom is initially in a state that interacts strongly with states of other manifolds, the energy separations at avoided crossings are large. There is then a high probability that, if the rate of increase, i.e., slew rate, of the electric field is sufficiently slow, these avoided crossings will be traversed adiabatically. The atom will follow paths to ionization such as those indicated by the heavy solid lines in Fig. 9.1 in which the atom successively assumes the character of several different states. All atoms in $m_l=0$ states in the $n=15$ manifold that follow adiabatic paths to ionization cross the saddle-point line V_{SAD} between points A and B. To the right of this line, the atoms experience a dramatic increase in ionization probability because of interactions with the lowest states of higher Stark manifolds, which are highly unstable against ionization in this region. They therefore ionize at field strengths close to those predicted by saddle-point theory in the range marked "adiabatic" in Fig. 9.1.

The intermanifold interactions are highly $|m_l|$ dependent: the larger the value of $|m_l|$, the smaller the interactions between states of the same $|m_l|$ at level crossings.[25] States of higher n also exhibit less interaction at level crossings. Thus the probability of diabatic passage to ionization increases both with $|m_l|$ and with n.

Selective field ionization (SFI)

Because different Rydberg states ionize at relatively distinct electric field strengths, it is, in principle, possible to identify Rydberg states using field ionization. To accomplish this, an increasing electric field may be applied across the region containing the Rydberg atoms and the resulting ion signal measured as a function of applied field, a technique referred to as selective field ionization (SFI).[38] The interpretation of SFI data must be approached with caution, however, as a given ionization feature can only be uniquely correlated with a particular Rydberg state or set of states if the nature of the path to ionization is completely known.

As an example, consider Fig. 9.4 (*top*) in which are shown SFI data for sodium 34d $^2D_{3/2,5/2}$ states populated in zero electric field by two-step photoexcitation via the intermediate 3p $^2P_{3/2}$ state.[36] The exciting lasers were plane polarized perpendicular to the direction of the ionizing field, resulting in the excitation of states with all possible $|m_j|$ values and hence intermediate field states with $|m_l|=0, 1, 2$. Two distinct ionization features are present in the SFI spectrum.

The high field feature is attributed to diabatic ionization of states with $|m_l|=2$ because, when the exciting lasers are polarized parallel to the direction of the ionizing field so that no $|m_l|=2$ states can be formed, the high-field ionization signal can be eliminated.[39] Further, the high-field ionization threshold is consistent with that predicted by QMT calculations[28] for the

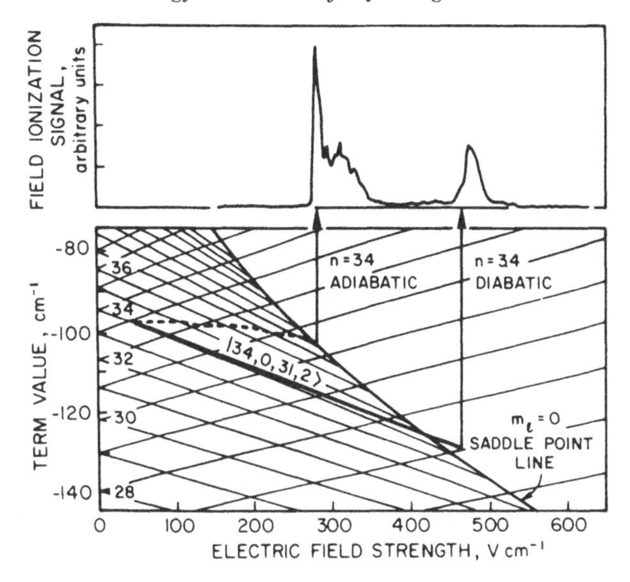

Fig. 9.4. Interpretation of SFI data. *Top,* selective field ionization data for sodium 34d atoms[36] obtained at a slew rate of $\sim 10^9$ V cm^{-1} s^{-1}. *Bottom,* the diagonal lines are the extreme members of the $m_l = 0$ hydrogenic Stark manifolds. The broken line indicates the adiabatic path to ionization of 34d $m_l = 0$ states, the heavy solid line the diabatic ionization of 34d $|m_l| = 2$ states.

lowest member of the $n = 34$, $|m_l| = 2$ Stark manifold, which corresponds to the 34d $|m_l| = 2$ state.

Adiabatic paths to ionization can be approximated using the construction discussed in Ref. 36 and, for $|34, 1, 32, 0\rangle$, is shown by the dashed line in Fig. 9.4 (*bottom*). The electric field strength at which this path intersects the saddle-point line is in good agreement with the experimentally observed low field-ionization threshold. Because states with $|m_l| = 1$ also ionize adiabatically at similar field strengths,[27] the low field feature is ascribed to predominantly adiabatic passage to ionization of $|m_l| = 0, 1$ states.

The small signal at electric fields between the adiabatic and diabatic thresholds is attributed to atoms in states that exhibit a combination of adiabatic and diabatic passage through avoided crossings.[36] As n increases, the ratio of this intermediate signal to the diabatic signal drops rapidly.

Studies of field ionization of xenon nf states with $n \simeq 30$ at slew rates of $\sim 10^9$ V cm^{-1} s^{-1} have shown them to ionize predominantly adiabatically,[41,42] even though nf states yield high field states with $|m_l| \leqslant 3$. This indicates that there is a stronger mutual interaction between states of given $|m_l|$ in xenon than in sodium. However, SFI analysis of states populated through n- and l-changing collisions, to be discussed later, shows that xenon states with $|m_l| > 3$ ionize predominantly diabatically.

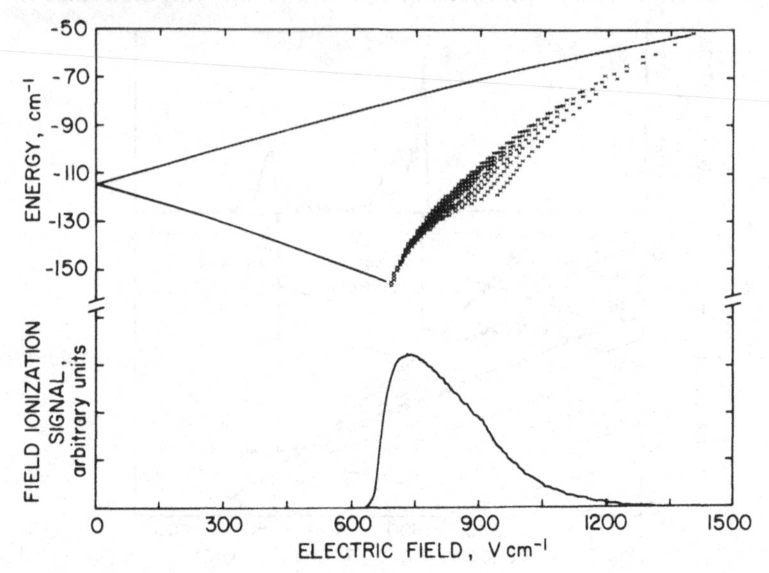

Fig. 9.5. Calculated SFI profile for diabatic ionization. *Top,* extreme members of the $n=31$, $|m_l|=3$ Stark manifold. The crosses represent the points at which each $|m_l|\geqslant 3$ Stark state achieves an ionization rate of $10^9\ \mathrm{s}^{-1}$. *Bottom,* calculated SFI profile for diabatic ionization of a mixture containing equal numbers of atoms in each $|m_l|\geqslant 3$ Stark level for $n=31$ at a slew rate of $10^9\ \mathrm{V\ cm^{-1}\ s^{-1}}$.

Synthesis of SFI spectra

Additional information concerning the field-ionization process and the state distribution of collision products can be obtained by comparing experimental SFI spectra with those derived theoretically.[38]

Knowledge of the electric field slew rate, together with the field dependence of the ionization rate, permits the calculation of SFI profiles for assumed Rydberg population distributions.[38] Consider a ramped ionization field F_t that is switched on at time t_0. If $N(t_0)$ atoms are present at t_0 in a particular Stark state whose ionization rate is $I(F_t)$, then, at a later time t, the number $N(t)$ of atoms remaining is

$$N(t)=N(t_0)\exp\left\{-\int_{t_0}^{t} I(F_t)\,dt\right\} \tag{12}$$

The field-ionization signal $ds(t)/dt$ corresponds to the number of ionization events per unit time and is given by

$$\frac{ds(t)}{dt}=I(F_t)N(t)=N(t_0)I(F_t)\exp\left\{-\int_{t_0}^{t} I(F_t)\,dt\right\} \tag{13}$$

Thus from a knowledge of $F(t)$ and $I(F_t)$, the SFI signal can be calculated. The total field-ionization signal appropriate to ionization of a mixture of initial states can then be calculated by summing the separate contributions from each state, weighted by the initial population in each state. For example, the profile shown in Fig. 9.5 (*bottom*) is that calculated for diabatic ionization, at a slew rate of 10^9 V cm^{-1} s^{-1}, of a mixture containing equal numbers of atoms in each $|m_l| \geqslant 3$ Stark level for $n=31$. In this case each atom remains in its original Stark state during passage to ionization and the ionization rate is obtained directly from QMT theory.[28] For comparison purposes, the extreme members of the $|m_l|=3$ manifold are shown in Fig. 9.5 (*top*) together with a series of points, each of which represents the electric field strength at which one of the $|m_l| \geqslant 3$ Stark states achieves an ionization rate of 10^9 s^{-1}.

Adiabatic ionization may be modeled by tracing the path of the initial state, as the field is increased, through avoided crossings with states from other manifolds until, at high fields, avoided crossings occur with states that are highly ionizing.[38] Knowledge of the electric field slew rate, coupled with the ionization rate for each state along the path, permits calculation of SFI profiles for adiabatic ionization.

The value of synthesized SFI profiles in interpreting experimental data will become evident during the discussion of experimental results.

9.2c. Other experimental considerations

Recent experiments demonstrate that Rydberg atoms are highly efficient absorbers of infrared radiation.[43] Thus effects due to interactions with background 300-K radiation must be considered in the analysis of experimental data. For instance, even in the absence of target gas, photoabsorption and stimulated emission can populate states other than the parent state at an appreciable rate.[44] Photoionization of Rydberg atoms by background radiation is also possible.[45]

Selective field-ionization spectra may also change in time as a result of quantum-beat phenomena or precession in stray off-axis fields.[46,47]

9.3. Experimental results: Rydberg atoms with $n \gtrsim 20$

9.3a. Polar targets

In collisions between a high-Rydberg atom and a polar molecule, rotational deexcitation can provide the energy necessary to further excite or ionize the Rydberg atom.[14,15] Recent experimental studies demonstrate that a complex array of reactions is possible. Only those targets that were investigated in detail, taking into account the interplay of these various reactions, will be discussed.

Ammonia

Collisions involving Xe(nf) and ammonia were studied by Smith et al.[41] and Kellert et al.[42] The rotational energies are given, neglecting inversion doubling, by

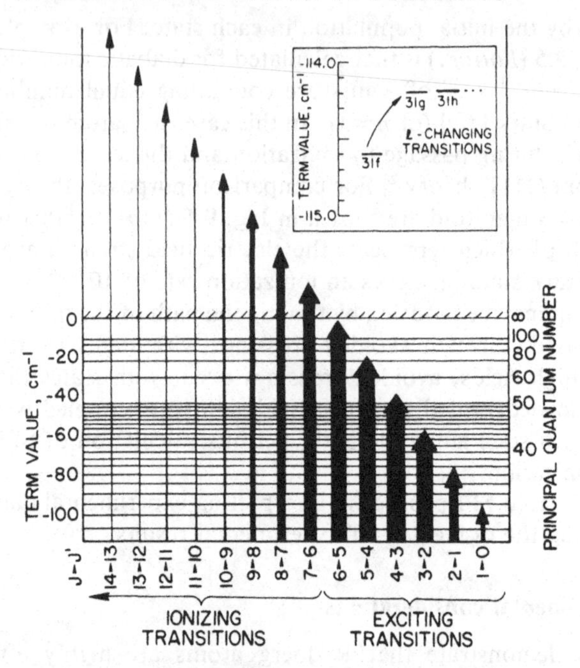

Fig. 9.6. Abbreviated term diagram for xenon showing the effect of Xe(31f)–NH_3 collisions. The lengths of the arrows show the energies released in the NH_3 rotational transitions indicated. The widths of the arrows are proportional to the room-temperature populations of the upper rotational levels involved.

$$E(J,K) = BJ(J+1) + (A-B)K^2 \tag{14}$$

where J is the total angular momentum quantum number, K its projection along the figure axis, and A and B molecular constants. In dipole-allowed deexcitations ($J \rightarrow J-1$, $K \rightarrow K$), the energies released are

$$\Delta E_J = E(J,K) - E(J-1,K) = 2BJ \tag{15}$$

The result of transfering these amounts of energy to a high-Rydberg atom is illustrated for Xe(31f) in Fig. 9.6, which includes a partial term diagram for xenon together with a series of arrows whose lengths correspond to the energies released in the rotational transitions indicated. The widths of the arrows are proportional to the room-temperature populations in the upper rotational levels.

As is evident from Fig. 9.6, energy transfer accompanying rotational deexcitation from levels with $J \geqslant 7$ results in collisional ionization via the reaction

$$\mathrm{Xe(31f)} + \mathrm{NH_3}(J) \xrightarrow{k_i(31f)} \mathrm{Xe}^+ + \mathrm{NH_3}(J-1) + e, \qquad J \geqslant 7 \tag{16}$$

Rotational deexcitation from levels with $J \leqslant 6$ leads to further excitation of the

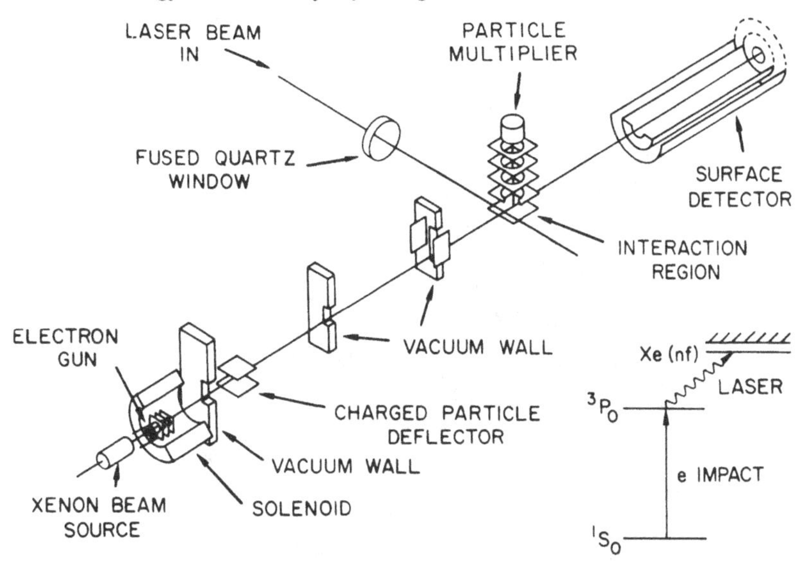

Fig. 9.7. Schematic diagram of the apparatus used to study Xe(nf) collisions.

atom to selected groups of high-lying excited states in n-changing collisions of the type

$$\mathrm{Xe(31f)} + \mathrm{NH_3}(J) \xrightarrow{k_n(\mathrm{31f})} \mathrm{Xe}(n'l') + \mathrm{NH_3}(J-1) \qquad J \leqslant 6 \tag{17}$$

In addition, as shown in the inset, the Rydberg atom can undergo l-changing collisions

$$\mathrm{Xe(31f)} + \mathrm{NH_3}(J) \xrightarrow{k_l(\mathrm{31f})} \mathrm{Xe}(31l') + \mathrm{NH_3}(J) \tag{18}$$

in which the J value of the target molecule and the principal quantum number of the Rydberg atom remain unchanged. The small amount of energy necessary for such l-changing transitions is furnished by the relative translational energy of the colliding particles. Each of these collision processes will contribute to the total collisional depopulation of Xe(31f).

Xenon(nf)–NH_3 collisions were studied using the apparatus shown schematically in Fig. 9.7.[42] A beam of ground-state xenon atoms is first formed by effusion from a multichannel array. A fraction of these atoms is then excited to the $^3P_{0,2}$ metastable levels by electron impact. The beam then enters a zero ($\lesssim 0.2$ V cm^{-1}) electric field region between two parallel grids where it is crossed by the output of a N_2-pumped dye laser tuned to excite the metastable atoms to a selected nf level. During collision experiments, NH_3 target gas is introduced into this interaction region at a pressure of a few microtorr. After a known interaction time following the laser pulse, charged collision products may be extracted from the interaction region by application of a weak electric field ($\lesssim 5$ V cm^{-1}) and detected by the particle multiplier. Alternately, the

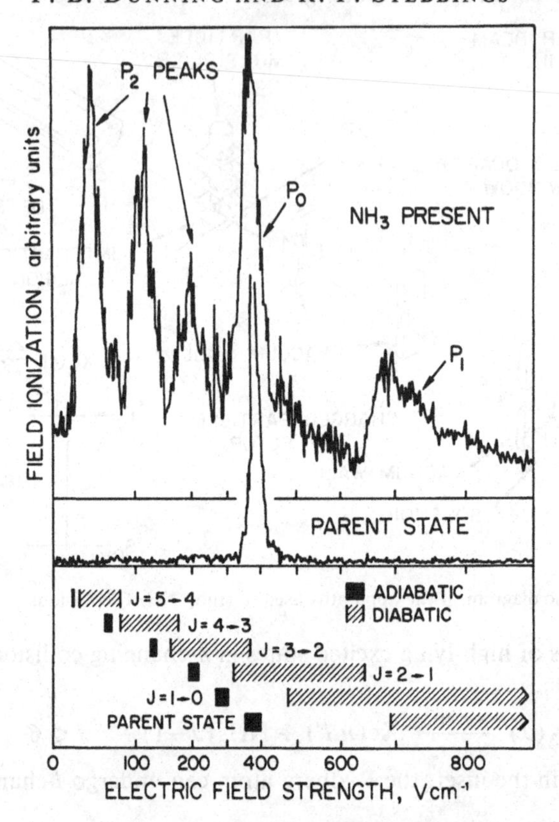

Fig. 9.8. Interpretation of SFI features observed after allowing Xe(31f) to interact with NH_3 for 20 μs. A typical SFI spectrum for 31f atoms is also included for comparison. The adiabatic and diabatic ionization ranges for states populated by the various possible rotational transitions in NH_3 are as indicated.

excited collision products may be analyzed utilizing SFI by application of a ramped potential to the lower interaction region grid. The resulting electrons are detected using the particle multiplier.

In Fig. 9.8 is shown the SFI spectrum obtained for ionization of Xe(31f) atoms together with that obtained following collisions of these atoms with NH_3. The various ionization features can be identified by reference to the horizontal bars below the data, which indicate the range of field strengths over which atoms in states with selected principal quantum numbers ionize diabatically and adiabatically. As is evident from Fig. 9.8, parent 31f atoms ionize predominantly adiabatically. The peak labeled P_0 results from ionization of remaining parent 31f atoms together with the products of l-changing collisions that ionize adiabatically, and P_1 comprises products of l-changing collisions that ionize diabatically. The location and profile of P_1 is consistent with that (see Fig. 9.5) for diabatic ionization of a mixture containing equal numbers of atoms in each of the $n=31$, $|m_l|>3$ Stark states. This suggests that collisions result in l-changed products having a wide range of values of l and $|m_l|$ although some mixing among states of very high l might result directly from the application of the ionizing field.

Furthermore, atomic precession about any residual off-axis field in the interaction region could also lead to changes in $|m_l|$. To test for the importance of such effects, SFI spectra were also recorded under conditions where, immediately following excitation in zero field, a weak electric field was applied parallel to the ionizing field direction. Collisions then occurred in this field that serves to prevent changes in $|m_l|$ from precession and results in a greater separation between adjacent levels so that in subsequent field ionization there is a reduced chance of field-induced l mixing. These experiments demonstrated that collisions are indeed responsible for the production of atoms having a broad range of values of l and $|m_l|$. However, it remains to be determined whether the observed distribution results from single or multiple collisions.

The peaks labeled P_2 result from ionization of the products of n-changing collisions. Each can be identified as arising from predominantly diabatic ionization of states populated by a specific rotational transition in NH_3. A given rotational transition may populate more than one manifold because of the effects of inversion doubling and because energy and momentum can be conserved in collisions in which there is a small interchange of translational and excitation energy.[48] The widths and shapes of the P_2 features again indicate that collisions result in the formation, either directly or through multiple collisions, of n-changed products with a broad range of values of l and $|m_l|$. The P_2 peaks are sufficiently close to one another that small features associated with adiabatic ionization of low-$|m_l|$ collision products are not resolvable. In addition, the ionization of states populated by $2 \rightarrow 1$ and $1 \rightarrow 0$ transitions is obscured by P_0 and P_1.

The total rate constant for collisional depopulation of a parent nf state, $k_d(n\mathrm{f})$, can be determined by measuring the time dependence of that part of P_0 that results from nf ionization. This can be identified by taking high-resolution SFI spectra of the P_0 peak using a low electric field slew rate.[42] If $N(0)$ 31f atoms are initially excited by the laser, the number $N(t)$ remaining at a later time t is given by

$$N(t) = N(0)\exp[-b(31\mathrm{f})t] \tag{19}$$

where

$$b(31\mathrm{f}) = \frac{1}{\tau(31\mathrm{f})} + \rho k_d(31\mathrm{f}) \tag{20}$$

and ρ is the NH_3 density and $\tau(31\mathrm{f})$ the effective radiative lifetime of the parent state.[49] Determination of $N(t)$ yields $b(31\mathrm{f})$ and, hence, $k_d(31\mathrm{f})$. Values of $k_d(n\mathrm{f})$ are shown in Fig. 9.9. Cross sections are assigned by assuming that the cross section varies slowly with velocity when

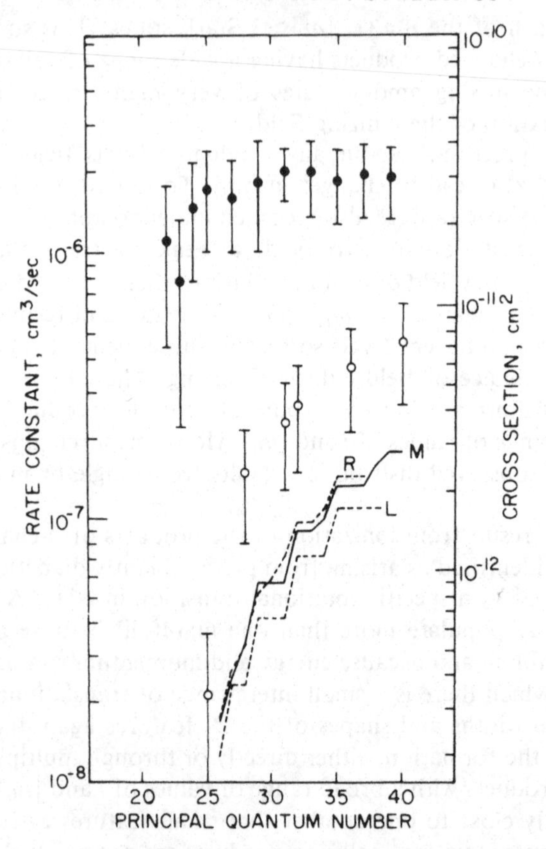

Fig. 9.9. Principal quantum number dependence of the experimental measurements and theoretical calculations for Xe(nf)–NH_3 collisions: filled circles, measured total depopulation rate constant $k_d(nf)$; open circles, measured collisional ionization rate constant $k_i(nf)$; lines, calculated ionization rate constants $k_i(nf)$; R, Rundel;[51] L, Latimer;[50] and M, Matsuzawa.[13,15]

$$\sigma_d(nf) = k_d(nf)/\bar{v}_{rel} \tag{21}$$

where $\bar{v}_{rel}$ is the average relative collision velocity. The measured rate constants are among the largest ever observed for thermal-energy neutral–neutral collisions.

Rate constants for collisional ionization $k_i(nf)$ can be determined from measurements of the Xe^+ ion signal. However, quantitative measurements for a specific Xe(nf) state must be approached with caution because the rate constants for l- and n-changing collisions are so large that the observed ion signal will result in part from ionization of such collision products. The complexity of these reaction sequences is such that they have not yet been treated in detail, although they can be treated collectively by postulating that state-changing collisions populate a "reservoir state" of Rydberg atoms.[42] Subsequent state-changing collisions then merely redistribute the Rydberg population within the reservoir state but do not significantly deplete it. Atoms are, however, lost from the reservoir through collisional ionization or radiative decay. The reaction sequence considered in analyzing the Xe^+ signal is

$$\begin{array}{lll} \mathrm{Xe}(n\mathrm{f}) + \mathrm{NH_3} & \xrightarrow{k_\mathrm{i}(n\mathrm{f})} \mathrm{Xe^+} & \\ \downarrow & & \\ \tau(n\mathrm{f}) & \xrightarrow{k_l(n\mathrm{f}) + k_n(n\mathrm{f})} \mathrm{Xe}(r) + \mathrm{NH_3} & \xrightarrow{k_\mathrm{i}(r)} \mathrm{Xe^+} \\ & \qquad\qquad \downarrow & \\ & \qquad\qquad \tau(r) & \end{array} \qquad (22)$$

Depopulation of the laser-excited nf state occurs through radiative processes, with an effective lifetime $\tau(n\mathrm{f})$, collisional ionization, and collisional n- and l-changing with rate constants $k_n(n\mathrm{f})$ and $k_l(n\mathrm{f})$. The collisionally populated reservoir state Xe(r) is depleted by radiative processes with lifetime $\tau(r)$ and by collisional ionization with rate constant $k_\mathrm{i}(r)$. Detailed analysis of the Xe^+ signal using this model yields the values of $k_\mathrm{i}(n\mathrm{f})$ shown in Fig. 9.9.[42]

Also shown in Fig. 9.9 are the results of various theoretical investigations.[13,15,50,51] Rate constants or cross sections are determined from a treatment of the collision between the Rydberg electron and NH_3 target utilizing e–NH_3 scattering parameters derived on the basis of a simple dipole interaction. A steplike dependence of the calculated rate constant on n is evident in all calculations. This results because as n increases the number of rotational transitions that yield sufficient energy to cause ionization increases discontinuously, enabling more of the NH_3 population to contribute to ionization. Rundel's calculation shows less-pronounced step structure than the others as a consequence of his inclusion of energy and momentum conservation and of inversion doubling.

It is evident from Fig. 9.9 that, although the experimental data do follow the trend of the theoretical calculations, i.e., $k_\mathrm{i}(n\mathrm{f})$ increases with n, the quantitative agreement is poor. This suggests that consideration of only the Rydberg electron–NH_3 interaction may not be entirely adequate for detailed quantitative predictions or that the e–NH_3 interaction is not well characterized by inclusion of only a dipole term. The experimental uncertainties are such that no steplike structure can be discerned.

It is apparent that, for the range of n included in Fig. 9.9, depopulation occurs primarily through n- and l-changing collisions, and analysis of the relative sizes of the corresponding features in SFI spectra enables their relative importance to be assessed. For Xe(31f), for example, $k_l(31\mathrm{f}) \simeq 3k_n(31\mathrm{f})$. There are few theoretical treatments of state-changing collisions.[48,52,53] Recently, Rundel,[48] using Flannery's semiquantal approach,[14] calculated individual reaction rate constants for n changing associated with each of the possible rotational transitions. The experimental data are approximately a factor of three above these calculated values.

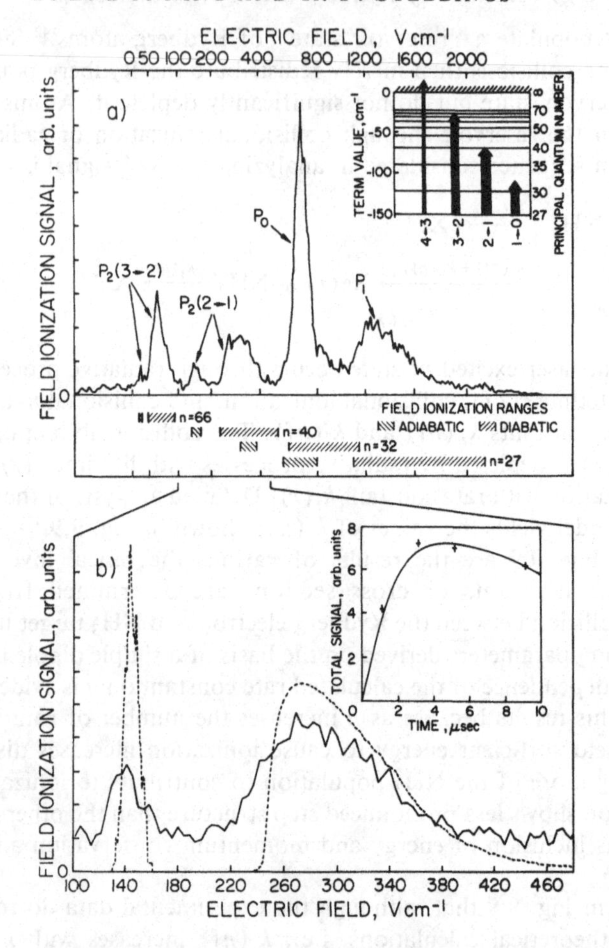

Fig. 9.10. Interpretation of SFI features observed following Xe(27f)–HF collisions. (*a*) SFI spectrum observed after collision with HF. The features can be identified by consideration of the inset and the indicated field-ionization ranges. (*b*) Detailed comparison of the SFI feature that results primarily from adiabatic and diabatic ionization of the products of reaction (23) with the profile calculated for ionization of a mixture containing equal numbers of atoms in each $n=40$ Stark state, assuming states with $|m_l| \leq 3$ ionize adiabatically, $|m_l| > 3$ diabatically.

Hydrogen chloride, hydrogen fluoride

Collisional mixing in Xe(nf)–HF and Xe(nf)–HCl collisions was studied by Higgs et al.[54,55] and representative SFI data obtained following Xe(27f)–HF collisions are shown in Fig. 9.10. The data are in many respects similar to those for Xe(nf)–NH_3 collisions. However, as evident from Fig. 9.10*a*, the large spacing of the rotational levels in HF means that collisions now populate fewer groups of excited states. In consequence, they are moderately well resolved in the SFI spectrum and it is possible, in certain cases, to identify separately the products of n-changing collisions that ionize adiabatically and diabatically. For example, the first two peaks in Fig. 9.10a result from adiabatic and diabatic ionization of states populated via $J=3 \rightarrow 2$ transitions. The second two peaks result primarily from ionization of states produced by the reaction

$$\mathrm{Xe(27f)} + \mathrm{HF}(J=2) \xrightarrow{k_n^{2\rightarrow 1}(27\mathrm{f})} \mathrm{Xe}(n'l') + \mathrm{HF}(J=1) \quad (23)$$

where $n' \simeq 40$. This portion of the spectrum is shown on an expanded scale in Fig. 9.10b, which also includes the SFI profile calculated for the ionization of a mixture containing equal numbers of atoms in each $n=40$ Stark state, assuming states with $|m_l| \leqslant 3$ ionize adiabatically, $|m_l| > 3$ diabatically. The generally good agreement between the experimental and calculated profiles indicates that collisions with HF result in n-mixed products with a wide range of values of l and $|m_l|$. The somewhat greater width of the experimentally observed adiabatic ionization feature relative to that calculated arises because the calculated feature is for a single manifold ($n=40$) whereas adjacent manifolds may be collisionally populated.

Xenon(23f)–HF collisions were also studied[54] because the energy match for the process

$$\mathrm{Xe(23f)} + \mathrm{HF}(J=0) \rightarrow \mathrm{Xe}(21l) + \mathrm{HF}(J=1) \quad (24)$$

in which electronic energy is transferred from the Rydberg atom to excite the target molecule rotationally is favorable. The data provide evidence of such reactions.

Rate constants for collisional depopulation and n- and l-changing collisions were determined by observing the time evolution of particular SFI features. The time dependence of the population $P(t)$ associated with a given collisionally induced SFI feature is given, neglecting second-order effects, by

$$dP(t)/dt = \rho k_x(n\mathrm{f})N(t) - b_\mathrm{p} P(t) \quad (25)$$

where $k_x(n\mathrm{f})$ is the rate constant for population of the feature, and b_p its total depopulation rate. Thus, if the integral form of Eq. (25) is fitted to the experimental measurements of $P(t)$, both $k_x(n\mathrm{f})$ and b_p can be determined. The time dependence of the population of the SFI features resulting from reaction (23) is shown in the inset in Fig. 9.10b together with the line representing the best fit obtained using the integral form of Eq. (25). Analysis of these, and other, SFI data yields the rate constants listed in Table 9.1, which also includes rate constants for collisional ionization. Similar data for HCl are also presented in the table. Because the values of k_d for both HF and HCl are equal, within experimental error, to the corresponding sums of the rate constants for collisional ionization and collisional mixing, this indicates that there are no other major, unidentified, reaction processes contributing to the collisional depopulation of the parent Rydberg state. As is evident from Table 9.1,

Table 9.1. Rate constants pertaining to Xe(nf) collisions with polar targets

Target species	Dipole moment (Debye)	n	Rate constant (10^{-7} cm^3 s^{-1}) k_d	k_l	k_n	k_i	$\Sigma k_i + k_n + k_l$
NH_3	1.3	31	20 (5)	~13	~4	2.3 (1.0)	—
HCl	1.08	31	11.3 (3)	4.8 (2.4)	5.5 (2.5)	0.9 (0.4)	11.2 (3)
HF	1.92	27	11.3 (3.2)	4.3 (2.2)	$k_n^{2\to 1}$, 3.0 (1.5) $k_n^{3\to 2}$, 2.2 (1.1)	1.5 (0.8)	11.0 (3)

Note: The numbers in parentheses are the experimental uncertainties.

the measured rate constants for NH_3, HF, and HCl are all similar in magnitude, a not unexpected result as these molecules have similar dipole moments and simple rotational structure.

Collisional ionizaton in collisions between krypton Rydberg atoms and HF and HCl was studied by Chupka.[10, 11] The Rydberg atoms were produced in well-defined states by direct photoexcitation from the ground state and their population was monitored by observing mass spectroscopically the SF_6^- signal from the electron-transfer process

$$Kr^{**} + SF_6 \rightarrow Kr^+ + SF_6^- \tag{26}$$

Addition of a small amount of HF or HCl results in competition between this process and collisional ionization via, say,

$$Kr^{**} + HCl(J) \rightarrow Kr^+ + e + HCl(J-1) \tag{27}$$

Those Rydberg states ionized by a collision with a polar molecule were effectively removed from the process of SF_6^- formation because the detached electron was immediately accelerated by an applied field, when the cross section for its capture by SF_6 is negligibly small. The influence of HCl on SF_6^- production is shown in Fig. 9.11. Also shown are the thresholds at which an additional rotational transition $J \rightarrow J-1$ provides sufficient energy to cause ionization. The data give a hint of steplike structure at these thresholds, although a number of effects, including n-changing collisions followed by ionization of these products, would tend to obscure these steps.

9.3b. Attaching targets

The majority of experimental studies of collisions between Rydberg atoms and molecules that attach free thermal-energy electrons focused on those reactions that result in negative-ion formation.[8, 9, 56–61] Such processes were investigated theoretically by Matsuzawa[16] utilizing the "essentially free" electron model in which negative-ion formation is viewed as resulting from the attach-

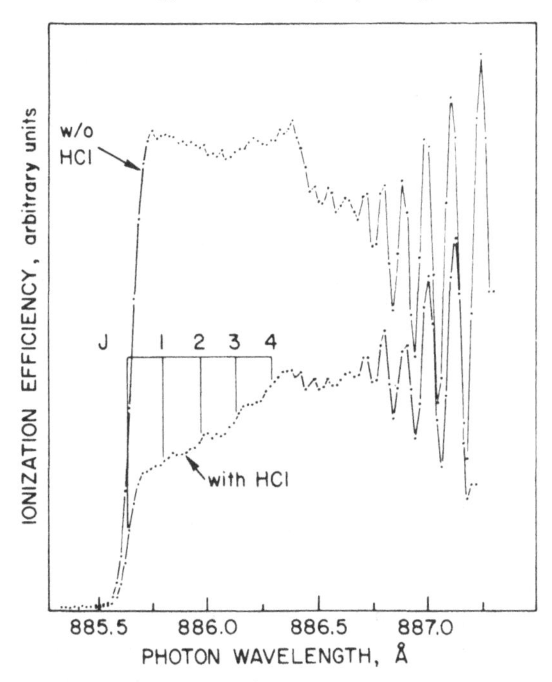

Fig. 9.11. SF_6^- ion production in Kr–SF_6 mixtures as a function of exciting photon wavelength with and without an admixture of HCl.

ment of the Rydberg electron to the target molecule. Matsuzawa predicted, on the basis of this model, that the rate constants for the transfer of bound electrons from high-Rydberg atoms should equal the rate constants for the attachment of free electrons having the same energy distribution.

Stockdale et al.[8] investigated the interaction of Ar** and Kr** atoms, excited by electron impact, with CH_3NO_2, CH_3CN, CH_3Br, and CH_3I. They identified the principal negative-ion products and compared them with those for the free-electron attachment to the same molecules. In addition, they observed that the rate constants for negative-ion formation were large; for example, in the case of CH_3NO_2, they estimated a value of $\sim 10^{-6}$ cm^3 s^{-1}. Similar studies involving interactions between He** atoms and CCl_4, CF_3Br, and CF_3I were reported by Klots[9] who observed only the dissociation products Cl^-, Br^-, and I^-. The Rice group[57-59] measured absolute rate constants for Xe^+ formation in collisions between Xe(nf) atoms and a variety of attaching targets. The identities of the primary negative-ion species were also determined. Negative-ion formation in Ar**-SF_6 collisions was investigated by Astruc et al.[56] and Dimicoli and Botter.[60] Collisions of laser-excited sodium ns and nd atoms with attaching targets were studied by Hill et al.[61]

Representative data for these studies are presented in Table 9.2 together

Table 9.2. Representative data for Rydberg and free-electron interactions with attaching targets

	Rydberg collisions			Thermal-electron collisions	
	Collisional ionization rate constant (10^{-7} cm^3 s^{-1})[a]		Principal negative-ion species observed	Rate constant for attachment (10^{-7} cm^3 s^{-1})	Principal negative-ion species observed
Target	k_i(26f)	k_i(36f)			
SF_6	4.3 (0.9)	4.3 (0.9)	SF_6^- [a,b]	2.7[d]	SF_6^- [d]
CCl_4	2.2 (0.4)	4.0 (0.8)	Cl^- [a,b]	3.6[e]	Cl^- [e]
CCl_3F	~4.1	5.8 (1.2)	Cl^- [a]	—	Cl^- [d]
CH_3I	2.7 (0.7)	4.2 (1.1)	I^- [a–c]	0.25[a]	I^- [c]
C_7F_{14}	3.1 (0.8)	1.3 (0.3)	$C_7F_{14}^-$ [a]	0.4 → 1.0[a]	$C_7F_{14}^-$ [f]

Note: The numbers in parentheses are the experimental uncertainties.
[a]Refs. 57–59, and references therein; [b]Ref. 9; [c]Ref. 8; [d]Ref. 62; [e]Ref. 63; [f]Ref. 64.

with data pertaining to free thermal-electron attachment. These data support the "essentially free" electron model because Rydberg collisions and free-electron attachment result in the formation of the same negative-ion species and the rates for such processes are comparable. In the following sections, detailed data relating to certain of the target species included in Table 9.2 will be presented.

Sulfur hexafluoride

A number of investigations of collisions between Rydberg atoms and SF_6 have been reported. Most concentrated on processes yielding ionic products. However, Kellert et al.[65] recently showed, using SFI, that collisions with SF_6 also result in l changing yielding states with a broad range of values of l and $|m_l|$, although no evidence of n-changing collisions was detected. The rate constants for collisional depopulation of the parent state and for l changing were measured. For Xe(31f), values $k_d(31\text{f}) \simeq 8\times10^{-7}$ cm^3 s^{-1} and $k_l(31\text{f}) \simeq 2.1\times10^{-7}$ cm^3 s^{-1} were obtained. The rate constant for l changing, although sizable, is less than those observed for polar targets but comparable to that measured for Xe(31f)–Xe collisions.[66]

Absolute measurements[57–59] of electron transfer in Xe(nf)–SF_6 collisions were undertaken using the apparatus shown in Fig. 9.7. Xenon atoms were excited, under field-free conditions, to selected nf states, and allowed to interact with SF_6 for a known time $t \lesssim 5$ μs. At the end of this period, Xe^+ ions resulting from collisional ionization via the electron-transfer reaction

$$\text{Xe}(n\text{f}) + SF_6 \xrightarrow{k_i(n\text{f})} \text{Xe}^+ + SF_6^- \tag{28}$$

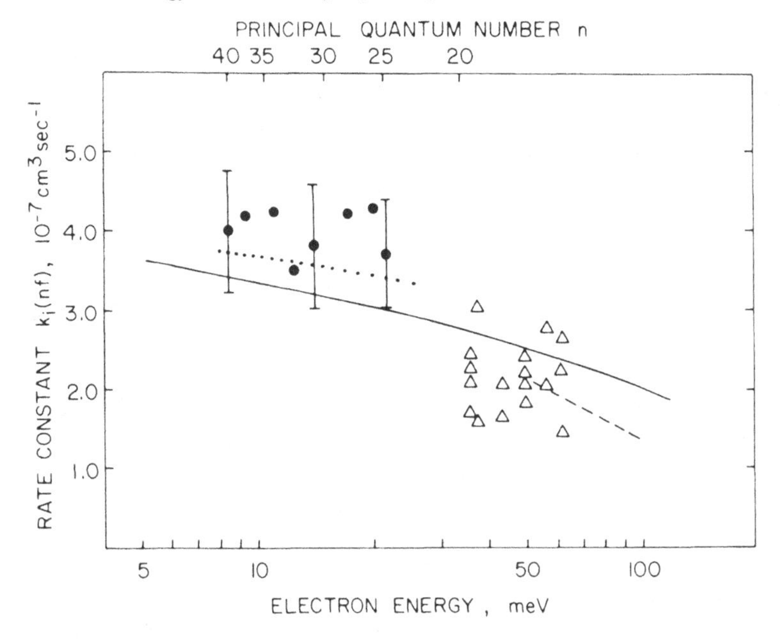

Fig. 9.12. Rate constants for electron attachment from Xe(nf) to SF_6 as a function of average electron kinetic energy: circles, Rydberg collision data;[57,58] dashed line, electron swarm experiment;[62] triangles, flowing afterglow experiment;[67] calculated values: solid line, Maxwellian velocity distribution;[68] dotted line, velocity distribution appropriate to Xe(nf) Rydberg electron.[68]

were extracted from the interaction region by application of a small extraction field and detected by the particle multiplier. The $k_l(n\text{f})$ is sufficiently small that only a very small fraction of the laser-excited nf atoms undergo l changing during the interval t, and the majority of the ion signal thus results from ionization of the parent state. Effects due to mixing can therefore be neglected in the data analysis.

The measured rate constants are shown in Fig. 9.12 and are located on the energy axis by ascribing to the Rydberg electrons their time-averaged kinetic energies. Also included are rate constants for free-electron attachment to SF_6 derived using the relationship

$$k = \int_0^\infty v\sigma_a(v) f(v)\, dv \tag{29}$$

where v is the electron velocity and $\sigma_a(v)$ the theoretical cross section[68] for free-electron attachment derived by assuming that the interaction is dominated by the polarization potential and that the process is purely s-wave capture. If Maxwellian velocity distributions $f(v)$ are used, the agreement with all the experimental data is reasonably good. If the $f(v)$ are taken as the velocity distribution of electrons in hydrogenic orbits of $l=3$, the resultant theoretical rate constants are seen to be in excellent accord with the Rydberg data. Hill et al.[61] recently determined the cross section for electron transfer in Na(10d)–SF_6 interactions. The measured value of $\sim 1\times10^{-11}$ cm^2 is comparable to that deduced from the Xe(nf) data, $\sim 1.4\times10^{-11}$ cm^2.[69] The velocity dependence of the electron-transfer cross section in Ar**–SF_6 collisions was also investigated.[60]

The success of the "essentially free" electron model in explaining the data indicates that Rydberg atoms may, in certain circumstances, be used to provide information on electron scattering at energies below 20 meV, a region virtually inaccessible by other techniques. For example, the data in Fig. 9.12 show that for Xe(nf) the rate constant for electron transfer to SF_6 is approximately independent of n in the range $25\leqslant n\leqslant 40$. This requires that the rate constant for Rydberg electron attachment to SF_6 be essentially independent of the Rydberg electron-velocity distribution and, hence, suggests that the free-electron-attachment rate constant is also independent of electron velocity at low electron energies. As is evident from Eq. (29), this requires that the free-electron-attachment cross section be inversely proportional to electron velocity at low electron energies, in good agreement with theoretical predictions.[68]

The primary negative-ion species produced in Xe(nf)–SF_6 collisions was identified as SF_6^- using a time-of-flight technique.[58] No free-electron production was observed and the SF_6^- produced in high-Rydberg collisions must therefore have a much longer lifetime against autodetachment than that reported for SF_6^- formed by free-electron attachment.[70] Klots[9] previously observed a similar effect in studies involving high-Rydberg helium atoms. He suggested that the SF_6^- might be stabilized by an interaction with the He^+ core in which the negative ion loses part of its internal energy, which appears as kinetic energy.

Astruc et al.[56] studied ion production in Ar**–SF_6 collisions as a function of SF_6 temperature. The product ions were analyzed mass-spectroscopically and the data are shown in Fig. 9.13. The Ar^+ signal, and hence electron-transfer rate, was found to be independent of SF_6 temperature. At room temperature, only SF_6^- ions were observed. Increasing the SF_6 temperature resulted in a marked decrease in SF_6^- production and the growth of a SF_5^- signal. A similar increase in SF_5^- formation with increasing temperature was noted in free-electron-attachment studies.[67,71] However, free-electron studies yield SF_6^- autodetachment lifetimes that are much less than the 40-μs ion flight times in the experiments of Astruc et al., who might thus have been expected to observe essentially no SF_6^- ions.

The data in Fig. 9.13 are interpreted using a model in which Rydberg electron attachment results in the initial formation of excited SF_6^- ions. These ions may then be stabilized by an interaction with the Rydberg core, undergo autodetachment, or, if they have enough internal energy, dissociate yielding SF_5^-. Clearly, stabilization becomes less effective and autodetachment increasingly important, as the SF_6 temperature is raised because, although the electron-

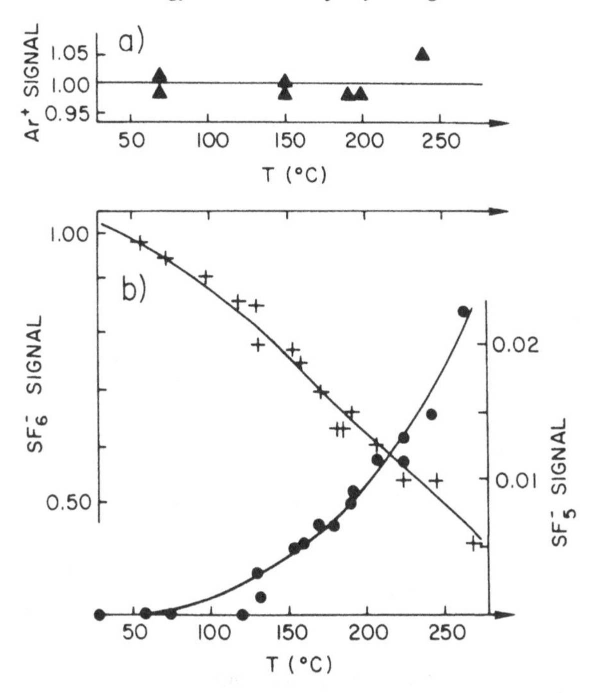

Fig. 9.13. Ar**-SF_6 collisions. Ion counting rate as a function of temperature:[56] triangles, Ar^+ production; crosses, SF_6^- production; circles, SF_5^- production.

transfer rate is independent of temperature, the SF_6^- and total $SF_6^- + SF_5^-$ signals decrease markedly. The observation of stabilization suggests that interactions with the Rydberg core may be important.

Carbon tetrachloride

Collisional ionization by CCl_4 proceeds via dissociative transfer reactions of the type[58, 60, 61]

$$\mathrm{Xe}(nf) + \mathrm{CCl_4} \xrightarrow{k_i(nf)} \mathrm{Xe^+} + \mathrm{CCl_3} + \mathrm{Cl^-} \tag{30}$$

In contrast to SF_6, the collisional ionization rate constant for CCl_4 increases markedly with increasing n. This is, for example, evidenced by the data of Hill et al.,[61] who studied Na(ns, nd)-CCl_4 collisions. In these experiments, the ratio of the cross sections σ_{Cl^-} and $\sigma_{SF_6^-}$ for Cl^- and SF_6^- production via the reactions

$$\mathrm{Na}(ns, nd) + \begin{cases} \mathrm{CCl_4} \rightarrow \mathrm{Na^+} + \mathrm{CCl_3} + \mathrm{Cl^-} \\ \mathrm{SF_6} \rightarrow \mathrm{Na^+} + \mathrm{SF_6^-} \end{cases} \tag{31}$$

was determined as a function of n. The measured ratios are shown in Fig. 9.14. On the assumption that, as for Xe(nf) atoms, $\sigma_{SF_6^-}(ns)$ and $\sigma_{SF_6^-}(nd)$ are es-

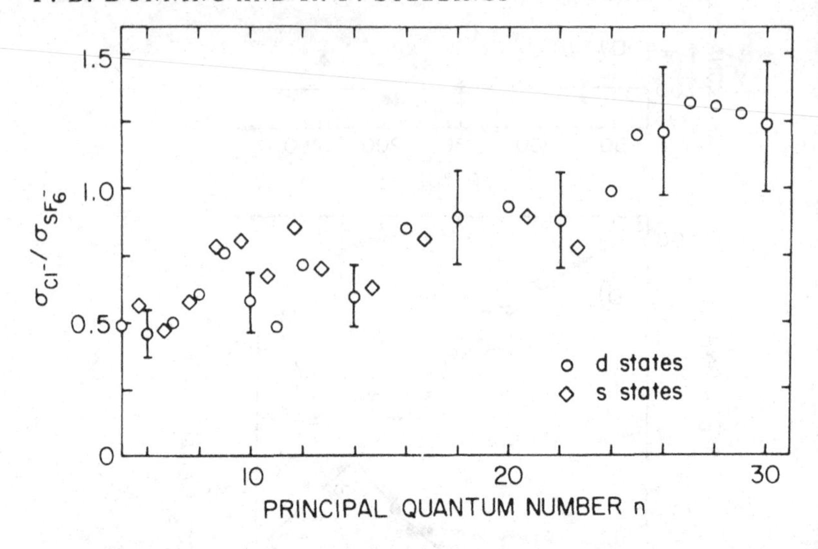

Fig. 9.14. Ratio of cross sections for the production of Cl^- and SF_6^- ions in Na(ns, nd)-CCl_4 and Na(ns, nd)-SF_6 collisions, respectively, as function of principal quantum number.

sentially independent of n, these data indicate that both $\sigma_{Cl^-}(ns)$ and $\sigma_{Cl^-}(nd)$ increase significantly with n.

Perfluoromethylcyclohexane

In the case of C_7F_{14}, both $C_7F_{14}^-$ and prompt free-electron production were noted,[59] indicating that collisional ionization results from the two reactions

$$\mathrm{Xe}(nf) + C_7F_{14} \begin{array}{l} \nearrow \mathrm{Xe}^+ + C_7F_{14}^- \\ \searrow \mathrm{Xe}^+ + e + C_7F_{14} \end{array} \quad (32)$$

The rate constants associated with each of these processes were found to be approximately equal, and $C_7F_{14}^-$ autodetachment was not observed.

The negative-ion production was observed to decrease with decreasing Rydberg electron energy in a manner similar to that observed in free-electron attachment.[64] A similar decrease in negative-ion production with increasing n was also recently noted in Na(nd)-$CHCl_3$ collisions.[61]

Hexafluorobenzene

Xenon(nf)-C_6F_6 collisions result in the formation of $C_6F_6^-$ ions that rapidly autodetach, pointing to the reaction sequence

$$\mathrm{Xe}(nf) + C_6F_6 \rightarrow \mathrm{Xe}^+ + C_6F_6^{-*} \searrow C_6F_6 + e \quad (33)$$

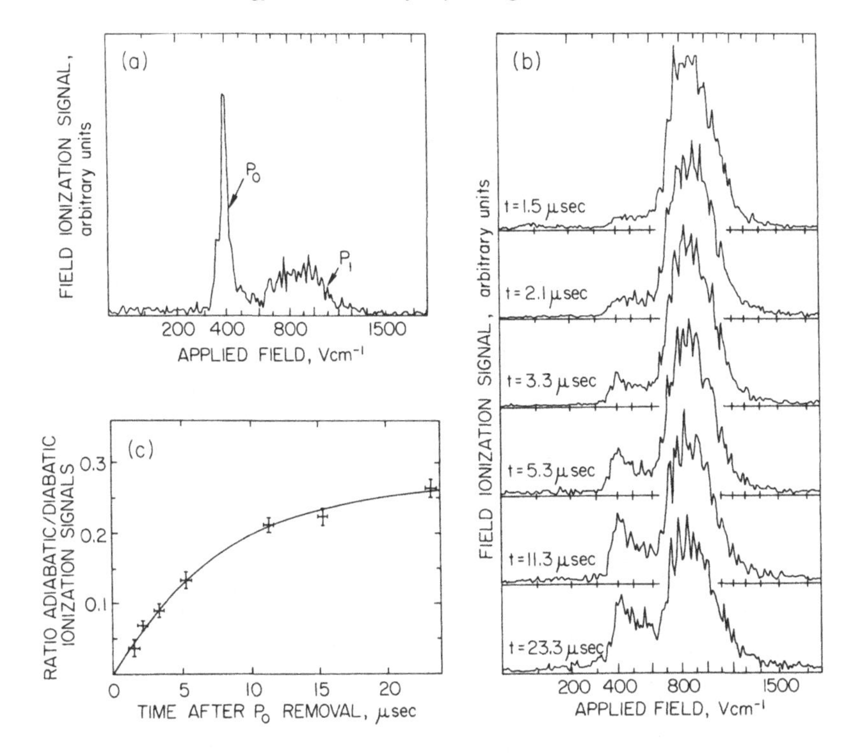

Fig. 9.15. Collisions with CO_2. (*a*) SFI spectrum obtained after allowing Xe(31f) to interact with CO_2. (*b*) Time development of the Rydberg population distribution resulting from collisions with CO_2 following removal of the low $|m_l|$ adiabatically ionizing states. (*c*) Time dependence of the ratio of adiabatic to diabatic ionization signals. The solid line shows the model fit to the data.

This is consistent with the work of Naff et al.[64] who observed that the ions formed by free low-energy electron attachment to C_6F_6 had an autodetachment lifetime of ~12 μs.

The autodetachment of the $C_6F_6^-$ ions[60] is in marked contrast to the behavior of the SF_6^- ions formed in room-temperature Xe(nf)–SF_6 collisions.

9.3c. Other targets

CO_2, CO, N_2

Higgs et al.[66] recently observed l changing in collisions between Xe(nf) atoms and CO_2, CO, and N_2. Typical SFI data obtained following Xe(31f)–CO_2 collisions are shown in Fig. 9.15*a*. The presence of P_1 and the shape of P_0 provide direct evidence of l-changing collisions. The profile of P_1 suggests that collisions populate states with a broad range of l and $|m_l|$. No SFI features

Table 9.3. *l* changing in Xe(*nf*)-CO_2, CO, and N_2 collisions

Target	n	$k_l(nf)$ [a] (10^{-7} cm^3 s^{-1})	$\sigma_l(nf)$ [a] (cm^2)	$\sigma_{ET}(\bar{v}_{rel})$ [b] (cm^2)	$\sigma_{CT}(\bar{v}_{rel})$ [b] (cm^2)	$\sigma_{ET}+\sigma_{CT}$ (cm^2)
CO_2	24	2.1 (0.5)	5.0 (1.3)$\times 10^{-12}$			
	26	1.9 (0.5)	4.6 (1.2)$\times 10^{-12}$			
	30	1.7 (0.5)	4.0 (1.1)$\times 10^{-12}$	2.5×10^{-12}	4.4×10^{-13}	2.9×10^{-12}
	34	1.2 (0.4)	2.8 (1.0)$\times 10^{-12}$			
	37	1.2 (0.4)	2.8 (1.0)$\times 10^{-12}$			
CO	37	0.11 (0.06)	2.2 (1.2)$\times 10^{-13}$	8.8×10^{-14}	3×10^{-13}	3.9×10^{-13}
N_2	37	0.04 (0.0025)	7.8 (4.9)$\times 10^{-14}$	2.6×10^{-14}	2.7×10^{-13}	3×10^{-13}

Note: The numbers in parentheses are the experimental uncertainties.
[a] Experimental data.[66]
[b] Theoretical[17] upper bounds to the *l*-changing cross sections σ_{ET} and σ_{CT} due to electron-target and core-target interactions, respectively.

associated with *n*-changed collision products are evident in Fig. 9.15*a*, indicating that the rate constant for this process is small. In a subsidiary experiment, no significant ion signal due to collisional ionization was detected.

The rate constants for *l*-changing collisions were determined by measuring the time evolution of P_0 and P_1 and analyzing the data by use of a single-reservoir model similar to that described previously. The measured rate constants are included in Table 9.3 together with the corresponding cross sections $\sigma_l(nf)$. The *l*-changing rate constant for CO_2 is larger than that for CO despite the fact that CO has a small permanent dipole moment.

Also included in Table 9.3 are upper bounds to the total Xe(*nf*)-target cross sections and, hence, to the *l*-changing cross sections calculated by the use of expressions[17] that provide upper limits on the separate contributions to the total cross section due to the Rydberg electron-target and core-target interactions.

In the case of CO_2, where the Rydberg electron-target interaction is dominant, the experimental values of $\sigma_l(nf)$ are in fair accord with the calculated values in Table 9.3 and with the results of an approximate scaling formula based on the Rydberg electron-target interaction derived by Hickman.[72] However, in the case of CO and N_2, the measured values of $\sigma_l(nf)$ appear too great to be accounted for solely on the basis of the Rydberg electron-target interaction. Inclusion of the core-target interaction provides results that are more consistent with the experimental data.

In a subsidiary experiment,[73] the rate constant for $|m_l|$ changing in collisions between the high $|m_l|$ states comprising the P_1 feature and CO_2 was determined. In this experiment, Xe(31f) atoms were allowed to interact with CO_2 for sufficient time to achieve a sizable population in the P_1 feature. The remaining parent atoms and the low $|m_l|$ collision products were then selectively ionized by application of a short-duration ($\lesssim 1$-μs) pulsed electric field of strength sufficient to ionize these states, but not the states comprising P_1. The remaining high $|m_l|$ atoms were then allowed to undergo further collisions with CO_2 in the presence of a weak (~ 5-V cm^{-1}) electric field applied along the ionizing field direction. This prevents effects due to atomic precession about any residual off-axis fields. The time dependence of the Rydberg state population distribution was then determined by use of SFI and is shown in Fig. 9.15*b*. Immediately following application of the pulsed field, no adiabatic ionization is observed, indicating the absence of low $|m_l|$ states. However, at later times, the adiabatic feature reappears, showing the collisional repopulation of low $|m_l|$ states. The time development of this feature is presented in Fig. 9.15*c*, which shows the ratio of the signals corresponding to adiabatic and diabatic ionization as a function of time after application of the pulsed field. The ratio of these signals tends to an equilibrium value of 0.27, in good agreement with that expected assuming that collisions result in a statistical distribution of l, $|m_l|$ states and that states with $|m_l| \leqslant 3$ ionize adiabatically, $|m_l| > 3$ diabatically.

The rate constant for $|m_l|$-changing collisions derived by analysis of the data in Fig. 9.15, taking into account that fact that only a fraction of the products of such collisions have values of $|m_l| \leqslant 3$, and thus ionize adiabatically, is $k_{|m_l|}(31) \simeq 2 \times 10^{-7}$ cm^3 s^{-1}. This value is comparable to the rate constant for l changing in Xe(31f)–CO_2 collisions, $k_l(31\text{f}) \simeq 1.5 \times 10^{-7}$ cm^3 s^{-1}, which was observed in both zero and weak electric fields. Thus, because the rate constants for l and $|m_l|$ changing are similar, this suggests that collisions with CO_2 result in simultaneous changes in l and $|m_l|$.

9.4. Experimental results: Rydberg atoms with $n \simeq 5$–15

Collisions between atoms in the lower-lying Rydberg states, i.e., $n \simeq 5$–15, and a variety of simple target molecules have been studied using the technique of laser-induced fluorescence. Data obtained in this manner will now be discussed.

9.4a. Resonant vibrational-energy transfer

Gallagher et al.[74] observed sharply resonant collisional-energy transfer from electronic excitation of Na(*n*s) states to vibrational excitation in CH_4 and CD_4. The Na(*n*s) atoms were excited by two-step laser-induced photoexcitation via the intermediate 3p state, as illustrated for Na(6s) states in the insert in Fig. 9.16. The sodium atoms, and targets CH_4 or CD_4, were contained in a heated Pyrex cell. The decay of the *n*s population was determined by monitoring the *n*s $\rightarrow$ 3p fluorescence. The decay rate $1/\tau(n\text{s})$, of the *n*s state is given by

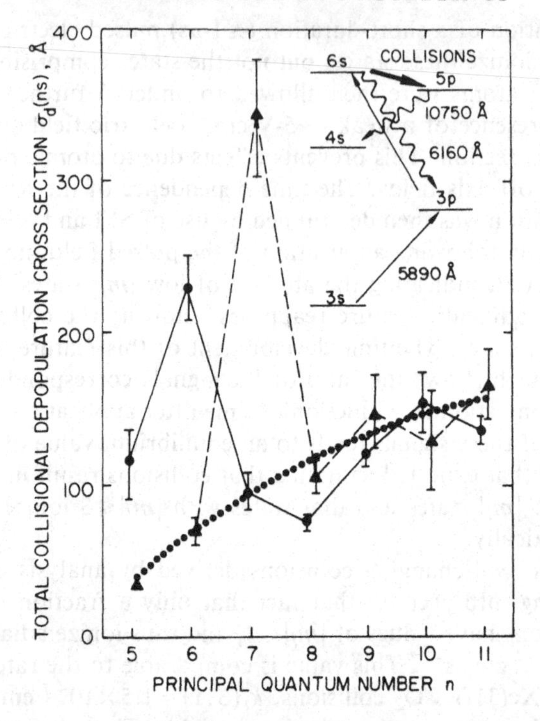

Fig. 9.16. Total collisional depopulation cross sections $\sigma_d(ns)$ for sodium ns states with CH_4 (circles and solid line) and CD_4 (triangles and dashed line).[74] The smooth dotted curve shows the expected depopulation cross section in the absence of resonant e–v transfer. The inset shows the energy levels relevant to the study of sodium 6s and 5p states. The straight arrows indicate the laser excitation steps, the heavy arrow collisional transfer, the wavy arrows the observed fluorescence.

$$\frac{1}{\tau(ns)} = \frac{1}{\tau_0(ns)} + \sigma_d(ns)\,\bar{v}_{rel}\,\rho \tag{34}$$

where $\tau_0(ns)$ is the radiative lifetime, $\bar{v}_{rel}$ the average relative collision velocity, and $\sigma_d(ns)$ the velocity-averaged total collisional depopulation cross section. Thus measurement of the Na(ns) decay rate as a function of CH_4 or CD_4 number density ρ yields $\sigma_d(ns)$.

The measured values of $\sigma_d(ns)$ are shown in Fig. 9.16. Marked increases in σ_d are observed at certain values of n. As indicated in Table 9.4, these enhanced cross sections can be explained as arising from close energy resonances between electronic transitions from these states and vibrational transitions in CH_4 or CD_4. Resonances are observed at different values of n for CH_4 and CD_4 because these molecules have different vibrational energy level spacings. To confirm that the transitions listed in Table 9.4 were responsible for the enhanced values of σ_d, the time dependence of the lower state populated in each transition was determined. In the case of 6s atoms, for example,

Table 9.4. Parameters pertaining to transitions resulting in enhanced depopulation cross sections

		CH_4, CD_4 transitions			State-specific cross sections (Å^2)	
Na transition	Energy (cm^{-1})	Species	Mode, branch	Branch center frequency (cm^{-1})	CH_4	CD_4
5s → 4p	2930	CH_4	ν_3, P	2940	103 (22)	10 (14)
6s → 5p	1331	CH_4	ν_4, R	1340	135 (22)	9 (4)
7s → 5d	975	CD_4	ν_4, P	965	12 (4)	215 (33)

Note: The numbers in parentheses are the experimental uncertainties.

measurement of the time dependence of the 5p → 4s fluorescence indicated that collisionally induced 6s → 5p transitions were indeed responsible for the increased value of $\sigma_d(6s)$ and yielded the state-specific cross sections $\sigma(6s \rightarrow 5p)$ for this process. This and other state-specific cross sections that were determined in a similar fashion are included in Table 9.4. As might be expected, the magnitudes of these cross sections are approximately equal to the amount by which the total depopulation cross section exceeds that expected in the absence of resonant electronic to vibrational energy transfer. As is evident from Table 9.4, resonant energy transfer occurs in cases where the Rydberg electron undergoes changes in angular momentum of $\Delta l=1$ and $\Delta l=2$. Because the cross sections in these two cases are comparable, this suggests that the interaction does not occur at long range, relative to the radius of the sodium atom, because this would favor $\Delta l=1$ transitions. This, as well as the size of the observed cross sections, implies that the interaction must take place at about the mean radius of the electron orbit. This result is consistent with the "essentially free" electron model. Interestingly, as discussed in Ref. 74, not all energy resonances or near resonances lead to enhanced depopulation cross sections.

9.4b. Nitrogen

The collisional depopulation of sodium *n*s states by N_2 has also been studied by the SRI group.[75,76] A fluorescence technique was again employed and the total collisional depopulation cross section obtained by measurement of the *n*s decay rate as a function of N_2 density. These are shown in Table 9.5 for several *n*s states together with earlier data of Czajkowski et al.[77] obtained using a somewhat less direct sensitized-fluorescence technique.

Bauer et al.[78] developed a theoretical model for collisional depopulation by N_2 that may be understood qualitatively by reference to the potential energy diagram shown in Fig. 9.17, which shows approximate potentials for the Na–N_2 and Na^+–N_2^- systems as a function of their separation *r* for collinear col-

Table 9.5. Collisional depopulation cross sections

	$Na–N_2$ (Å^2)					$He–N_2$ (Å^2)[a]	
State	Humphrey et al.[76]	Czajkowski et al.[77]	Curve-crossing model	Na–He[81] (Å^2)	Na–Ar[81] (Å^2)	Hitachi et al.[82]	Curve-crossing model
5s	86 (8)	27 (5)	75	—	—	58 (+9, −6)	102
6s	84 (8)	41 (8)	102	0.62 (0.15)	<0.12	—	121
7s	82 (9)	16 (3)	123	3.8 (0.6)	0.48 (0.15)	—	—
8s	91 (8)	95 (19)	135	5.5 (0.9)	5.3 (0.8)	42 (+16, −9)	145
9s	104 (10)	55 (11)	147	7.3 (1.0)	18.9 (3.0)	48 (+14, −9)	153
5d manifold	43 (7)	7.3 (1.4)	111	—	—	67 (+12, −9)	—
6d manifold	36 (8)	10 (2)	128	—	—	—	—
7d manifold	32 (7)	45 (9)	140	—	—	—	—
8d manifold	18 (6)	25 (5)	150	—	—	40 (+7, −4)	—
9d manifold	—	—	—	—	—	41 (+6, −4)	—

Note: The numbers in parentheses are the experimental uncertainties.
[a] Helium *n*s and manifold state data are for triplet states at 600 K.

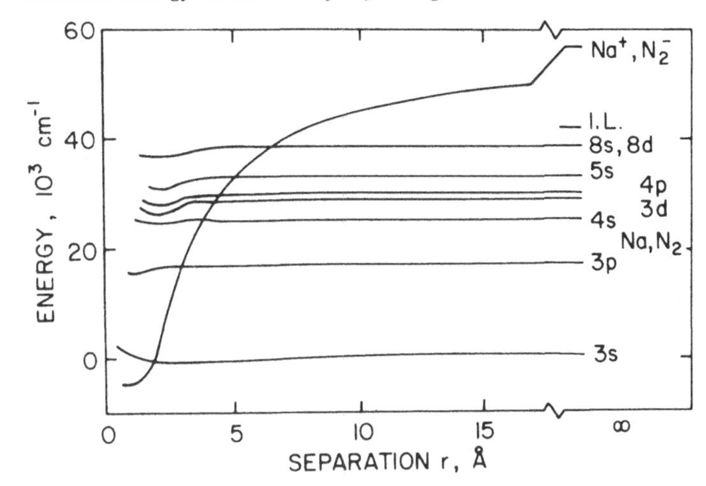

Fig. 9.17. Approximate potential curves for the Na-N_2 and Na^+-N_2^- systems in the $v=0$ vibrational state. (After Ref. 76)

lisions. Above each of the covalent curves lies a series of potential curves corresponding to excited vibrational states of N_2. A similar set of curves lies above the ionic curve corresponding to the excited vibrational states of N_2^-; these were omitted for clarity. At some radius r_c, each excited Na-N_2 curve crosses the ionic Na^+-N_2^- curve. Thus, if during a collision the sodium atom and N_2 molecule pass within r_c of each other, the system will pass through many curve crossings. If these crossings are all traversed adiabatically or diabatically, the system will follow a unique path through these curve crossings, and after the collision, the system, and hence the sodium atom, will be left in its initial state. However, if some of the crossings are traversed partially adiabatically and partially diabatically, the system can be in many different states by the time the repulsive wall is reached at small separations. The system then has a high probability of being in other than the initial state after the collision. It is reasonable to expect that some crossings will indeed be traversed partially diabatically and partially adiabatically.[75] Thus some of the excited sodium atoms that come within r_c of a target N_2 molecule will, after the collision, end up in a different excited state. The collisional depopulation cross section should therefore be given approximately by πr_c^2. Cross sections determined using this model are included in Table 9.5 and are seen to be in general agreement with the data of Humphrey et al.[76]

Collisions between laser-excited sodium *n*d atoms and N_2 were also investigated using time-resolved fluorescence. In this case, *l*-changing collisions provide the dominant depopulation mechanism.[79] Measured cross sections are shown in Fig. 9.18 (*right*) together with results obtained using a two-state quantum close-coupling method.[79] In Fig. 9.18 (*left*) are shown similar data

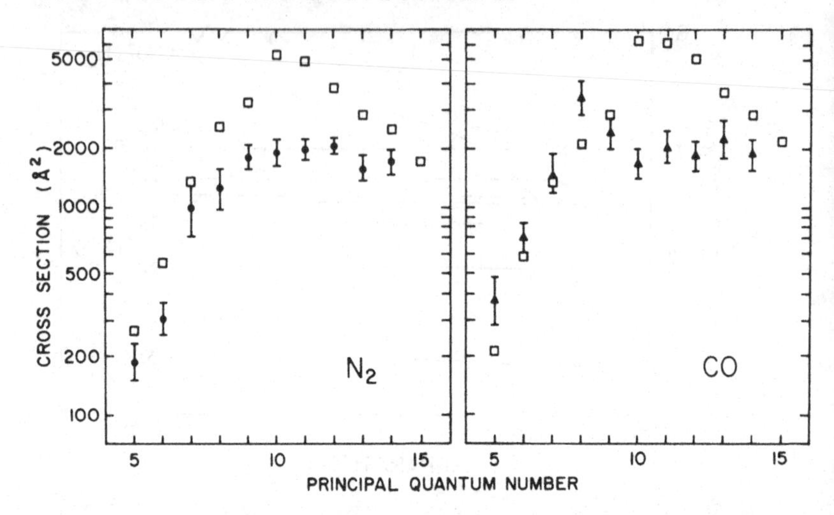

Fig. 9.18. Cross sections for *l* mixing in collisions of sodium *n*d atoms with N_2 and CO as a function of principal quantum number *n*.[79] Circles, experimental data N_2; triangles, experimental data CO; squares, calculated values.

obtained for CO. The *l*-changing cross sections for CO and N_2, although large, are comparable to those for helium, indicating that the presence of energetically accessible rotational transitions does not necessarily increase the cross section.[80] For N_2 this is not surprising because it is a homonuclear molecule and thus has no permanent dipole moment to help induce rotational transitions. The observation that CO exhibits a cross section comparable to that for N_2 and helium suggests that long-range coupling between the CO permanent dipole moment and the excited sodium is not important for *l*-changing collisions.

At high target gas densities, the laser-excited *n*d states are rapidly mixed to higher-*l* states and an equilibrium established where the rate of depopulation of the *n*d states through collisional mixing equals the rate of repopulation through collisions involving higher-*l* states. Under these conditions, the fluorescence from the *n*d state provides a measure of the population of the resultant mix of $l \geqslant 2$ states, referred to as a reservoir or manifold state.[76] Thus determination of the time dependence of the *n*d fluorescence as a function of N_2 pressure, at relatively high nitrogen pressures, yields the total collisional depopulation cross section for the mix of $l \geqslant 2$ states. The measured cross sections are included in Table 9.2. The data of Humphrey et al.[76] exhibit a steady decrease with *n* in this depopulation cross section, in disagreement both with the curve crossing model and with the results of Czajkowski et al.[77]

Interestingly, although N_2 and the rare gases yield similar *l*-changing cross sections in collisions with sodium *n*d states, this is not the case for collisional depopulation of *n*s states. As shown in Table 9.5, the depopulation cross sections observed with helium and argon are very much smaller than those for N_2.[81] Thus, as suggested by the curve-crossing model, the presence of energetically accessible rotational and vibrational levels in N_2 is important.

Hitachi et al.[82] studied the collisional depopulation of helium Rydberg *n*s and *n*-manifold states by N_2. The helium atoms were excited by the use of a pulsed electron gun and emissions from the $n\mathrm{S} \rightarrow 2\mathrm{P}$ and $n\mathrm{D} \rightarrow 2\mathrm{P}$ transitions were monitored to obtain the time dependence of the populations in the *n*S and manifold states, respectively. Collisional depopulation cross sections were obtained by determining the excited-state decay rate as a function of N_2 density. The measured cross sections are included in Table 9.5 and are of the same order of magnitude as those deduced using the curve-crossing model and observed by Humphrey et al.[76]

9.5. Future trends

A wide variety of reaction processes have been identified in studies of thermal collisions between Rydberg atoms and molecules. However, many details of these processes require further study. For instance, no systematic study of *n*- and *l*-changing collisions has been undertaken to assess the influence of such parameters as target dipole moment, polarizability, density of vibrational and rotational states, etc., or the energy separation between Rydberg levels.[83] It might be possible, for example, to use Stark tuning to enhance the mixing rate between particular initial and final states.

Quantitative measurements have indicated that the "essentially free" electron model is basically correct but that effects ascribable to the core can be observed. The available data point to the future utilization of Rydberg atoms to study low-energy electron–molecule scattering at energies of a few millielectronvolts, below those readily accessible using alternate techniques. Although such investigations can be undertaken using low-*l* states, high-*l* states are to be preferred because they have a narrower Rydberg electron momentum distribution and hence afford better energy resolution. A number of low-energy electron-scattering resonances should be accessible for study in this manner.[16]

The study of collisions between Rydberg atoms and molecules also holds the promise of providing information concerning thermal-energy ion–molecule reactions because products resulting from the target–core interaction should be separately identifiable, especially at high *n*. In this regard, a Rydberg atom beam can be viewed as a space-charge neutralized low-energy ion beam. An advantage of using Rydberg atoms to study ion chemistry is that the state of the core ion is well defined.

It is thus evident that collisions between high-Rydberg atoms and molecules require much further study, and if past experiments provide any indication, many new and interesting phenomena remain to be discovered.

Acknowledgment

The research by the authors and their colleagues reported in this chapter was supported by the National Science Foundation under Grant No. PHY78-09860 and by the Robert A. Welch Foundation.

References and notes

1. Recent reviews discussing the properties of Rydberg atoms include: F. B. Dunning and R. F. Stebbings, *Comments At. Mol. Phys. 10,* 9 (1980); H. J. Metcalf, *Nature 284,* 127 (1980); R. F. Stebbings, *Adv. At. Mol. Phys. 15,* 77 (1979); S. A. Edelstein and T. F. Gallagher, *Adv. At. Mol. Phys. 14,* 365 (1978); K. Takayanagi, *Comments At. Mol. Phys. 6,* 177 (1977); R. F. Stebbings, *Science 193,* 537 (1976).
2. V. Čermák and Z. Herman, *Coll. Czech. Chem. Commun. 29,* 953 (1964).
3. S. E. Kupriyanov, *Sov. Phys. JETP 21,* 311 (1965); *24,* 674 (1967); G. A. Surskii and S. E. Kupriyanov, *Sov. Phys. JETP 27,* 61 (1968).
4. H. Hotop and A. Niehaus, *Z. Phys. 215,* 395 (1965).
5. H. Hotop and A. Niehaus, *J. Chem. Phys. 47,* 2506 (1967).
6. T. Shibata, T. Fukuyama, and K. Kuchitsu, *Mass Spectrosc. 21,* 217 (1973); *Chem. Lett. 1974,* 75 (1974).
7. T. Sugiura and K. Arakawa, in *Recent Developments in Mass Spectroscopy, Proceedings of the International Conference on Mass Spectroscopy (Kyoto, 1969),* eds. K. Ogata and T. Hayakawa (Baltimore: University Park Press, 1971), p. 848.
8. J. A. Stockdale, F. J. Davis, R. N. Compton, and C. E. Klots, *J. Chem. Phys. 60,* 4279 (1974).
9. C. E. Klots, *J. Chem. Phys. 66,* 5240 (1977).
10. W. A. Chupka, *Bull. Am. Phys. Soc. 19,* 70 (1974).
11. M. Matsuzawa and W. A. Chupka, *Chem. Phys. Lett. 50,* 373 (1977).
12. E. Fermi, *Nuovo Cimento 11,* 157 (1934).
13. For a review of theoretical calculations using the "essentially free" electron model, see M. Matsuzawa, in Invited Papers and Progress Reports, *XI ICPEAC,* eds. N. Oda and K. Takayanagi (Amsterdam: North-Holland Publ., 1980).
14. M. R. Flannery, *Ann. Phys. (N.Y.) 79,* 480 (1973); *61,* 465 (1970).
15. M. Matsuzawa, *J. Chem. Phys. 55,* 2685 (1971); *J. Electron. Spectrosc. Rel. Phen. 4,* 1 (1974).
16. M. Matsuzawa, *J. Phys. Soc. Japan 32,* 1088 (1972); *J. Phys. B 8,* L382 (1975); *8,* 2114 (1975); *10,* 1543 (1977).
17. M. R. Flannery, *J. Phys. B 13,* L657 (1980); *Phys. Rev. A 22,* 2408 (1980).
18. See, for example, J. A. Schiavone, D. E. Donohue, D. R. Herrick, and R. S. Freund, *Phys. Rev. A 16,* 48 (1977); S. M. Tarr, J. A. Schiavone, and R. S. Freund, *Phys. Rev. Lett. 44,* 1660 (1980).
19. P. M. Koch and J. E. Bayfield, *Phys. Rev. Lett. 34,* 448 (1975).
20. R. R. Freeman and D. Kleppner, *Phys. Rev. A 14,* 1614 (1976); C. Fabre, P. Goy, and S. Haroche, *J. Phys. B 10,* L183 (1977).
21. See, for example, F. Devos, J. Boulmer, J.-F. Delpeche, *J. Phys. (Paris) 40,* 215 (1979); F. Gounand, M. Hugon, P. R. Fournier, and J. Berlande, *J. Phys. B 12,* 547 (1979); *12,* 2707 (1979); L. R. Pendrill and G. W. Series, *J. Phys. B 11,* 4049 (1978); and other references in this chapter.
22. See, for example, P. M. Koch, *Phys. Rev. Lett. 41,* 99 (1978); J. E. Bayfield and P. M. Koch, *Phys. Rev. Lett. 33,* 258 (1974).

23. W. E. Cooke and T. F. Gallagher, *Opt. Lett. 4,* 173 (1979).
24. H. A. Bethe and E. E. Salpeter, *Quantum Mechanics of One- and Two-Electron Atoms* (New York: Springer-Verlag, 1957).
25. M. L. Zimmerman, M. G. Littman, M. M. Kash, D. Kleppner, *Phys. Rev. A 20,* 2251 (1979).
26. H. J. Silverstone, *Phys. Rev. A 18,* 1853 (1978).
27. W. E. Cooke and T. F. Gallagher, *Phys. Rev. A 17,* 1226 (1978).
28. R. J. Damburg and V. V. Kolosov, *J. Phys. B 12,* 2637 (1979); *11,* 1921 (1978); *9,* 3149 (1976).
29. M. G. Littman, M. L. Zimmerman, T. W. Ducas, R. R. Freeman, and D. Kleppner, *Phys. Rev. Lett. 36,* 788 (1976).
30. M. G. Littman, M. L. Zimmerman, and D. Kleppner, *Phys. Rev. Lett. 37,* 486 (1976).
31. T. F. Gallagher, L. M. Humphrey, R. M. Hill, and S. A. Edelstein, *Phys. Rev. Lett. 37,* 1465 (1976).
32. T. F. Gallagher, L. M. Humphrey, W. E. Cooke, R. M. Hill, and S. A. Edelstein, *Phys. Rev. A 16,* 1098 (1977).
33. T. F. Gallagher and W. E. Cooke, *Phys. Rev. A 19,* 694 (1979).
34. J. L. Vialle and H. T. Duong, *J. Phys. B 12,* 1407 (1979).
35. M. G. Littman, M. M. Kash, and D. Kleppner, *Phys. Rev. Lett. 41,* 103 (1978).
36. T. H. Jeys, G. W. Foltz, K. A. Smith, E. J. Beiting, F. G. Kellert, F. B. Dunning, and R. F. Stebbings, *Phys. Rev. Lett. 44,* 390 (1980).
37. J. R. Rubbmark, M. M. Kash, M. G. Littman, and D. Kleppner, *Phys. Rev. A 23,* 3107 (1981). See also J. H. M. Neijzen and A. Dönszelmann, *J. Phys. B 15,* L87 (1982).
38. F. G. Kellert, T. H. Jeys, G. B. McMillian, K. A. Smith, F. B. Dunning, and R. F. Stebbings, *Phys. Rev. A 23,* 1127 (1981).
39. T. H. Jeys, G. B. McMillian, K. A. Smith, F. B. Dunning, and R. F. Stebbings, *Phys. Rev. A 26,* 335 (1982).
40. I. V. Komarov, T. P. Grozdanov, and R. K. Janev, *J. Phys. B 13,* L573 (1980).
41. K. A. Smith, F. G. Kellert, R. D. Rundel, F. B. Dunning, and R. F. Stebbings, *Phys. Rev. Lett. 40,* 1362 (1978).
42. F. G. Kellert, K. A. Smith, R. D. Rundel, F. B. Dunning, and R. F. Stebbings, *J. Chem. Phys. 72,* 3179 (1980).
43. See, for example, P. R. Koch, H. Hieronymous, A. F. J. Van Raan, and W. Raith, *Phys. Lett. 75A,* 273 (1980); T. F. Gallagher and W. E. Cooke, *Appl. Phys. Lett. 34,* 369 (1979); T. F. Gallagher and W. E. Cooke, *Phys. Rev. Lett. 42,* 835 (1979); E. J. Beiting, G. F. Hildebrandt, F. G. Kellert, G. W. Foltz, K. A. Smith, F. B. Dunning, and R. F. Stebbings, *J. Chem. Phys. 70,* 3551 (1979); T. W. Ducas, W. P. Spencer, A. G. Vaidyanathan, W. H. Hamilton, and D. Kleppner, *Appl. Phys. Lett. 35,* 382 (1979).
44. G. F. Hildebrandt, E. J. Beiting, C. Higgs, G. J. Hatton, K. A. Smith, F. B. Dunning, and R. F. Stebbings, *Phys. Rev. A 23,* 2978 (1981); J. W. Farley and W. H. Wing, *Phys. Rev. A 23,* 2397 (1981).
45. W. P. Spencer, A. G. Vaidyanathan, D. Kleppner, and T. W. Ducas, *Phys. Rev. A 26,* 1490 (1982).
46. T. H. Jeys, K. A. Smith, F. B. Dunning, and R. F. Stebbings, *Phys. Rev. A 23,* 3065 (1981).
47. G. Leuchs and H. Walther, *Z. Phys. 293,* 93 (1979); G. Leuchs and S. J. Smith, in

Laser Spectroscopy, eds. H. Walther and K. W. Rothe (New York: Springer-Verlag, 1979), vol. IV.

48. R. D. Rundel, in Abstracts of Contributed Papers, *XI ICPEAC*, eds. N. Oda and K. Takayanagi (Amsterdam: North-Holland Publ., 1980), p. 936. [These data are too large by a factor of π; R. D. Rundel, private communication.]
49. R. F. Stebbings, C. J. Latimer, W. P. West, F. B. Dunning, and T. B. Cook, *Phys. Rev. A 12,* 1453 (1975).
50. C. J. Latimer, *J. Phys. B 10,* 1889 (1977).
51. R. D. Rundel, private communication.
52. M. Matsuzawa, *Phys. Rev. A 20,* 860 (1979).
53. M. Matsuzawa (private communication) recently calculated cross sections for l changing resulting from inversion transitions during Xe(nf)–NH_3 collisions.
54. C. Higgs, K. A. Smith, G. B. McMillian, F. B. Dunning, and R. F. Stebbings, *J. Phys. B 14,* L285 (1981).
55. R. F. Stebbings, F. B. Dunning, and C. Higgs, *J. Electron. Spectrosc. Rel. Phen. 23,* 333 (1981).
56. J. P. Astruc, R. Barbé, J. P. Schermann, *J. Phys. B 12,* L377 (1979).
57. W. P. West, G. W. Foltz, F. B. Dunning, C. J. Latimer, and R. F. Stebbings, *Phys. Rev. Lett. 36,* 854 (1976).
58. G. W. Foltz, C. J. Latimer, G. F. Hildebrandt, F. G. Kellert, K. A. Smith, W. P. West, F. B. Dunning, and R. F. Stebbings, *J. Chem. Phys. 67,* 1352 (1977).
59. G. F. Hildebrandt, F. G. Kellert, K. A. Smith, F. B. Dunning, and R. F. Stebbings, *J. Chem. Phys. 68,* 1349 (1978).
60. I. Dimicoli and R. Botter, *J. Chem. Phys. 74,* 2346 (1981); *74,* 2355 (1981).
61. R. M. Hill, private communication.
62. L. G. Christophorou, D. L. McCorkle, and J. G. Carter, *J. Chem. Phys. 54,* 253 (1971).
63. F. J. Davis, R. N. Compton, and D. R. Nelson, *J. Chem. Phys. 59,* 2324 (1973).
64. W. T. Naff, C. D. Cooper, and R. N. Compton, *J. Chem. Phys. 49,* 2784 (1968).
65. F. G. Kellert, C. Higgs, K. A. Smith, G. F. Hildebrandt, F. B. Dunning, and R. F. Stebbings, *J. Chem. Phys. 72,* 6312 (1980).
66. C. Higgs, K. A. Smith, F. B. Dunning, and R. F. Stebbings, *J. Chem. Phys. 75,* 745 (1981).
67. F. C. Fehsenfeld, *J. Chem. Phys. 53,* 2000 (1970).
68. C. E. Klots, *Chem. Phys. Lett. 38,* 61 (1976).
69. The average relative collision velocities used to reduce the rate constant data in Refs. 57 and 58 to cross sections were slightly in error. Improved average relative collision velocities are 2.7×10^4 cm s^{-1} for both SF_6 and CCl_4.
70. D. Edelson, J. E. Griffiths, and K. B. McAfee, Jr., *J. Chem. Phys. 37,* 917 (1962); R. N. Compton, L. G. Christophorou, G. S. Hurst, and P. W. Reinhardt, *J. Chem. Phys. 45,* 4634 (1966); P. W. Harland and C. J. Thynne, *J. Phys. Chem. 75,* 3517 (1971).
71. C. L. Chen and P. J. Chantry, *Bull. Am. Phys. Soc. 15,* 418 (1970).
72. A. P. Hickman, *Phys. Rev. A 23,* 87 (1981).
73. C. Higgs, M. P. Slusher, K. A. Smith, F. B. Dunning, and R. F. Stebbings, *J. Chem. Phys. 76,* 5303 (1982).
74. T. F. Gallagher, G. A. Ruff, and K. A. Safinya, *Phys. Rev. A 22,* 843 (1980).
75. T. F. Gallagher, W. E. Cooke, and S. A. Edelstein, *Phys. Rev. A 17,* 125 (1978).
76. L. M. Humphrey, T. F. Gallagher, W. E. Cooke, and S. A. Edelstein, *Phys. Rev. A 18,* 1383 (1978).

77. M. Czajkowski, L. Krause, and G. M. Skardis, *Can. J. Phys. 51,* 1582 (1973).
78. E. Bauer, E. R. Fisher, and F. R. Gilmore, *J. Chem. Phys. 51,* 4173 (1969).
79. T. F. Gallagher, R. E. Olson, W. E. Cooke, S. A. Edelstein, and R. M. Hill, *Phys. Rev. A 16,* 441 (1977).
80. T. F. Gallagher, S. A. Edelstein, and R. M. Hill, *Phys. Rev. Lett. 35,* 644 (1975); *Phys. Rev. A 15,* 1945 (1977).
81. T. F. Gallagher and W. E. Cooke, *Phys. Rev. A 19,* 2161 (1979).
82. A. Hitachi, T. A. King, S. Kubota, and T. Doke, *Phys. Rev. A 22,* 863 (1980).
83. M. P. Slusher, C. Higgs, K. A. Smith, F. B. Dunning, and R. F. Stebbings, *Phys. Rev. A 26,* 1350 (1982).

10

High-Rydberg molecules

ROBERT S. FREUND

10.1. Introduction

In this chapter, we shall address the properties and behavior of high-Rydberg molecules (HRMs), that differ from those of high-Rydberg atoms. For the purposes of this discussion, a "high-Rydberg" state is defined as one with principal quantum number $n \gtrsim 10$. This is high enough for the Rydberg electron to be considered independent of the core and for molecular properties to be near their asymptotic high-n limits.

The main differences between HR atoms and molecules result from vibration and rotation of the molecular core ion and from its dissociation. Three cases can be distinguished. In one, the core ion is electronically excited with enough energy to dissociate while the HR electron is far enough away to have no influence on the dissociation process; this situation will be expanded upon in Sect. 10.2a. Another case is when the core ion is in its electronic ground state but is vibrationally excited; the Rydberg electron, on one of its infrequent passes through the core region, collides superelastically with the core and gains sufficient kinetic energy at the expense of vibrational energy to become ionized. This is called vibrational autoionization. The third case is again when the Rydberg electron passes through a ground-state core but jumps to a potential curve (or surface) of a repulsive valence state. The molecule then dissociates into non-Rydberg state fragments by a process called predissociation. These three cases can deactivate most HRMs, which raises the question of whether any HRMs are as stable as HR atoms, that is, with lifetimes close to radiative lifetimes. This point will be addressed in Sect. 10.3, where examples of stable HRMs will be discussed.

Another way in which HRMs differ from atoms is in their high-resolution optical spectra. Vibration and rotation lead to a great many levels and there are interactions between these levels. In the simplest approximation, there is a Rydberg series converging to every rotational–vibrational level of the core ion. For sufficiently high n, this means that electronic levels are more closely spaced than vibrational (and even rotational) levels of the core. The recent application of multichannel quantum-defect theory (MQDT) accurately described much of the behavior of Rydberg series in several molecules, as will be discussed in Sect. 10.3d.

To be sure, HR molecules and atoms are similar in many ways. Each consists of a core ion and a HR electron at a large mean distance from the core. For many purposes, the core can be approximated by a point charge, so the hydrogen atom is a useful prototype of a HR atom or molecule. The coarse energy-level structure is given by

$$E_n = -13.6/(n - \delta_l)^2 \text{ eV} \tag{1}$$

Small deviations result from interaction of the Rydberg electron with the structured core; they are expressed as quantum defects δ_l, which are functions of the orbital angular momentum l. Rydberg series of molecular energy levels are well known, although they are not as extensive as those observed in atoms. Energy levels lying sufficiently close to the ionization potential are susceptible to electric field ionization, although no molecule has yet been treated in much detail.

This chapter will be divided into two major sections. The first will deal with dissociation of HRMs into fragments, one of which is in a HR state. Understanding of this process is based on knowledge about dissociation of the molecular ion that constitutes the core of the HRM. Conversely, the study of HRM dissociation is one of very few methods that has provided information on repulsive electronic states of molecular ions. It has also provided some of the most accurate data on translational energy distributions of dissociation fragments. The other major section of this chapter will consider the conditions necessary for stability of HRMs and describe a number of observations.

10.2. Dissociation of high-Rydberg molecules

Since the first report by Kupriyanov[1] in 1965, many measurements have been made on dissociation of molecules to give high-Rydberg fragments. All of these studies used electron-impact excitation, in part because the necessarily high excitation energy is more difficult to obtain with photon sources and in part because the electron-impact process was itself of interest. Most of these experiments used translational spectroscopy as the primary diagnostic method and nearly all used measurements of the electron-impact thresholds or excitation cross section. We shall begin this section with a discussion of the core ion model, follow with a discussion of the experimental methods, and then conclude, in Sect. 10.2c, with a survey of typical experimental results.

10.2a. The core ion model of dissociation

The main principle for understanding the behavior of a high-Rydberg molecule is that it consists of a molecular core ion surrounded by a distant electron in a high-Rydberg orbital.[2-4] The HR electron spends most of its time far from

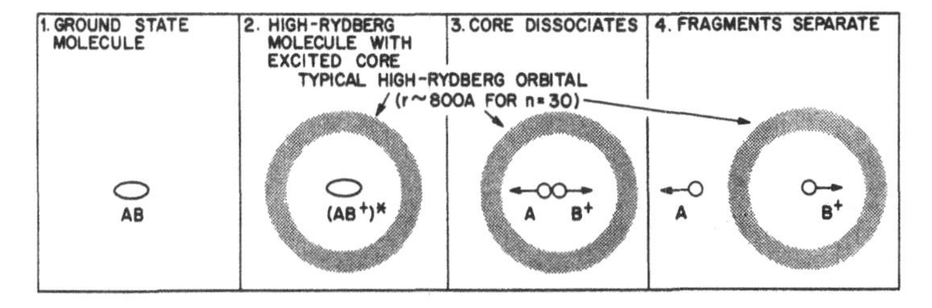

Fig. 10.1. The core ion model: schematic illustration of dissociation of molecule AB to give a HR fragment B(HR).

the core with an average distance proportional to n^2, but occasionally it penetrates the core with a probability proportional to n^{-3}.

The core ion model is illustrated in Fig. 10.1. In the first step, the ground-state molecule is excited to a HR state by an electronic transition, which, according to the Franck–Condon principle, takes place in a time so short that the nuclei move negligibly. Such excitation to a HR state represents only a small fraction of the possible excitation processes, which include ionization and excitation to low-lying states.

The excited electron may have been the most loosely bound electron in the ground-state molecule, in which case the core ion of the HRM is left in its ground electronic state. Its vibrational energy distribution is determined by the Franck–Condon factors between the ground-state molecule and the ground-state ion. If, however, the excited electron was originally more tightly bound, then the core ion is left in an excited electronic state. If the core ion forms in a sufficiently excited state, it can dissociate with no help from the HR electron. The fragments can separate to an appreciable distance and still remain completely within the HR orbital. Only at large separation must the electron choose whether to remain bound to the ionic fragment (at a large distance and therefore as a HR electron) or to follow neither fragment and leave an ionic fragment behind. No theoretical treatment of this process has yet been carried out, so a major unanswered question is how the HRM quantum numbers correlate with those of the HR fragments.

Another way to look at the core ion model is by potential energy curves (Fig. 10.2). At any separation of the nuclei, the HR electron is bound to the core ion by the energy given in Eq. (1). Thus molecular high-Rydberg states are described by potential curves parallel to those of each core ion state but lying slightly below. From this point of view it is clear that the appearance potential (AP), the energy of the dissociation limit, and the kinetic energy of the fragments are nearly identical for a HR fragment and for the ion from the corresponding dissociative ionization process.

For the correspondence between HR dissociation and dissociative ionization to be observed, it is necessary that the HR fragment be sufficiently stable to reach a detector. If the fragment is molecular, then it is subject to decay by

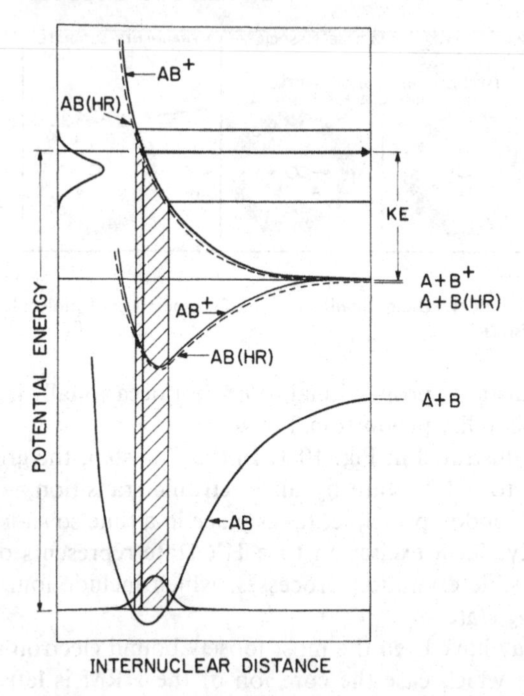

Fig. 10.2. Potential curve description of the core ion model. Excitation from the ground state in the Franck–Condon region (shaded) leads either to a stable HR around an AB^+ core or a replusive HR state. The fragments separate with released kinetic energy (KE) and a KE distribution determined approximately by the reflection of the ground state $v=0$ vibrational wave function in the replusive potential curve.

autoionization and predissociation. If the fragment is atomic, with an electronically excited core ion, it can usually autoionize very quickly.

The potential energy curve interpretation of the core ion model also predicts a similarity in behavior between high Rydbergs and low Rydbergs with $n=3$ or 4, which are usually observed by their emitted radiation. For $n=3$, the appearance potentials and dissociation limits lie about 1.4 eV below the corresponding high-Rydberg energies. To the extent that an $n=3$ or 4 Rydberg electron does not influence the molecular bonding, the resulting molecular potential curves are parallel to the HR potential curves. As a consequence, dissociation limits for emission from low-n fragments are found just below the dissociative ionization energies, and kinetic energies measured by the Doppler shift closely resemble those measured for HR fragments from the same molecule.

Another application of the concepts of the core ion model is to the potential energy curves of high-Rydberg states in singly charged diatomic molecular ions. These states, which are responsible for some of the observations that will

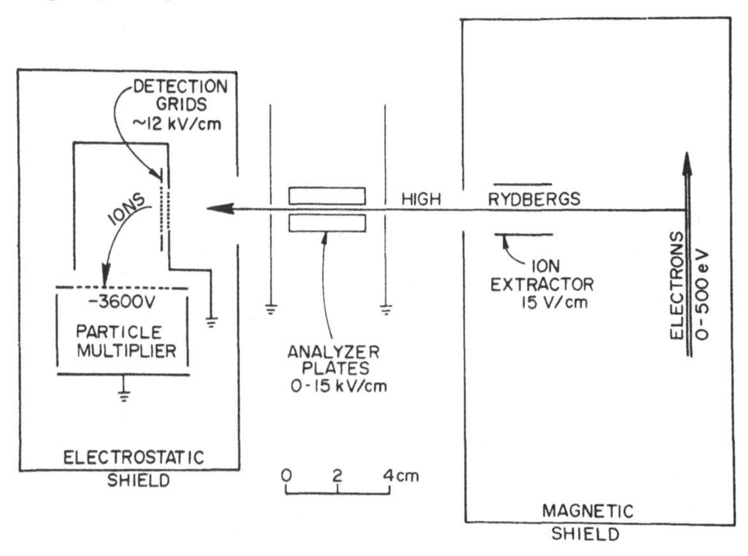

Fig. 10.3. Typical apparatus used for the study of HRs excited by electron impact.[9] The detector electric field ionizes all HRs with $n > 15$. The analyzer plates are used to determine n distributions. Time-of-flight measurements and electron-impact excitation cross sections are used to characterize further the HRs and their formation processes.

be presented later, are describable as a Rydberg-electron bound to a doubly charged molecular core ion. The ground state of a doubly charged core ion dissociates to two singly charged ions and so is determined by Coulomb repulsion at a large internuclear distance.[5,6] For a triply ionized core, the Coulomb repulsion is even stronger, thus giving higher thresholds and higher kinetic energy fragments.

10.2b. Experimental methods

Time-of-flight distributions

Translational spectroscopy refers to any method of measuring the translational kinetic energy distribution of an atom or molecule. When dealing with charged particles, the kinetic energy or momentum can be measured electrically. For neutral particles, however, it is more practical to measure time of flight (TOF) and then transform to kinetic energy. Another consideration is that the energies of low-energy ions, on the order of several electron volts or less, can be shifted by stray electric fields, whereas neutrals are not affected.

A typical apparatus, used for the TOF spectroscopy of HRs, is shown in Fig. 10.3. In this and similar apparatuses, both static gas sources, at pressures between 10^{-4} and 10^{-6} Torr, and molecular-beam sources have been used. Molecular-beam sources permit higher resolution by detecting the fragment at

90° to the parent molecule velocity. A static gas source permits pressure measurements for quantitative cross section determinations.

Time-of-flight measurements depend on generating a short pulse of HR fragments. This has been accomplished by various simple electron guns that produce electron pulse widths down to 0.1 μs at energies from below threshold to several hundred electron volts. Flight paths for the HRs from the electron beam to the detector have ranged from 0.5 to 41 cm, with the resulting TOF ranging from several microseconds to several milliseconds.

Three basic kinds of HR detectors have been used, those depending on the Auger process,[7,8] those depending on electric field ionization (EFI),[4,9,10] and those depending on collisional ionization in a gas.[3,11,12] The earliest experiments used dual-source mass spectrometers.[1,13] High-Rydbergs were ionized by accelerating electric fields and by the electric fields due to the image charges as a HR passed near slit edges or grid wires in the second chamber. Later, EFI detectors created electric fields of known magnitude up to 15 kV cm^{-1} between a pair of grids and then collected the ions.[9] The Auger process is a two-electron radiationless transition that occurs when an excited atom or molecule is in close proximity to a metal surface. The observable result is that potential energy of the excited atom or molecule is converted to kinetic energy of an electron, which escapes into free space. Auger detectors have long been used to detect low-lying metastable states and have been found more recently to respond to HRs also. They have several disadvantages compared with EFI detectors: they detect both low-lying metastables and HRs so that the signal is a composite; their absolute efficiency for HR detection is probably several times smaller than an EFI detector; and the EFI detector produces free ions that can be directed to a mass spectrometer.

Ions or electrons from any of these forms of detector are routed to a particle multiplier or channeltron and the output pulse to the appropriate electronics. Commercial multichannel scalers rarely have time resolutions better than 10 μs $channel^{-1}$ and so have not been used often. Some homemade attachments to minicomputers have provided resolutions down to 0.1 μs. The fastest method remains using a time-to-amplitude converter (TAC) followed by a pulse-height analyzer (PHA), which provides resolutions from several nanoseconds to several microseconds with commercial equipment.

The TOF resolution is a function of a number of factors, the most important of which are:[4]

1. The effect of thermal energy, by which the component of parent molecule velocity parallel to the direction toward the detector is added to or subtracted from the center-of-mass velocity of the fragment.
2. The recoil of the scattered electron due to conservation of momentum, which can add to or subtract from the fragment velocity.
3. The lengths of the excitation and detection regions, which introduce a range of distances l to the detector.
4. The duration of the excitation pulse and the width of the detector channels.

Another factor that distorts experimental TOF measurements, but does not come under the heading of resolution, is radiative decay during flight to the detector. The lowest principal quantum number that reaches the detector is determined by the lifetime. At least some HRs decay in flight, and this attenuates the signal detected at longer times more than at shorter times. If the fragments consisted of a single state with a single lifetime, the radiative decay could be accurately accounted for by an exponential decay term. With a multitude of different lifetimes present, a practical approach is to describe the decay by a single exponential with an effective lifetime, but this cannot be a completely accurate description. The effect of radiative decay on fragments in flight was displayed by several workers who measured TOF for different principal quantum numbers.[14,15]

A final distortion of TOF curve shapes was reported by workers who used Auger detectors.[16] Metastable species with radiative lifetimes comparable to the TOF emitted photons that could eject electrons from the detector. The result was a decaying contribution to the detected signal at short times but extending beyond the end of the electron pulse.

Kinetic energy distributions

It is usually desirable to convert TOF distributions to velocity or translational kinetic energy distributions. In this form the data can be directly related to molecular potential energy curves or surfaces and to dissociation limits. The data are also directly applicable to physical situations in which knowledge of the kinetic energies or velocities of particles is desired.

Transformation from a TOF to a KE distribution is based on[4,8,17]

$$F'(E_1)\,dE_1 = F(t)\,dt \tag{2}$$

$$v_1 = d/t \tag{3}$$

$$E_1 = m_1 d^2/2t^2 \tag{4}$$

where the subscript 1 refers to the detected particle, E is the translational (as opposed to vibrational or rotational) kinetic energy, t the TOF to the detector a distance d away, and m the mass. The kinetic energy distribution of the observed particle then becomes

$$F'(E_1) = F(t)t^3/m_1 d^2 \tag{5}$$

and the sum of the kinetic energies of the two particles, by conservation of momentum, becomes

$$F'(E_r) = \left(1 + \frac{m_2}{m_1}\right)F'(E_1) \tag{6}$$

The subscript 2 refers to the undetected particle and E_r is the released kinetic energy. In practice, the transformation consists of multiplying the amplitude of each point in the TOF distribution by t^3 and plotting the result against the corresponding kinetic energy.

In order to carry out this procedure, it is necessary to know the mass of the detected HR fragment. The most direct way is to use a mass spectrometer, but only one laboratory has used this approach.[10, 18, 19] When the fragment is hydrogen, its TOF is so short that no heavier fragment could be responsible. As to long TOF peaks thought to be due to H, the shift upon deuteration can provide a convincing assignment. Another method, effective in some cases, is based on Eq. (6). For diatomic molecules, the slope of a plot of the *released* KE against the observed appearance potential must be unity. If the masses m_1 and m_2 are incorrectly assigned, the slope will deviate from unity and indicate the incorrect assignment.

Perhaps the major limitation to obtaining accurate KE distributions from TOF is the extreme sensitivity of the signal at small KE to the precise placement of the TOF baseline. This results because of the double problem that many TOF points at long times map into a narrow KE range and are multiplied by t^3 in the transformation. Thus the largest differences among nominally identical KE measurements from different laboratories are usually the amplitude of the signal at low KE and the minimum reported KE. An effective procedure for identifying the baseline level has been to delay the electron pulse about 10 channels after the multichannel scaler is triggered. The count level during those channels is taken as the baseline because it occurs at the longest time possible following the preceding electron pulse, and any signal from slow particles has had the most practical time to decay.

Another important source of differences among KE distributions reported by different laboratories is the TOF resolution for the channels with shortest TOF. These are the channels that map into the largest KE. The distorted KE results have an excess of signal at high KE. When data with different resolutions were reported, those with the better resolution usually showed sharper KE peaks and sharper drops at the highest KE.

An interesting consequence of the transformation is that some readily observed shoulders or even peaks in the TOF distribution become unresolved in the KE distribution and other features that are unresolved in TOF appear clearly in the KE distribution. This is a result of the t^3 factor, which clearly has a large effect on the shape. Usually, each feature seen in TOF or KE can be confirmed by appearance potential measurements.

Another possible cause of distortion is radiative decay, which weakens the signal at long TOFs. Although an attempt was made to correct for this effect, no single lifetime could be determined to fit the data.[4] The result of radiative decay is to weaken the signal at low KE. It is likely, however, that this effect is smaller than the uncertainty due to the baseline error, so its neglect is probably of little importance.

Finally, it should be mentioned that, with only a few exceptions, KE measurements on HR dissociation were made with the detector at 90° to the electron beam. Anisotropy may be present, however, and the measured KE distribution may vary with angle. This effect was demonstrated by Van Brunt[20] with measurements on dissociative ionization of H_2 and by Misakian, Mumma, and Faris[7] with measurements on CO_2.

Principal quantum number distributions

In order to better understand HR measurements and, certainly, to make quantitative measurements, it is necessary to know the principal quantum number n of the HR being detected. In the absence of any decay process other than radiation, the intensity I_n of HRs excited by electron impact to principal quantum number n and surviving for t seconds is proportional to

$$n^{-3}\exp(-t/\tau_n) \tag{7a}$$

where the electronic angular momentum (l) dependence is omitted and

$$\tau_n = \tau_1 n^x \tag{7b}$$

where τ_1 is a constant and x is often, but not always, 3. The shape of this function at high n, where τ_n is long, is proportional to n^{-3}. At low n, there is exponential attentuation. Measurements of the n-distribution utilized parallel-plate electrodes with electric fields up to 15 kV cm^{-1}.[9,21-23] The HRs pass between the plates and to a HR detector. As the electric field is increased, field ionization takes place for lower and lower n. Although the field-ionization probability does depend on l (or more precisely on the Stark levels) as well as on n, the error in ignoring this dependence is unlikely to be more than one or two n units. The function that relates the critical n, which is the lowest n ionized by an electric field F_c, to that electric field is

$$F_c \simeq \frac{6 \times 10^8}{n^4} \text{ V cm}^{-1} \tag{8}$$

where the constant in the numerator applies to a mixture of l states.[9]

The measurement yields a transmission function, not the n distribution of interest. Thus a transformation is necessary. The procedure that was used transformed Eq. (8) to a function with F as independent variable and then fit the data by the method of least squares.[9,21,22] The determined constants τ_1 and x are then substituted in Eq. (7) to give the n distribution.

Electron-impact cross sections

Although a general treatment of electron-impact excitation cross sections is not appropriate here, several topics are important to the present discussion. Specifically, these are the excitation function, which includes the electron energy calibration, the threshold energy, and absolute values of the cross section.

For the dissociation processes discussed here for producing HR fragments, the shape of the cross section as a function of impact energy (often called the excitation function or relative cross section) generally rises from zero at threshold to a peak in the vicinity of 100 eV and then falls slowly, approaching zero asymptotically. The observed excitation functions are superpositions of cross sections for dissociation of many molecular states, however, so much information is obscured. A considerable separation of different processes can be achieved by measuring the excitation function with the detector gated to accept signal arriving in a narrow TOF interval and therefore in a narrow KE interval. This was done in several laboratories either by gating the detector for a particular TOF interval or by using an on-line computer to record the signal from multiple TOF intervals simultaneously. Simultaneous measurement assures the absence of energy scale shifts for the different TOF intervals.

For sufficiently high values of n, the absolute magnitude of the cross section $\sigma_n(E)$ for exciting a state with principal quantum number n by electrons of energy E is given approximately by[9, 12, 22]

$$\sigma_n(E) = \sigma_1(E)/n^3 \tag{9}$$

where $\sigma_1(E)$ in turn must be derived from measurements of the cross section $\sigma_T(E)$ for excitation of the total range of detected HRs. The measured signal is given by

$$I_T(E) = \eta L \omega \rho i \sigma_T(E)/4\pi e \tag{10}$$

where η is the detector efficiency, L the electron-beam length viewed by the detector, ω the detector solid angle, ρ the gas density, i the electron current, and e the electron charge. All of these quantities except $\sigma_T(E)$ are measurable. The quantity $\sigma_1(E)$ is derived from

$$\sigma_T(E) = \sum_n \sigma_1(E) n^{-3} \exp(-t/\tau_n) \tag{11}$$

Tables 10.1 and 10.3 give the measured values of σ_1 for dissociation of a number of molecules and for excitation of several molecules and rare gas atoms.

Information on the states of dissociation fragments and on molecular states can be obtained from threshold energies. When sufficiently narrow TOF regions are detected, corresponding to narrow ranges of translational energy, well-defined thresholds can be identified. Straight-line approximations to the data above and below a threshold are used by most workers, with the threshold being taken at the intersection of the lines. A useful way to treat the data is to plot the released kinetic energy [Eq. (6)] against the measured appearance potential for a range of detected TOF regions. As Fig. 10.2 illustrates for a diatomic molecule, the threshold should increase exactly as much as the released KE. The resulting plot thus consists of a set of lines with unit slope. Extrapolation of these lines to zero released KE gives the potential energies of the dissociation limits.

Table 10.1. List of molecules observed to dissociate into HR fragments. The electron-impact cross section can be obtained from the value of σ_1 given here and from Eq. (9). Values of σ_1 are from Ref. 22 and are for excitation by 100-eV electrons. The rare gas atoms are included for comparison.

Molecule	HR	σ_1 (Å^2)	References
He	He	0.67	9, 21
Ne	Ne	0.61	21
Ar	Ar	1.5	21
Kr	Kr	2.0	21
Xe	Xe	4.6	21
H_2	H	0.14	20, 22, 31, 44, 45
HD	H, D	—	45
D_2	D	0.13	22, 31, 45
N_2	N	1.6	2, 4, 8, 10, 14, 22
O_2	O	—	2, 3, 17
CO	C, O	1.6	2, 16, 22, 58
Cl_2	Cl	—	60
N_2O	N, O	—	18
H_2O	H, O	—	11, 23, 31
NH_3	H, N	—	61
CO_2	O	3.0	7, 10, 15, 22, 33
OCS	?	—	59
CH_4	H	1.7	22, 24, 32, 39
CH_4	C	0.36	22, 24, 32, 39
CH_4	CH_3	—	38
CD_4	D	—	39
CH_3CN	H, C, N	—	55–57
CCl_2F_2	C, Cl, F	—	19
CCl_3F	C, Cl, F	—	19
C_2H_2	H, C	—	31
C_2H_4	H	2.5	22, 24
C_2H_4	C	—	22, 24
C_2H_6	H	2.9	22, 24
C_2H_6	C	—	22, 24
C_2H_6	C_2H_4	—	38
C_3H_4	H	3.0	22
(Allene)	C	—	22
C_3H_6	H	1.3	22
(Cyclopropane)	C	—	22
C_3H_8	H	2.4	22, 30
(Propane)	C	—	22, 30
C_4H_6	H	2.3	22
(1, 3-Butadiene)	C	—	22
C_6H_{14}	H	1.6	22
(*n*-Hexane)	C	—	22

For dissociation of polyatomic molecules, some energy may end up as internal (vibrational and rotational) energy of a fragment, in addition to the translational energy. This has been treated in two somewhat conflicting ways. If a constant fraction β of the excess energy goes into translation, then plots of KE vs. threshold energy display a slope of β. Some measurements have been treated successfully this way.[24] The other treatment identified each electronic vibrational-rotational state of the fragments as a dissociation limit and argued that the released KE vs. threshold plot should be linear with a slope of unity from each such limit.[19] This too was found to fit some data.[18,19]

An interesting phenomenon seen primarily for the dissociation of polyatomic molecules is the occurrence of an infinite slope (a vertical segment) in a plot of KE vs. AP.[18,19,24-28] Three possible interpretations have been offered.[18] One is that a bound state of the molecular core ion is predissociated by a number of repulsive states that produce molecular fragments with a range of internal energies.[24,25] A second is that simultaneous dissociation into three or more fragments leads to a vertical threshold.[18] This can be thought of as related to the first interpretation, in that the relative motion of the two undetected fragments is one form of "internal energy." The third interpretation, and one that can apply also to dissociation of diatomic molecules, is that the vertical threshold is caused by autoionization in conjunction with dissociation.[18,19,26,27] This is really related to the second interpretation, with one of the fragments being an electron. The feature common to these three interpretations is that there are at least two unobserved particles with a variable energy relative to each other.

No measurement of threshold energy is better than the accuracy of the electron energy scale calibration. The collision energy differs from the voltage between the cathode and collision chamber walls because of contact potentials and space charge.[29] These effects can be measured for a particular electron gun and set of operating conditions, but changes in conditions can invalidate such a calibration; the degree of space-charge neutralization by positive ions is a function of pressure, and of course space charge is a function of both electron current and energy. More reliable ways to calibrate the energy scale are to observe some different threshold process in the molecule under study or to mix in a small quantity of another atom or molecule and measure one of its thresholds simultaneously. High-Rydberg states of the rare gases or H_2 show sharp thresholds suitable as references,[21] and TOF can be used to separate the signals from the calibrant and the molecule under study. The rare gases have metastable states with well-known excitation functions that can often be observed with a HR detector. Various optical emission spectra can be observed with an interference filter and photomultiplier which view the electron-beam region. Emission from states such as the b $^3\Sigma$ state of CO or the C $^3\Pi_u$ state of N_2, which have sharp excitation functions and narrow or resonant peaks, can be especially useful. Experience has been that thresholds can be measured

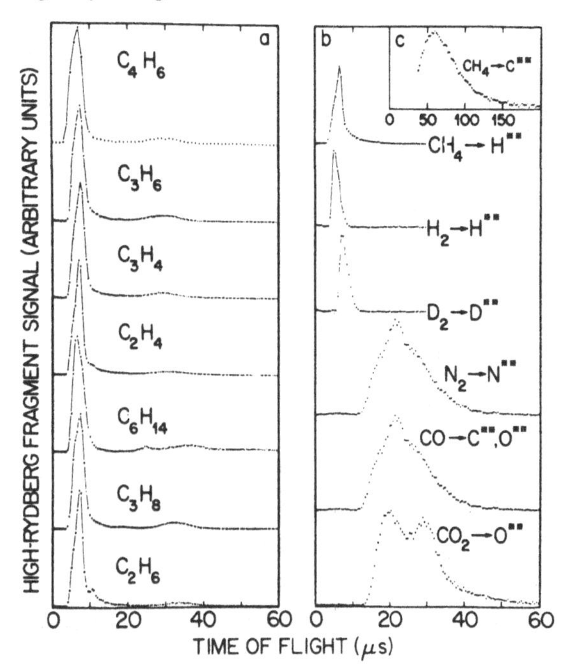

Fig. 10.4. TOF distributions for the HR fragments from a number of molecules excited by 100-eV electrons. The flight path for these measurements was 20.1 cm.

reproducibly to an accuracy of about ±0.2 eV, using an electron beam with energy half-width of 0.3 eV.

10.2c. Experimental results

Overview

The previous sections presented the principles and methods used to study dissociation of high-Rydberg molecules. This section will discuss the experimental results. Studies of only five diatomic, four triatomic, and about a dozen polyatomic molecules have been reported, as listed in Table 10.1. The first observations of the majority of these molecules were made by Kupriyanov.[1,2,30-33] He used a mass spectrometer that formed both ions and HRs by electron impact but had slits with potentials to prevent the ions from reaching the mass analyzer. This apparatus demonstrated the formation of HR fragments by means of low-resolution (~2-eV) excitation functions. These excitation functions displayed thresholds at the lowest dissociative ionization energy, second thresholds near the energy for formation of the double ion, and peaks in the vicinity of 100 eV.

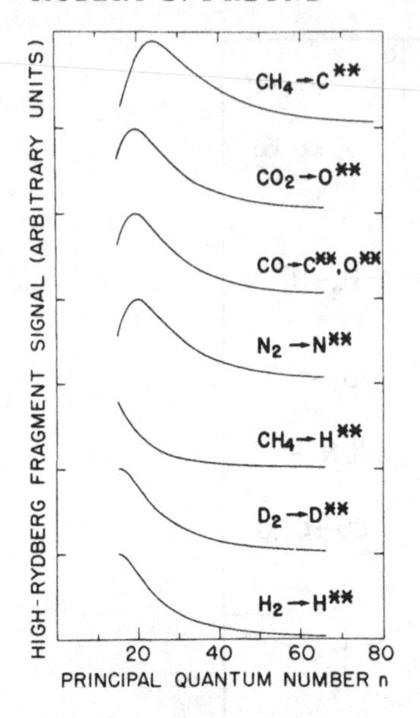

Fig. 10.5. Measured principal quantum number distributions for fragments from a number of molecules.

Quantitative measurements were made a decade later by Schiavone et al.[22] The four different measurements were TOF, n distributions, excitation functions, and absolute cross sections. Figure 10.4 shows TOF distributions measured with 100-eV electrons. For the 20.1-cm flight path, hydrogen-containing species yield HR hydrogen atoms with a TOF around 4–10 μs. Heavier atoms show peaks in the 15–50 μs range. Principal quantum number distributions (Fig. 10.5) are found from $n=15$, the lower limit of the detector sensitivity, to an upper limit of 70 or 80, set by the electric field used to deflect ions. Hydrogen HRs show little radiative decay during their short TOF and so have maximum population at $n=15$. Heavier fragments peak at higher values of n, consistent with a larger t in Eq. (7a). The determination of n distributions also leads to values for τ_1 and x in Eq. (7b). These values, although not highly accurate and not straightforward to interpret, do suggest that the fragment atoms form with electronic angular momentum l greater than about 4. Excitation functions are shown in Fig. 10.6. The absolute magnitudes were determined by Eqs. (10) and (11) and the resulting values of σ_1 (100 eV) are included in Table 10.1. The remarkable point is that σ_1 varies by only a factor of two for the majority of the molecules, and H_2 and D_2 dissociations give appreciably smaller values for reasons that will be explained later. It is also plausible that the C(HR) formation from hydrocarbons has a small σ_1 because of its small statistical probability and the large degree of dissociation required to obtain C atoms.

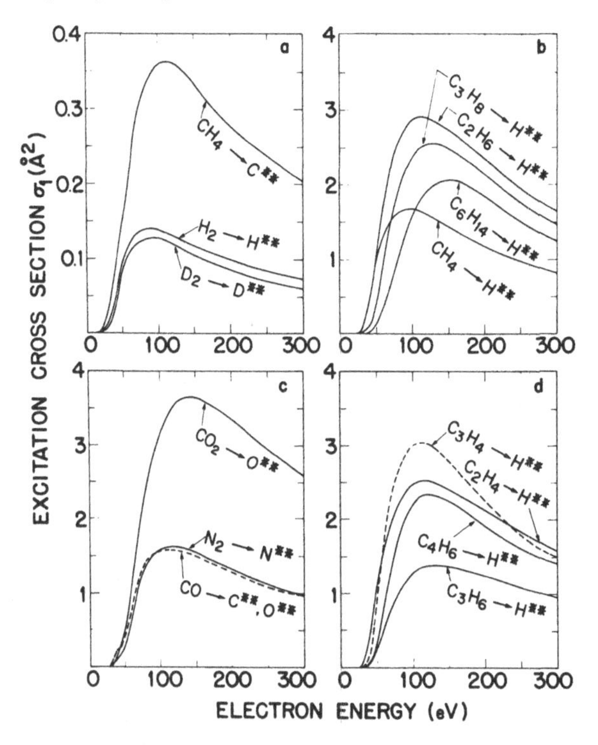

Fig. 10.6. Measured electron-impact excitation cross sections expressed as the parameter $\sigma_1(E)$, which, together the Eq. (9), give quantitative cross sections for all fragments with principal quantum number n.

Rather than attempt to present results on each of the molecules studied, we shall give, in the next sections, detailed discussions of two representative molecules and mention several noteworthy points about others.

Example: N_2

The most extensive investigations of N_2 were carried out by Smyth, Schiavone, and Freund[4] and by Wells, Borst, and Zipf.[8] Both groups used TOF, transformation to KE, and excitation functions to identify four principal dissociation processes. These four processes are indicated on the data of Smyth et al.[4] in Fig. 10.7–10.10. Wells et al.[8] reported essentially the same results, although some differences will be mentioned later. Allcock and McConkey,[10] in a paper on CO_2, commented that they tested their apparatus on N_2 and obtained results identical to those of Smyth et al. Thus, the experimental results from these three laboratories appear to be in good agreement.

The core ion model is useful in interpreting the four dissociation processes

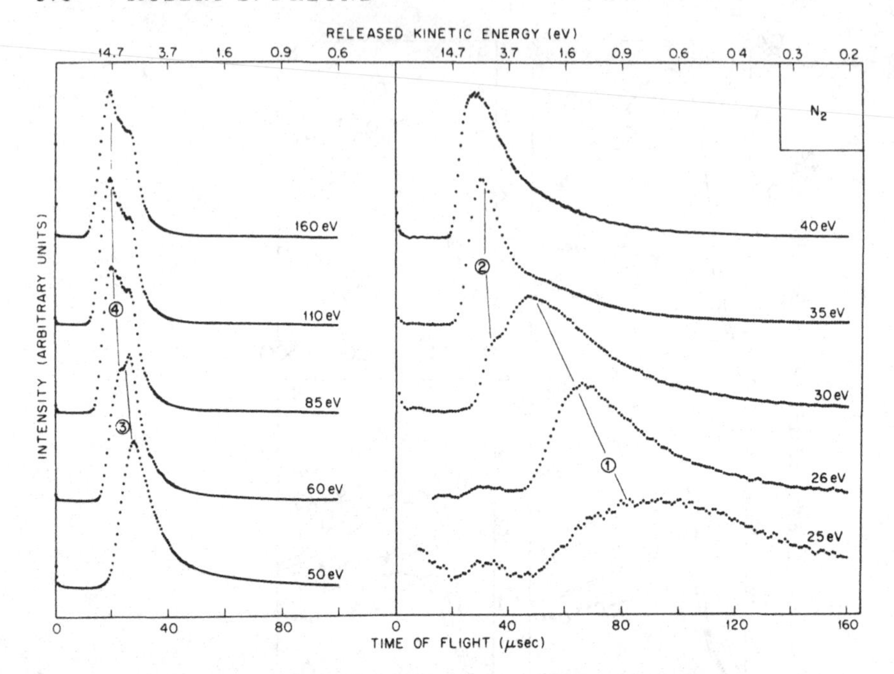

Fig. 10.7. TOF distributions of HR N atoms from dissociation of N_2, excited by electrons of the indicated energies. Corresponding released KE is given at the top of the diagram. Four processes are indicated by circled numbers and are discussed in the text.

noted in the figures. The detected fragment is a HR atom and the undetected fragment appears in essentially all states of the atom or its ionic states. State identifications, dissociation limit energies, and released KE are given in Table 10.2. Process 1 has the longest TOF, the smallest KE, and appearance potentials that extrapolate back at zero KE to the potential energy of the lowest dissociative ionization limit $N(^4S) + N^+(^3P)$. Process 2 represents dissociation to the limit where the detected HR atom is the same as for process 1 but the undetected atom is in its first excited state $N(^2D)$. Wells et al.[8] identified an additional process just above process 2, which they assigned to the $N(^2P)$ state. Process 3 is really a group of processes in which one fragment is in a $N(^3P)$ HR state as for processes 1 and 2 and the other is in any Rydberg state of the atom from the lowest $N(3s\,^4P)$ to the series limit N^+. This set of closely spaced limits obscures individual sharp appearance potentials. Process 4 is seen most clearly in the TOF curves and also appears in the KE transform. Wells et al.[8] observed two appearance potentials for process 4 and identified the undetected fragment as N^{2+} in its ground and first excited state. In addition to these HR processes, Wells et al.[8] followed a suggestion by Fairchild et al.[14] that the $2s(2p)^3 3s\,^6S_{5/2}$ state of the nitrogen atom is metastable and assigned three dissociation limits to this state. Because it should be subject to

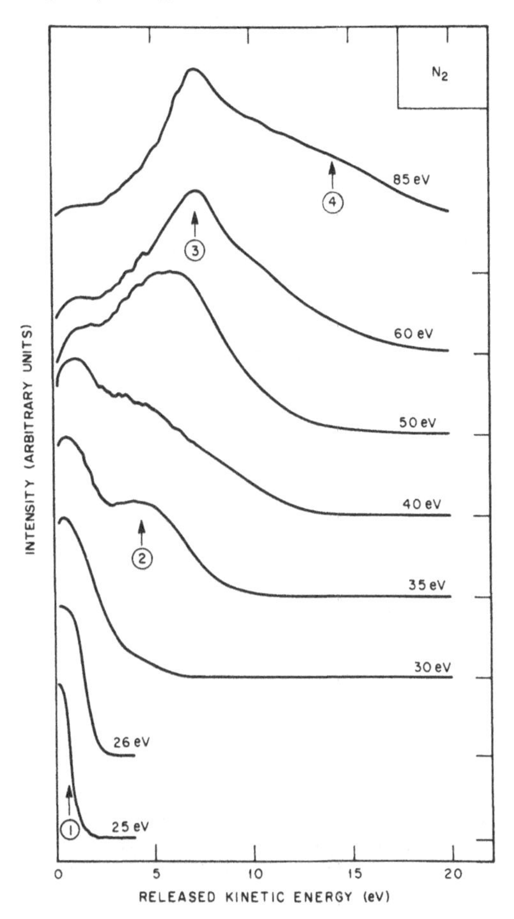

Fig. 10.8. KE distributions of HR N atoms from N_2 calculated from the TOF distributions of Fig. 10.7.

autoionization in the presence of an electric field, the ^6S state would behave similarly to N(HR) in many ways but would be distinguishable by its electric field behavior and dissociation limits.

The core ion model also predicts "identical" kinetic energy distributions for HR fragments and for ions from dissociative ionization.[34-37] Comparison of the available HR and dissociative ionization measurements are made in Fig. 10.11. Agreement between the two HR experiments is excellent. Agreement with the results of dissociative ionization is not as good, although the shapes, the number, and the positions of peaks are qualitatively the same. The largest differences are for low KE, which is the most difficult region for both kinds of experiments to measure.

It is difficult to assign those molecular states of the core ion that are initially

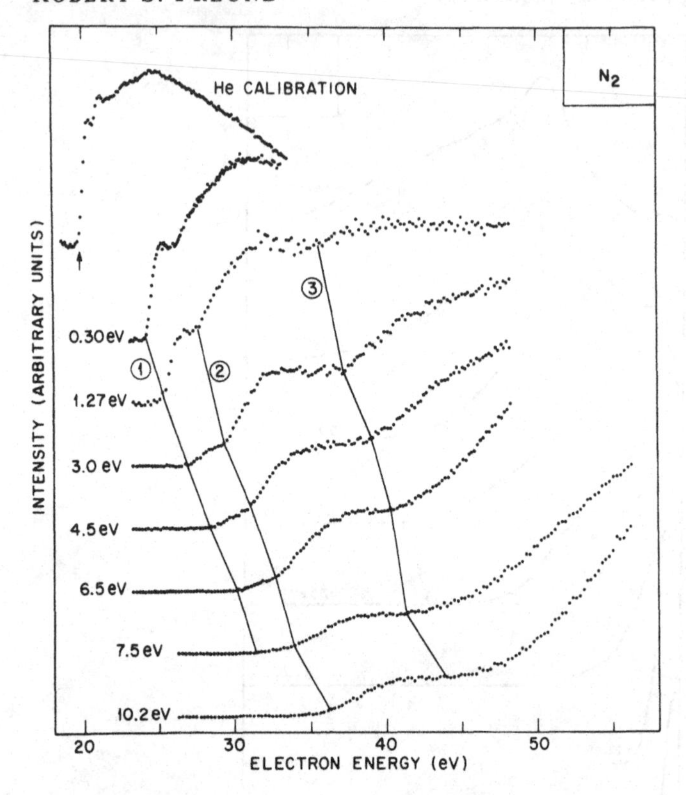

Fig. 10.9. Excitation functions for HR N atoms from N_2. Energies to the left of each curve (from 0.30–10.2 eV) indicate the released KE that was detected. Electron energy calibration is based on the excitation function for metastable helium, with a threshold at 19.8 eV. The threshold for process 4 is above the energy of this diagram.

excited and dissociate to the observed fragments. The state or states of process 1 were discussed by Smyth et al.[4] and three candidates identified: predissociation of the $C\,^2\Sigma_u^+$ state, direct dissociation of the $D\,^2\Pi$ state, or dissociation of an unknown quartet state (Fig. 10.12). Predissociation might have been expected to show vibrational structure in the translational spectrum, but because of limited resolution, the absence of observed structure cannot rule out the $C\,^2\Sigma_u^+$ state. Wells et al.[8] constructed a segment of a potential curve from the kinetic energy distribution of process 1 and found that it lies close to the $C\,^2\Sigma_u^+$ potential curve. Another argument that process 1 is due to C-state predissociation is that the sharp peak of its excitation function could be due to direct high-*l* excitation near threshold (see Sect. 10.3c). If the HR electron were in a low-*l* orbital around the C state, autoionization could take place faster than predissociation of the core and no HR would be observed. For a high-*l* HR electron, however, autoionization would become slower and predissociation of the $C\,^2\Sigma_u^+$ core would lead to the observed HR fragments. These

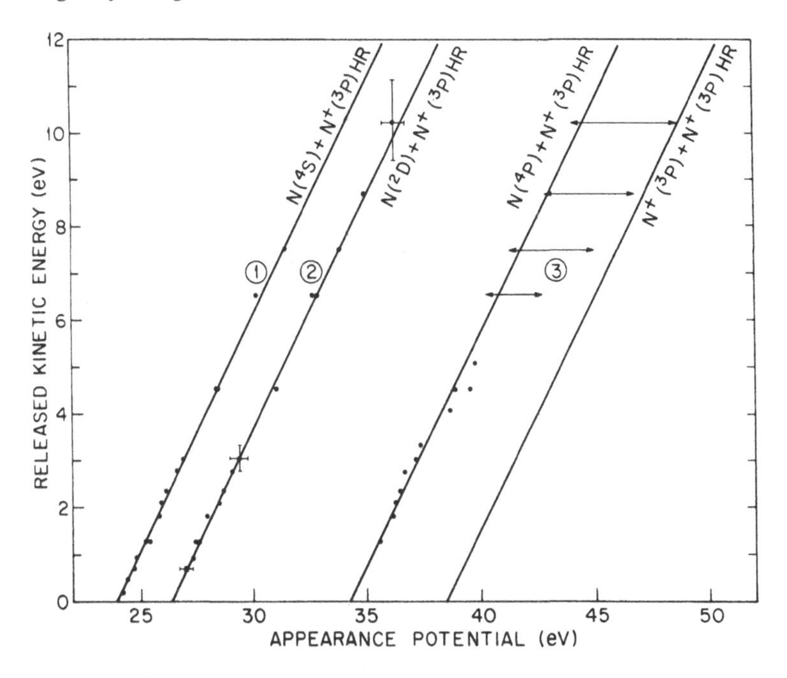

Fig. 10.10. Plot of AP vs. KE for N_2, showing a point for each measured AP and the theoretically expected lines of unit slope.

Table 10.2. Summary of HR dissociation processes observed for N_2

Process number	Dissociation limit: States	Dissociation limit: Energy	Released KE (eV): Peak	Released KE (eV): Range
1	$N(^4S) + N(HR)$	24.3	0.5	0–7.5
2	$N(^2D) + N(HR)$	26.7	5	0–20
3	$N(3s\,^4P) + N(HR)$	34.6		
	⋮	⋮	7.3	1.3–20
	$N^+(^3P) + N(HR)$	38.8		
4	$N^{2+}(^2P) + N(HR)$	68.4	15	—
	$N^{2+}(^4S°) + N(HR)$	91.6		

Source: Data from Smyth, Schiavone, and Freund.[4]

arguments strongly suggest that some process 1 fragments are due to predissociation of the $C\,^2\Sigma_u^+$ state. Those process 1 fragments with combined kinetic and potential energies above the $N(^4S) + N^+(^1D)$ dissociation limit are more likely to come from other states.

Little can be said about the potential curves leading to process 2. As the

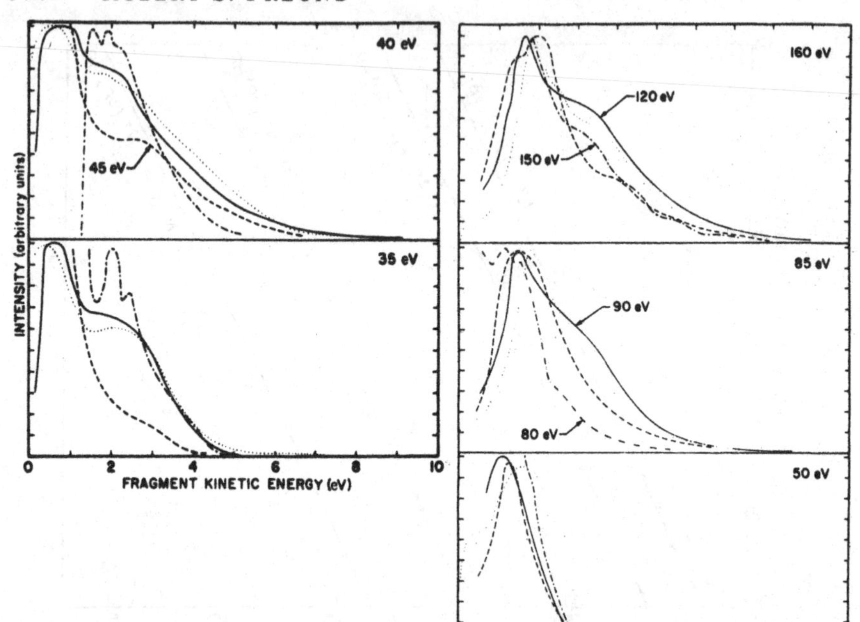

Fig. 10.11. Comparisons of fragment KE (=1/2 released KE) for N_2 measured by two HR experiments (solid lines, Ref. 8; dotted lines, Ref. 4) and three dissociative ionization experiments (dot–dash lines, Ref. 37; dashed lines, Refs. 34 and 35; dash–double-dot lines, Ref. 36).

excitation function is sharply peaked, typical of a spin–exchange collision, process 2 may result from excitation of a quartet state core ion.

Estimated potential curves leading to process 3 are included in Fig. 10.12. On the horizontal scale of this figure the HR electron is far away. The potential curves therefore are those of N_2^{2+}, displaying Coulomb repulsion at the larger internuclear distance but showing potential minima at smaller internuclear distance where the electron clouds overlap and lead to bonding. Because of the potential maxima (shown near 2 Å), no zero KE fragments are possible. However, released KE over 10 eV should be obtainable from excitation within the Franck–Condon region.

Little is known about potential curves of N_2^{3+}. We expect behavior similar to that of N_2^{2+} but the Coulomb replusion should be twice as strong because of the increased nuclear charge. This could be the cause of the higher KE observed for process 4.

Example: CH_4

Unlike N_2, methane contains two different atomic species. Their signals are readily distinguishable by the large difference in TOF (Fig. 10.13), with

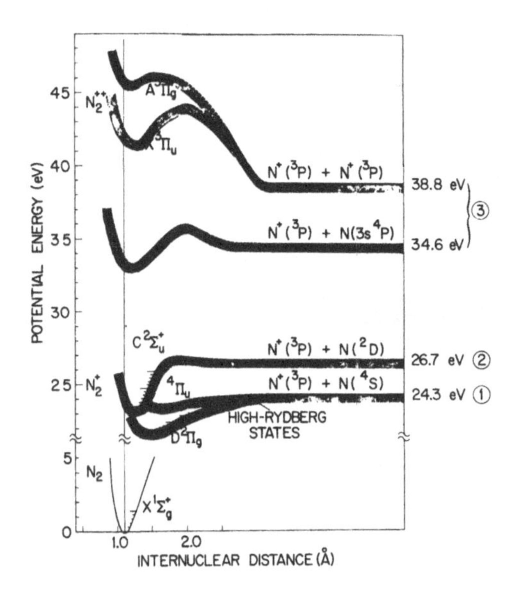

Fig. 10.12. Potential energy curves for N_2. The shaded areas represent HR states lying just below ionic states.

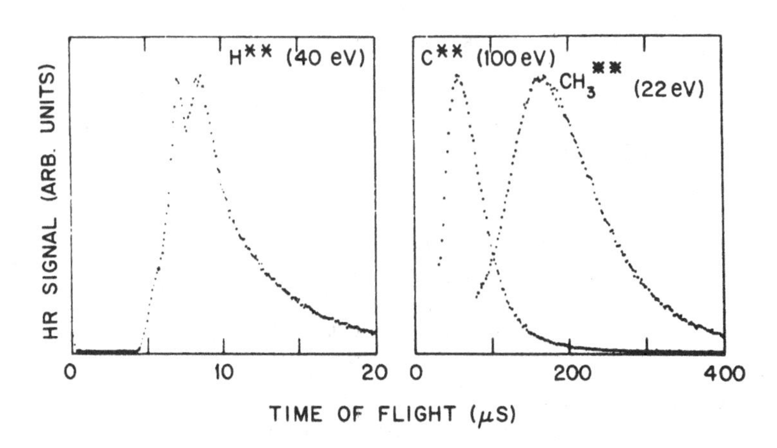

Fig. 10.13. TOF distributions for HR H and C atoms and CH_3 molecules from CH_4.

H(HR) peaking around 10 μs and C(HR) around 50 μs. In addition, a stable CH_3(HR) fragment forms (and is identified by its low threshold energy).[38] Its TOF is almost as slow as thermal CH_4 because the mass of the undetected H atom is so small. There is no explanation for why H_2(HR), a species frequently observed under other conditions, is not seen as a CH_4 dissociation product.

Kinetic energy distributions are shown in Fig. 10.14. The HR data of Finn et al.[39] and Schiavone et al.[24] generally agree with each other, although the

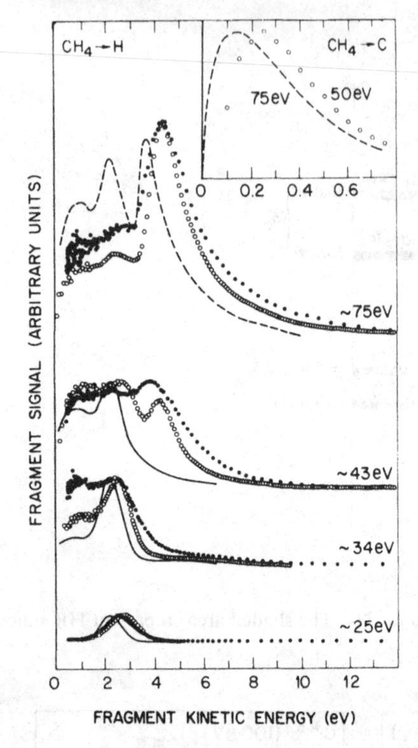

Fig. 10.14. KE distributions for HR H and C atoms from CH_4 (filled circles, Ref. 39; open circles, Ref. 24) compared with those for dissociative ionization (solid lines, Ref. 25; dashed lines, Ref. 40).

Schiavone et al. data are systematically lower at low and high energies, a result possibly of better resolution at short TOF and differences in the baseline correction before transformation from TOF to KE. As with N_2, there is qualitative agreement between KE from HRs and dissociative ionization,[25,40] but there are some distinct differences that may be attributable to autoionization, insensitivity of the ion experiments to very slow ions, and errors in the energy-scale calibration. Agreement with the more recent dissociative ionization data of Locht and Momigny[28] (not shown in Fig. 10.14) is somewhat better.

Excitation functions (Fig. 10.15) and the corresponding plots of KE vs. AP (Fig. 10.16) reveal six processes corresponding to H(HR), one giving C(HR), and one giving CH_3(HR). At the lowest excitation energy, there is a sharply peaked function with a 14.3-eV threshold that proves its assignment as $H + CH_3$(HR). A process at 23 eV is unusual in that it exhibits a single KE for a variety of APs. This phenomenon, which was discussed in Sect. 10.2b, probably results from a CH_4 state at 23 eV, which is predissociated by another

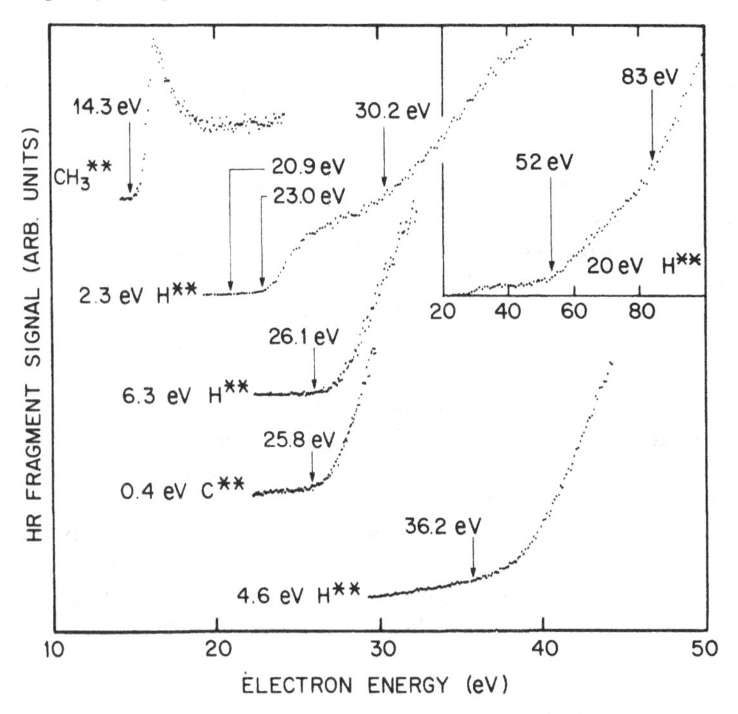

Fig. 10.15. Excitation functions for HR H, C, and CH_3 from CH_4.

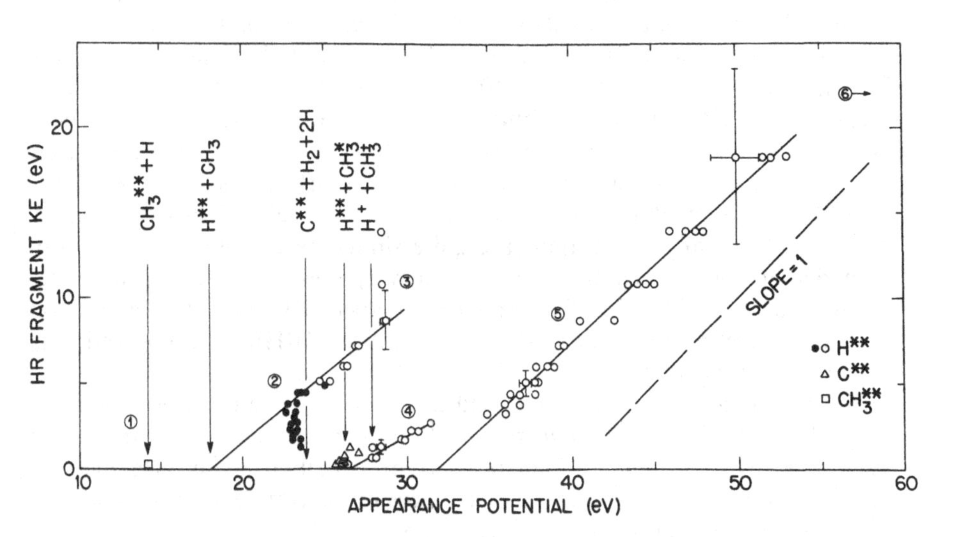

Fig. 10.16. Plots of AP vs. KE for HR fragments from CH_4.

state leading to CH_3 fragments with a range of internal energies. The strength of this process could be attributed to a 23 eV one-electron excitation from the $2a_1$ orbital to a HR orbital. Process 3 extrapolates to 18.2 eV, the minimum energy for forming H(HR). Process 5 very likely corresponds to dissociation of a state with a doubly charged ionic core into H(HR) and CH_3^+ or excited states of CH_3 lying just below CH_3^+. At the highest energy, process 6 (which lies off the scale of this figure but is shown in Ref. 24) may be analogous to process 4 in N_2, dissociation into H(HR) and a doubly charged CH_3 ion, or to a higher degree of dissociation.

One process is clearly identified as a carbon HR by its TOF. The minimum thermodynamic energy for a triple dissociation is 19.6 eV if the unobserved fragments are two ground-state H_2 molecules. If the unobserved fragments are four hydrogen atoms, the required energy would be 28.6 eV, higher than the observed thresholds of 25.7 or 26.6 eV.[24,39] Thus, at least one unobserved fragment must be H_2. This is an example of a dissociation process in which a rearrangement takes place to form a new chemical bond.

A final point to note in the methane results is that some energy can end up in internal (mostly vibrational) modes of a fragment. Thus, a plot of released kinetic energy vs. appearance potential is not necessarily linear and many not have a slope of unity, behavior clearly illustrated by process 4.

Examples: other molecules

For the two molecules discussed here in detail, most of the important methods and principles were mentioned. For two additional molecules, figures were published comparing KE distribution from HR dissociation to those from dissociative ionization. These molecules are H_2[20] and O_2,[41–43] and the comparisons are given in Figs. 10.17 and 10.18. For H_2, two independent measurements of the H(HR) KE distribution are in excellent agreement.[44,45] There is also a substantial similarity between the HR and dissociative ionization kinetic energies.[20,46] It had been thought that the differences were major, but Van Brunt[20] pointed out the importance of the angle of observation. In Fig. 10.17, the 90° distributions are quite similar, except for a high-energy shoulder on the dissociative ionization curves, presumably owing to direct dissociation of $H_2^+(2p\sigma_u)$. The similarity at lower KE suggests that for the electron-impact-excited states, which dissociate to H(HR), autoionization is relatively unimportant.

Kinetic energy of HR from HD and D_2 show significant differences from H_2.[45] In particular, there appears to be a larger contribution from doubly excited states near $2p\sigma_u$. Carnahan and Zipf[45] attribute these differences to mass differences and therefore to differences in the competition between dissociation and autoionization for various states.

A phenomenon observed in H_2 but not yet in any other molecule is excitation of a two-electron excited state of the neutral molecule lying above the

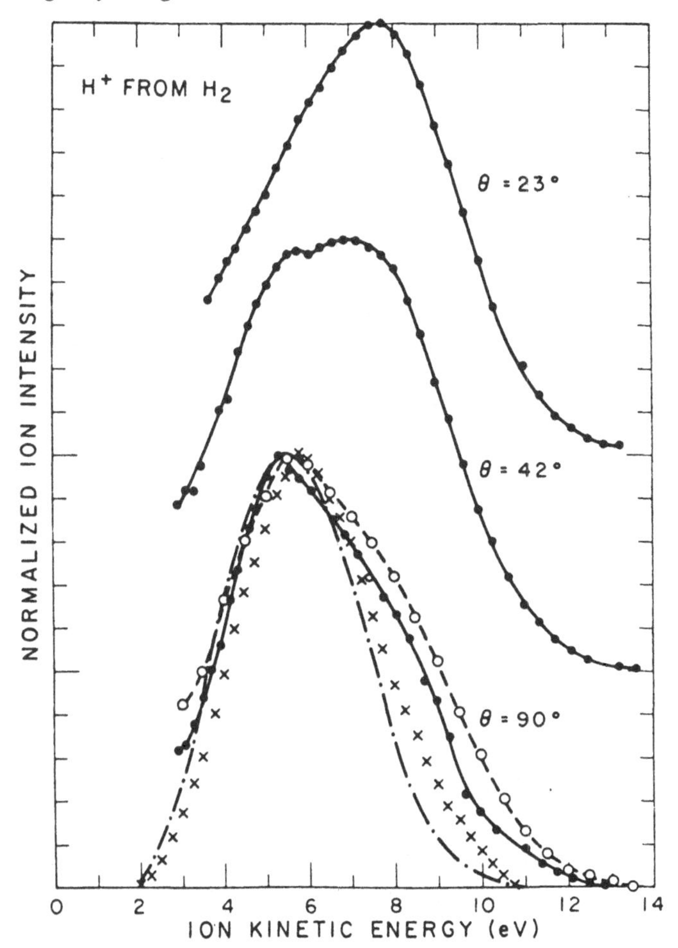

Fig. 10.17. Comparison of fragment kinetic energy distributions for HR H atoms observed at 90° to the electron beam (dash-dot lines, Ref. 33; crosses, Ref. 43) and for dissociative ionization (solid circles, Ref. 20; open circles, Ref. 46).

ground state of the ion. Such a state can therefore autoionize but it can also dissociate; the two processes are competitive. The identities of such states (often called resonances), their energies, and their autoionization and dissociation behavior are at present only partially solved problems and are the focus of much activity.[26,47-54] Such doubly excited states must of course exist in all molecules, but their role is especially prominent in H_2 because there are no one-electron excited states in the same energy region. This is because ground-state H_2 with a $(1s\sigma_g)^2$ electron configuration has no inner orbitals. For H_2, it is therefore impossible to obtain excited states well above the ionization potential by promoting one electron from an inner orbital, a relatively

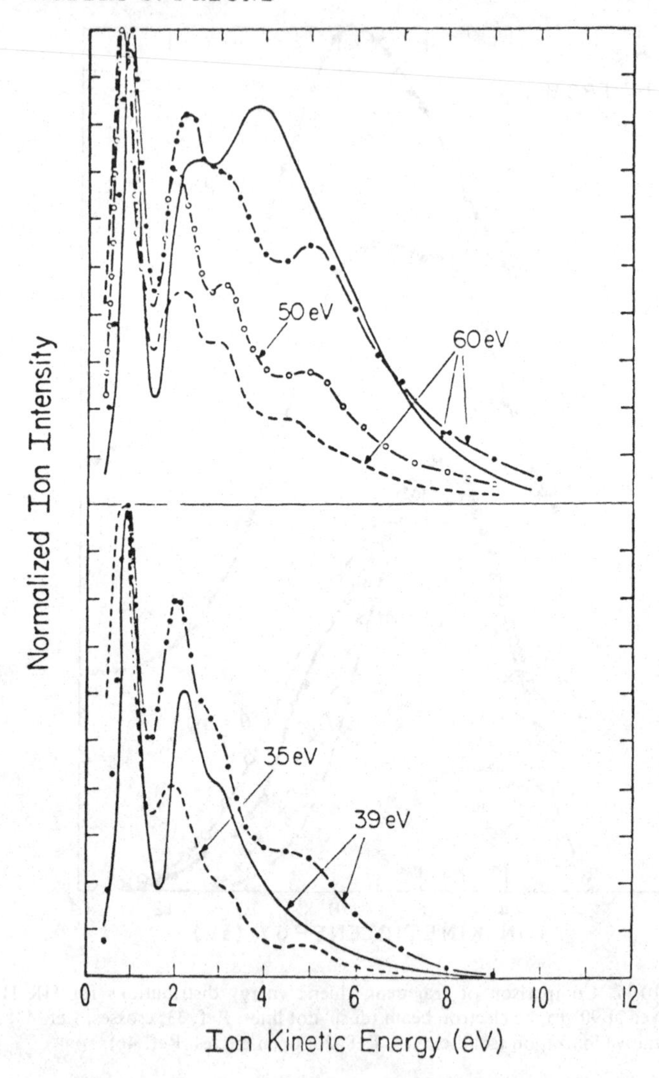

Fig. 10.18. Kinetic energy distributions for HR O atoms from dissociation of O_2 (dashed lines, Ref. 3) and for O^+ from dissociative ionization of O_2 (solid circles, Ref. 46; open circles, Ref. 43; dashed lines, Ref. 42).

probable process in other molecules. The partially forbidden character of these two-electron transitions accounts for the small values of σ_1 given in Table 10.1.

For several molecules, CO_2,[10] N_2O,[18] CCl_2F_2,[19] and CCl_3F,[19] it was crucial that a mass spectrometer follow the HR detector to identify the species. An interesting result is that about 98% of the HR fragments from CO_2 were found to be oxygen atoms.[10] References to other studies of HRM dissociation are given in Table 10.1.[55-61]

10.3. Stable high-Rydberg molecules

The previous section dealt with HRMs that dissociate into HR fragments. In this section, we shall address HRMs that are stable on the time scale of an experiment. They must have potential curves with potential minima (bound states) or they would dissociate almost immediately. A bound excited state may be observed with a lifetime as short as $\sim 10^{-11}$ s when detection is by absorption spectroscopy to that state, where the only requirement is that the predissociation lifetime not excessively broaden a level. At the other extreme, a HRM may exhibit a lifetime as long as 10^{-3} s, which is a typical TOF of a thermal molecule in a molecular beam. A relatively small number of molecular HRs has been found to live long enough to undergo collisions in the gas phase ($\sim 10^{-9}$ s) or to be formed into molecular beams. We shall therefore begin by examining the mechanisms that lead to HRM decay.

10.3a. Decay of high-Rydberg molecules

All Rydberg states are unstable, of course, in that they can radiate their excitation energy. Radiative lifetimes are, however, quite long on the time scale of many experiments, especially for states with large values of the quantum numbers n and l. Our concern, therefore, is with the competing decay processes, autoionization and predissociation. Their rates are often found to be in the range of $10^{10}n^{-3}$ to $10^{12}n^{-3}$ s^{-1},[62] whereas radiative decay rates are closer to $10^{8}n^{-3}$ s^{-1}.

Electronic autoionization occurs when the core is electronically excited and the HR electron, on one of its infrequent traverses of the core, collides superelastically; the core gives up its excitation energy to translational energy of the electron.[63] Vibrational autoionization occurs when a electronic ground-state core gives up vibrational energy to the HR electron. Both of these processes depend on the HR electron penetrating the core and so have rates that scale as n^{-3}. When the core is in $v=0$ of its ground electronic state, autoionization is energetically impossible for many HR levels and there is often sufficient rotational energy to autoionize only the very highest HR levels.

Predissociation differs from autoionization in that $v=0$ of the ground-state core ion for almost every molecule lies above the energy of one or more dissociation limits. Repulsive valence states, or the replusive inner walls of bound valence states, can cross the Rydberg states and predissociate them. Such valence–Rydberg interactions are familiar from the low-Rydberg spectroscopy of NO[64] and N_2[65] and theoretical computations on O_2.[66-69] At low n, the interaction between valence state and Rydberg state shows up as an avoided curve crossing, but the strength of the interaction between them should scale as n^{-3}, so for high n the potential curves should be affected very little. Predissociation of HRMs by valence states in this way is expected to occur in most molecules.

A notable exception is known in the H_2 molecule; the $v=0$ vibrational level (and maybe $v=1$) is not crossed by any replusive curve. Thus, as for HR atoms with ground-state cores, its only decay mechanism is by radiation. This situation has permitted the many observations of H_2(HR). Apparently a similar situation exists in the potential curves of He_2.[70] In both of these molecules, the first excited dissociation limit lies nearly as high as the energy of the lowest molecular ionization potential, decreasing the number of replusive states that might pass near the lowest HRM levels. Similar situations might be looked for in molecules containing other atoms with high-lying first excited states such as the diatomic halogens.

10.3b. Examples of long-lived HRMs

Low-Rydberg states of many diatomic and polyatomic molecules have been observed by absorption spectroscopy in the vacuum ultraviolet region.[71,72] For many of these molecules, several members of one or more Rydberg series were seen, and in a few cases the series was followed to $n=15$ or slightly higher.[71] It is rare, however, that rotational structure has been resolved and studied for series members above the lowest. This is probably ascribable to two differences between atoms and molecules: the density of levels in a molecule is very high, because of rotation and vibration, and individual levels can be lifetime broadened by predissociation.[73] Series converging to excited states of the core ion are of course subject to lifetime broadening by autoionization also. Past observations of Rydberg series in absorption therefore show that bound Rydberg states exist in many or most molecules but provide no evidence for lifetimes close to radiative lifetimes.

Experiments with built-in time scales are those based on collisional ionization of HRMs. The HRM lifetime must be comparable to the collision time in a low-pressure gas in order to be detected. Revealing experiments were performed by Klots[74] and Person et al.[75] A number of organic molecules were photoexcited to energies just below the lowest ionization potential where the range of quantum numbers was above $n \simeq 18$. Collisional ionization of HRs with ground-state molecules and buffer gases at a total pressure around 10^{-1} Torr produced a measurable current across a pair of parallel plates. The kinetic scheme describing this arrangement is

$$M + h\nu \rightarrow M^*$$

$$M^* \xrightarrow{K_1} \text{neutral predissociation products}$$

$$M^* + M \xrightarrow{K_2} \text{ionic products}$$

$$M^* + M \xrightarrow{K_3} \text{neutral products}$$

Variation of pressure during the measurements and solution of the kinetic equations gives values for the rate constants K_1, K_2, and K_3. The physical basis of this approach is a direct comparison of the bimolecular rate constants K_2 and K_3 to the unimolecular decay constant K_1. Person[75] found K_1 to be 10^8–10^{10} s^{-1} for several molecules, corresponding to lifetimes of 10^{-8} to 10^{-10} s, and Klots[74] deduced lifetimes somewhat longer than 10^{-7} s. These values are shorter than radiative lifetimes for $n > 18$ and so must represent predissociation.

A number of other examples of collisional ionization of HR molecules in mixed gases have been reported. Chupka and Berkowitz,[76] in studies of H_2 photoionization, found that ions formed at photon energies below the ionization potential. Koyano et al.[77] reported a similar observation for photoexcitation of N_2. Most recently, Panock et al.[78] studied laser excitation of He_2 from a metastable state to HR states, with subsequent collisional ionization. In these experiments, pressures ranged from $\sim 10^{-3}$ to 10 Torr, so the corresponding lifetimes must be similar to those reported by Person[75] and Klots:[74] 10^{-10}–10^{-7} s.

Another effective way to determine lifetimes is to observe radiative emission from the HRM. This is well known to be difficult for high n because the transition probabilities decrease as n^{-3} and this in addition permits more time for diffusional or collisional loss. Emission spectra of He_2 have been seen from n up to 17, indicating lifetimes in the 10^{-6} s range.[70] Emission has been seen also from relatively low members of a number of other diatomic molecules, but there are no reports of emission from polyatomic Rydberg molecules.[79] This may be the result of rapid deactivation processes or of insufficient observational efforts.

The most definitive method to prove a long lifetime is to use a molecular beam and procedures similar to those used to study dissociation. The earliest observations of HRMs, H_2 and D_2 by Kupriyanov[1,80] and by Hotop and co-workers,[81,82] were done in mass spectrometers in which the electron beam was separated from the ion optics by 1 or 2 cm. Those experiments proved that neutral molecules traveled this distance and therefore had lifetimes near 10^{-5} s. At about the same time, Barnett and co-workers[83-87] and Solov'ev et al.[88] began experiments on the formation of H_2 and H_3 HR by charge-transfer neutralization of fast beams of H_2^+ and H_3^+, respectively. The TOFs in these experiments proved lifetimes $\sim 10^{-7}$ s. More recently, Hiraishi et al.[12,89] produced H_2 (HRs) by electron impact and allowed them to travel 6 cm to where they were collisionally ionized and mass-spectrometer detected. We refer, finally, to the observation of long-lived HRMs other than H_2 by Tarr et al.[90,91] Electron impact was used to excite CO and N_2 in addition to H_2 and D_2. Direct TOF measurements showed lifetimes on the order of 10^{-4} s. Other references to stable HRMs are included in Table 10.3.[31,32,44,92]

Table 10.3. List of molecules that have been observed in stable HR states

Molecule	σ (Å^2)	References
H_2	0.21	1, 12, 31, 32, 44, 76 81–85, 87–92
D_2	0.082	31, 80, 82
He_2	—	70, 78
N_2	0.035	77, 90, 91
CO	0.023	91
H_3	—	86
CH_3	—	38
CH_3Br	—	75
CH_3OH	—	75
C_2H_2	—	75
C_2H_4	—	38, 75
C_2H_6	—	74
CH_3CHO	—	75
C_2H_5Br	—	75
C_3H_4 (propyne)	—	75
C_3H_6 (cyclopropane)	—	74
C_3H_6O (acetone)	—	75
CH_3COOCH_3 (methyl acetate)	—	75
C_4H_6 (ethylacetylene)	—	74

Note: The electron impact cross section can be obtained from the value of σ_1 given here and from Eq. (9). Values of σ_1 are from Ref. 91 and are for excitation by electrons with energy corresponding to the peak of the cross section.

10.3c. High-Rydberg molecules in high-l states

The frequency with which H_2(HR) have been observed is due in part to the accidental absence of predissociation of their $v=0$ levels. The other, and more general reason, is that electron impact and charge transfer can form HRMs in states of high electronic angular momentum (l). High-l electrons have very small probabilities for penetrating the core, so rates for all three decay processes, autoionization, predissociation, and radiation, are greatly reduced because they depend on Rydberg-core interactions. High-l states are not formed by photoexcitation from the ground state because of the $\Delta l = \pm 1$ selection rule. Once high-n, low-l states are formed, they can be collisionally converted to high l, but in the molecular-beam experiments that have been done, pressures are low enough that collisional Δl processes are unlikely.

The first suggestion of high l as the mechanism for stabilizing H_2(HR) was by Band,[93] who proposed that high-l states were formed in the charge-transfer experiments of Barnett et al.[83-86] Band showed that autoionization (and presumably predissociation) of all vibrational levels was slowed greatly for $l \geqslant 3$.

High-l states can form also be electron impact with a high-n, low-l molecule (see chap. 12 by Delpech).[9] The necessary condition is that the high-n, low-l

state formed initially survives decay long enough for a second electron to give the Δl process. This delay may be on the order of 10^{-6} s for typical electron beams. This *stepwise* mechanism of exciting high-l states was observed for the rare gas atoms[9,21] but for only one molecule, H_2 in its $v=0$ level.[90,91]

Another general mechanism for production of high-l states was proposed by Fano.[94] Just above threshold in an electron-impact ionization process, the two outgoing electrons move slowly and are highly correlated, both radially and angularly. If the correlation is high enough, both electrons escape and ionization results. If they are less well correlated, one electron may remain bound in a HR orbital. As these two slow electrons separate, they exchange angular momentum, so the HR forms in high-l states. Because only one incident electron is involved, this is called *direct* high-l excitation, to distinguish it from stepwise.

Direct excitation of high l was recently observed in the excitation of H_2, D_2, N_2, and CO.[90,91] It appears as a sharp peak at threshold in the excitation function (Fig. 10.19). In the case of H_2 (Fig. 10.20), there is a clear contribution at higher energies from stepwise excitation, but the two contributions can be approximately separated because they produce different distributions of l states and, therefore, produce different radiative lifetimes. The only other known example of direct high-l excitation is process 1 in the dissociation of N_2 to N(HR) for which a threshold peak was interpreted in Sect. 10.2c to result from direct high-l excitation with a $C\,^2\Sigma^+$ core.

One last mechanism to consider for the production of high l is the process of molecular dissociation. As the fragments separate, the axial electric field mixes l states to form Stark states, and at large internuclear separation, the Stark states resolve back to a range of l states, including high l. We can also use a classical description in which a Rydberg electron in a position perpendicular to the axis of dissociation must, during dissociation, acquire angular momentum about the ionic core. Evidence for high-l production was found in the dissociation of CH_4 to give H(HR).[22] Another example appears to be the observation of a CH_3(HR) fragment from CH_4 and C_2H_4(HR) from C_2H_6.[38] One would expect both of these HRM fragments to predissociate were they not in high-l states.

10.3d. Multichannel quantum defect theory

In this discussion of stable HRMs, we deemphasized optical spectroscopy. One topic of recent interest cannot be omitted, however, multichannel quantum-defect theory (MQDT).[95,96] This theory has proven to yield remarkably accurate descriptions of Rydberg series even when interactions between overlapping series perturb energy levels and intensities so strongly that Rydberg series are no longer apparent in the spectrum. The theory has been applied to numerous atoms, but so far it has been applied to only four molecules: H_2, He_2, and, most recently, N_2,[97] and CH_3I.[98]

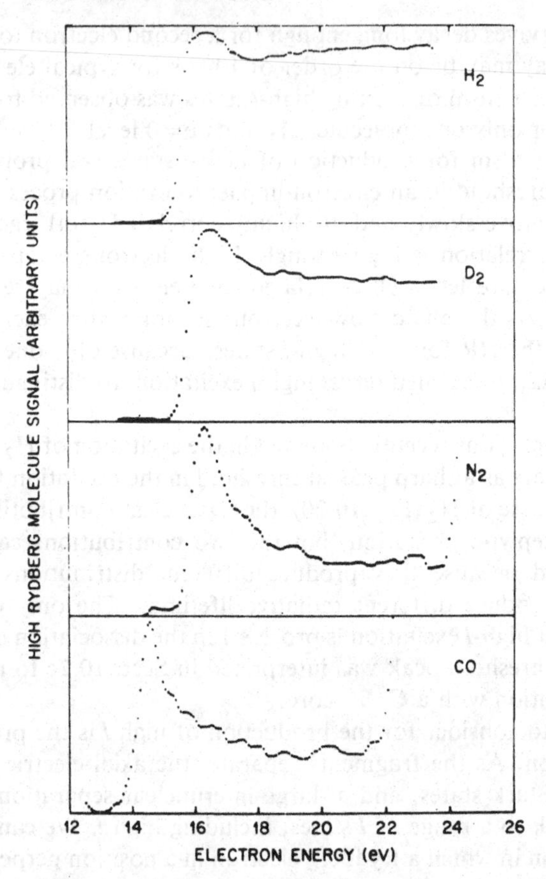

Fig. 10.19. Excitation functions for the formation of stable high-Rydberg molecules. The prominent threshold peaks are due to direct excitation of high-l electrons.

As might be expected, the Rydberg states of H_2 have been analyzed in greatest detail. The lowest states were exhaustively measured by Dieke.[99] Subsequently, Mulliken[100] presented a thorough interpretation of molecular Rydberg states, resting most heavily on H_2 data. Measurements to high n were published almost simultaneously by Herzberg,[101] Takezawa,[102,103] and Chupka and Berkowitz.[76] It was two years later that MQDT was first applied to a molecule by Herzberg and Jungen.[104] Subsequently, Jungen and Atabek,[105] Dehmer and Chupka,[106] and Jungen and Dill[107] presented further analyses of H_2(HR) by means of MQDT. The absorption spectra that have been studied involve excitation from the $(1s\sigma_g)^2$ ground state to the np Rydberg states consisting of two interacting series of + symmetry, $(1s\sigma_g, np\sigma_u)\,^1\Sigma_u^+$ and $(1s\sigma_g, np\pi_u)\,^1\Pi_u^+$ and a third noninteracting series $(1s\sigma_g, np\pi_u)\,^1\Pi_u^-$. Multichannel quantum defect theory has produced remarkably accurate analyses of

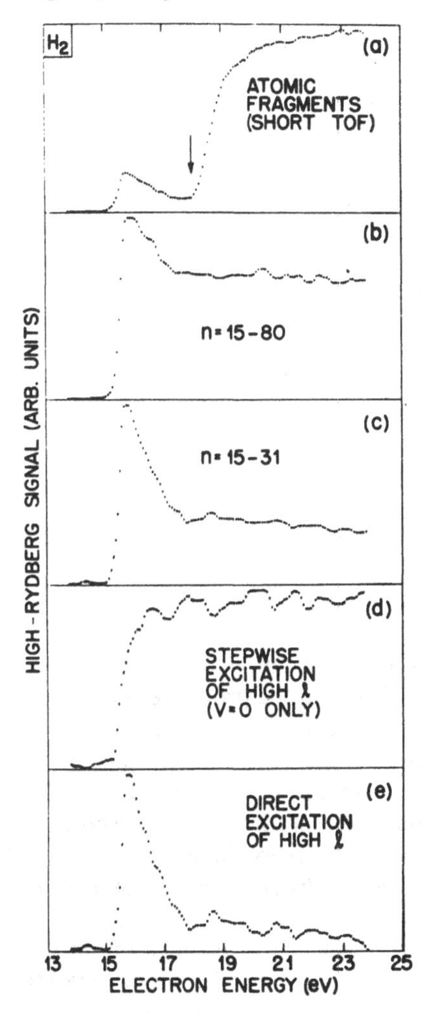

Fig. 10.20. Excitation functions for the formation of stable HR H_2 molecules showing in (*a*) the contribution from dissociation by process 1, in (*b*) and (*c*) the signals for different ranges of principal quantum numbers and therefore different lifetimes, and in (*d*) and (*e*) the separated stepwise and direct excitation functions obtained by subtraction of (*b*) and (*c*).

the two interacting + states, whereas conventional methods are awkward at best.

The He_2 molecule has also been well studied. Unlike H_2, however, its ground state is unstable. Measurements, therefore, have been by emission spectroscopy[70] or by absorption spectroscopy from a metastable state.[78] Ginter and Ginter[70] recently reported the $(np\pi)\,^3\Pi_g$ and $(np\sigma)\,^3\Sigma_g^+$ states to $n=16$ and applied MQDT to interpret the interacting Σ^+ and Π^+ levels.

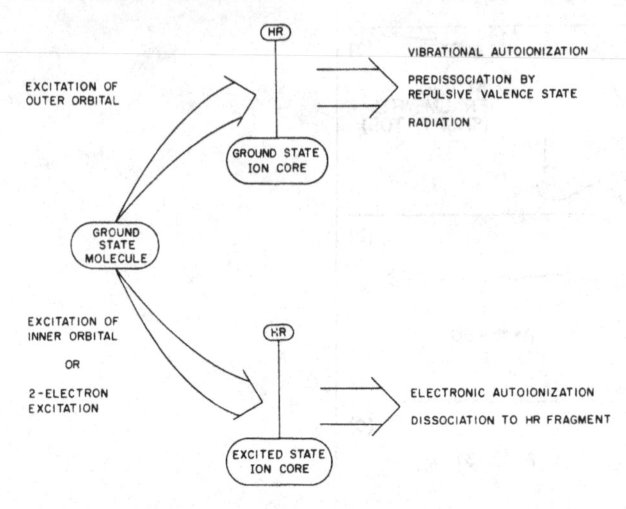

Fig. 10.21. Summary of the most prominent decay processes of HRMs following either inner or outer shell excitation.

10.4. Conclusions

This final section will summarize our understanding of high-Rydberg molecules. There are extensive similarities between the properties of HR atoms and molecules because of the small size of the core ion with respect to the size of a high-Rydberg orbital. High-Rydberg molecules display a number of distinct behaviors, however, because the core ion can vibrate, rotate, and dissociate. The more dominant processes are indicated in Fig. 10.21. The core ion model is generally very accurate in describing dissociation limits and gives good qualitative agreement between kinetic energy distributions from HR TOF and from dissociative ionization.

Although the number of molecules studied for dissociation into HR fragments is relatively small, it appears that all molecules do include HR atoms among their dissociation fragments. The remaining fragments of such processes are found in many different states and, presumably, all states occur to some degree. Several categories of dissociation processes appear prominent in the measurements: (1) Dissociation of every molecule (with one exception) has been observed to give a HR atomic fragment at the minimum possible energy, that is, the other fragment is in its ground state. (2) There is generally a cluster of dissociation limits near the lowest and with data not always clearly distinguishable. (3) Strong, often well-resolved signals correlate with limits just below the double-ionization limit. These result from dissociation to an atomic HR with the other fragment in a Rydberg state or an ionic state. (4) At the very highest excitation energies, 70–100 eV, there is usually a very strong TOF peak corresponding to high kinetic energy atoms. Although not well characterized, such peaks may result from dissociation of HR molecules with triply ionized cores.

Attempts to characterize the molecular states leading to the observed fragments have been largely unsuccessful. If a fragment resulted from dissociation by a single repulsive state, it would be possible to use the measured kinetic energy distribution to determine the potential curve. Only that portion of the curve in the Franck–Condon region would be determined, however, and available resolution could not distinguish between a purely repulsive curve and predissociation. The major practical limit has been that kinetic energy distributions from dissociation by several repulsive states invariably overlap, so any derived curve is some unknown composite.

A major result of HR dissociation work has been the accumulation of experimental data on kinetic energy distributions of fragment atoms. Direct measurements have been made by very few other methods and probably none with more accuracy. Measurements of ion kinetic energy, as made in many mass spectrometers, are prone to discrimination, either for or against slow ions, and this may be the cause of many discrepancies in Figs. 10.11, 10.17, and 10.18.

Stable HRMs have been studied less extensively than dissociative HRMs. The work that has been done suggests that any molecule with a stable ground-state ion should have a long-lived HR state, provided the orbital angular momentum is high enough. Threshold peaks in electron-impact excitation, resulting from direct excitation of high-l states, may therefore be common, although weak and difficult to observe.

Thus, a consistent description of HRMs can now be presented. To be sure, it rests on measurements of relatively few molecules. Further measurements could provide the data to test and extend this description. The increased use of synchrotron radiation and multiphoton laser methods may be especially useful in this regard. Such use of photon absorption could provide more accurate information than electron-impact excitation on the shapes and symmetries of repulsive states. Alternately, repulsive states of molecular ions might be characterized by various ion spectroscopic techniques now being developed such as absorption of laser radiation by ions in a discharge[108] or a matrix,[109] laser photofragmentation of an ion beam,[110] and photoexcitation combined with ion cyclotron resonance.[111] Finally, there is a need for more theoretical work on topics such as computations of repulsive potential energy curves, valence-Rydberg state interactions, the nature and positions of two-electron excited states, and the competition between dissociation and autoionization.

References and notes

1. S. E. Kupriyanov, *Zh. Eksp. Teor. Fiz. 48,* 467 (1965) [*Sov. Phys. JETP 21,* 311 (1965)].
2. S. E. Kupriyanov, *Zh. Eksp. Teor. Fiz. 55,* 460 (1968) [*Sov. Phys. JETP 28,* 240 (1969)].
3. R. S. Freund, *J. Chem. Phys. 54,* 3125 (1971).
4. K. C. Smyth, J. A. Schiavone, and R. S. Freund, *J. Chem. Phys. 59,* 5225 (1973).

5. A. C. Hurley and V. W. Maslen, *J. Chem. Phys. 34,* 1919 (1961).
6. A. C. Hurley, *J. Mol. Spectrosc. 9,* 18 (1962).
7. M. Misakian, M. J. Mumma, and J. F. Faris, *J. Chem. Phys. 62,* 3442 (1975).
8. W. C. Wells, W. L. Borst, and E. C. Zipf, *Phys. Rev. A 14,* 695 (1976).
9. J. A. Schiavone, D. E. Donohue, D. R. Herrick, and R. S. Freund, *Phys. Rev. A 16,* 48 (1977).
10. G. Allcock and J. W. McConkey, *J. Phys. B 9,* 2127 (1976).
11. T. Shibata, T. Fukuyama, and K. Kuchitsu, *Chem. Lett.* (*Japan*) 75 (1974).
12. H. Hiraishi, T. Kondow, T. Fukuyama, and K. Kuchitsu, *J. Phys. Soc. Japan 46,* 1628 (1979).
13. V. Čermák and Z. Herman, *Coll. Czech. Chem. Commun. 29,* 953 (1964).
14. C. E. Fairchild, H. P. Garg, and C. E. Johnson, *Phys. Rev. A 8,* 796 (1973).
15. J. A. Schiavone, *J. Chem. Phys. 70,* 2236 (1979).
16. W. C. Wells, W. L. Borst, and E. C. Zipf, *Phys. Rev. A 17,* 1357 (1978).
17. W. L. Borst and E. C. Zipf, *Phys. Rev. A 4,* 153 (1971).
18. G. Allcock and J. W. McConkey, *Chem. Phys. 34,* 169 (1978).
19. G. Allcock and J. W. McConkey, *J. Phys. B 11,* 741 (1978).
20. R. J. Van Brunt, *Phys. Rev. A 16,* 1309 (1977).
21. J. A. Schiavone, S. M. Tarr, and R. S. Freund, *Phys. Rev. A 20,* 71 (1979).
22. J. A. Schiavone, S. M. Tarr, and R. S. Freund, *J. Chem. Phys. 70,* 4468 (1979).
23. T. Shibata, T. Fukuyama, and K. Kuchitsu, *Bull. Chem. Soc. Japan 47,* 2883 (1974).
24. J. A. Schiavone, D. E. Donohue, and R. S. Freund, *J. Chem. Phys. 67,* 759 (1977).
25. J. Appell and C. Kubach, *Chem. Phys. Lett. 11,* 486 (1971).
26. K. Köllman, *J. Phys. B 11,* 339 (1978).
27. R. Locht, J. L. Oliver, and J. Momigny, *Chem. Phys. 43,* 425 (1979).
28. R. Locht and J. Momigny, *Chem. Phys. 49,* 173 (1980).
29. D. W. O. Heddle, *Proc. Phys. Soc. London 90,* 81 (1967).
30. S. E. Kupriyanov, *Khimiya Vysokikh Energii 2,* 113 (1968) [*High-Energy Chem. 2,* 94 (1968)].
31. S. E. Kupriyanov, *Khimiya Vysokikh Energii 3,* 17 (1969) [*High-Energy Chem. 3,* 13 (1969)].
32. S. E. Kupriyanov and A. A. Perov, *Khimiya Vysokikh Energii 3,* 306 (1969) [*High-Energy Chem. 3,* 278 (1969)].
33. S. E. Kupriyanov, *Khimiya Vysokikh Energii 4,* 304 (1972) [*High-Energy Chem. 6,* 266 (1973)].
34. J. T. Tate and W. W. Lozier, *Phys. Rev. 39,* 254 (1932).
35. L. J. Kieffer and R. J. Van Brunt, *J. Chem. Phys. 46,* 2728 (1967).
36. R. Locht, J. Schopman, H. Wankenne, and J. Momigny, *Chem. Phys. 7,* 393 (1975).
37. L. Deleanu and J. A. D. Stockdale, *J. Chem. Phys. 63,* 3898 (1975).
38. S. M. Tarr, J. A. Schiavone, and R. S. Freund, unpublished.
39. T. G. Finn, B. L. Carnahan, W. C. Wells, and E. C. Zipf, *J. Chem. Phys. 63,* 1596 (1975).
40. R. Fuchs and R. Taubert, *Z. Naturforsch., Teil A 19,* 494 (1964).
41. R. J. Van Brunt, G. M. Lawrence, L. J. Kieffer, and J. M. Slater, *J. Chem. Phys. 61,* 2032 (1974).
42. J. Schopman and R. Locht, Proceedings, VIII ICPEAC, Belgrade (1973), p. 424.
43. J. A. D. Stockdale and L. Deleanu, *Chem. Phys. Lett. 22,* 204 (1973).

44. J. A. Schiavone, K. C. Smyth, and R. S. Freund, *J. Chem. Phys. 63,* 1043 (1975).
45. B. L. Carnahan and E. C. Zipf, *Phys. Rev. A 16,* 991 (1977).
46. W. W. Lozier, *Phys. Rev. 36,* 1285 (1930).
47. B. van Wingerden, Ph. E. van der Leeuw, F. J. de Heer, and M. J. van der Wiel, *J. Phys. B 12,* 1559 (1979).
48. M. Glass-Maujean, K. Köllmann, and K. Ito, *J. Phys. B 12,* L453 (1979).
49. M. D. Burrows, L. C. McIntyre, Jr., S. R. Ryan, and W. E. Lamb, Jr., *Phys. Rev. A 21,* 1841 (1980).
50. J. P. Johnson and J. L. Franklin, *Int. J. Mass. Spectrosc. Ion Phys. 33,* 393 (1980).
51. C. Bottcher and K. Docken, *J. Phys. B 7,* L5 (1974).
52. C. Bottcher, *J. Phys. B 9,* 2899 (1976).
53. A. U. Hazi, *Chem. Phys. Lett. 25,* 259 (1974); *J. Chem. Phys. 60,* 4358 (1974).
54. M. Landau, R. I. Hall, and F. Pichou, *J. Phys. B 14,* 1509 (1981).
55. T. Sugiura and K. Arakawa, in *Recent Developments in Mass Spectroscopy,* eds. K. Ogata and T. Hayakawa (University of Tokyo Press, 1970), p. 848.
56. T. Shibata, T. Fukuyama, and K. Kuchitsu, *Bull. Chem. Soc. Japan 47,* 2573 (1974).
57. J. A. Stockdale, F. J. Davis, R. N. Compton, and C. E. Klots, *J. Chem. Phys. 60,* 4279 (1974).
58. K. C. Smyth, J. A. Schiavone, and R. S. Freund, *J. Chem. Phys. 60,* 1358 (1974).
59. R. J. Van Brunt and M. J. Mumma, *J. Chem. Phys. 63,* 3210 (1975).
60. W. C. Wells and E. C. Zipf, *J. Chem. Phys. 66,* 5828 (1977).
61. B. L. Carnahan, W.-W. Kao, and E. C. Zipf, *J. Chem. Phys. 74,* 5149 (1981).
62. R. S. Berry, *Rec. Chem. Prog. 31,* 9 (1970).
63. U. Fano, C.N.R.S. Colloque International No. 273, 1977.
64. E. Miescher and K. P. Huber, in *International Review of Science, Physical Chemistry,* ser. 2, vol. 3: *Spectroscopy,* ed. D. A. Ramsay (London: Butterworth, 1976), p. 37.
65. A. Lofthus and P. H. Krupenie, *J. Phys. Chem. Ref. Data 6,* 113 (1977).
66. R. J. Buenker and S. D. Peyerimhoff, *Chem. Phys. 8,* 324 (1975).
67. R. J. Buenker and S. D. Peyerimhoff, *Chem. Phys. Lett. 34,* 225 (1975).
68. R. J. Buenker and S. D. Peyerimhoff, *Chem. Phys. Lett. 36,* 415 (1975).
69. R. P. Saxon and B. Liu, *J. Chem. Phys. 73,* 870, 876 (1980).
70. D. S. Ginter and M. L. Ginter, *J. Mol. Spectrosc. 82,* 152 (1980).
71. A. B. F. Duncan, *Rydberg Series in Atoms and Molecules* (New York: Academic Press, 1971).
72. M. B. Robin, *Higher Excited States of Polyatomic Molecules* (New York: Academic Press, 1974).
73. J. P. Byrne and I. G. Ross, *Aust. J. Chem. 24,* 1107 (1971).
74. C. E. Klots, *J. Chem. Phys. 62,* 741 (1975).
75. J. C. Person, R. L. Watkins, and D. L. Howards, *J. Phys. B 9,* 1811 (1976).
76. W. A. Chupka and J. Berkowitz, *J. Chem. Phys. 51,* 4244 (1969).
77. I. Koyano, H. Yamada, and I. Tanaka, *Int. J. Mass. Spectrosc. Ion Phys. 21,* 113 (1976).
78. R. Panock, R. R. Freeman, R. H. Storz, and T. A. Miller, *Chem. Phys. Lett. 74,* 203 (1980).
79. A. Tramer, *J. Chim. Phys. Phys.-Chim. Biol. 77,* 41 (1980).
80. S. E. Kupriyanov, Proceedings, V ICPEAC, Leningrad (1967), p. 571.
81. H. Hotop and A. Niehaus, *J. Chem. Phys. 47,* 2506 (1967).

82. H. Hotop, F. W. Lampe, and A. Niehaus, *J. Chem. Phys. 51,* 593 (1969).
83. C. F. Barnett and J. A. Ray, Proceedings, IV ICPEAC, Quebec (1965), p. 250.
84. C. F. Barnett, J. A. Ray, R. A. Langley, and A. Russek, Proceedings, V ICPEAC, Leningrad (1967), p. 17.
85. C. F. Barnett, J. A. Ray, and A. Russek, *Phys. Rev. A 5,* 2110 (1972).
86. C. F. Barnett and J. A. Ray, *Phys. Rev. A 5,* 2120 (1972).
87. T. J. Morgan, C. F. Barnett, J. A. Ray, and A. Russek, *Phys. Rev. A 20,* 1062 (1979).
88. E. S. Solov'ev, R. N. Il'in, V. A. Oparin, and N. V. Fedorenko, *Zh. Eksp. Teor. Fiz. 53,* 1933 (1967) [*Sov. Phys. JETP 26,* 1097 (1968)].
89. H. Hiraishi, T. Kondow, T. Fukuyama, and K. Kuchitsu, *Chem. Phys. Lett. 66,* 9 (1979).
90. S. M. Tarr, J. A. Schiavone, and R. S. Freund, *Phys. Rev. Lett. 44,* 1660 (1980).
91. S. M. Tarr, J. A. Schiavone, and R. S. Freund, *J. Chem. Phys. 74,* 2869 (1981).
92. W. A. Chupka, M. E. Russell, and K. Refaey, *J. Chem. Phys. 48,* 1518 (1968).
93. Y. B. Band, *J. Phys. B 7,* 2072 (1974).
94. U. Fano, *J. Phys. B 7,* L401 (1974).
95. K. T. Lu and U. Fano, *Phys. Rev. A 2,* 81 (1970).
96. U. Fano, *Phys. Rev. A 2,* 353 (1970).
97. A. Giusti-Suzor and H. Lefebvre-Brion, *Chem. Phys. Lett. 76,* 132 (1980).
98. J. A. Dagata, G. L. Findley, S. P. McGlynn, J.-P. Connerade, and M. A. Baig, *Phys. Rev. A 34,* 2485 (1981).
99. G. H. Dieke, *J. Mol. Spectrosc. 2,* 494 (1958).
100. R. S. Mulliken, *J. Am. Chem. Soc. 86,* 3183 (1964); *88,* 1849 (1966); *91,* 4615 (1969).
101. G. Herzberg, *Phys. Rev. Lett. 23,* 1081 (1969).
102. S. Takezawa, *J. Chem. Phys. 52,* 2575 (1970).
103. S. Takezawa and Y. Tanaka, *J. Chem. Phys. 56,* 6125 (1972).
104. G. Herzberg and Ch. Jungen, *J. Mol. Spectrosc. 41,* 425 (1972).
105. Ch. Jungen and O. Atabek, *J. Chem. Phys. 66,* 5584 (1977).
106. P. M. Dehmer and W. A. Chupka, *J. Chem. Phys. 65,* 2243 (1976).
107. Ch. Jungen and D. Dill, *J. Chem. Phys. 73,* 3338 (1980).
108. T. A. Miller and V. E. Bondybey, *J. Chim. Phys. 77,* 695 (1980).
109. V. E. Bondybey, T. A. Miller, and J. H. English, *J. Chem. Phys. 72,* 2193 (1980).
110. J. Moseley and J. Durup, *Ann. Rev. Phys. Chem. 32,* 53 (1981).
111. R. C. Dunbar, in *Mass Spectroscopy,* ed. R. J. W. Johnstone (London: Chemical Society, 1981), vol. VI.

11

Theory of Rydberg collisions with electrons, ions, and neutrals

M. R. FLANNERY

11.1. Classification of treatments for Rydberg collisions

In this chapter we shall present theory suitable for the description of collisions between Rydberg atoms $B(n)$ in quantum state n and electrons, ions, and neutrals denoted, in general, by A, with detailed emphasis on electron impact. The various methods fall quite naturally into three broad, overlapping, classifications - quantal, semiquantal-semiclassical, and classical - each characterized by certain properties peculiar to the collision and by certain validity criteria satisfied by the particular process under consideration. Various quantal-classical combinations within the continuous classification can be used to describe both the relative A-$B(n)$ *external* motion in the continuum and the *internal* bounded motion of the Rydberg electron e attached to its parent core B^+ in state n, such that the overall treatment is hybrid in nature.

The term *semiclassical* is used in the sense of the JWKB approximation to external and internal wave functions, and *semiquantal* refers to an impulse treatment wherein the Rydberg electron-projectile A scattering is described by full quantal scattering technology and the associated energy and momentum changes are prescribed by classical mechanics. This semiquantal method[1] is equivalent[2] to the full quantal impulse approximation with a plane-wave for the final state of the Rydberg electron.

As n is increased, the Rydberg electron eventually behaves as a classical particle in the sense that it becomes increasingly localized in phase space, where the quantal imprecisions Δr_n and Δv_n associated with its position r_n relative to the core B^+ and with its orbital speed v_n about the core are $\ll r_n$ and v_n, respectively. Quantal characteristics remain as exhibited, for example, by radio-frequency line emissions between neighboring levels,[3] with n as high as ~ 250. Because electronic quantal wave functions for bound excited states tend, in the limit of high n, to semiclassical JWKB wave functions, which lead naturally to the Bohr-Sommerfeld quantization rule and hence to the Bohr correspondence principle for line emission and absorption between highly excited levels (Sects. 11.4a and 11.4b), the essential quantum mechanics underlying internal motion is preserved and the correct quantal-classical connection or correspondence is provided by semiclassical internal wave functions. Also, for heavy-particle collisions at impact energies above the well depth of A–B(n) attraction and for high-energy electron impact, the more familiar semiclassical description of external relative motion is accurate. Cross sections, transition amplitudes, etc., require knowledge of matrix elements M_{ij} of some kernel M (such as the scattering operator and electrostatic interaction, for example) averaged over electronic wave functions for a given (i,j) pair of excited states i and j. Direct quantal evaluation of M_{ij} is laborious because each highly oscillatory quantal wave function (with $\sim n$ nodes for each direction) requires specification at more than n^3 points and is, apart from the end result, unrewarding. Semiclassical wave functions not only expose much beauty in the quantal–classical connection (Sect. 11.4e) but also lead directly to the Heisenberg correspondence principle (Sect. 11.4b), which expresses any general M_{ij} as the sth Fourier component ($s=|i-j|\ll i,j$) of M evaluated along a mean internal classical orbit $\mathbf{r}(t)$ for the Rydberg electron. This principle represents a very efficient and powerful method for evaluation of matrix elements with $i,j\gg s$.

Many kinds of A–B(n) collisions involve strong coupling between many states strongly connected to the initial and final states under consideration such that perturbation-based full quantal procedures developed originally with ground-state B in mind are, in general, unrealistic, except in the weak-coupling limit. For collisional transitions between strongly coupled neighboring or adjacent, equally spaced levels, the probability amplitudes for $n\rightarrow n'$ transitions can be exactly determined,[4-6] with the aid of Heisenberg's correspondence principle (or semiclassical internal wave functions) within a semiclassical description of the collision, to give what is termed the *equivalent oscillator theorem* (Sects. 11.4d and 11.4e) or, alternatively, the *strong-coupling correspondence principle.*[5] Although its one- and three-dimensional forms can be derived from several related avenues,[4-7] a new, direct, and rigorous derivation based on the action-angle representation will be presented in Sect. 11.4d. These amplitudes may then be inserted into an appropriate treatment (such as the semiclassical multichannel eikonal treatment[8] or the classical path treatment given in Sect. 11.3) for the collision cross section. Overall, semiclassical methods properly developed for Rydberg collisions are not alternatives to full quantal procedures; rather they provide the most efficient methods for application of quantum mechanics to A–B(n) collisions.

Application of purely quantal methods would prove too cumbersome in that the density of states for transitions between excited levels is too large for individual attention to each state. Only for scattering by a dipole in the weak-coupling limit can summation over the n^2 degenerate states associated with a given hydrogenic level n be carried out analytically. The standard quantal close-coupling (QCC) method, based on a perturbation expansion, is useful only when a limited small number of states are accessible (as in collisions that mix only the angular momentum states l of a given n). Also the standard difficulties associated with the convergence of partial cross sections for each partial angular momentum L wave of relative motion would be considerably amplified over those evident for processes with atoms initially in ground states because, in e–B(n) collisions, the effect of the induced dipole is very strong and long range, and the cross section is, in general, dominated by scattering in the forward direction. This difficulty is greatly reduced in an impact-parameter representation of the scattering amplitude, as in the multichannel eikonal treatment[8] such that convergence is rapidly obtained. Use of a restricted atomic basis set, however, limits close-coupling procedures (whether quantal QCC or semiclassical SCC) to cases of weak coupling with the neglected states, a circumstance that becomes increasingly rare even for moderate values of n. For high n, the Rydberg target behaves more like two distinct scattering centers such that expansion in terms of functions that describe external relative motion distorted by the averaged atomic field is not the most effective or even desirable procedure for a limited basis set. For example, a product of Coulomb functions centered at the Rydberg electron e and its core B^+ is obviously a better zero-order choice for relative motion in ion or electron impact than is a plane wave distorted by the averaged atomic field.

A full quantal method more effective than any close-coupling (CC) procedure, QCC or SCC, in that it is not based on a perturbation expansion, has been developed specifically for A–B(n) collisions.[9] Here A moves in a field given, not by the static interaction averaged over the internal electronic motion, the basis of CC, but rather by the superposition of two strong (A–e and A–B^+) fields that remain essentially instantaneous when the speed v_n of the Rydberg electron about B^+ is very much less than the speed v_{AB} of relative motion of the projectile A. This combination reduces to the averaged field only for distant encounters (a weak-coupling situation). When the effect of the Rydberg core is neglected, this "fixed-center" quantal method (QFC) reduces (Sect. 11.2) to the quantal and semiquantal (or plane-wave) impulse approximations[2] and then to the Born first-order result in the limit of weak coupling. The classical binary-encounter method for charged-particle–atom collisions can then be deduced (Sect. 11.5) by assuming a plane-wave final electronic state and the Rutherford cross section for projectile–Rydberg electron scattering and by neglecting the momentum distribution of the Rydberg electron.

Collisional transitions between neighboring levels ($n, n' \gg |n-n'|$) demand strong coupling theories, such as the equivalent oscillator theorem (EOT) for the $n \rightarrow n'$ transition amplitudes within a semiclassical description of the cross section or the quantal fixed-center treatment for $nlm \rightarrow n'l'm'$ transitions. For adiabatic collisions (distant encounters) and sudden collisions (close encounters), EOT reduces to the Born and sudden approximations, respectively, to be used where and when appropriate.

For $n' \gg n$ and for ionization, the quantal and semiquantal impulse treatments[2] (Sect. 11.5) and the pure classical binary-encounter method are valid, provided that the energy and momentum changes are much greater than the energy and momentum imparted to the Rydberg electron by its parent core B^+

during the time of close collision. Classical Monte Carlo computer simulations,[10,11] based on a solution of the equations of motion for the three-body system, are effective and reliable for ionization.

Various classical-quantal correspondences and theories of charged-particle-Rydberg collisions were already discussed by Percival and Richards.[12] In this chapter we shall attempt to provide a comprehensive and unified formulation of collision treatments suitable for Rydberg collisions with electrons, ions, and neutrals. In so doing, new material will be developed and conditions for validity of each method will become more apparent. It will also become apparent that considerable power lies in solution of the problem in either the action-angle variable representation of quantum mechanics or the (perturbed) Hamilton–Jacobi equation of classical mechanics for the action S so as to yield a semiclassical wave function with its automatic built-in quantization, which is very natural for Rydberg states. The former procedure is very akin to the Heisenberg matrix formulation of quantum mechanics,[13,14] and the first part of the latter procedure awakens interest in various perturbative techniques,[15,16] based on the Hamilton–Jacobi equation, which were highly developed in connection with problems in celestial mechanics and which remained dormant after the introduction of quantum mechanics. The atomic bound problem[15] is simpler, in that only a discrete number of motions is selected by quantization from the whole continuum of classical solutions, and yet more difficult, in that the replusion between two electrons can be greater than or comparable to the force of attraction with the core, whereas the perturbation between two planets is extremely small in comparison to attraction with the sun.

11.2. Quantum theory of Rydberg collisions

11.2a. Close-coupling procedures

The wave function $\Psi(\mathbf{r},\mathbf{R})$ for the scattering of a Rydberg atom A and some atomic or molecular species B by their mutual interaction $V(\mathbf{r},\mathbf{R})$ at relative separation $\mathbf{R}$ and internal electronic coordinates $\mathbf{r}$ is, in their c.m. reference frame, the appropriate outgoing-wave solution of the stationary-state Schrödinger equation,

$$H\Psi \equiv \left[-\frac{\hbar^2}{2\mu}\nabla_{\mathbf{R}}^2 + H_0(\mathbf{r}) + V(\mathbf{r},\mathbf{R})\right]\Psi(\mathbf{r},\mathbf{R}) = E\Psi(\mathbf{r},\mathbf{R}) \tag{1}$$

where μ is the reduced mass and E the total energy of the collision system. The internal structure of the isolated species with internal coordinates collectively denoted by $\mathbf{r}$ is described by the Hamiltonian H_0 with eigenenergies ϵ_n and eigenfunctions $\{\phi_n\}$ assumed as known solutions of

$$H_0(\mathbf{r})\phi_n(\mathbf{r}) = \epsilon_n\phi_n(\mathbf{r}) \tag{2}$$

such that, in terms of the wave number k_n for relative motion, the total energy

$$E = \epsilon_n + (\hbar^2 k_n^2 / 2\mu) \tag{3}$$

is constant for all channels n. The scattering solution of Eq. (1) can be expanded in terms of $\{\phi_n\}$ as

$$\Psi(\mathbf{r}, \mathbf{R}) = \mathop{\mathrm{S}}_{n} F_n(\mathbf{R}) \phi_n(\mathbf{r}) \tag{4}$$

where the unknown F_n for the relative motion can be shown, with the aid of (2) and (3), to satisfy the standard set of coupled, three-dimensional, differential equations

$$[\nabla_{\mathbf{R}}^2 + K_f^2(\mathbf{R})] F_f(\mathbf{R}) = (2\mu/\hbar^2) \mathop{\mathrm{S}}_{n \neq f} V_{fn}(\mathbf{R}) F_n(\mathbf{R}) \tag{5}$$

where

$$K_f(\mathbf{R}) = [k_f^2 - (2\mu/\hbar^2) V_{ff}(\mathbf{R})]^{1/2} \tag{6}$$

are the *local* wave numbers for relative motion under the static interaction $V_{ff}(R)$, the diagonal elements of the matrix

$$V_{fi}(\mathbf{R}) = \langle \phi_f(\mathbf{r}) | V(\mathbf{r}, \mathbf{R}) | \phi_i(\mathbf{r}) \rangle_{\mathbf{r}} \tag{7}$$

The usual quantal procedure converts (5) via a partial-wave analysis to analogous sets of coupled one-dimensional equations that require solution for each conserved angular momentum quantum number L, the vector sum of internal and relative motions. This procedure is the standard quantal close-coupling method (QCC), exact in principle but limited in practice to situations that involve only a small number of internal states n and values $L \lesssim 10$–20, the number depending on both the impact energy and effective range of interaction. When the basis set is, however, limited to all $|lm\rangle$ angular momentum states appropriate to principal quantum number n remaining fixed, then the further quantal development from Eq. (5) by Arthurs and Dalgarno[17] for collisional-rotational transitions in molecules is directly applicable to $nl \rightarrow nl'$ transitions in Rydberg atoms (see Chap. 6) and is feasible at low thermal energies.

The general application of Eq. (5) when decomposed to the corresponding set of one-dimensional equations is plagued by slow convergence of the partial-wave expansion and becomes prohibitively difficult particularly for higher-energy Rydberg collisions involving changes in both n and l. Semiclassical close-coupling methods[8, 18] (multichannel eikonal and orbital treatments in Sect. 11.3), based on initial approximation of the three-dimensional set of Eq. (5) to a set that can then be solved exactly in three dimensions without recourse to angular momentum decomposition of the relative motion, are extremely powerful for Rydberg collisions. Here, convergence of transition amplitudes in an "impact-parameter" representation (not necessarily purely classical) is more rapidly achieved than in the angular momentum representation, and the methods can be applied to situations prohibitive to a pure QCC treatment, e.g., Rydberg collisions in general, particularly at higher impact energies.

Because electronic wave functions ϕ_n for higher n contain many oscillations (with n^3 nodes in three dimensions), direct evaluation of those matrix elements in Eq. (7) that enter in quantal and semiclassical close-coupling treatments is, in general, not feasible or even recommended. More effective methods (Sect. 11.4) based on the Heisenberg correspondence principle or on the use of semiclassical (hydrogenic) wave functions, represent powerful techniques.

However, an obvious disadvantage, i.e., the limited extent of the basis set (4) taken in practice, remains in all close-coupling treatments, whether quantal or semiclassical, of Rydberg collisions. With the aid of the Bohr and Heisenberg correspondences (Sect. 11.4b) for high-Rydberg atoms, transition amplitudes can, however, be obtained exactly in the semiclassical version under certain conditions. This exact solution is designated here as an equivalent oscillator theorem (Sect. 11.4d).

A second quantal approach, which bears no obvious relationship to close-coupling methods characterized by (5)–(7), has been developed specifically for Rydberg collisions and will now be outlined.[9]

11.2b. Scattering by a fixed two-centered Rydberg target

A useful quantal treatment specifically designed for collisions involving a range of excited states has been developed from the operator formalism of scattering to allow greater transparency than (5) to the inclusion of various important effects and to permit construction of the resulting expressions in a form suitable for effective and interesting approximations.[9] The Lippmann-Schwinger operator equation describing the outgoing scattering of two atomic collision partners by their mutual interaction V is, in terms of the Green's resolvent G and transition operator T for the collision, given by

$$\Psi_i^+ = \Phi_i + G_0^+ V \Psi_i^+ = \Phi_i + G^+ V \Phi_i = \Phi_i + G_0^+ T \Phi_i \tag{8}$$

in the center-of-mass system. The solution Ψ_i^+ of (1) with appropriate outgoing, spherically scattered waves can be written as

$$\Psi_i^+(\mathbf{r},\mathbf{R}) = \phi_i(\mathbf{r}) \exp(i\mathbf{k}_i \cdot \mathbf{R}) + \iint d\mathbf{r}'\, d\mathbf{R}'\, G_0^+(\mathbf{r},\mathbf{R};\mathbf{r}',\mathbf{R}') V(\mathbf{r}',\mathbf{R}) \Psi_i^+(\mathbf{r}',\mathbf{R}') \tag{9}$$

The Green's function G_0^+ satisfies

$$(E_i - \mathcal{H}_0 + i\epsilon) G_0^+(\mathbf{r},\mathbf{R};\mathbf{r}',\mathbf{R}') = \delta(\mathbf{r}-\mathbf{r}')\delta(\mathbf{R}-\mathbf{R}') \tag{10}$$

and can therefore be expanded in terms of the complete set

$$\Phi_i(\mathbf{r},\mathbf{R}) = \phi_i(\mathbf{r})\exp(i\mathbf{k}_i\cdot\mathbf{R}) \tag{11}$$

which are free-particle eigenfunctions of $\mathcal{H}_0$ $(\equiv H-V)$, i.e., (1) in the absence of the interaction V, to give

$$G_0^+(\mathbf{r},\mathbf{R};\mathbf{r}',\mathbf{R}') = -\frac{1}{4\pi}\frac{2\mu}{\hbar^2}\,\underset{n}{\mathrm{S}}\,\frac{\exp(ik_n|\mathbf{R}-\mathbf{R}'|)}{|\mathbf{R}-\mathbf{R}'|}\phi_n^*(\mathbf{r})\phi_n(\mathbf{r}') \tag{12}$$

By considering the asymptotic $(R\to\infty)$ behavior of (12) in (9), the transition matrix T and associated scattering amplitude f_{if} can be defined with elements

$$\begin{aligned} T_{fi} = \langle\phi_f|T|\phi_i\rangle &= \langle\phi_f(\mathbf{r})\exp(i\mathbf{k}_f\cdot\mathbf{R})|V(\mathbf{r},\mathbf{R})|\Psi_i^+(\mathbf{r},\mathbf{R})\rangle \\ &= -(4\pi\hbar^2/2\mu)f_{if}(\mathbf{k}_i,\mathbf{k}_f) \end{aligned} \tag{13}$$

Because exact solutions to (8) for use in (13) do not, in general, exist, various methods for constructing the T matrix (or Ψ_i^+ and G^+), based on a perturbation expansion in the interaction V (assumed weak), give rise to close-coupling schemes (Sect. 11.2a) that are exact in principle but limited in practice to a restricted basis set. There is, however, another alternative that involves the approximation of G_0^+ in (12) with respect to k_n, rather than G^+ in (8) with respect to V. In Rydberg collisions at intermediate and high impact energies, for example, it is a good approximation to write $k_n \simeq k_i$ in (12), which reduces, with the aid of the closure formula

$$\underset{n}{\mathrm{S}}\,\phi_n^*(\mathbf{r})\phi_n(\mathbf{r}') = \delta(\mathbf{r}-\mathbf{r}') \tag{14}$$

for the complete set of internal target states to

$$G_0^+(\mathbf{r},\mathbf{R};\mathbf{r}',\mathbf{R}') = -\frac{1}{4\pi}\frac{2\mu}{\hbar^2}\frac{\exp(ik_i|\mathbf{R}-\mathbf{R}'|)}{|\mathbf{R}-\mathbf{R}'|}\delta(\mathbf{r}-\mathbf{r}') \tag{15}$$

with the result that the integration over $\mathbf{r}'$ in (9) can be performed. Thus this closure approximation, *valid when the energy of the incident projectile is large compared with the internal energy-level spacings of the A–B system,* replaces the many-particle Green's function, in effect by the free-particle Green's function, with the result that the total scattering function is

$$\begin{aligned} \Psi_i^+(\mathbf{r},\mathbf{R}) &= \phi_i(\mathbf{r})\exp(i\mathbf{k}_i\cdot\mathbf{R}) \\ &\quad -\frac{1}{4\pi}\frac{2\mu}{\hbar^2}\int d\mathbf{R}'\frac{\exp(ik_i|\mathbf{R}-\mathbf{R}'|)}{|\mathbf{R}-\mathbf{R}'|}V(\mathbf{r},\mathbf{R}')\Psi_i^+(\mathbf{r},\mathbf{R}') \end{aligned} \tag{16}$$

a form that suggests the following substitution:

$$\Psi_i^+(\mathbf{r},\mathbf{R}) = \phi_i(\mathbf{r})\chi_i^+(\mathbf{r},\mathbf{R}) \tag{17}$$

where the new function χ_i^+ satisfies the integral equation

$$\chi_i^+(\mathbf{r},\mathbf{R}) = \exp(i\mathbf{k}_i\cdot\mathbf{R}) - \frac{1}{4\pi}\frac{2\mu}{\hbar^2}\int d\mathbf{R}'\frac{\exp(ik_i|\mathbf{R}-\mathbf{R}'|)}{|\mathbf{R}-\mathbf{R}'|}V(\mathbf{r},\mathbf{R}')\chi_i^+(\mathbf{r},\mathbf{R}') \tag{18}$$

This equation describes the potential scattering of a fictitious projectile of original wave number k_i by a fixed, multicentered, electrostatic interaction $V(\mathbf{r},\mathbf{R})$. The transition matrix (13) for the A–B collision is therefore

$$T_{fi} = \langle\phi_f(\mathbf{r})\exp(i\mathbf{k}_f\cdot\mathbf{R})|V(\mathbf{r},\mathbf{R})|\phi_i(\mathbf{r})\chi_i^+(\mathbf{r},\mathbf{R})\rangle \tag{19}$$

which may be alternatively written as

$$T_{fi} = \langle\phi_f(\mathbf{r})|T_e(\mathbf{k}_i,\mathbf{k}_f;\mathbf{r})|\phi_i(\mathbf{r})\rangle_{\mathbf{r}} \tag{20a}$$

where

$$T_e(\mathbf{k}_i,\mathbf{k}_f;\mathbf{r}) = \langle\exp(i\mathbf{k}_f\cdot\mathbf{R})|V(\mathbf{r},\mathbf{R})|\chi_i^+(\mathbf{r},\mathbf{R})\rangle_{\mathbf{R}} \tag{20b}$$

is the T matrix for scattering by the fixed structureless potential $V(\mathbf{r},\mathbf{R})$ and is evaluated both on $(k_i=k_f)$ and off $(k_i\neq k_f)$ the energy shell. Thus (20a) emphasizes directly the unique role of elastic scattering in inelastic collisions and involves, as the only unknown (20b) or, alternatively, the full solution to the equation

$$[-(\hbar^2/2\mu)\nabla_{\mathbf{R}}^2 + V(\mathbf{r},\mathbf{R})]\chi_i^+(\mathbf{r},\mathbf{R}) = E_i\chi_i^+(\mathbf{r},\mathbf{R}), \qquad E_i = \hbar^2k_i^2/2\mu \tag{20c}$$

subject to the usual outgoing scattering condition. Note that all the information obtained in general by numerical integration of (20c) is used in T_{fi}. The scheme is therefore efficient in that the work entailed in the solution χ_i^+ to (20b) is not redundant, as opposed to other methods based on perturbation series for which a solution is integrated out from the origin in an effort to obtain only its asymptotic behavior. The full knowledge of χ_i^+ for all $\mathbf{R}$ is, of course, associated with the fact that the full T matrix (20b), with elements on and off the energy shell, is required.[9] Moreover, once $\chi_i^+(\mathbf{r},\mathbf{R})$ is obtained for a given scattering system, then it is preset for examination of all transitions within the system, i.e., χ_i^+ needs to be determined only once over the effective $(\mathbf{r},\mathbf{R})$ range of a given system.

Thus the inelastic scattering of composite structures is reduced to the solution of elastic scattering by fixed centers of potential, which are, in general, multiple. For A–B(n) collisions,

$$V(\mathbf{r},\mathbf{R}) = V(\mathbf{r}_1,\mathbf{r}_3) = V_{13}(\mathbf{r}_{13}) + V_{C3}\left(\mathbf{r}_3 + \frac{M_1}{M_1+M_2}\mathbf{r}_1\right) \tag{21}$$

where the Rydberg electron, its core C, and the projectile are denoted by 1, 2, and 3, respectively, and the position vector of each particle $\mathbf{r}_i$ is relative to the center of mass, which may be taken as the core of the Rydberg atom, except when electronic transitions arise from A–C encounters (see Sect. 11.5).

A full hierarchy of approximations based on the quantal fixed-center treatment (20) has been constructed for Rydberg collisions, of which only a few will be outlined next.[9]

11.2c. Born, Coulomb fixed-center, and impulse treatments

Born treatment

In the limit of zero V in (20c), χ_i^+ tends to a plane-wave $\exp(i\mathbf{k}_i \cdot \mathbf{R})$ and (20a) reduces, for inelastic transitions, to the Born approximation written as

$$T_{fi}^{(B)}(\mathbf{K}) = F_{fi}(\mathbf{K})\, T_{el}^{(B)}(\mathbf{K}), \qquad \mathbf{K} = \mathbf{k}_i - \mathbf{k}_f \tag{22}$$

which is a product of the Born T-matrix element

$$T_{el}^{(B)}(\mathbf{K}) = \int V_{13}(\mathbf{r}_{13}) \exp(i\mathbf{K} \cdot \mathbf{r}_{13})\, d\mathbf{r}_{13} \tag{23}$$

for potential scattering by V_{13} and of

$$F_{fi}(\mathbf{K}) = \langle \phi_f(\mathbf{r}) | \exp(i\mathbf{K} \cdot \mathbf{r}) | \phi_i(\mathbf{r}) \rangle \tag{24}$$

the inelastic form factor for $i \to f$ transitions in the Rydberg atom A. This product (22) is written to emphasize the underlying role in inelastic collisions of elastic collisions, characterized by $T_{el}^{(B)}$, between the projectile and the Rydberg electron. It can also be recast in the usual and less revealing way as

$$\begin{aligned} T_{fi}^{(B)} &= \langle \phi_f(\mathbf{r}) \exp(i\mathbf{k}_f \cdot \mathbf{R}) | V(\mathbf{r}, \mathbf{R}) | \phi_i(\mathbf{r}) \exp(i\mathbf{k}_i \cdot \mathbf{R}) \rangle \\ &\equiv \int V_{fi}(\mathbf{R}) \exp(i\mathbf{K} \cdot \mathbf{R})\, d\mathbf{R} \end{aligned} \tag{25}$$

which is simply the Fourier transform of the coupling interaction (7). The Born integral cross section for $\mathbf{n}(nlm) \to \mathbf{n}'(n'l'm')$ collisional transitions at impact energy E is therefore

$$\sigma_{\mathbf{n},\mathbf{n}'}^{(B)}(E) = \frac{1}{8\pi k_f^2} \left(\frac{2M_{AB}}{\hbar^2} \right)^2 \int_{(k_i - k_f)}^{(k_i + k_f)} |F_{\mathbf{n},\mathbf{n}'}(K)|^2 |T_{el}^{(B)}(K)|^2 K\, dK \tag{26}$$

where M_{AB} is the reduced mass of the A–B(n) collision system. For e–B(n) collisions, $T_{el}^{(B)}$ in (26) is $4\pi e^2/K^2$, which, because of its singularity as $K \to 0$, exerts a dramatic influence on F_{fi} only in the "optical" limit of vanishing K when it effectively amplifies any dipole term in F_{fi}. This dipole then dominates the remaining multipoles at high E with the result that the cross section $\sigma_{\mathbf{nn}'}$ tends to the Bethe asymptotic limit with its characteristic $f_{\mathbf{nn}'} \ln E/E$ dependence in terms of the oscillator strength $f_{\mathbf{nn}'}$ for $\mathbf{n} \to \mathbf{n}'$ transitions. With increasing n and n', however, the onset of this limit is pushed further into regions of much higher E (relative to the location of the cross-sectional peak) such that, for a wide range of E (up to 1000 eV for $2 \to 3$ transitions in e-He($2\,^3$S) collisions), the $l \to l \pm 1$ dipole transitions are no longer dominant.[8, 19, 20]

For H(1s) projectiles remaining in the ground state,

$$T_{\text{el}}^{(\text{B})}(K) = \frac{4\pi e^2}{K^2}\,[1 - F_{1s1s}^2(K)] = \frac{4\pi e^2}{K^2}\,\frac{K^2(8 + K^2)}{(4 + K^2)^2} \tag{27}$$

is nonsingular as $K \to 0$ such that the corresponding $\sigma_{\mathbf{nn}'}$ will correlate more closely than e-atom collisions with any systematic trends (Sect. 11.2d) exhibited in $F_{\mathbf{nn}'}$ as $\mathbf{n}$ and $\mathbf{n}'$ are varied.

Coulomb fixed-center treatment

When (21) is a sum of Coulomb interactions

$$V = -(Ze^2/r_{13}) + (ZZ_c e^2/R) \tag{28}$$

as for particles of charge Ze incident on $B(n)$ of core charge $Z_c e$, Flannery[9] provided the exact solution $\Lambda(\lambda)M(\mu)\exp(im\Phi)$ for χ_i^+ in (20c), in terms of prolate–spheroidal coordinates (λ, μ, Φ) such that (20a) is then given by

$$T_{fi} = \langle \phi_f(\mathbf{r})\exp(i\mathbf{k}_f \cdot \mathbf{R}) \left| \frac{Ze^2}{R} - \frac{Ze^2}{|\mathbf{R}-\mathbf{r}|} \right| \phi_i(\mathbf{r})\, T_{\mathbf{r}}\Lambda(\lambda\,|\,\mathbf{r})M(\mu\,|\,\mathbf{r})\exp(im\Phi)\rangle \tag{29}$$

where $T_{\mathbf{r}}$ denotes a rotation operator that transforms χ_i from a coordinate axis in which $\mathbf{r}$ is fixed to a space-fixed frame. To zero order, this solution χ_i^+ tends to the unperturbed product of Coulomb functions centered about each scattering center, respectively. Then, by following an analysis similar to that of Vainshtein et al.,[21] the cross section reduces to

$$\sigma_{nn'}(E) = (8\pi/k_i^2 a_0^2)(M_{\text{AB}}/m_e)^2 \int_{(k_n - k_{n'})}^{(k_n + k_n')} |F_{\mathbf{nn}'}(\mathbf{K})|^2 [f(\nu, \chi)]^2\, dK/K^3 \tag{30a}$$

where m_e is the electronic mass. Eq. (30a) is identical with the plane-wave Born approximation (26) except for introduction of

$$f(\nu, \chi) = (\pi\nu/\sinh \pi\nu)F(-i\nu, i\nu, 1, \chi) \tag{30b}$$

where F is a hypergeometric function with arguments

$$\chi = (2\epsilon_{n'n} + K^2)/(2\epsilon_{n'n} + 3K^2), \qquad \nu = k_n^{-1} \text{ or } [k_n + (2\epsilon_n)^{1/2}]^{-1} \tag{30c}$$

in which $\epsilon_{n'n}$ is $(\epsilon_{n'} - \epsilon_n)$, the transition energy in atomic units, and the second value of ν is so designed as to account for the fact that the Rydberg electron is bound to give an effective-charge effect. The result, Eq. (30), derived originally from a first-order treatment[21] replaces the (zero-order) plane-wave $\exp(i\mathbf{k}_i \cdot \mathbf{R})$ for relative motion of the charged projectile in the Born approximation (25) by a product of Coulomb functions that represent the unperturbed relative motion in the field of the fixed Rydberg e and its core. The overall treatment, Eq. (29), of Flannery,[9] which tends to (30) in the first-order limit, is termed the *Coulomb fixed-center approximation* to remind us of the

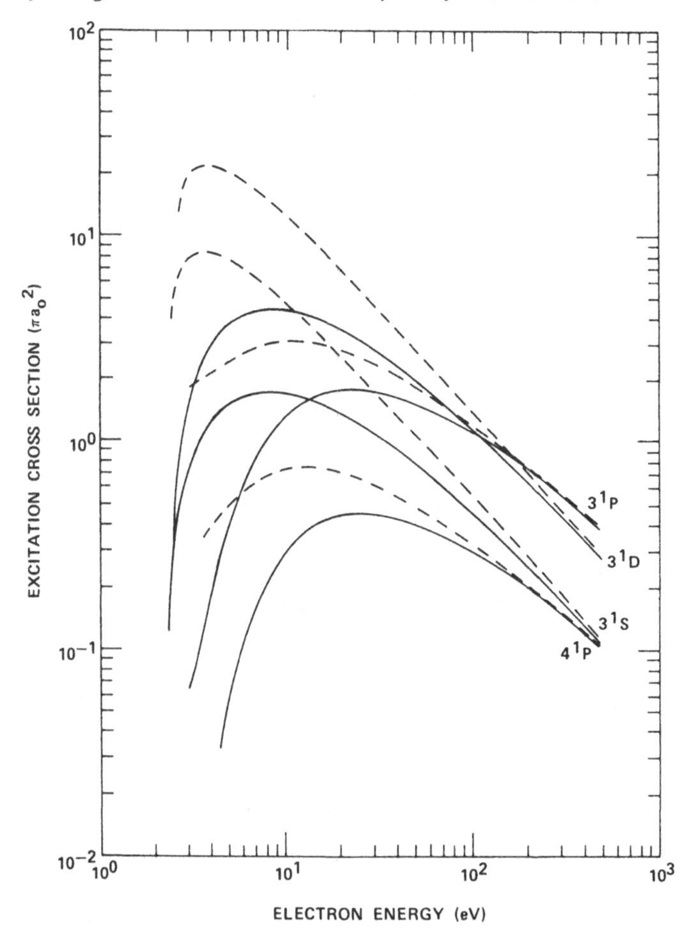

Fig. 11.1. Cross sections (πa_0^2) for the $3\,^1S$, $3\,^1P$, $3\,^1D$, and $4\,^1P$ excitations in e-He($2\,^1S$) collisions. (From Flannery et al.[19]) Broken curves: Born approximation; solid curves: equation (30) generalized to include electron exchange and effective charge. (From Flannery et al.[19])

underlying assumptions and provides systematic improvement to the first-order result of Vainshtein et al.[21]

In charged-particle–Rydberg collisions, the fixed-center treatment is based on the fact that the incident particle is subjected not to the averaged field of the Rydberg electron and its core – the basis of close-coupling perturbation treatments of Sect. 11.2a – but actually to two strong Coulomb-type fields that reduce to the averaged field only for distant encounters.

Flannery et al.[19] investigated (26) and (30) for $2\,^{1,3}S$–$n\,^{1,3}L$ transitions in e-He($2\,^{1,3}S$) collisions through the use of highly accurate wave functions for He. Representative cross sections based on generalization of (30a) to include electron exchange are displayed in Fig. 11.1. The main feature is that the optically forbidden $2\,^1\mathrm{S} \rightarrow 3\,^1\mathrm{D}$ transition, and $2\,^1\mathrm{S} \rightarrow 3\,^1\mathrm{S}$ to a lesser extent, is greater than the $2\,^1\mathrm{S}$–$3\,^1\mathrm{P}$ optically allowed transition over a large E range. This feature is but an example of a more systematic trend exhibited in collisional transitions between excited states in general.[20] Because of its slower $E^{-1}\ln E$ asymptotic dependence, the optically allowed transition will eventually dominate.

Impulse approximation treatment

When interaction V_{C3} between the core C and projectile A is neglected and the Rydberg wave function $\phi_i(\mathbf{r}_1)$ in (20) is expressed in terms of its momentum wave function or amplitude $g_i(\mathbf{k}_1)$ associated with the Rydberg electron of momentum $\mathbf{k}_1$ by

$$\phi_i(\mathbf{r}_1) = (2\pi)^{-3/2} \int g_i(\mathbf{k}_1) \exp(i\mathbf{k}_1 \cdot \mathbf{r}_1)\, d\mathbf{k}_1 \tag{31}$$

it can be shown[22] that the T matrix (20a) for inelastic scattering of projectile 3 from wave vector $\mathbf{k}_3$ to $\mathbf{k}_3'$ relative to the target core reduces after some analysis to

$$T_{fi}^{\mathrm{I}}(\mathbf{k}_3, \mathbf{k}_3') = \int g_f^*(\mathbf{k}_1') g_i(\mathbf{k}_1) T_{13}(\mathbf{k}, \mathbf{k}')\, d\mathbf{k}_1 \{\delta[\mathbf{P} - (\mathbf{k}_1' - \mathbf{k}_1)]\, d\mathbf{k}_1'\} \tag{32a}$$

where

$$T_{13}(\mathbf{k}, \mathbf{k}') = \langle \exp(i\mathbf{k}' \cdot \mathbf{r}) | V_{13}(\mathbf{r}) | \psi(\mathbf{k}, \mathbf{r}) \rangle \tag{32b}$$

is the exact T matrix and ψ is the exact wave function for potential scattering by V_{13} in the (1–3) center of mass such that $\chi_i^+(\mathbf{r}_1, \mathbf{r}_3)$ in (20c) is $\exp(i\mathbf{K}_0 \cdot \mathbf{R}) \times \exp(-i\mathbf{k}_1 \cdot \mathbf{r}_1)\psi(\mathbf{k}, \mathbf{r})$. The δ function in (32a) implies conservation of total linear momentum $\mathbf{K}_0 = (\mathbf{k}_1 + \mathbf{k}_3)$ before and after the (1–3) collision, which changes the momentum of particle i of mass M_i from $\mathbf{k}_i$ to $\mathbf{k}_i'$ such that the momentum change is

$$\mathbf{P} = \mathbf{k}_3 - \mathbf{k}_3' = \mathbf{k}_1' - \mathbf{k}_1 = \mathbf{k}' - \mathbf{k} \tag{32c}$$

where

$$\mathbf{k}^{(\prime)} = \frac{M_3}{M}\mathbf{k}_1^{(\prime)} - \frac{M_1}{M}\mathbf{k}_3^{(\prime)}, \qquad M = M_1 + M_3 \tag{32d}$$

is the initial (unprimed) or final (primed) momentum of (1–3) relative motion. Expressions (32) represent the quantal impulse approximation (QIA), which can be derived from many different directions.[23] Although the present method of derivation has exposed several underlying assumptions in QIA, the improvement and establishment of rigorous validity criteria are best achieved from the exact two-potential expression, a procedure natural for Rydberg collisions (Sect. 11.5).[2]

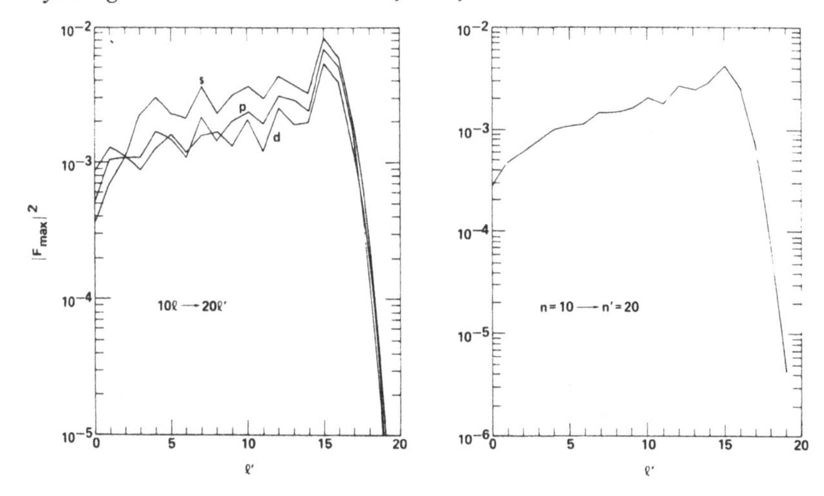

Fig. 11.2. Variation with final angular momentum l' of the peak of the inelastic form factor, Eq. (24), unaveraged and averaged over l for $10l \to 20l'$ transitions. (From Flannery and McCann[20])

11.2d. Systematic trends in inelastic form factors for $nl \to n'l'$ collisional transitions

Many sets of quantal Born and related calculations of cross sections for $[e, H^+, H(1s)]$–$H(nl)$ collisions exist (see, for example, Table V of Percival and Richards[12] for partial list and Refs. 24–26 for some additional work). However, by analogy with transitions from ground states, it is tempting to suggest that dipole transitions $nl \to n'(l \pm 1)$ dominate the remaining multipole collisional transitions between any two excited levels (n, n') and thereby to restrict calculation to only a few transitions with $\Delta l = |l' - l| = 0, 1, 2$, and 3, as in the case of the applications just referenced. As Fig. 11.1 suggests and as Flannery and McCann[20] explicitly showed, this analogy is quite restrictive and is without foundation for $n \to n'$ transitions in general.

Not only is the inelastic form factor F_{fi} [Eq. (24)] basic to Born's approximation and derivatives, but it is also key to more elaborate efforts such as the impulse, fixed-center, and close-coupling treatments. Because any systematic behavior of F_{fi} with variation in quantum numbers (nlm) is, in general, reflected in Born[20] and more elaborate cross sections and because the maximum in $\sigma^{(B)}_{\mathbf{n},\mathbf{n}'}(E)$ originates in (26) from K integration over the largest peak F_{max} in $F_{fi}(K)$, we may initially consider the variation of F_{max} with respect to the initial and final quantum numbers.

$nl \to n'l'$ Transitions $(n' \gg n)$

Figure 11.2 shows that F_{max} for various l oscillate on an increasing background as l' is increased until a unique value of l'_{max} is attained, after which they

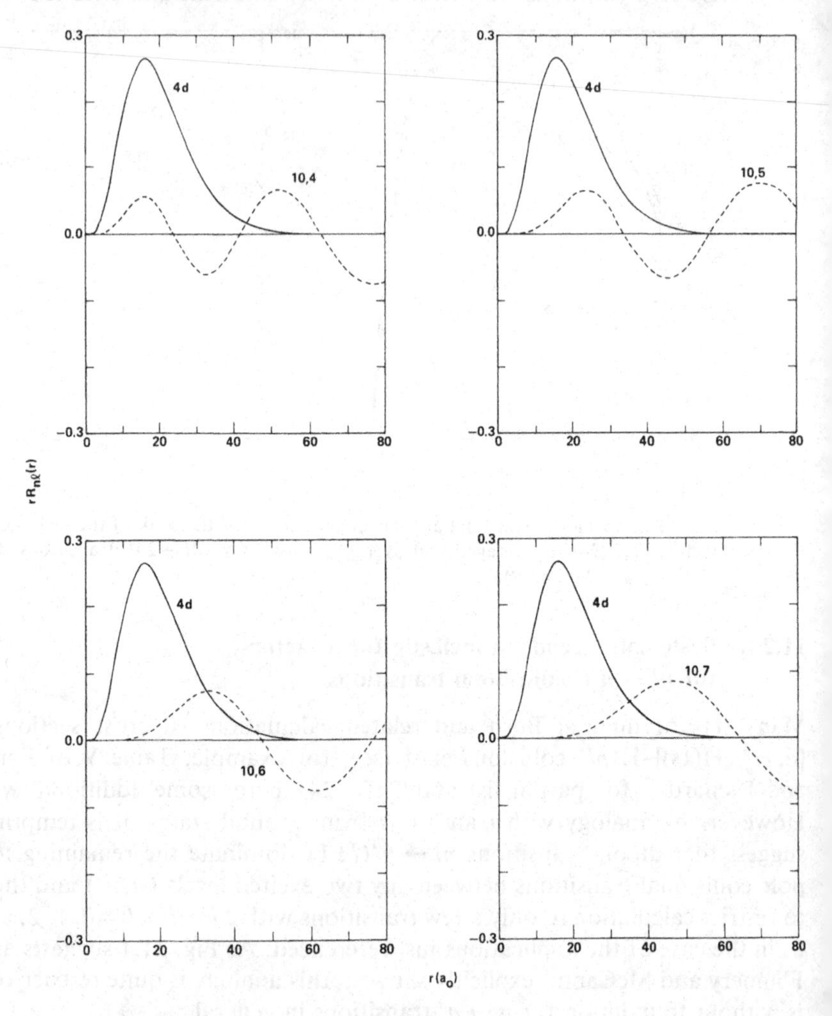

Fig. 11.3. The 4d and $10(l'=4\text{–}7)$ hydrogen radial orbitals R_{nl} times r. Note the outward shift in the $10l'$ orbital and the consequent variation in overlap with increase of l'. Here maximum overlap is obtained for $l'=l'_{\max}=6$.

exhibit a rapid monotonic decrease. The value $l'_{\max}$ is strongly dependent on the initial value of the principal quantum number n, is relatively insensitive to changes in the initial angular momentum quantum number l, and is given by[20]

$$l'_{\max} = \min\left\{(n'-1), \sim\left[n\left(\frac{2(n+3)}{(n+1)}\right)^{1/2} - \frac{1}{2}\right]\right\} \tag{33}$$

The chief contribution to the population of the final level n' in electron–atom and atom–atom collisions arises from the $n[l=0,1,\ldots,(n-1)] \rightarrow n'l'_{\max}$ array of transitions. This array may include some with dipole character, i.e., $l=l'_{\max}\pm 1$; and these dipole transitions tend to be somewhat more probable.

However, the important feature is that l'_{max} is primarily n dependent and, as such, may have a value inaccessible to dipole transitions. For example, the strongest collisional transitions in the $10l \to 20l'$ array (with variation of l and l') are the set $10l \to 20(l'_{max} \simeq 15)$, none of which possess dipole character. Because $n' \gg n \gg 1$, the magnitude and range of the initial and final orbits are quite different, as shown in Fig. 11.3 for $(4 \to 10)$ transitions (because of the relatively large energy separation between the levels), and the expression (33) for l'_{max} is determined[20] from a consideration of the overlap of the initial radial orbital with the first lobe of the final orbital, the region yielding any significant overlap.

The dramatic drop in F_{max} for $l > l'_{max}$ will not be evident unless the final n' is sufficiently large to accommodate that unique l'_{max} fixed by n. The final decline becomes more emphasized by making n' as large as in Fig. 11.2. By averaging over l before varying K and l', the oscillations can be suppressed and the key issues - the rise, the peak, and the rapid decline - become more marked. These trends are preserved in the Born cross sections for transitions in Rydberg atoms induced by e and H(1s) impact (see Ref. 20 for examples for many different $(n \to n')$ sets of transitions) and also in the more elaborate close-coupling eikonal treatment of even e-He($2\,^{1,3}$S) collisions.[8]

$n \to n+1$ Transitions

For $n \to n+1$ transitions, the radial orbitals are almost similar in both magnitude and range (in contrast to that in Fig. 11.3 for $n' \gg n$). Maximum overlap is obtained when all the *innermost* lobes are in phase and almost coincide and each outermost lobe is out of place, as happens when $l \simeq l'$. As l' is increased, the lobe of the final n' orbital moves inward with increasing phase difference between the orbitals. Maximum overlap is again attained when the *outermost* oscillations of both n and n' orbitals are almost out of phase by π or in phase as for the overlap of $n'=11$, $l'=6, 7$ with 10s, respectively.[20] With further increase of l', the number of oscillations in the n' orbitals reduces and cancellation occurs, thereby yielding small overlap with the initial state. Low initial l is therefore characterized by two well-separated peaks, one at $l' \simeq l$ and the other at higher l'. As l is increased, these two peaks merge and yield relatively larger form factors and cross sections (see Ref. 20 for detailed examples).

The main conclusions here are that dipole transitions do not, in general, provide the main contribution to $n \to n'$ collisional transitions unless the initial value of l is $\sim(l_{max} \pm 1)$ and that accurate cross sections require at least evaluation of the full array $nl \to n'l'_{max}$ of collisional transitions.

11.3. Semiclassical theory of Rydberg collisions

Semiclassical collision treatments in general focus on the transition amplitude $a_{nn'}(\mathbf{R})$ as a function of the relative external vector $\mathbf{R}$, whereas quantal treatments concentrate on scattering amplitudes or eigenphases associated with each partial wave of total angular momentum of the complete collision system. This feature can be exploited in Rydberg collisions for which $a_{nn'}$ can be exactly determined for transitions involving a large number of equally spaced levels. Semiclassical methods, properly constructed, are based on a stationary-state treatment of the collision process and furnish the scattering amplitude $f_{nn'}(\theta)$ as an $\mathbf{R}$ integral over $a_{nn'}(\mathbf{R})$, or functions thereof, modulated by a Bessel function (Sect. 11.3a). For heavy-particle collisions, a finite number of classical trajectories $R_i(\theta)$ provide the dominant contribution to $f_{nn'}$, which then reduces to $a_{nn'}(R_i(\theta))$ times the amplitude for classical elastic scattering (Sect. 11.3a). In the heavy-particle–high-energy limit, when the scattering is mainly in the forward direction, the integral cross section reduces to the transition probability $|a_{nn'}|^2$ integrated over $2\pi\rho\, d\rho$, where ρ is the impact parameter. Here, the amplitudes $a_{nn'}$ satisfy equations identical to the time-dependent set in Dirac's method of variation of constants for the response of a system to a time-dependent perturbation (Sect. 11.3a).[27] The "time-dependent" semiclassical method (Sect. 11.3b) is therefore valid only in the heavy-particle–high-energy limit and the full stationary-state description (Sect. 11.3a) extends the validity of the semiclassical treatment down to much lower impact energies.

11.3a. Stationary-state treatment

The function F_n for relative motion in (4) is decomposed as

$$F_n(\mathbf{R}) = A_n(\mathbf{R}) \exp[iS_n(\mathbf{R}) - \chi_n(\mathbf{R})] \tag{34}$$

where S_n is real and is identified with the eikonal for relative motion under the static interaction $V_{nn}(\mathbf{R})$ so that

$$S_n(\mathbf{R}) = \int_{\mathbf{R}_0}^{\mathbf{R}} \mathbf{K}_n \cdot d\mathbf{R} + \mathbf{k}_n \cdot \mathbf{R}_0 \tag{35}$$

is the solution of the Hamilton–Jacobi equation,[15] where $\mathbf{R}_0$ is the initial point on the trajectory $\mathbf{R}_n$ at which $K_n = k_n$. When the $\mathbf{R}$ variation in F_n is mainly controlled by S_n, then various second-order terms $\nabla^2(A_f, \chi_f)$, $(\nabla\chi_f)$, and $(\nabla A_f \cdot \nabla\chi_f)$ can be neglected, so that the substitution of (34) into (5) yields the first-order differential equations in three dimensional space:[8, 18]

$$i\mathbf{K}_f \cdot \nabla_{\mathbf{R}} A_f(\mathbf{R}) = \left(\frac{\mu}{\hbar^2}\right) \underset{n \neq f}{\mathrm{S}} A_n(\mathbf{R}) V_{fn}(\mathbf{R}) \exp[i(S_n - S_f)] \exp[-(\chi_n - \chi_f)] \tag{36}$$

where χ_n satisfies

$$\nabla^2_{\mathbf{R}} S_n - 2(\nabla_{\mathbf{R}} S_n) \cdot (\nabla_{\mathbf{R}} \chi_n) = 0 \tag{37}$$

When S_n and χ_n are purely real, then, for unit A_n, Eq. (34) describes motion of a classical ensemble of particles with momentum ∇S_n and flux density $\exp(-2\chi_n)$. The classical current J_n (number of particles per unit area per second) in this "elastic" channel n is therefore $\exp(-2\chi_n)|\nabla S_n|/\mu$ such that (37) is equivalent to current conservation $(\nabla \cdot \mathbf{J}_n)=0$ in channel n. With the initial $\mathbf{k}_n$ motion directed along the Z axis with impact parameter $\boldsymbol{\rho}(\rho, \phi)$, then $\chi_n(\mathbf{R}_0)$ is, from (34) and (35), identically zero and the scattered flux across an elemental area $d\mathbf{S}$ normal to the direction $\mathbf{R}(R, \theta, \phi)$ of the outgoing "elastic" scattered wave is, therefore,

$$\lim_{R\to\infty} \{\mu^{-1}\exp[-2\chi_n(\mathbf{R})]\mathbf{K}_n \cdot d\mathbf{S}\} = \mu^{-1}\mathbf{k}_n \cdot d\boldsymbol{\rho} \tag{38}$$

such that $\exp(-2\chi_n)$ as $R \to \infty$ is $[\partial(\tau_n)/\partial(\tau_{n0})]$, where τ_n is the infinitesimal volume along a classical path that evolved from τ_{n0}. Flux is, therefore, lost only by transitions with probability controlled by A_f.

Multichannel eikonal treatment

For collisions of electrons and heavy particles with Rydberg atoms, scattering is mainly in the forward direction (because of the very large internal dipole associated with atoms in a state n high enough for large effective dipole yet low enough so that the Rydberg electron and core do not behave as separate scatterers). The trajectories in (35) and (36) can be assumed identical $(\chi_n - \chi_f \simeq 0)$ and rectilinear $(\chi_n \simeq 0)$ along the initial Z direction so that the matrix element for direct $i \to f$ transitions is

$$T_{fi}(\mathbf{k}_f, \mathbf{k}_i) = \langle \phi_f(\mathbf{r}) \exp(i\mathbf{k}_f \cdot \mathbf{R}) | V(\mathbf{r}, \mathbf{R}) | \mathop{\mathrm{S}}_n A_n(\mathbf{R})\phi_n(\mathbf{r}) \exp iS_n(\mathbf{R}) \rangle \tag{39}$$

where the transition amplitudes are solutions of

$$iK_f\, \partial A_f(\mathbf{R})/\partial Z = (\mu/\hbar^2) \mathop{\mathrm{S}}_n A_n(\mathbf{R}) V_{fn}(\mathbf{R}) \exp i(S_n - S_f) \tag{40}$$

These equations represent the multichannel eikonal treatment of Flannery and McCann,[8] which was developed more formally via linearization of the Green's function propagator (see last listing in Ref. 8). The method proved to be very reliable for both heavy-particle collisions[28] at energies where forward scattering predominates and for electron-(ground-state) atom inelastic scattering[8] by $\theta \lesssim 40°$ at impact energies E greater than a few times the ionization threshold. Its application to e–He($2\,^{1,3}$S) collisions represents, as yet, the only direct application of any stationary-state closely coupled procedure to e-excited atom collisions (see last listing in Ref. 8). It is applicable to detailed collisional transitions $(nl \to n'l')$, particularly between close levels $n \simeq n'$, by electron, ion, and neutral impact. The matrix elements V_{fn} can be determined from correspondence principles (Sect. 11.4).

Development of (39) and (40) follows by setting

$$a_f^{(E)}(\boldsymbol{\rho}, Z) = A_f(\boldsymbol{\rho}, Z) \exp\left[i \int_{-\infty}^{Z} (K_f - k_f)\, dZ\right] \exp[i\Delta\Phi] \tag{41}$$

where $\mathbf{R}(R,\theta,\Phi)\equiv\mathbf{R}(\rho,\Phi)$, where Δ is the change (m_f-m_i) in the magnetic quantum number of the Rydberg atom, and $a_f^{(E)}(\rho,-\infty)=\delta_{fi}$ ensures that the system wave function $\Psi\simeq\phi_i(\mathbf{r})\exp(ik_iZ)$ as $Z\to-\infty$. Hence (40) reduces to the set

$$i\frac{\hbar^2}{\mu}K_f\frac{\partial a_f^{(E)}}{\partial Z}+\left[\frac{\hbar^2}{\mu}K_f(K_f-k_f)+V_{ff}\right]a_f^{(E)}$$

$$=\sum_{n=1}^{N}a_n^{(E)}(\rho,Z)V_{fn}(\rho,Z)\exp i(k_n-k_f)Z \tag{42}$$

The scattering amplitude on Φ-integration in (39) becomes[8]

$$f_{if}^{\mathrm{ME}}(\theta,\phi)=-i^{\Delta+1}\int_0^\infty J_\Delta(K'\rho)[I_1(\rho,\theta)-iI_2(\rho,\theta)]\rho\,d\rho \tag{43}$$

where K' is the momentum increase $k_f\sin\theta$ in the direction $\hat{\rho}$, J_Δ the Bessel functions of integral order, and

$$I_1(\rho,\theta;\alpha)=\int_{-\infty}^{\infty}K_f\left(\frac{\partial a_f^{(E)}}{\partial Z}\right)\exp(i\alpha Z)\,dZ,\qquad \alpha=k_f(1-\cos\theta) \tag{44}$$

is the main contributor because

$$I_2(\rho,\theta;\alpha)=\int_{-\infty}^{\infty}[K_f(K_f-k_f)+(\mu/\hbar^2)V_{ff}]a_f^{(E)}\exp(i\alpha Z)\,dZ \tag{45}$$

contributes only at lower impact energies.

Equations (42)–(45) represent the full multichannel eikonal (ME) treatment.[8] Approximation follows when

$$K_f\simeq k_f-(\mu/\hbar^2k_f)V_{ff}(\mathbf{R})\qquad\text{for}\quad k_f^2\gg(2\mu/\hbar^2)V_{ff} \tag{46}$$

such that I_2 and the second term on the left-hand side of (42) vanish. For higher energies $K_n\simeq k_n$, in the high-energy–heavy-particle limit of relative speed v_i,

$$k_f=k_i-(\epsilon_{fi}/\hbar v_i)[1+(\epsilon_{fi}/2\mu v_i^2)+\cdots] \tag{47}$$

such that (43) reduces to

$$f_{if}^{(1)}(\theta,\phi)=-i^{\Delta+1}k_i\int_0^\infty\rho\,d\rho\,J_\Delta(K'\rho)\left[\int_{-\infty}^{\infty}(\partial a_f/\partial Z)\exp(i\alpha'Z)\,dZ\right] \tag{48}$$

where $\alpha'=(k_i-k_f)-(\epsilon_{fi}/\hbar v_i)$ and the amplitudes a_f satisfy

$$i\hbar\frac{\partial a_f}{\partial t}=\sum_{n=1}^{N}a_f(\rho,Z)V_{fn}(\rho,Z)\exp(i\omega_{fn}t);\qquad \hbar\omega_{fn}=\epsilon_{fn} \tag{49}$$

in which the dummy variable t in $Z=v_it$ specifies variation along the straight-line path. When α' can be neglected, as in the high-energy limit, then (48) reduces to

$$f_{if}^{(2)}(\theta, \phi) = -i^{\Delta+1}k_i \int_0^\infty J_\Delta(K'\rho)[a_f(\rho, \infty) - \delta_{if}]\rho\, d\rho \tag{50}$$

In the sudden approximation [i.e., when the collision time $\tau_{\text{coll}} \ll 2\pi/\omega_{fi} \simeq T/(f-i)$, where the classical period T of the orbital Rydberg electron has been introduced by Bohr's correspondence (Sect. 11.4a)], then set (49) can be solved exactly to give

$$a_f^{(s)}(\rho, Z) = \left\langle \phi_f(\mathbf{r}) \middle| \exp\left[-(i/\hbar)\int_{-\infty}^{t} V(\mathbf{r}, \mathbf{R})\, dt\right] \middle| \phi_i(\mathbf{r}) \right\rangle, \qquad Z = v_i t \tag{51}$$

The Glauber approximation is recovered by substitution of (51) into (50).[8] The differential cross section for $i \to f$ collisional transitions given by the full ME treatment (42)–(45) and by its approximation I [Eqs. (48) and (49)] and II [Eq. (50)] is

$$\sigma_{if}(k_i) = 2\pi(k_f/k_i)\int_0^\pi |f_{if}^{(\text{ME},1)}(\theta, \phi)|^2\, d(\cos\theta) \tag{52}$$

which reduces for approximation II (50) to

$$\sigma_{if}(k_i) = 2\pi(k_f/k_i)\int_0^\infty |a_f(\rho, \infty) - \delta_{if}|^2 \rho\, d\rho \tag{53}$$

Thus (53) is a valid representation only in the heavy-particle–high-energy limit as previously outlined, whereas (52) extends the validity down to intermediate and lower energies. For example, (52) is accurate for 1s–2p and $1\,^1$S–$2\,^1$P transitions in e–H(1s) and e–He($1\,^1$S) collisions for impact energies $E \gtrsim 2I$, twice the ionization energies, whereas (53) tends to accurate values only for $E \gtrsim 10I$.[8]

Multistate orbital treatment

We note that the basic expression (39) for a given scattering angle θ provides contributions from all regions of $\mathbf{R}$ space and not just those $\mathbf{R}$ associated with an external classical trajectory $R_n(\theta)$. For heavy-particle collisions at lower energies, when the main contribution to both the phase and magnitude of the scattering amplitude arises from the trajectories $R_n(\theta)$, which are not in the forward direction, then the multistate orbital treatment can be adopted for Rydberg collisions.[18] From (38),

$$\lim_{R\to\infty} \exp[-2\chi_n(\mathbf{R})] = \left(\frac{\partial \tau_{no}}{\partial \tau_n}\right) = \left[\frac{\rho\, d\rho\, d\phi}{R^2\, d(\cos\theta)\, d\phi}\right] = \frac{1}{R^2}\left(\frac{d\sigma_{\text{CL}}^{(n)}}{d\Omega}\right) \tag{54}$$

where $\sigma_{\text{CL}}^{(n)}$ is the classical cross section for elastic scattering by angle θ under V_{nn} in channel n. The scattered amplitude $f_{ni}(\theta)$ for $i \to n$ transitions then follows from the outgoing spherical wave of (4) with the aid of (54) in (34) to yield

$$f_{ni}(\theta) = \lim_{R\to\infty} A_n(\mathbf{R})\{\exp[-iS_n'(\mathbf{R})]\}[\sigma_{\mathrm{el}}^n(\theta)]^{1/2} \tag{55}$$

where S_n' is the classical action along the nth path measured relative to the action along the path of undeflected particles with the same wave number K_n, i.e.,

$$S_n'(R) = -(1/\hbar)\int_{\mathbf{R}_0}^{\mathbf{R}} \mathbf{R}_n \cdot \mathbf{P}_n\, dt; \qquad \mathbf{P}_n \equiv \hbar\mathbf{K}_n \equiv \mu\, d\mathbf{R}_n/dt \tag{56}$$

in terms of the external relative momentum $\mathbf{P}_n$. The amplitudes $A_n(\mathbf{R})$ are solutions of (36), where it proves convenient to write

$$i\mathbf{K}_f \cdot \nabla_{\mathbf{R}} A_f(\mathbf{R}) \equiv i\mathbf{K}_f \cdot \hat{\mathbf{s}}_f \partial A_f(\rho, s_f)/\partial s_f = (i\mu/\hbar)\,\partial A_f(\rho, t)/\partial t \tag{57}$$

in terms of $\mathbf{s}_f$ the vector distance traveled along the trajectory $\mathbf{R}_f$ with initial impact parameter ρ and where "time" t is merely a dummy variable invoked only to classify variation along the trajectories specified by $\mathbf{R}_f$. When the momentum $\mathbf{P}$ is the mean $\hbar(\mathbf{K}_n + \mathbf{K}_f)/2$, directed along some averaged trajectory $\mathbf{R}(t)$ common to all channels n, then $\chi_n \simeq \chi_f$ in (36) and the difference $(S_n - S_f)\hbar$ of classical action in channels n and f evolves as

$$S_n(\mathbf{R}) - S_f(\mathbf{R}) = \omega_{fn} t + (1/\hbar)\int_{t_0}^{t} [V_{ff}(\mathbf{R}(t) - V_{nn}(\mathbf{R}(t))]\, dt \tag{58}$$

such that (49) is recovered from (40) via (57) and the transformation $a_f = A_f \exp[-(i/\hbar)\int_{t_0}^{t} V_{ff}\, dt]$. The scattering amplitude (55) is therefore

$$f_{ni}^{(j)}(\theta) = a_n(\rho_j, t\to\infty)\exp\left[\frac{i}{\hbar}\int_{t_0}^{t}\{V_{nn}[\mathbf{R}(t)] + \mathbf{R}\cdot\mathbf{P}(t)\}\, dt\right]\left[\frac{d\sigma_{\mathrm{CL}}(\theta)}{d\Omega}\right]^{1/2} \tag{59}$$

where one impact parameter ρ_j yields scattering angle θ (which, in terms of the deflection angle Θ, is $|\Theta \pm 2\pi m|$ for integer m). The differential cross section is simply

$$(d\sigma_{if}/d\Omega) = |a_f(\rho_j, t\to\infty)|^2 (d\sigma_{\mathrm{CL}}/d\Omega) \tag{60}$$

When more than one trajectory j yields a given θ, then interference occurs between the phases associated with each trajectory and a three-dimensional generalization[18] of the usual one-dimensional analysis yields

$$f_{ni}(\theta) = -i\sum_{j=1}^{N} \alpha_j \beta_j f_{ni}^{(j)}, \qquad \alpha_j, \beta_j = \exp(\pm i\pi/4) \tag{61}$$

where the exponent of α_j is positive or negative, depending on whether the scattered particle emerges on the same side of the axis it entered (as in an overall repulsive collision) or on the opposite side (as in an overall attractive collision) and where the exponent of β_j is positive or negative as $d\Theta/d\rho$ is, respectively, positive or negative (see second listing in Ref. 18). When these trajectories coalesce, as for caustics (e.g., a rainbow), representation by classical trajectories becomes invalid and a (uniform) solution, which tends to (61) for well-separated trajectories, is then obtained in terms of Airy functions.[18]

The orbit common to all channels n is determined by the averaged Hamiltonian[18]

$$\bar{H} = \langle \Psi | H | \Psi \rangle = \sum_{k=1}^{3} (1/2\mu) P_k^2(\mathbf{R}) + \mathcal{V}_{\text{opt}}(\mathbf{R}) \tag{62}$$

where Ψ is defined by (4) with (34) and H by (1). The first term on the right-hand side of (62) is the kinetic energy of relative motion (obtained by assuming that $S_n(\mathbf{R})$ in (34) contains the major $\mathbf{R}$ variation). The second term

$$\mathcal{V}_{\text{opt}}(\mathbf{R}) = \underset{n}{\mathrm{S}} \, [|a_n(t)|^2 \epsilon_n + \underset{f}{\mathrm{S}} \, a_f^*(t) a_n(t) V_{fn}[\mathbf{R}(t)] \exp(i\omega_{fn} t)] \tag{63}$$

behaves like an optical potential that effectively couples response (4) provided by the transition amplitudes a_f to the interaction $V(\mathbf{r}, \mathbf{R})$ back to the relative motion and vice versa. The common trajectory is then given by the solution of Hamilton's equations of motion:

$$\begin{aligned} \frac{\partial Q_k}{\partial t} &= \frac{\partial \bar{H}}{\partial P_k} \equiv \frac{1}{\mu} P_k(t) \\ \frac{\partial P_k}{\partial t} &= -\frac{\partial \bar{H}}{\partial Q_k} = -\frac{\partial V_{\text{opt}}}{\partial Q_k} = -\underset{n}{\mathrm{S}} \, \underset{f}{\mathrm{S}} \, a_f^*(t) a_n(t) \frac{\partial V_{fn}(Q_k)}{\partial Q_k} \exp(i\omega_{fn} t) \end{aligned} \tag{64}$$

for the variation with time t of the generalized coordinates $Q_k \equiv (X, Y, Z)$ for $\mathbf{R}$ and their conjugate momenta P_k. The solutions of (64) are therefore coupled to the complex solutions a_f of (49), thereby resulting in $(2N+4)$ coupled equations rather than $2N$ for rectilinear trajectories. This semiclassical procedure reproduces[18] with remarkable accuracy differential and integral collisions, together with detailed oscillatory structure due to interference between the action S_j' phases peculiar to contributing classical paths.

A valuable feature of scattering by $\mathcal{V}_{\text{opt}}$ is that total energy E of the collision system is conserved, which is confirmed by showing, with the aid of (63) and (64), that $(d\bar{H}/dt) = (\partial \bar{H}/\partial t) = 0$.[18]

In this multistate orbital treatment, energy is therefore continually being redistributed between the relative motion and the internal degrees of freedom as motion along the trajectory proceeds, which is a unique and valuable asset, particularly for Rydberg collisions with ions, atoms, and molecules at thermal and higher energies.

When this coupling between internal and external motions is ignored by assuming some averaged spherical interaction $\bar{V}(R)$ independent of transition amplitudes $a_n(t)$, then the time variation of $\mathbf{R}(R, \Theta, \Phi)$ given by (64) is

$$dR/dt = \pm \bar{v}[1 - (\rho^2/R^2) - \bar{V}(R)/E]^{1/2} \qquad d\Theta/dt = \bar{v}\rho/R^2 \tag{65}$$

where $\bar{v}$ is some averaged speed and Φ is a constant.

11.3b. "Time-dependent" treatment: classical path, Magnus, and sudden approximations

Next attention will be focused only on the response of the target system to a time-dependent interaction generated by the passing projectile. Thus (1) is replaced by the time-dependent Schrödinger equation

$$H\Phi = [H_0(\mathbf{r}) + V(\mathbf{R}(t), \mathbf{r})]\Phi(\mathbf{r}, t) = i\hbar\, \partial\Phi(\mathbf{r}, t)/\partial t \tag{66}$$

and the response is expanded as

$$\Phi(\mathbf{r}, t) = \mathop{\mathrm{S}}_{n} a_n(t)\phi_n(\mathbf{r}) \exp(-i\omega_n t) \tag{67}$$

where the transition amplitudes are determined by

$$i\hbar\, \partial a_f(t)/\partial t = \mathop{\mathrm{S}}_{n} a_n(t) V_{fn}(\mathbf{R}(t)) \exp(i\omega_{fn} t) \tag{68}$$

subject to $a_f(t \to -\infty) = \delta_{ni}$, as given by Dirac's method of variation of constants.[27] Although in this treatment the inelastic integral cross sections are

$$\sigma_{if} = \int_0^\infty |a_f(\boldsymbol{\rho}, \infty)|^2\, d\rho \tag{69}$$

the treatment provides no details on either integral elastic or differential scattering cross sections without recourse to the preceding stationary-state analysis.

Upon use of a straight-line trajectory

$$\mathbf{R} = \boldsymbol{\rho} + \mathbf{v}t \tag{70}$$

(69) reduces to the standard impact-parameter result. Because (68) is identical to (49) obtained as a high-energy approximation to the basic stationary-state analysis [Eqs. (42)–(45)], the transition amplitudes a_f computed from (68) may be inserted directly into (48) or (50)–(53) for the various scattering amplitudes and cross sections.[8,28] By comparison, we note that (69) with a straight-line path (70) provides a valid representation of the inelastic cross sections *only in the high-energy–heavy-particle collision limit,* which has been used[12,29] for e–H(n) inelastic cross sections that are claimed to be accurate for impact energies $E > 4I_n$, a claim that may be optimistic because (69) for H(1s–2p) collisional excitation becomes accurate only for $E > 10I_1$. Validity to lower energies can be extended through the use of a_f in (50) or, even better, by solution of the full ME equations (42)–(45) for use in (52).

Valuable information on the transition amplitudes for high-Rydberg atoms can be obtained, however, from (68), which becomes considerably simplified through the use of the Bohr and Heisenberg correspondence principles (Sect. 11.4b) such that an exact solution (the equivalent oscillator theorem; Sect. 11.4d) is obtained.

In the preceding Schrödinger picture, the time variation of Φ is developed by the full Hamiltonian (H_0+V) in (66). A complete hierarchy of approximation to (1) becomes more transparent in the interaction picture,[30] where the transformed interaction

$$V_\mathrm{I}(t) = \exp\left(\frac{i}{\hbar} H_0 t\right) V(t) \exp\left(-\frac{i}{\hbar} H_0 t\right) \tag{71}$$

controls the full-time evolution of the transformed wave function

$$\Phi_\mathrm{I}(\mathbf{r},t) = \exp\left(\frac{i}{\hbar} H_0 t\right)\Phi(\mathbf{r},t) \equiv \sum_{p=0}^{\infty} \frac{1}{p!}\left(\frac{i}{\hbar} H_0 t\right)^p \Phi(\mathbf{r},t) \tag{72}$$

which, with the aid of (66) satisfies,

$$i\hbar\, \partial\Phi_\mathrm{I}/\partial t = V_\mathrm{I}(\mathbf{r},t)\Phi_\mathrm{I}(\mathbf{r},t) \tag{73}$$

Thus, in this representation, the free-particle states that remain independent of t are simply $\phi_n(\mathbf{r})$, such that the system wave function in (73) can be expanded as

$$\Phi_\mathrm{I}(\mathbf{r},t) = \mathop{\mathrm{S}}_{n} a_n(t)\phi_n(\mathbf{r}) \tag{74}$$

to give, in matrix notation,

$$i\hbar\dot{\mathbf{a}} = \mathbf{V}_\mathrm{I}(t)\mathbf{a} \tag{75}$$

where $\mathbf{a}\equiv\{a_n\}$ and the matrix elements of the anti-Hermitian matrix $\mathbf{V}_\mathrm{I}$ $(\equiv -\mathbf{V}_\mathrm{I}^\dagger)$ of (71) are

$$(V_\mathrm{I})_{mn} = \langle\phi_m(\mathbf{r})|V_\mathrm{I}(\mathbf{r},t)|\phi_n(\mathbf{r})\rangle \equiv V_{mn}\exp(i\omega_{mn}t) \tag{76}$$

so that (75) and (68) are equivalent. Then we introduce the operator $\mathbf{U}$, which translates Φ_I in time as

$$\Phi_\mathrm{I}(t) = \mathbf{U}(t,t_0)\Phi_\mathrm{I}(t_0) \tag{77}$$

Because $\mathbf{a}$ provides the full-time development of Φ_I, then

$$\mathbf{a}(t) = \mathbf{U}(t,t_0)\mathbf{a}(t_0) \tag{78}$$

The unitary property of $\mathbf{U}$ (i.e., $\mathbf{U}\mathbf{U}^\dagger=\mathbf{I}$) ensures conservation of transition probability and $\mathbf{U}(t_0,t_0)$ is the unit operator $\mathbf{I}$. From (75)

$$i\hbar\, d\mathbf{U}(t,t_0)/dt = \mathbf{V}_\mathrm{I}(t)\mathbf{U}(t,t_0) \tag{79}$$

which, on integration, yields,

$$\mathbf{U}(t,t_0) = \mathbf{I} - (i/\hbar)\int_{t_0}^{t} \mathbf{V}_\mathrm{I}(t_1)\mathbf{U}(t_1,t_0)\, dt_1 \tag{80}$$

such that iteration provides the expansion series

$$\mathbf{U}(t,t_0)=\mathbf{I}-\frac{i}{\hbar}\int_{t_0}^{t}\mathbf{V}_{\mathrm{I}}(t_1)\,dt_1+\left(\frac{i}{\hbar}\right)^2\int_{t_0}^{t}dt_1\left[\int_{t_0}^{t_1}\mathbf{V}_{\mathrm{I}}(t_1)\mathbf{V}_{\mathrm{I}}(t_2)\,dt_2\right]+\cdots \tag{81}$$

$$\equiv \overleftarrow{T}\left[\exp\left(-\frac{i}{\hbar}\right)\int_{t_0}^{t}\mathbf{V}_{\mathrm{I}}(t)\,dt\right] \tag{82}$$

The chronological operator $\overleftarrow{T}$ generates in (81) the time ordering that is essential because $\mathbf{V}_{\mathrm{I}}(t_1)$ and $\mathbf{V}_{\mathrm{I}}(t_2)$ in the interaction picture will, in general, not commute for $t_1 \neq t_2$. Although (81), known as the *the Dyson expansion,*[31] is similar in form and in difficulty of calculation to the Born expansion in stationary-state theory, it can be suitably rearranged[32] so that truncation of the resulting series retains the unitarity of **U** at each level of approximation. *The Magnus expansion* is therefore written as

$$\mathbf{U}(t,t_0)=\exp\mathbf{A}(t,t_0)=\mathbf{I}+\mathbf{A}(t,t_0)+\frac{1}{2!}\mathbf{A}^2(t,t_0)+\cdots \tag{83}$$

where

$$\mathbf{A}=\sum_{k=1}^{\infty}(i\hbar)^{-k}\mathbf{A}^{(k)}(t,t_0)/k! \tag{84a}$$

of which the first few terms are

$$\mathbf{A}^{(1)}(t,t_0)=\int_{t_0}^{t}\mathbf{V}_{\mathrm{I}}(t_1)\,dt_1 \tag{84b}$$

$$\mathbf{A}^{(2)}(t,t_0)=\int_{t_0}^{t}dt_1\left\{\int_{t_0}^{t_1}dt_2\,[\mathbf{V}_{\mathrm{I}}(t_1),\mathbf{V}_{\mathrm{I}}(t_2)]\right\} \tag{84c}$$

where $[\mathbf{V}_{\mathrm{I}}(t_1),\mathbf{V}_{\mathrm{I}}(t_2)]$, the commutator of the interaction $\mathbf{V}_{\mathrm{I}}$, is, as $\mathbf{V}_{\mathrm{I}}$, also anti-Hermitian. Thus **U** of (83) retains its unitarity [because $(\exp\mathbf{A})^{\dagger}=\exp(-\mathbf{A})$] in (84b, c) and in all higher terms, as

$$\mathbf{A}^{(3)}(t,t_0)=\int_{t_0}^{t}dt_1\left\{\int_{t_0}^{t_1}dt_2\int_{t_0}^{t_2}dt_3\,[[\mathbf{V}_{\mathrm{I}}(t_1),\mathbf{V}_{\mathrm{I}}(t_2)],\mathbf{V}_{\mathrm{I}}(t_3)]\right.$$

$$\left.+\,[[\mathbf{V}_{\mathrm{I}}(t_3),\mathbf{V}_{\mathrm{I}}(t_2)],\mathbf{V}_{\mathrm{I}}(t_1)]\right\} \tag{84d}$$

We shall show in Sect. 11.4d that an exact solution (expressed as an equivalent oscillator theorem) to (79) can then be obtained from (83) for high-energy Rydberg collisions upon transformation to an action-angle (**J**, **w**) representation for the unperturbed functions of $H_0(\mathbf{J})$ and the use of the Bohr and Heisenberg correspondences such that $\mathbf{V}_{\mathrm{I}}(t)$ then commutes at all times.

In the full quantal theory (Sect. 11.2a), the transition probability is given by quantal matrix elements as

$$P_{fi}^{(Q)} = |S_{fi}^{(Q)}|^2 = |\langle \Psi_f^- | \Psi_i^+ \rangle|^2$$
$$= |\langle \phi_f(\mathbf{r}) \exp(i\mathbf{k}_f \cdot \mathbf{R}) | S^{(Q)} - 1 | \phi_i(\mathbf{r}) \exp(i\mathbf{k}_i \cdot \mathbf{R}) \rangle_{\mathbf{r},\mathbf{R}}|^2 \quad (85)$$

where $\Psi_i^\pm$ are the appropriate solutions, with outgoing (+) and incoming (−) scattered waves, of the stationary-state wave equation (1) and where $S_{fi}^{(Q)}$ are elements of the scattering matrix $\mathbf{S}^{(Q)}$ associated with the quantal scattering operator $S^{(Q)}$. The corresponding probability in the "time-dependent" semiclassical formulation is

$$P_{fi}^{(SC)}(t) = |\langle \phi_f | \Phi_I(t) \rangle|^2 = |\langle \phi_f | U(t, -\infty) | \phi_i \rangle|^2 \quad (86)$$

so that, by analogy with (85), a time-dependent semiclassical scattering operator can be defined as

$$\bar{S}^{(SCH)}(t) = U(t, -\infty) + 1 \quad (87)$$

with matrix elements $\bar{S}_{fi}^{(SCH)}$ given simply by the transition amplitude $a_{fi}(t)$ where the added subscript i denotes solution of (68) appropriate to $a_{fi}(t \to -\infty) = \delta_{fi}$. Note that $\bar{S}_{fi}^{(SCH)}$ is only a heavy-particle–high-energy approximation to the actual quantal $\mathbf{S}_{fi}^{(Q)}$ scattering matrix. True semiclassical time-independent scattering matrix elements $S_{fi}^{(SC)}$ are defined via the multichannel eikonal treatment (Sect. 11.3a) by

$$|\langle \phi_f(\mathbf{r}) \exp(i\mathbf{k}_f \cdot \mathbf{R}) | S^{(SC)} - 1 | \phi_i(\mathbf{r}) \exp(i\mathbf{k}_i \cdot \mathbf{R}) \rangle|^2 = -2\pi i \delta(E_i - E_f) T_{fi}$$
$$= (i\hbar/\mu) f_{if}^{(ME)}(\mathbf{k}_i, \mathbf{k}_f) \quad (88)$$

where $f_{if}^{(ME)}$ is given by (43), which tends, in the high-energy limit, to (50). This distinction between the true semiclassical $S^{(SC)}$ scattering operator and its high-energy limit $\bar{S}^{(SCH)}$ is important. A bar as in $\bar{S}$ is used to denote this distinction.

Because each component matrix $\mathbf{A}^{(k)}$ in (84a) remains anti-Hermitian, the semiclassical transition probability $\mathbf{a}(t)\mathbf{a}^\dagger(t)$ is conserved at all times. When $\mathbf{V}_I(t)$ commutes at all times, an exact solution from (84) to (79) is

$$\bar{\mathbf{S}}^{(1)} = \exp\left[-(i/\hbar) \int_{-\infty}^{t} \mathbf{V}_I \, dt \right] \quad (89)$$

Because

$$e^{\mathrm{A}} \mathrm{B} e^{-\mathrm{A}} \equiv e^{[\mathrm{A} \cdot]} \mathrm{B}$$
$$= \mathrm{B} + [\mathrm{A}, \mathrm{B}] + \frac{1}{2!} [\mathrm{A}, [\mathrm{A}, \mathrm{B}]] + \frac{1}{3!} [\mathrm{A}, [\mathrm{A}, [\mathrm{A}, \mathrm{B}]]] + \cdots \quad (90)$$

where the commutator operator $[\mathrm{A} \cdot]$ simply produces the commutator, then

$$V_I = e^{(i/\hbar) H_0 t} V e^{-(i/\hbar) H_0 t}$$

$$= V + \left(\frac{it}{\hbar}\right)[H_0, V] + \frac{1}{2!}\left(\frac{it}{\hbar}\right)^2 [H_0, [H_0, V]] + \cdots \tag{91}$$

reduces to the first term V, provided H_0 and V commute, which is also a necessary condition for $V_I(t_1)$ and $V_I(t_2)$ to commute. Hence from (89)

$$\bar{S}_{mn}^{\text{EXACT}} = \left\langle \phi_m \left| \exp\left[-(i/\hbar)\int_{-\infty}^{t} V\,dt\right] \right| \phi_n \right\rangle, \qquad [H_0, V] = 0 \tag{92}$$

is exact, provided H_0 and V commute, i.e., when $V_{mn}(\epsilon_m - \epsilon_n)$ vanishes.

Under particular conditions fulfilled in general by Rydberg atoms, $[H_0, V]$ is zero, so that (92) is then the exact scattering matrix at *high impact energies*. This circumstance is also fulfilled in general when $(\omega_{fn}t)$ in (76) is effectively zero, i.e., when the collision time $\tau_{\text{coll}} \ll 2\pi/\omega_{mn} \simeq T/(m-n)$, where T is the orbital classical internal period. Equation (92) is the *sudden approximation.* When inserted for a_m in (50) and (51), (92) yields

$$f_{mn} = -\frac{ik_m}{2\pi}\int e^{i\mathbf{K}\cdot\boldsymbol{\rho}} S_{mn}(\boldsymbol{\rho}, t \to \infty)\, d\boldsymbol{\rho} \tag{93}$$

for the scattering amplitude which reduces, with the aid of the straight-line trajectory (70) to the Glauber approximation.[8]

When H_0 and V do not commute, then (89) is the solution obtained from (83) to zero order in the commutator $[V_I(t_1), V_I(t_2)]$ and to all powers in V_I alone. Expanding (89) to first order in V_I then yields

$$\bar{S}_{fi}^{\text{B}} = -(i/\hbar)\int_{-\infty}^{t} V_{fi}(\mathbf{R}(t)) \exp(i\omega_{fi}t)\, dt \tag{94}$$

which is known as the "Born impact-parameter" transition amplitude when the straight-line trajectory (70) is appropriate for the external motion but which obviously holds for a general classical trajectory $\mathbf{R}(t)$.

In conclusion, differential and, hence, integral cross sections for electron, ion, and neutral collisions with Rydberg atoms are given by the multichannel eikonal treatment (42)–(45) and (52) when scattering is mainly in the forward direction ($\lesssim 40°$). As the impact energy is increased, the cross sections tend to those calculated from the solutions a_f of (68) inserted in (50) and (52) and in the high-energy–heavy-particle limit to (53).

When scattering into larger angles becomes important, as for thermal-energy ion and atom impacts, then the multistate orbital treatment (59) is valuable, with $\sigma_{\text{CL}}(\theta)$ in (59) based on either the optical potential $\mathcal{V}_{\text{opt}}$ of (63) or on some spherical average $\bar{V}(R)$ as (65).

11.4. Quantal–classical correspondences

11.4a. Bohr–Sommerfeld quantization

As the principal quantum number n of a Rydberg atom is increased, the electron in the highly excited state eventually behaves as a classical particle in the sense that it is becoming increasingly localized in (p, q) phase space, where the quantal imprecisions Δq_n and Δp_n associated with its generalized coordinates q_n and conjugate momenta p_n are much less than q_n and p_n, respectively. However, quantal characteristics remain evident, as exhibited, for example, by the detection in H I and H II regions of hot stars of radio-frequency line emissions between neighboring levels n (up to ~250 at present). The link between a classical description of the bound electron and the observable quantal properties is provided by the generalized Bohr–Sommerfeld quantization rule,[33] which states that the classical action

$$A_i = J_i \Delta w_i = \oint p_i \, dq_i$$
$$= 2\pi\hbar(n_i + \alpha_i), \qquad n_i = 0, 1, 2, \ldots, \qquad \alpha_i = \begin{cases} 0, & \text{rotation } q_i \\ \frac{1}{2}, & \text{libration } q_i \end{cases} \tag{95}$$

is equal to the quantum h of action times an integer n_i or half-integer $(n_i + \frac{1}{2})$, depending on whether q_i for periodic motion is a rotation or libration variable, respectively. When q_i and p_i have the same frequency, then the orbit in phase space is closed, which yields a motion (radial motion, harmonic oscillator) designated by the astronomical term *libration* and bounded by zeros in the kinetic energy. When p_i is some periodic function of q_i, which, on increase by a fixed period q_0, does not change the configuration of the system, then the phase-space orbit is not closed and the motion is designated as *rotation* (e.g., a rigid rotor). The q_i integration in (95) is over a complete period of oscillation or of rotation of q_i.

When the frequency of the classical periodic motion is highlighted (as in radiation theory), then the motion is more conveniently represented in terms of action-angle variables (J, w) than in terms of the classical trajectory of the particle in (p, q) phase space. Transformation between the two sets of variables (p, q) and (J, w) is achieved via knowledge of a generating function $S(q, J)$, which is such that[15,16]

$$p = \partial S(q, J)/\partial q, \qquad w = \partial S(q, J)/\partial J \tag{96}$$

Hamilton's equations of motion in the action-angle representation are then[15,16]

$$\begin{aligned} \dot{w}_i &= \partial \tilde{H}(J_i)/\partial J_i = \omega_i(\text{const}) = \nu_i \Delta w_i(\text{const}); & w_i &= \omega_i t + \text{const} \\ \dot{J}_i &= -\partial \tilde{H}(J_i)/\partial w_i = 0, & J_i &= \text{const} \end{aligned} \tag{97}$$

where the transformed Hamiltonian $\tilde{H}$, whose value E is preserved by the canonical transformation, depends only on the action variables J_i, which now become constants of the motion. The cyclic angle variables w_i are periodic over a range Δw_i with angular frequency ω_i and natural frequency $\nu_i = \omega_i/\Delta w_i$. When Δw_i extends to 2π such that w_i is identified as a real angle, the resulting action-angle set (J_i, w_i) with $\nu_i = \omega_i/2\pi$ is more suitable for modern quantal development than the set (J', w_i') with $\nu_i = \omega_i'$, where ω_i', with a $[0 \to 1]$ range, is customarily used in most treatises of classical mechanics.[15,16] The action integral A_i of (95) is of course invariant to either choice so that the corresponding action variables are related by $J_i = J_i'/2\pi$.

For spherically symmetric interactions $V(r)$ (central forces), the angular frequencies ω_θ and ω_ϕ, obtained from the action set (J_r, J_θ, J_ϕ) associated with $\mathbf{r}(r, \theta, \phi)$ in spherical polar coordinates, are degenerate such that $E \equiv E(J_r, J_\theta + J_\phi)$, where the sum $(J_\theta + J_\phi)$ is the total conserved angular momentum. For a Coulomb field - the Kepler problem[16] - the motion is fully degenerate $(\omega_r = \omega_\theta = \omega_\phi)$ with $E \equiv E(J_r + J_\theta + J_\phi)$ and the orbit is closed. This full degeneracy may be eliminated by transformation to a new set of "*proper*" action variables $\mathbf{J}(J_\phi, J_\phi + J_\theta, J_\phi + J_\theta + J_r)$ and angle variables $\mathbf{w}(w_\phi - w_\theta, w_\theta - w_r, w_r)$ peculiar to the Coulomb problem. Thus application of the Bohr quantization rule (95) to (J_r, J_θ, J_ϕ) associated with libration variables (r, θ) and rotation variable ϕ yields, with $\Delta w_i = 2\pi$, the proper $(\mathbf{J}, \mathbf{w})$ representation:

$$\begin{aligned} J_m &\equiv J_\phi = m\hbar, & w_m &= w_\phi - w_\theta = \phi_E(\text{const}) \\ J_l &\equiv (J_\phi + J_\theta) = (m + n_\theta + \tfrac{1}{2})\hbar \equiv (l + \tfrac{1}{2})\hbar, & w_l &= w_\theta - w_r = \psi_E(\text{const}) \\ J_n &= (J_\phi + J_\theta + J_r) = (m + n_\theta + n_r + 1)\hbar \equiv n\hbar, & w_n &= w_r = (\partial E/\partial J_n)\, t + \delta \\ J_m/J_l &= \cos\theta_E, & \epsilon^2 &= 1 - (J_l/J_n)^2 \end{aligned} \tag{98}$$

where $|m| = 0, 1, \ldots, l$; $l = 0, 1, \ldots, n$; $n = 1, 2, \ldots, \infty$; δ is a constant phase; and ϵ the eccentricity of the elliptical orbit. The new actions (J_n, J_l, J_m) can therefore be associated with quantum numbers (n, l, m), which are principal, angular momentum, and azimuthal, respectively, with the proper range of permitted values. For spherical symmetric $V(r)$, the orbit is confined to a plane. The action J_m is the constant component of the total angular momentum along a fixed Z axis and its conjugate angle w_m (fixed for constant m) is the Euler angle $\phi_E (0 \to 2\pi)$ between the line of nodes (the intersection of the plane of the orbit with the equatorial fixed XY plane) and the fixed X axis. The angle w_l conjugate to J_l is the Euler angle ψ_E between the line of nodes and some direction (in the orbit plane of the particle) usually taken as the direction of the perihelion (or perigee), which is fixed only for a Coulomb interaction. The remaining Euler angle θ_E is $\cos^{-1}(J_m/J_l)$, which is fixed for a given l and m. Hence the orientation of the orbit w.r.t. a fixed XYZ reference is defined by l and m. The remaining action J_n for a particle of reduced mass μ moving in $V(r)$ is[16]

$$J_n = \oint [2\mu(E - V(r)] - J_l^2/r^2]^{1/2}\, dr \tag{99}$$

such that $E(J_n, J_l)$. When V is $-e^2/r$, then (99) and (97) yield

$$E(J_n) = -\frac{\hbar^2}{J_n^2}\left(\frac{e^2}{2a_0}\right), \qquad \omega_n = \frac{\partial E}{\partial J_n} = n^{-3}\left(\frac{e^2}{\hbar a_0}\right) \tag{100}$$

such that for fixed n (and J_n) the particle moves in its bound orbit with a constant angular frequency ω_n. When $V(r)$ departs from pure Coulomb, a precession of the perihelion in the orbit plane occurs with angular frequency $\dot{\psi}_E$. Whereas the time-dependent $\mathbf{w}(t)$ are given by (97) and (100), the orbit $q(\mathbf{w})$ is obtained from (96).

The Bohr–Sommerfeld quantization rule (95) can now be exploited to establish the following correspondences between properties of classical motion and the quantum-mechanical observables:

1. It provides the general distribution of energy levels in the high-n spectrum, *the Bohr correspondence principle.*
2. It can be used to obtain *Heisenberg's correspondence principle*, which furnishes a powerful technique for the evaluation of various matrix elements involving highly excited states (nl) such as oscillator strengths, inelastic form factors, interaction matrix elements $V_{if}(\mathbf{R})$, S-matrix elements S_{ij}, etc., which occur in quantal and semiclassical treatments of scattering.

In the second case, the required quantal matrix element $\langle \phi_i(\mathbf{r})|V(\mathbf{r},\mathbf{R})|\phi_j(\mathbf{r})\rangle$, for example, is expressed as a given Fourier component of the static interaction $V(\mathbf{R}(t),\mathbf{r}(t))$ evaluated along a mean internal classical orbit $\mathbf{r}(t)$ for the Rydberg electron. When this classical $\mathbf{r}$ motion is periodic, the classical orbits $\mathbf{r}(\theta,\phi)$ and frequencies ω_i can be solved by the introduction of action-angle variables $(\mathbf{J},\mathbf{w})$ such that the internal $\mathbf{r}$ motion can then be quantized by replacing the action $\mathbf{J}$ by $2\pi(\mathbf{n}+\frac{1}{2})\hbar$, a procedure that provides, according to Sommerfeld, "the royal road to quantization," as quoted in Ref. 16 or else helps "sew the quantal flesh on classical bones."[34] Another advantage to the action-angle representation of the Rydberg atom in state $|nlm\rangle$ is that five of the six $(\mathbf{J},\mathbf{w})$ variables are constant, so that any small departures from a pure Coulomb field can be described very efficiently in terms of precession w.r.t. the angular variables.

11.4b. Bohr and Heisenberg correspondence principles

For large n, the energy separation ΔE between adjacent levels is small compared with the energies E_n, E_{n+1} of the levels, so that, for motion in one dimension, (95) applied to each level yields, upon subtraction,

$$\Delta E \oint (\partial p_i/\partial H_0)\, dq_i = \Delta E \oint dq_i/\dot{q}_i = \Delta E\, T_i \simeq 2\pi\hbar \tag{101}$$

where T_i is the classical period $(2\pi/\omega_i)$ for motion with angular frequency ω_i and where H_0 is the Hamiltonian. Hence, in a given range Δn of the quasi-classical spectrum, the neighboring levels are approximately equidistant by

$\hbar\omega_i$ such that the separation between any two levels n and $n+s$ within the range Δn is given by

$$E_{n+s} - E_n = h\nu_{n+s,n} \simeq s\hbar\omega_n \qquad (s = 1, 2, \ldots, \ll n) \tag{102a}$$

which is *Bohr's correspondence* between the line emission frequency $\nu_{n+s,n}$ and the angular frequency ω_n of classical orbital motion.

Use of Taylor's expansion yields, for quantal hydrogenic systems,

$$E_{n+s} - E_n = s\hbar\omega_n[1 - \tfrac{3}{2}(s/n) + 4(s/n)^2 + \cdots] \tag{102b}$$

which agrees with (102a) when $s \ll n$. This principle is easily extended to include D dimensions by adopting vector notation $\boldsymbol{\omega} \equiv \{\omega_i\}$, $\mathbf{n} \equiv \{n_i\} = (n, l, m, \ldots)$ and $\mathbf{s}(n'-n, l'-l, m'-m, \ldots)$. From (97), with angular range $\Delta\omega_i$ as 2π and (95)

$$\partial E/\partial n_i = \hbar\omega_i \tag{103}$$

such that the energy separation of levels $(n'l'm')$ and (nlm) is

$$\hbar\omega_{\mathbf{n}'\mathbf{n}} = E(\mathbf{n}') - E(\mathbf{n}) = \sum_{i=1}^{D} \Delta n_i\,(\partial E/\partial n_i) = \hbar\mathbf{s} \cdot \boldsymbol{\omega} \tag{104}$$

The number of states with quantum numbers in the range $\Delta\mathbf{n}$ is

$$\Delta N = \prod_{i=1}^{D} \Delta n_i = \prod_{i=1}^{D} (\Delta J_i\,\Delta w_i)/(2\pi\hbar)^D = \prod_{i=1}^{D} (\Delta p_i\,\Delta q_i)/(2\pi\hbar)^D \tag{105}$$

because the $(\mathbf{p}, \mathbf{q} \to \mathbf{J}, \mathbf{w})$ transformation is canonical, which, therefore, implies invariance of the corresponding elements of phase space. Each quantum state of a system with D degrees of freedom occupies a cell of phase volume $(2\pi\hbar)^D$, as expected.

The mean value $\bar{F}$ of any physical quantity $F(q)$ in any quantum state Ψ tends, in the classical limit, to the classical value of that quantity *provided* that the quantum state itself can be described in the same limit by classical motion of the particle in a definite path. On expanding Ψ in terms of the quantal stationary states Ψ_n of the system with expansion coefficients a_n,

$$\bar{F} = \langle\Psi|F|\Psi\rangle = \sum_n \sum_m a_m^* a_n F_{mn}^{(\mathrm{q})} \exp(i\omega_{mn} t) \tag{106}$$

where $F_{mn}^{(\mathrm{q})}$ are the required quantal matrix elements between the time-independent states ϕ_n. With correspondence (104) used for ω_{mn} in (106) and with $s = n - m$,

$$\bar{F} = \sum_{n=1}^{\infty} \sum_{s=-\infty}^{n-1} a_{n-s}^* a_n F_{n-s,n}^{(\mathrm{q})} \exp(-is\omega t) \tag{107}$$

Although the elements $F_{n-s,n}$ decrease rapidly as $|s|$ increases, they exhibit only a slow variation with n for s fixed within some small range $\Delta n \ll n$ about n, so that

$$\bar{F} = \sum_n |a_n|^2 \sum_{s=-\infty}^{\infty} F_s^{(q)} \exp(-is\omega t); \qquad F_s = F_{\tilde{n}-s,\tilde{n}} \tag{108}$$

where $\tilde{n}$ is some mean value of n in the range Δn. Because $\bar{F}$ in the classical limit tends to $F^c(t)$ and $\Sigma_n |a_n|^2 = 1$, the quantal matrix elements F_{mn} reduce in the classical limit to the components F_{n-m} obtained in the expansion of the classical function $F^c(t)$ as a Fourier series, i.e.,

$$F_{mn}^{(q)}(\mathbf{R}) = \int_0^\infty \phi_m^*(r) F(r, \mathbf{R}) \phi_n(r)\, dr \simeq F_s^{(c)}$$

$$= (\omega/2\pi) \int_0^{2\pi/\omega} F^{(c)}(r(t)) \exp(is\omega t)\, dt \tag{109}$$

where $r(t)$ is the classical variation of r with time t. When bounded motion for several degrees of freedom can be described by a separable Hamiltonian (e.g., Kepler problem), the classical position $\mathbf{r}$ is periodic in the angle variables $\mathbf{w}$ and separates as a Fourier series[16]

$$\mathbf{r}(t) = \sum_{\text{all } \mathbf{s}} \mathbf{r}_s(\mathbf{J}) \exp(-i\mathbf{s} \cdot \mathbf{w}), \qquad \mathbf{w} = \omega t + \delta \tag{110}$$

where the components of order $+s$ and $-s$ result in emission and absorption of radiation at frequency $s\omega$. Any classical function $F^c(\mathbf{r})$ is also periodic in $\mathbf{r}$, i.e.,

$$F^c(\mathbf{r}(t)) = \sum_{\mathbf{s}} F_{\mathbf{s}}^c(\mathbf{J}) \exp(-i\mathbf{s} \cdot \mathbf{w}) \tag{111}$$

where the Fourier coefficients in three dimensions yield the quantal matrix elements

$$F_{\mathbf{n}'\mathbf{n}}^{(q)} \simeq F_{\mathbf{s}}^c(\mathbf{J}) = (2\pi)^{-3} \int F^c(\mathbf{r}(\mathbf{J}, \mathbf{w})) \exp(i\mathbf{s} \cdot \mathbf{w})\, d\mathbf{w}, \tag{112}$$

in the classical limit, where $\mathbf{n} \equiv (n, l, m)$ and $\mathbf{s} = (\mathbf{n} - \mathbf{n}')$. Expressions (109) and (112) represent *the Heisenberg correspondence principle* for one and three internal degrees of freedom, respectively. When $\mathbf{s}$ is zero, the expectation value of F is then the average of F over the classical orbit associated with quantum numbers (nlm). A nondiagonal element is a similar average of $F \exp(i\Delta S/\hbar)$, where ΔS is the difference between the actions $J_n w_n$ and $J_{n'} w_{n'}$ associated with the initial and final classical orbits at a given time.

For example,

$$\sum_{m=-l}^{l} \sum_{m'=-l'}^{l'} |\langle n'l'm'|F(\mathbf{r})|nlm\rangle|^2$$

$$= (2\pi)^{-6} \sum_{m=-l}^{l} \sum_{\Delta m} \int F(\mathbf{J}, \mathbf{w}) F^*(\mathbf{J}, \mathbf{w}') \exp[i\mathbf{s} \cdot (\mathbf{w} - \mathbf{w}')]\, d\mathbf{w}\, d\mathbf{w}' \tag{113}$$

with $d\mathbf{w}=(dw_n, dw_l, dw_m)$ and $\mathbf{s}=-(\Delta n, \Delta l, \Delta m)$, which, on summation over Δm, reduces with the aid of (98) to

$$\sum_m \sum_{m'} |\langle n'l'm'|F(\mathbf{r})|nlm\rangle|^2$$

$$= \frac{(l+\frac{1}{2})}{(2\pi)^5} \int_{-1}^{+1} d(\cos\theta_E) \int_0^{2\pi} d\phi_E \left| \int F(\mathbf{J},\mathbf{w}) \exp[-i(\Delta n\, w_n + \Delta l \psi_E)\, dw_n\, d\psi_E \right|^2 \tag{114}$$

Similarly, on summing (114) over all l and l' and averaging over the n^2 initial states, we have

$$|\langle n'|F(\mathbf{r})|n\rangle|^2$$

$$= (2\pi)^{-4} \int_0^1 d\epsilon^2 \int_{-1}^{+1} d(\cos\theta_E) \int_0^{2\pi} d\psi_E \int_0^{2\pi} d\phi_E \left| \int F(\mathbf{r}(\mathbf{J},\mathbf{w})) \exp(-i\Delta n\, w_n)\, dw_n \right|^2 \tag{115}$$

which is an average over w_n in the plane of the orbit, over orientations (ϕ_E, different m or lines of nodes; ψ_E, different l or perihelions; θ_E, different planes of orbit for fixed line of nodes or directions of total angular momentum) of that plane, and over all possible eccentricities (shapes) ϵ^2 of the orbits. Thus (115) represents an average of the one-dimensional result (109) over a microcanonical distribution, which assumes equal weighting of each of the n^2 states (lm) and uniform distribution of ϵ^2 between 0 and 1. Evaluation of the w_n integral in (115) or (109) can be achieved for bounded motion under a pure Coulomb attraction via the following parameter form of the classical orbits:[35]

$$r = a_n(1-\epsilon\cos\chi), \quad w_n = \omega_n t + \delta = \omega_n \tau = \chi - \epsilon \sin\chi$$

$$x = a_n(\cos\chi - \epsilon), \quad y = a_n(1-\epsilon^2)^{1/2}\sin\chi, \quad \tan\tfrac{1}{2}w_n = \left(\frac{1+\epsilon}{1-\epsilon}\right)^{1/2} \tan\tfrac{1}{2}\chi \tag{116}$$

where χ is the eccentric anomaly that varies between 0 and 2π (0 at perihelion and π at aphelion) during a full revolution of the angle w_n and τ the epoch or time measured from the instant when the particle is at perihelion [perigee, $\chi=0$, $r=a_n(1-\epsilon)$]. The semimajor axis a_n, eccentricity ϵ, and angular frequency ω_n for the classical orbit appropriate to fixed n and l are

$$a_n = n^2 a_0, \qquad \epsilon = [1-(l+\tfrac{1}{2})^2/n^2]^{1/2}, \qquad \omega_n = n^{-3}(e^2/a_0\hbar) \tag{117}$$

11.4c. Correspondence approximations for atomic properties and collision amplitudes

By selection of different observables F, a variety of correspondence limits can be obtained and expressed as further correspondence principles. For example, the mean power spontaneously emitted in a radiative transition $n' \to n$ or the mean power absorbed from incident radiation due to induced upward and downward transitions $n' \leftrightarrow n$ can be identified, respectively, with the mean emitted or absorbed power determined by the classical Fourier components of order s $(=n-n')$ and $-s$ of the dipole moment.[13,36]

Formulae for the classical intensity of the components of dipole radiation are given in standard textbooks of classical radiation theory.[35,37] Also, the classical energy $\Delta E^{(c)}$ transferred to a Rydberg atom by an incident charged particle is, with the aid of (111) and (112),

$$\Delta E^{(c)} = \int_{-\infty}^{\infty} \mathbf{F} \cdot \mathbf{v}\, dt = \sum_{s=-\infty}^{\infty} \mathbf{v}_s \cdot \int_{-\infty}^{\infty} \mathbf{F}(t) \exp(-is\omega t)\, dt$$

$$= \sum_s \Delta E_s^{(c)} \simeq \sum_{n'} \overline{\Delta E_{n'n}^{(q)}} \tag{118}$$

which equals the mean energy transferred in quantal upward and downward transitions $n \leftrightarrow n'$ summed over all final states n'. Here $\mathbf{v}(\mathbf{r}(t))$ is the classical velocity of the Rydberg electron at $\mathbf{r}$ and $\mathbf{F}$ the electric force provided at the electron by the incident charge Ze of impact parameter ρ. When $\rho \gg r$, then F remains effectively constant over the atom (dipole approximation) and the preceding Fourier transforms (of momentum transfers) $\mathbf{F}_s$ of $\mathbf{F}$ can be readily obtained[37] in terms of modified Bessel functions $K_{0,1}$ as

$$\mathbf{F}_s(\omega) = \int_{-\infty}^{\infty} \mathbf{F}(t) \exp(-is\omega t)\, dt$$

$$= 2\left(\frac{Ze^2}{v\rho}\right)\left(\frac{\omega s \rho}{v}\right)\left[-iK_0\left(\frac{s\omega\rho}{v}\right)\hat{\mathbf{k}} + K_1\left(\frac{s\omega\rho}{v}\right)\hat{\boldsymbol{\rho}}\right] \tag{119}$$

where $\hat{\mathbf{k}}$ is the direction of incidence normal to the impact-parameter direction $\hat{\boldsymbol{\rho}}$. Equation (119) can then be used in (118) to determine the energy transferred to a charge bound harmonically[37] or by a Coulomb attraction for which $\mathbf{v}_s$, with the aid of (116), is[35]

$$\mathbf{v}_s = -is\omega_n \mathbf{r}_s = \omega_n a_n\{-iJ_s'(s\epsilon)\hat{\mathbf{i}} + [(1-\epsilon^2)^{1/2}/\epsilon]J_s(s\epsilon)\hat{\mathbf{j}}\} \tag{120}$$

where $\hat{\mathbf{i}}$, directed toward the perihelion, and $\hat{\mathbf{j}}$ denote directions of the XY axes in the orbit plane and J_s' the differentiation of the Bessel function J_s with respect to its argument. The mean energy transferred $\langle \Delta E^{(q)} \rangle$ in $n \leftrightarrow n'$ collisional transitions is therefore $F_s(\omega_n)v_s$, with $s=n-n'$ in the appropriate correspondence limit of weak interaction for simultaneous validity of both classical and quantal perturbation theory (small variation to classical orbit and small transition probability $P_{nn'}$, respectively). By paying attention to detailed balancing between upward and downward transition probabilities, Percival and Richards,[38] with the use of (118)–(120), were able to show that at asymptotic impact energies E the cross section for charged-particle–Rydberg collisions varies as $E^{-1} \ln E$, in contrast to the E^{-1} variation obtained from a classical theory (binary encounter, for example) based on the full classical energy transfer $\Delta E^{(c)}$. Correspondence is, therefore, preserved, provided classical probabilities are deduced from the Fourier components $\Delta E_s^{(c)}$ of $\Delta E^{(c)}$ rather than from $\Delta E^{(c)}$ alone.[38] For Rydberg collisions at lower energies, contributions from higher multipoles tend to dominate the cross sections (see Sect. 11.2). The $\ln E$ term is significant therefore only in the high-energy limit where the proper energy change is given by the individual components $\mathbf{F}_s \cdot \mathbf{v}_s$ of (118).

Atomic properties

The accuracy of results obtained from Heisenberg's principle can be tested, when possible, by direct comparison with quantal results. It follows from (120) that the dipole matrix element summed over all states m and m' and averaged over the $(2l+1)$ initial states is

$$|\langle n'l \pm 1|\mathbf{r}|nl\rangle|_c^2 = \frac{3}{2}\, a_n^2 \frac{1+(s/n)}{3s}\, [J_s'(s\bar{\epsilon}) \pm (\bar{\epsilon}^{-2}-1)^{1/2} J_s(s\bar{\epsilon})]^2 g_c(s) \tag{121}$$

where $\bar{\epsilon}$ is the eccentricity (117) appropriate to some averaged $\tilde{l}$ and $\tilde{n}$ and

$$g_c(s) = \frac{2n'}{n+n'}\,\frac{\tilde{n}}{n} \tag{122}$$

On averaging over all n^2 initial states $|lm\rangle$ and summing over all final states i.e., integrating (121) over ϵ^2 between 0 and 1, then

$$|\langle n'|\mathbf{r}|n\rangle|_c^2 = \frac{3}{2}\, a_n^2\left(1+\frac{s}{n}\right)\frac{4}{3s^3} J_s(s) J_s'(s) g_c(s) \tag{123}$$

to be compared with the quantal results of Menzel,[39] who gives (123) with g_c replaced by $1+(3s/2n)+[A(s)/n^2]$, where $A(s)$ is of order unity. For a mean orbit, $\tilde{n}=\frac{1}{2}(n+n')$ and g_c is then $1+(s/n)$. For $\tilde{n}=2n'^2/(n+n')$, which ensures detailed balance between the oscillator strengths

$$n^2 f_{nn'} = [\tfrac{2}{3}(m/\hbar^2)E_{n'n}]n^2|\langle n'|\mathbf{r}|n\rangle|^2 = -n'^2 f_{n'n} \tag{124}$$

for emission and absorption, then g_c is $1+(2s/n)+(2s^2/4n^2)$.[40] Use of asymptotic J_s in (123) for large s but $\ll n$ yields the Gaunt–Kramers result[41]

$$f_{nn'} \simeq (0.49/2n^2)/(nn'\omega)^3, \qquad 2\omega = n^{-2} - n'^{-2} \tag{125}$$

and provides a reasonable approximation for the oscillator strength between two excited levels even for small s. On extrapolating n' to a continuum state of energy ϵ,

$$df(n,\epsilon)/d\epsilon \simeq (0.49/2n^2)/(n\omega)^3 \tag{126}$$

Upon use of the oscillator strength sum rule, we note, therefore, that most of the oscillator strength from an excited level n is essentially exhausted by transitions to bound levels and that little remains ($\sim 0.49/n$) for transitions to the continuum (a feature responsible even for the abnormally small cross sections for the photoionization of metastable rare gas atoms).[42]

Further applications of the correspondence principle to determination of various matrix elements are given by Naccache[43] and de Prunelé.[44] In conclusion, agreement between quantal and correspondence results are, in general, obtained when $s \ll n$, $n' \gg 1$. However, de Prunele[44] pointed out that a further sufficient condition for validity is that $F(\mathbf{r}, \mathbf{p})$ must not significantly weight the classical inaccessible region. If this latter condition is not fulfilled, then an appropriate extension of the semiclassical procedure given in Sect. 11.4 is valuable.

Collision amplitudes

Use of the Bohr and Heisenberg correspondences (102) and (112), respectively, for ω_{fi} and V_{fi} in the first-order semiclassical result (94) yields

$$\begin{aligned} S_{fi}^{(1)} &= -\frac{i}{\hbar}\left(\frac{\omega}{2\pi}\right)\int_{-\infty}^{\infty} dt \int_{0}^{2\pi/\omega} V(\mathbf{R}(t), r(t_1)) \exp[is\omega(t_1 - t)]\, dt_1 \\ &= -\frac{i}{\hbar}\left(\frac{\omega}{2\pi}\right)\int_{0}^{2\pi/\omega} dt_e \int_{-\infty}^{\infty} V(\mathbf{R}(t), r(t + t_e)) \exp(is\omega t_e)\, dt, \qquad s = i - f \end{aligned} \tag{127}$$

in which the external (projectile) motion along a classical path $\mathbf{R}(t)$ is correlated to the classical internal motion of the Rydberg electron in the orbit $r(t_e)$. When $S_{fi}^{(1)}$ is substituted for a_f in (53) with its straight-line external trajectory, the resulting cross sections are valid only (Sect. 11.3) in the heavy-particle–high-energy limit for collisions of ions, neutrals, and electrons with Rydberg atoms.

For the sudden approximation (92), F in (112) is taken as $S^{(S)}$ such that

$$S_{fi}^{(S)} = \frac{\omega}{2\pi}\int_{0}^{2\pi/\omega} dt_e \exp\left[is\omega t_e - \frac{i}{\hbar}\int_{-\infty}^{\infty} V(\mathbf{R}(t), r(t_e))\, dt\right] \tag{128}$$

directly, where there is now no correlation between internal and external motions in accordance with the impulse assumption. It is tempting to replace $r(t_e)$ in (128) by $r(t+t_e)$ in order to obtain an expression that yields both the correct sudden and weak-interaction limits. Percival and Richards[5] introduced such correlations into (128) by substituting the sequential relation

$$U(t_N, t_0) = \prod_{j=0}^{N-1} \mathbf{U}(t_{j+1}, t_j), \qquad t_j = t_0 + j\,\Delta t \tag{129}$$

between the evolution operators for small time intervals Δt into (80), which they then solved to first order in V_I to give, upon use of Heisenberg's correspondence, the overall matrix element that reduces (after detailed analysis) to

$$U_{fi}(T,t_0) = \frac{\omega}{2\pi} \int_0^{2\pi/\omega} \prod_{j=0}^{N-1} \left[1 - \frac{i\,\Delta t}{\hbar} V(\mathbf{R}(t_j), r(t_j+\tau)) \right] \exp(is\omega\tau)\, d\tau \quad (130)$$

where $s=i-f$. Upon replacing the bracketed term by an exponential (equivalent to first order in Δt) and by letting $N\to\infty$ such that the product is converted to an integral, Percival and Richards[5] obtained the following collision amplitude:

$$S_{fi}^{(\mathrm{SC})} = \frac{\omega}{2\pi} \int_0^{2\pi/\omega} \exp\left[i(s\omega t_e) - \frac{i}{\hbar} \int_{-\infty}^{\infty} V(\mathbf{R}(t), r(t+t_e))\, dt \right] dt_e \quad (131)$$

which they termed the *strong correspondence principle*. This result indeed exhibits the desired characteristics of (127) for weak V and of (128) for sudden collision.

An interesting and rigorous theoretical generalization of (128) that includes diabatic effects and a three-dimensional electron orbit will now be presented.

11.4d. Equivalent oscillator theorem

The main advantage underlying the use of the action-angle representation in quantum mechanics is that the unperturbed Hamiltonian $H_0(\mathbf{J})$ depends only on $\mathbf{J}$, which is therefore conserved (because $\mathbf{w}$ is cyclic). The action operator $\hat{\mathbf{J}}$ is defined with the aid of its associated quantal commutator

$$[\hat{J}_i, w_i] = -i\hbar \quad (132)$$

and of (95) as

$$\hat{J}_i \phi_{n_i}(w_i) = \left[-i\hbar \frac{\partial}{\partial w_i} + \alpha_i \hbar \right] \phi_{n_i}(w_i) = J_i \phi_{n_i} = (n_i + \alpha_i)\hbar \quad (133)$$

such that

$$\phi_{\mathbf{n}}(\mathbf{w}) = (2\pi)^{-3/2} \exp(i\mathbf{n}\cdot\mathbf{w}) \quad (134)$$

in three dimensions. Hence the quantal matrix elements $F_{\mathbf{n}'\mathbf{n}}$ of $F(\mathbf{n})$ given by Heisenberg's correspondence principle (112) are simply the matrix elements of F in the action-angle representation. Because the total energy E_n and time t are also conjugate variables, the time-dependent wave function in the action-angle representation is

$$\phi_{\mathbf{n}}(\mathbf{w},t) = \phi_n(\mathbf{w}) \exp(-iE_n t/\hbar) \quad (135)$$

The "classical path" equation (66) in this representation is

$$[H_0(\hat{\mathbf{J}}) + V(\mathbf{J},\mathbf{w},t)]\Psi(\mathbf{w},t) = i\hbar\, \partial\Psi(\mathbf{w},t)/\partial t \quad (136)$$

for which the system wave function is expanded in terms of the unperturbed basis set $\{\phi_n(\mathbf{w},t)\}$ as

$$\Psi(\mathbf{w},t) = \mathop{\mathrm{S}}_{n} a_n(t)\phi_n(\mathbf{w}) \exp(-iE_n t/\hbar) \quad (137)$$

The transition amplitudes [or $S_{fi}(t\to\infty)$] then satisfy

$$i\hbar\,\partial a_f/\partial t = \mathop{\mathrm{S}}_{n} a_n(t)V_{fn}(t)\exp(i\omega_{fn}t) = \mathop{\mathrm{S}}_{d=-f} a_{d+f}(t)V_d(t)\exp(-id\omega t) \tag{138}$$

in which the Bohr and Heisenberg correspondence principles (102) and (112) have been invoked, respectively, for $\omega_{fn}=(f-n)\omega\equiv -d\omega$ and for V_{fn}, i.e.,

$$V_{fn}(t) = V_d(t) = (2\pi)^{-3}\int_0^{2\pi} V(\mathbf{w}',t)\exp(i\mathbf{d}\cdot\mathbf{w}')\,d\mathbf{w}', \qquad \mathbf{d}=\mathbf{n}-\mathbf{f} \tag{139}$$

Hence set (138) can be written as

$$i\hbar\,\partial a_f/\partial t = \mathop{\mathrm{S}}_{d=-f} a_{d+f}(t)V'_d(t) \tag{140}$$

where, on replacing $(\mathbf{w}-\omega t)$ by the angle variable $\mathbf{w}$,

$$V'_d(t) = \exp(-id\omega t)V_d(t) = (2\pi)^{-3}\int_0^{2\pi} V(\mathbf{w}+\omega t,t)\exp(i\mathbf{d}\cdot\mathbf{w})\,d\mathbf{w}$$
$$\equiv \langle\phi_f(\mathbf{w})|V(\mathbf{w}+\omega t,t)|\phi_n(\mathbf{w})\rangle \tag{141}$$

The operator $V(\mathbf{w}+\omega t,t)$ commutes at different times t_1 and t_2 such that set (140) can be solved exactly (Sect. 11.3a), subject to $a_f(t\to-\infty)=\delta_{fi}$, to yield

$$a_f(t) = \left\langle\phi_f(\mathbf{w})\left|\exp\left[-(i/\hbar)\int_{-\infty}^{t} V(\mathbf{w}+\omega t,t)\,dt\right]\right|\phi_i(\mathbf{w})\right\rangle \tag{142}$$

and hence the transition probability amplitude is

$$S_{\mathbf{n}'\mathbf{n}} = a_{\mathbf{n}'}(t\to\infty) = (2\pi)^{-3}\int_0^{2\pi} d\mathbf{w}\exp\left[i\mathbf{s}\cdot\mathbf{w} - (i/\hbar)\int_{-\infty}^{\infty} V(\mathbf{w}+\omega t,t)\,dt\right] \tag{143}$$

where $\mathbf{s}=\mathbf{n}-\mathbf{n}'$ for $\mathbf{n}(nlm)\to\mathbf{n}'(n'l'm')$ transitions. This exact expression follows directly from (136) under the conditions of Bohr and Heisenberg correspondence, which, therefore, expose the key – the Fourier index d in (139) being equal to the angular index d in $\exp(-id\omega t)$ of (138) – which is essential for an exact solution. Because w_n of $\mathbf{w}(w_n,w_l,w_m)$ is ωt_e, where t_e is the time describing the internal Rydberg electron in its orbit, we note that the action integral over the external classical path contains, via $(\mathbf{w}+\omega t)$, a time evolution that expresses coupling between the classical internal and external motions, i.e., diabatic effects are acknowledged.

In the limit of weak interaction, (i.e., when $V\tau_c\ll 1$ where τ_c is the collision time),

$$S^{(B)}_{\mathbf{n}',\mathbf{n}} = (2\pi)^{-3}(-i/\hbar)\int_0^{2\pi} d\mathbf{w}\exp(i\mathbf{s}\cdot\mathbf{w})\int_{-\infty}^{\infty} V(\mathbf{w}+\omega t,t)\,dt \tag{144}$$

which is the equivalent of the Born approximation to (138) and which also correlates the internal and external classical motions of the Rydberg electron and projectile, respectively.

In the impulse limit, i.e., when $\tau_c \ll w/\omega \simeq 2\pi/\omega = T_e$ the orbital period,

$$S^{(S)}_{\mathbf{n}',\mathbf{n}} = (2\pi)^{-3}\int_0^{2\pi} d\mathbf{w}\exp\left[i\mathbf{s}\cdot\mathbf{w} - (i/\hbar)\int_{-\infty}^{\infty} V(\mathbf{w},t)\,dt\right] \tag{145}$$

which is the equivalent of the sudden approximation that neglects ω_{fn} in (138). We note also that (145) contains no correlation between the internal and external motions in accordance with the impulse assumption that the Rydberg electron remains fixed during the encounter. When (145) is inserted for a_f in (50) and (53), then expressions identical with the Glauber approximation are obtained.[8]

Because the only assumptions underlying (143) are Bohr correspondence (equally spaced levels as in a harmonic oscillator) and Heisenberg correspondence for the interaction matrix elements, the basic expression (138) yields what we term the *equivalent oscillator theorem* for the collisional transition amplitudes. Similar or related expressions have also been obtained less directly by other means (see Sects. 11.4e and 11.4f).[4-7]

The classical limit to (143) is obtained by evaluating the integral by the method of stationary phase, which is located at $\mathbf{w}$, given by

$$\mathbf{s}\hbar = \frac{\partial}{\partial\mathbf{w}}\left[\int_{-\infty}^{\infty} V(\mathbf{J},\mathbf{w},t)\,dt\right] = -\int_{-\infty}^{\infty}\mathbf{F}(\mathbf{J},\mathbf{w},t)\,dt = (\mathbf{n}-\mathbf{n}')\hbar \tag{146}$$

where $\mathbf{F}$ is the generalized force $-\nabla_{\mathbf{w}} V$. Thus, in the classical limit, the change $(\mathbf{n}'-\mathbf{n})\hbar$ in the generalized momentum $\mathbf{J}$ is given by the impulse of the generalized force. The transition amplitude (143) is therefore

$$S_{\mathbf{n}'\mathbf{n}} = (2\pi)^{-3}\int_0^{2\pi} d\mathbf{w}\,\delta\left[\mathbf{s}\hbar + \int_{-\infty}^{\infty} F(\mathbf{J},\mathbf{w}+\omega t,t)\,dt\right] \tag{147}$$

in accordance with the classical perturbation theory.

11.4e. Semiclassical wave functions, correspondences, and perturbation theory

It is well known that the WKB-bound state wave functions for classical accessible and inaccessible regions when matched result directly in Bohr's quantization rule (95).[30] It is, therefore, of interest to develop the radial matrix elements appropriate to the radial WKB wave function

$$\chi_{nl}(r) = \left(\frac{1}{\pi}\frac{\partial p_r}{\partial J_r}\right)^{1/2}\exp\left[iS_{nl}(q_r,J_r)/\hbar + \frac{\pi}{4}\right] \tag{148}$$

where, for the classical accessible region, the radial generator of the canonical $(\mathbf{p},\mathbf{q}\to\mathbf{J},\mathbf{w})$ transformation that preserves the value E of the Hamiltonian $H(\mathbf{p},\mathbf{q}) = \tilde{H}(\mathbf{J}) = E$ in the action-angle $(\mathbf{J},\mathbf{w})$ representation is

$$S_{nl}(q_r, J_r) = \int p_r\, dq_r = \int_{r_1}^{r} \left\{2m[E_{nl} - V(r)] - \frac{(l+\frac{1}{2})^2\hbar^2}{r^2}\right\}^{1/2} dr \tag{149}$$

for a particle of mass m bound with energy E_{nl} by $V(r)$ between classical turning points $r_{1,2}$ given by $p_r^2(r_{1,2})=0$. This transformation is therefore governed by

$$\begin{aligned} \dot{w}_r &= \partial\tilde{H}(\mathbf{J})/\partial J_r \equiv \dot{q}_r\, \partial p_r/\partial J_r = \omega_n, & p_r &= \partial S(q_r, J_r)/\partial q_r \\ \dot{J}_r &= \partial\tilde{H}(\mathbf{J})/\partial w_r = 0, & w_r &= \partial S(q_r, J_r)/\partial J_r \end{aligned} \tag{150}$$

For each full revolution, the action phase (149) of χ_{nl} increases by the action integral

$$J_r\, \Delta w_r = 2\pi J_r = \oint p_r\, dq_r = S_{nl}(2\pi, J_r) \tag{151}$$

evaluated around a closed loop (denoted by 2π).

When the levels nl and $n'l'$ are close (i.e., $\Delta = (n_r - n_r') \ll n, n'$), then

$$S_{n'l'}(q_r, J_r') = S_{nl}(q_r, J_r) + (\partial S_{nl}/\partial J_r)\,(J_r' - J_r) + \cdots \tag{152}$$

which, with the aid of (98)

$$J_r = (n_r + \tfrac{1}{2})\hbar = (n - l - 1)\hbar, \qquad w_n = w_r \tag{153}$$

for a Rydberg atom, yields the eikonal difference

$$S_{nl}(q, J_r) - S_{n'l'}(q, J_r') = (\Delta n - \Delta l)\hbar\omega_n t, \qquad \Delta n = n' - n, \quad \Delta l = l' - l \tag{154}$$

Hence, with the aid of $\dot{w}_n$ in (148), the radial matrix element

$$\begin{aligned} F_{n'l', nl} &= \int_{r_1}^{r_1} \chi^*_{n'l'}(r) F(r) \chi_{nl}(r)\, dr \\ &= (1/2\pi) \oint F(r) \exp[i(\Delta n - \Delta l) w_n]\, dw_n \end{aligned} \tag{155}$$

such that the function

$$F(r) = \sum_{s=-\infty}^{\infty} F_{n'l', nl} \exp[-i(\Delta n - \Delta l) w_n] \tag{156}$$

which is equivalent to the Heisenberg correspondence principle for radial matrix elements. The angular portion of the three-dimensional matrix element $\langle \phi_{n'l'}(\mathbf{r}) | V(\mathbf{r}) | \phi_{nl}(\mathbf{r}) \rangle$ may be determined either quantum mechanically by standard angular momentum coupling theory involving $Y_{lm}(\hat{\mathbf{r}})$ or by the semiclassical angular equivalents to (148) and (155). Thus a full semiclassical analysis – (148) or its three-dimensional equivalent in (34) – leads directly to the equivalent oscillator theorem (143).

The semiclassical wave function over the classical accessible (q_1, q_2) and inaccessible regions are appropriate combinations of

$$\chi_{nl}(q) = \left(\frac{2\omega_n}{\pi^{1/2}p_r}\right)^{1/2} \sin\left[\frac{1}{\hbar}\int_{q_1}^{q} p\,dq + \frac{\pi}{4}\right], \qquad q \gg q_1$$

$$= \left(\frac{\omega_n}{2\pi^{1/2}|p_r|}\right)^{1/2} \exp\left[-\frac{1}{\hbar}\int_{q}^{q_1} |p|\,dq\right], \qquad q \ll q_1$$

$$= \pm\left(\frac{2\omega_n}{\pi^{1/2}p_r}\right)^{1/2} \sin\left[\frac{1}{\hbar}\int_{q}^{q_2} p\,dq + \frac{\pi}{4}\right], \qquad q \ll q_2$$

$$= \pm\left(\frac{\omega_n}{2\pi^{1/2}|p_r|}\right)^{1/2} \exp\left[-\frac{1}{\hbar}\int_{q_2}^{q} |p|\,dq\right], \qquad q \gg q_2 \tag{157}$$

where the inequalities are satisfied within a few wavelengths. The sum of phases for a bound state equals $n_r\pi$ and Bohr's quantization (98) is recovered. Matrix elements of a function $F(p,q)$ that weights the classical inaccessible region significantly may then be determined by using (155) for the classical accessible region (which contains all the difficult oscillatory portion of the bound-state wave function) and the exponential increasing and decreasing $\chi_{nl}(q)$ in (157) for the inaccessible regions.

In the action-angle representation, the normalized wave function is

$$\chi_{nl}(w_r) = (2\pi)^{-1/2}\exp(iJ_r w_r/\hbar) = (2\pi)^{-1/2}\exp i[n - (l + \tfrac{1}{2})]w_n \tag{158}$$

from which the wave function in the (p,q) representation is obtained from

$$\chi_{nl}(q) = \oint A_1(q, w_r)\chi_{nl}(w_r)\,dw_r \tag{159}$$

where the transform A_1 is unitary, i.e.,

$$\int A_1^*(q, w_r')A_1(q, w_r)\,dq = \delta(w_r - w_r') \tag{160}$$

to preserve orthonormality $\langle\chi_i|\chi_f\rangle = \delta_{if}$ in both representations. The quantal matrix elements

$$\langle\chi_{n'l'}(q)|F(q)|\chi_{nl}(q)\rangle = \oint \chi_{n'l'}^*(w_r)F(J_r, w_r)\chi_{nl}(w_r)\,dw_r \tag{161}$$

are therefore the corresponding matrix elements in the action-angle $(\mathbf{J}, \mathbf{w})$ representation. Note that the transform A_1 for hydrogenic systems is simply the Fourier transform of the WKB semiclassical wave function (148).

Not only does the action-angle representation permit wave functions (134), (135) and (158) to have particularly simple forms but it also permits construction of a systematic perturbation procedure based on the Hamilton–Jacobi equation[15]

$$(\partial S/\partial t) + H_0(\partial S/\partial \mathbf{w}) + V(\mathbf{w}^0, \partial S/\partial \mathbf{w}, t) = 0 \tag{162}$$

where $S(\mathbf{w}^0, \mathbf{J})$ is the generator of the canonical transformation from the "unperturbed" representation $(\mathbf{J}^0, \mathbf{w}^0)$ to the "perturbed" representation $(\mathbf{J}, \mathbf{w}^0)$ such that

$$\mathbf{J}^0 = \partial S(\mathbf{w}^0, \mathbf{J})/\partial \mathbf{w}^0, \qquad \mathbf{w} = \partial S(\mathbf{w}^0, \mathbf{J})/\partial \mathbf{w}^0$$
$$\dot{\mathbf{J}} = -\partial H(\mathbf{J}, \mathbf{w})/\partial \mathbf{w}, \qquad \dot{\mathbf{w}} = \partial H(\mathbf{J}, \mathbf{w})/\partial \mathbf{J} \tag{163}$$

where H, the total Hamiltonian $H_0(\mathbf{J}^0) + \lambda V(\mathbf{J}, \mathbf{w})$, is no longer a function of $\mathbf{J}$ alone. By expanding S as $S_0 + \lambda S_1 + \cdots$ and

$$H_0(\mathbf{J}) - H_0(\mathbf{J}^0) = (\partial H_0/\partial \mathbf{J})_{\mathbf{J}^0} \cdot (\mathbf{J} - \mathbf{J}_0) + \cdots$$
$$\equiv \lambda \sum_j (\partial H_0/\partial J_j)(\partial S_1/\partial w_j^0) + \cdots \tag{164}$$

then the first two action terms are

$$S_0 = \mathbf{J}^0 \cdot \mathbf{w}^0 - E_n^0 t \tag{165}$$

and, provided S_1 is separable in the angle variables,

$$S_1(\mathbf{w}, t) = -\int_{-\infty}^{t} V(\mathbf{w} + \omega(t' - t), t')\, dt' \tag{166}$$

which is the solution of

$$(\partial S_1/\partial t) + \omega(\partial S_1/\partial \mathbf{w}) = -V(\mathbf{w}, t) \tag{167}$$

On using the semiclassical wave function for the system, the transition probability is, therefore,

$$P_{nn'} = (2\pi)^{-1} \lim_{t \to \infty} |\langle \exp i\mathbf{J}_{n'}^0 \cdot \mathbf{w}_{n'}^0 | \exp iS(\mathbf{J}, \mathbf{w}^0)\rangle|^2 \tag{168}$$

which, with S taken as $S_0 + S_1$, reduces to

$$P_{nn'} = \left| (2\pi)^{-3} \int_0^{2\pi} d\mathbf{w} \exp\left[i\mathbf{s} \cdot \mathbf{w} - (i/\hbar) \int_{-\infty}^{\infty} V(\mathbf{w} + \omega t, t)\, dt \right] \right|^2 \tag{169}$$

in agreement with (143). This result is essentially the three-dimensional generalization of the result obtained by Beǐgman et al.[4] for one degree of internal freedom.

Because semiclassical wave functions with Bohr's correspondence yield the Heisenberg correspondence directly and, when substituted into a classical path semiclassical treatment (34), yield the equivalent oscillator theorem (143), it is therefore consistent that the preceding perturbation treatment, based on the Hamilton–Jacobi equation, which is intimately related to (34) (see Sect. 11.3b), should yield similar results. Presynakov and Urnov[6] obtained for one degree of internal freedom a related "equally spaced levels" approximation

$$S_{n',n} = (2\pi)^{-1} \int_0^{2\pi} dw \exp\left[isw - (1/\hbar) \sum_k \int_{-\infty}^{\infty} V_k(t) \exp ik(\omega t + w)\, dt \right] \tag{170}$$

which reduces to the one-dimensional equivalent oscillator.[7]

From (130), or by expanding (169) to first order in V taken as the dipole interaction

$$V(\mathbf{R},\mathbf{r}) = -Z_{\mathrm{p}}e^2(\mathbf{R}\cdot\mathbf{r}/R^3) \tag{171}$$

of a Rydberg atom of nuclear charge $Z_{\mathrm{c}}e$ with a projectile of charge $Z_{\mathrm{p}}e$, the collisional probability for $n \to n'$ transitions reduces, for large-impact parameters ρ, to[29]

$$P_{nn'}(\rho,v) = \left[n\left(\frac{Z_{\mathrm{p}}}{Z_{\mathrm{c}}}\right)\left(\frac{v_n}{v}\right)^2\right]^2 \frac{8J_s(s)J_s'(s)}{3s}\left\{\left[K_1\left(\frac{s\omega\rho}{v}\right)\right]^2+\left[K_0\left(\frac{s\omega\rho}{v}\right)\right]^2\right\} \tag{172}$$

and the contribution to the cross section from impact parameters $\geqslant \rho_1$ is[29]

$$\sigma_{nn'}(v,\rho_1) = \left(\frac{2Z_{\mathrm{p}}}{Z_{\mathrm{c}}}\frac{n}{v}\right)^2\left\{\frac{4\pi}{3}\frac{J_s(s)J_s'(s)}{s^3}\right\}\frac{\rho_1 s}{v}K_0\left(\frac{s\rho_1}{v}\right)K_1\left(\frac{s\rho_1}{v}\right) \tag{173}$$

The probability (172) is valid for ρ much larger than that associated with unit $P_{nn'}$, which is obtained for small $\omega\rho/v \ll 1$, the sudden limit, when[29]

$$(v/v_n)(\rho/a_n) = n(Z_{\mathrm{p}}/Z_{\mathrm{c}}) \tag{174}$$

where a_n is the radius of the Rydberg orbital. Perturbation weak-coupling methods can therefore be taken as applicable in the region to the right of the inverse curve (174) in (ρ, v) space. To the left of (174), the effectively exact equivalent oscillator result (EOT), Eq. (169), must be used.

The line

$$(s\hbar\omega)(\tau_{\mathrm{c}}/\hbar) \equiv s\omega_{\mathrm{p}}/v \equiv (\rho/a_n)(sv_n/v) = 1 \tag{175}$$

in (ρ, v) space separates the sudden and adiabatic regions, which lie, respectively, above and below the diagonal (175). Sudden collisions, $\rho s v_n/a_n v \ll 1$, involve small ρ, large v, and, from (118) and (119), large energy transfers, whereas adiabatic collisions, $\rho s v_n/a_n v \gg 1$, involve large ρ, small v, and vanishingly small energy changes often less than the energy separation between adjacent levels (the classical inaccessible region). The probability of an adiabatic transition is usually small in comparison to that of a sudden transition such that the chief contribution to $\sigma_{nn'}$ arises from "sudden" impact parameters $\rho \ll \rho_{\mathrm{S}}/s\omega$.

Weak coupling approximations to (169) are essentially valid for all $\rho \gtrsim a_n$ at relative speeds v and impact energies E_i, which satisfy[12]

$$v > v_{\mathrm{A}} = (nv_n)(Z_{\mathrm{p}}/Z_{\mathrm{c}}), \qquad E_i > Z_{\mathrm{c}}^2 R_1 \tag{176}$$

where R_1 is the Rydberg unit of energy, and are independent of the binding energy of the Rydberg electron. For lower impact speeds and energies in the ranges

$$v_{\mathrm{A}} > v > v_{\mathrm{B}} = (snZ_{\mathrm{p}}/Z_{\mathrm{c}})^{1/2}, \qquad Z_{\mathrm{c}}^2 R_1 > E_1 > (2sZ_{\mathrm{c}}Z_{\mathrm{p}}/n)R_1 \tag{177}$$

where v_{B} is the intersection of (174) and (175), the sudden approximation (145) to (169) is valid for small ρ and the weak-coupling approximation (144) to (169) for large ρ. The full EOT (169) is valid for all ρ. For lower speeds and energies in the range

$$v_n \ll v \lesssim v_B, \qquad 4I_n \lesssim E_i \lesssim (2sZ_cZ_p/n)R_1 \tag{178}$$

where I_n is the binding energy of the Rydberg electron with orbital speed v_n, the adiabatic region is extended down to much smaller ρ and the sudden region becomes contracted. Here EOT or, better still, the multichannel eikonal treatment MET (43) with (143) for the transition amplitudes in (44) and (45) is valid over all ρ.

At yet lower speeds $v \simeq v_n$, there is, in this section, no satisfactory perturbation-based treatment for transitions between close levels with $s \ll n$ in charged-particle–Rydberg collisions. The fixed two-centered scattering analysis[9] outlined in Sect. 11.2b is valid under the sole condition that

$$k_i = k_f + \frac{2\mu}{\hbar^2}\frac{(\epsilon_f - \epsilon_i)}{(k_i + k_f)} \simeq k_f\left[1 + \frac{s\omega}{k_f\bar{v}}\right] \simeq k_f \tag{179}$$

where $\bar{v}$ is some averaged projectile speed. This condition is satisfied provided $k_f\bar{v} \gg s\omega$, i.e., when the impact energy is much greater than the transition-energy $s\hbar\omega$. For ionization and excitational processes involving large changes in principle quantum number, none of the methods in this section based on the Heisenberg correspondence principle are satisfactory. Classical procedures, e.g., the Monte Carlo method of Abrines and Percival[10] and Abrines et al.[10] based on the solution of the equation of motion for the three-body system are applicable. For neutral collisions with Rydbergs, methods based on various simplifications of the analyses in Sects. 11.2b for scattering by a fixed two-centered Rydberg target, e.g., the impulse treatment, are applicable.

The previous correspondence methods - EOT and its sudden and weak-interaction limits - within MET (43) or its high-energy limit (53) are applicable to $nl \to n'l'$ collisional transitions in hydrogenic systems. Other Rydberg atoms exhibit an additional interesting feature, which is of special significance. In Na, for example, the energy splitting $\Delta\epsilon_l$ between successive lower-l levels associated with a given n correspond roughly to the energy separation between adjacent n levels. High-l levels become essentially indistinguishable because $\Delta\epsilon_l \simeq l^{-6}$. Thus the initial and final levels a and b in the transition $nl \to n'l'$ may well be embedded within clusters $c_{i,f}$ of levels. The energy separation of each cluster is approximately equal to the transition energy so that c_i and c_f may be regarded as being in resonance with a and b, respectively. Transitions within each cluster can be determined by the sudden approximation (145). The probability for a $\to c_i$ and b $\to c_f$ transitions is in effect unity for small-impact parameters $\rho \ll \rho_s$, the sudden limit $(v/s\omega)$, such that the probability for a $\to$ b transitions can be satisfactorily determined by the first-order treatment (144). For intermediate $\rho \simeq \rho_s$, direct a $\to$ b transitions and a $\to c_{i,f}$ transitions are strongly coupled and compete. Transitions to $c_{i,f}$ may be regarded as loss mechanisms because the probability within a cluster of returning from many l'' levels to a specified l' is relatively small. Percival and Richards[45] presented an interesting probability analysis of the situation and obtained an expression for the a → b transition probability in terms of the probabilities for the resonance transitions a → c_i, b → c_f and the first-order probability for a → b direct transitions. They illustrated their results by considering s → p collisional transitions, 10s → 10p(9d) and 11s(11p) → 12p(11d), where the resonant levels are enclosed in parentheses. The cross section for the first (single-resonance) collisional transition is greater at all electron-impact energies E than the latter (double-resonance) transition because the transition energy is smaller and the radial matrix element (123) in (172) is larger. Because of resonant depletion of the lower 11s state, the cross section for 11s → 12p transitions becomes markedly reduced at lower E, in contrast to that for the single-resonant (10s → 10p) case.

11.5. Quantal impulse and semiquantal methods

The quantal impulse approximation (QIA) was originally developed[46] by Chew, was extended by Chew and Wick and Chew and Goldberger for high-energy neutron scattering by deuterons, and has been derived from various directions.[23] The fixed-center analysis of Sect. 11.2b provides yet another new derivation of Eqs. (32) basic to QIA. However, the three basic assumptions underlying QIA of A–B(n) collisions become fully transparent from a derivation based on the exact two potential formula that are as follows:[2]

1. The interactions V_{12} and V_{32} of Rydberg electron 1 and projectile 3 with the B^+ core 2 are switched off during the 1–3 collision time and V_{12} is invoked only to establish the initial and final quantal states of $B(n)$.
2. The distortion of the motion of projectile 3 in the field V_{32} due to core 2 is neglected when interacting with both 2 and with 1.
3. Inelastic transitions in B are prohibited in direct A–B^+ encounters.
4. Although not essential to QIA, "on-the-energy-shell" 1–3 encounters are tacitly assumed, a procedure valid only in the high-energy or weak-binding limit.

These assumptions imply important considerations of special significance to A–B(n) collisions and will be discussed in Sect. 11.5b. Although conditions underlying the four items listed may be justified for A–B(n) direct collision processes at sufficiently high n and collision speeds $v_3 \gg v_1$ of the Rydberg electron, they may be seriously violated, particularly at thermal energies when $v_3 \ll v_1$.[2,47]

Flannery[1] derived, from classical-quantal principles, a *semiquantal* version of QIA suitable for A–B(n) ionizing collisions at high-impact energies, a version that was recently derived directly from QIA by replacing the final wave function $\phi_f(r)$ in (32) by a plane wave.[2] The analysis is sufficiently general so as to include a general (1–3) scattering amplitude $f_{13}(\mathbf{k}, \mathbf{k}')$ in contrast to the other QIA derivatives (Sect. 11.5a and Chap. 8) that assume f_{13} either constant or a function only of the momentum change $\mathbf{P}=(\mathbf{k}'-\mathbf{k})$, as in Born's approximation.

Rydberg collisions at thermal energies involve, in addition to e–A encounters, a contribution from direct A–B$^+$ encounters.[2,47] Theory[2] of this nuclear-recoil effect may not only be applicable to l-changing and other quasi-elastic–elastic processes but may complement expressions for the shift and shape of spectral lines originating from highly excited levels $B(n)$ collision broadened by neutral perturbers A. This shift was predicted by Fermi[48] on the basis of S-wave scattering in slow e–A collisions and was generalized within the framework of QIA by Alekseev and Sobel'man[49] who also assumed that e–A collisions alone were responsible.

The quantal impulse treatment, its derivatives, and the validity criteria (which are easily violated[47] by several recent models for A–B(n) collisions) are discussed more fully by Flannery,[2] to which the reader is referred.

11.5a. Basic formulae in impulse treatments

From the basic QIA transition matrix element (32), the integral cross section for scattering of projectile 3 by the bound (1, 2) Rydberg system in the center of mass of the 3–(1, 2) system of reduced mass M_{AB} is

$$\sigma_{if}(k_3) = \left(\frac{M_{\mathrm{AB}}}{M_{13}}\right)^2 \frac{k_3'}{k_3} \int |\langle g_f(\mathbf{k}_1+\mathbf{P})|f_{13}(\mathbf{k},\mathbf{k}')|g_i(\mathbf{k}_1)\rangle_{\mathbf{k}_1}|^2 \, d\hat{\mathbf{k}}_3' \tag{180}$$

where $\mathbf{P}$ is the momentum change (32c) suffered in the (1–3) collision characterized by

$$f_{13}(\mathbf{k},\mathbf{k}') = \frac{1}{4\pi}\left(\frac{2M_{13}}{\hbar^2}\right) T_{13}(\mathbf{k},\mathbf{k}'); \qquad \mathbf{k}^{(\prime)} = \frac{M_3}{M}\,\mathbf{k}_1^{(\prime)} + \frac{M_1}{M}\,\mathbf{k}_3^{(\prime)} \tag{181}$$

the (on-and-off-the-energy-shell) amplitude for $\mathbf{k}\to\mathbf{k}'$ scattering of 1 and 3 in the center of mass of the (1–3) system with reduced mass M_{13} and by the conservation of momentum $\mathbf{k}_1+\mathbf{k}_3$. At high energy $k_3 \simeq k_3'$ such that (180), with the aid of closure, yields the total elastic and inelastic cross section arising from (1–3) collisions as[2]

$$\sigma_{\mathrm{tot}}(k_3) = (M_{\mathrm{AB}}/M_{13})^2 \int |g_i(\mathbf{k}_1)|^2 |f_{13}(\mathbf{k},\mathbf{k}')|^2 \, d\mathbf{k}_1 \, d\hat{\mathbf{k}}_3' \tag{182}$$

When distortion in the scattering of 3 by the core C is neglected in the incident wave, which is then plane, the only contribution from (3–C) collisions is elastic and is given by Born's approximation. When the final state f in (180) is described by a plane wave, it can then be shown that the total cross section for all elastic and inelastic events arising from (1–3) collisions satisfies[47]

$$\sigma_{\mathrm{tot}}(k_3) = (1/v_3) \int |g_i(\mathbf{k}_1)|^2 [v_{13}\sigma_{13}^{\mathrm{T}}(v_{13})] \, d\mathbf{k}_1 \tag{183}$$

where σ_{13}^{T} is the total cross section for (1-3) scattering at relative speed v_{13} and v_3 the speed of the incident atom A. Cross section (183) is an upper limit to any collision process satisfying specific criteria (Sect. 11.5b) for validity of the e-A impulse approximation and shows that the rate $v_3\sigma_{tot}$ for all A-B(n) elastic and inelastic processes is limited by the total rate of free Rydberg e-A collisions with free momentum amplitude $g_i(\mathbf{k}_1)$ specified by the initial state of $B(n)$.

Plane-wave final state

When the final state of the Rydberg electron is described by a plane wave

$$\sigma_f(\mathbf{r}_1) = (2\pi)^{-3/2}\exp(i\kappa_1'\cdot\mathbf{r}_1); \qquad g_f(\mathbf{k}_1') = \delta(\mathbf{k}_1' - \kappa_1') \tag{184}$$

then the differential cross section for scattering into unit solid angle $d\hat{\mathbf{k}}_3'$ and for the final momentum of Rydberg electron to be within an interval $d\mathbf{k}_1'$ about $\mathbf{k}_1'$ is, with the aid of (180),

$$\left[\frac{d\sigma_{if}}{d\hat{\mathbf{k}}_3'\, d\mathbf{k}_1'}\right] = \left(\frac{M_{AB}}{M_{13}}\right)^2 \frac{k_3'}{k_3} |g_i(\mathbf{k}_1)|^2 |f_{13}(\mathbf{k},\mathbf{k}')|^2 \tag{185}$$

which, in addition to the impulse requirement, assumes that the core is acknowledged only in the preparation of the initial state i and is "switched off" thereafter. It is this quantal result that yields[2] the semiquantal treatment previously developed[1] from a classical–quantal basis, which specified the cross section per unit of energy-change interval $d\epsilon$ about

$$\epsilon = (k_1'^2 - k_1^2)/2m_1 \tag{186}$$

per unit momentum-change interval dP about P, and per unit initial momentum interval $d\mathbf{k}_1$ about $\mathbf{k}_1(k_1, \theta, \phi_1)$. The polar angle θ_1 of $\mathbf{k}_1$ relative to $\hat{\mathbf{k}}_3$ along the Z axis can be expressed in terms of the momentum k of relative motion and of k_1 and k_3. Thus (185), in terms of these "classical" variables $(\epsilon, P, \mathbf{k}_1)$ rather than the set $[\hat{\mathbf{k}}_3'(\cos\theta_3', \phi_3'), \mathbf{k}_1', \theta_1', \phi_1')]$ natural to full quantal treatments, is

$$\frac{d\sigma_{if}}{(d\epsilon\, dP\, dk_1\, dk\, d\phi_1)} = \frac{k_1'^2}{J_{55}}\left(\frac{k_3'}{k_3}\right)\left(\frac{M_{AB}}{M_{13}}\right)^2 |g_i(\mathbf{k}_1)|^2 |f_{13}(\mathbf{k},\mathbf{k}')|^2 \tag{187}$$

where the Jacobian of the five-dimensional transformation is[2]

$$J_{55} \equiv \frac{\partial(P, \epsilon, k_1, k, \phi_1)}{\partial(\cos\theta_3', \phi_3', k_1', \cos\theta_1', \phi_1')}$$
$$= \frac{M_3}{(M_1+M_3)^2}\,\frac{k_1'^3(k_3k_3')^2}{k_1kP}\sin\theta_1'\sin\theta_3'\sin(\phi_1'-\phi_3') \tag{188}$$

With the aid of conservation of momentum and energy, (187) with (188) eventually yields the semiquantal integral cross section[2]

$$\sigma_{SQ}(v_3) = \frac{1}{M_{13}^2 v_3^2} \int_{\epsilon_1}^{\epsilon_2} d\epsilon \int_{v_{10}}^{\infty} \frac{F_{nl}(v_1)\, dv_1}{v_1} \int_{P^-}^{P^+} dP \int_{g_-}^{g_+} \frac{|f_{13}(P,g)|^2\, dg^2}{[(g_+^2 - g^2)(g^2 - g_-^2)]^{1/2}} \tag{189}$$

where $g_\pm(P, v_1, \epsilon; v_3)$ and $P^\pm(v_1, \epsilon; v_3)$ denote limits to the relative (1–3) speed g and momentum change P consistent with fixed (P, v_1, ϵ) and (v_1, ϵ), respectively, where $\frac{1}{2}M_1[1+(M_1/M_2)]v_{10}^2$ is $\max(0,\epsilon)$ and $\epsilon_{1,2}$ the limits to ϵ. The initial distribution in speed v_1 of the Rydberg electron in state nl is

$$F_{nl}(v_1)\, dv_1 = \frac{2}{2l+1} \left(\int \sum_{m=-l}^{l} |g_{nlm}(\mathbf{k}_1)|^2\, d\phi_1 \right) k_1^2\, dk_1 \tag{190}$$

which can be determined from the initial wave function $\phi_{nlm}(\mathbf{r})$. In situations where the (1–3) scattering amplitude $f_{13}(\mathbf{k}, \mathbf{k}')$ is expressed as $f_{13}(g, \psi)$, where ψ is the scattering angle, then (189) can be rewritten alternatively as[1]

$$\sigma_{SQ}(v_3) = \frac{1}{v_3^2} \int_{\epsilon_1}^{\epsilon_2} d\epsilon \int_{v_{10}}^{\infty} \frac{F_{nl}(v_1)\, dv_1}{v_1} \int_{g_-}^{g_+} \frac{g\, dg}{S(v_1, g; v_3)} \times \int_{\psi^-}^{\psi^+} \frac{|f_{13}(g,\psi)|^2\, d(\cos\psi)}{[(\cos\psi^+ - \cos\psi)(\cos\psi - \cos\psi^-)]^{1/2}} \tag{191}$$

where $\psi^\pm(g, v_1, \epsilon; v_3)$ are certain angular limits[1,2] to the scattering and ensure a given energy change ϵ for fixed g, v_1, and v_3 and where S is also a known function.[1,2] Cross sections are more efficiently calculated from the semiquantal formula (189) or (191) than from their quantal equivalent (185), which gives

$$\sigma_{if} = \left(\frac{M_{AB}}{M_{13}} \right)^2 \frac{k_3'}{k_3} \int |g_i(\mathbf{k}_1)|^2 |f_{13}(\mathbf{k}, \mathbf{k}')|^2\, d\mathbf{k}_1'\, d\hat{\mathbf{k}}_3' \tag{192}$$

for the integral cross section.

When f_{13} is taken as the Born amplitude, then the SQ treatment for

$$\mathrm{H(1s) + H(1s)} \rightarrow \left\{ \begin{matrix} \mathrm{H}(n) \\ \mathrm{H}(\Sigma) \end{matrix} \right. + \mathrm{H}^+ + e \tag{193}$$

at high-impact energies reproduces[1] the quantal Born results of Omidvar and Kyle.[50] Semiquantal methods have also been applied to[51]

$$\mathrm{H(1s) + H(1s)} \rightarrow \left\{ \begin{matrix} \mathrm{H(1s)} + \mathrm{H}^+ + e \\ \mathrm{H}(1,2) + \mathrm{H}(n') \end{matrix} \right. \tag{194}$$

with successful results and to[52]

$$\mathrm{H}(nl) + \mathrm{H}(n_0 l_0) \rightarrow \mathrm{H}(n'l') + \mathrm{H}^+ + e \tag{195}$$

where the interesting systematic trends discussed in Sect. 11.2d were exhibited. We recall that any treatment, such as SQ, based on QIA is valid only when momentum and energy changes imparted in e–A collisions are greater than any momentum and energy imparted to e by its core during the collision (see Sect. 11.5b).

Further approximations will follow from (180) upon the use of closure or the peaking approximation.

$f_{13}(\mathbf{P})$

In particular, when f_{13} in (180) or implicitly in (32) is a function *only* of momentum change $\mathbf{P}=(\mathbf{k}'-\mathbf{k})$, then

$$T_{fi}(\mathbf{k}_3, \mathbf{k}_3') = T_{13}(\mathbf{P})F_{fi}(\mathbf{P}) \tag{196}$$

such that when the free (1–3) scattering is described by the Born approximation, then

$$T_{fi}^{(B)}(\mathbf{k}_3, \mathbf{k}_3') = T_{13}^{(B)}(\mathbf{P})F_{fi}(\mathbf{P}) \tag{197}$$

which is simply the Born result (22) for A–B(n) collisions. This is the form of QIA exploited extensively by Matsuzawa (see Chap. 8) and used by Matsuzawa[53] in situations where it may not describe the actual state of affairs either in validity or in mechanism.[47]

The full impulse expression (180) and its semiquantal derivatives (185), (189), or (191), which are all equivalent, provide, via a general $f_{13}(\mathbf{k}, \mathbf{k}')$, a scattering description much more general than (193), which is valid only at high-impact energies. Moreover, the semiquantal cross section for all elastic and inelastic events yields a result identical with (183) given by the full QIA.

$f_{13} = A$

When f_{13} can be taken as a constant scattering length A, then

$$\sigma_{if}(\mathbf{k}_3) = \frac{2\pi A^2}{k_3^2}\left(\frac{M_{AB}}{M_{13}}\right)^2 \int_{(k_3-k_3')}^{(k_3+k_3')} |F_{fi}(\mathbf{P})|^2 P\,dP \tag{198}$$

such that (183) for the total cross section gives

$$\sigma_{tot}(k_3) = \frac{4\pi A^2}{v_3}\int v_{13}|g_i(\mathbf{k}_1)|^2\,d\mathbf{k}_1 \rightarrow \begin{cases} 4\pi A^2, & v_3 \gg v_1 \quad (199a) \\ \dfrac{\langle v_1\rangle}{v_3} 4\pi A^2, & v_3 \ll v_1 \quad (199b) \end{cases}$$

where $\langle v_1\rangle$ is the mean speed of the Rydberg electron. However, at high v_3, T_{13} is never constant so that (199a) is never attained in practice, whereas at low v_3, (199b) must be augmented by contributions from direct collision with the Rydberg core. Note that $v_3 \ll v_1$ for principal quantum numbers[2]

$$n \ll 870 M_{AB}^{1/2}(300/T)^{1/2} \tag{200}$$

where the reduced mass is in atomic mass units and where T is the temperature such that (200) is satisfied in most cases of interest for thermal atoms A. The total cross section (113) for this case is then

$$\sigma_{\text{tot}}^{13}(v_3) = \frac{1}{v_3}\int |g_i(\mathbf{k}_1)|^2 v_1 \sigma_{13}^{\text{T}}(v_1)\, dv_1$$
$$\equiv \frac{1}{v_3}\langle v_1 \sigma_{13}^{\text{T}}(v_1)\rangle; \qquad v_1 \gg v_3 \tag{201}$$

the average rate for e–A collisions divided by the projectile speed v_3, a rate that may be augmented by the contribution σ_{tot}^{23} from 2–3 collisions. The minimum temperatures T corresponding to $\frac{3}{2}kT$ energy of relative motion required for $n \to n+s$ excitation and ionization of H(n) are

$$T_{\text{E}}(K) = (10/n)^3 300\text{s} \tag{202a}$$

$$T_{\text{I}}(K) = (23/n)^2 300 \tag{202b}$$

respectively.

11.5b. Validity criteria

Rigorous criteria for validity of QIA of A–B(n) collisions have been properly traced via a derivation of QIA from the exact two-potential formula.[2] The assumptions basic to QIA imply the following conditions for A–B(n) collisions.[47]

Condition A

Switching off the core interactions $V_{12}+V_{32}$ during the (1–3) collision time τ_c implies that energy can be controlled only to within imprecision $\Delta E_{\text{I}} \simeq \hbar/\tau_c$ during the collision, i.e., $\sigma_{13}(v_{13})$ must not exhibit too rapid a variation with v_{13} such as would occur, e.g., in the neighborhood of either an A^- resonance or a Ramsauer minimum evident for e–Ar, Kr, and Xe scattering. This implication has been ignored in several studies (as in Matsuzawa[53] and Hickman[54]). When $v_1 \gg v_3$, $\tau_c \simeq A_1 n$ (a.u.), where $A_1(a_0)$ is the e–A interaction distance such that, during τ_c, $\Delta E_{\text{I}} \simeq (A_1 n)^{-1}$ a.u. is comparable with the small impact energy $\frac{1}{2}v_1^2$. For $v_3 \gg v_1$, $\Delta E_{\text{I}} \simeq v_3/A_1$, which is much less than $\frac{1}{2}v_3^2$ the relative energy over which σ_{13}, in general, varies slowly.

Condition B

The momentum P transferred (impulsively) to 1 during τ_c must be much greater than the momentum imparted to 1 during τ_c via the force F associated with core interaction V_{12}, i.e.,

$$P \gg \int_{\tau_c} F\,dt = |\langle \phi_{nl}| -\nabla V_{12} |\phi_{nl}\rangle| \tau_c \simeq \tau_c / n^3(l+\tfrac{1}{2}) \tag{203a}$$

such that in terms of the orbital period T_n,

$$\tau_c \ll T_n(l+\tfrac{1}{2})P, \qquad T_n \simeq n^{-3} \text{ (a.u.)} \tag{203b}$$

If V_{12} varies sufficiently slowly (but need not be necessarily small!) over the range A_1 of the collision interaction V_{13} such that the force F ($\equiv -\nabla V_{12}$) due to the core is small in comparison with the force ($-\nabla V_{13}$) due to the Rydberg electron–projectile interaction, then (203) is satisfied; in this sense, V_{12} can be regarded as "quasi-classical."

For ionizing collisions, $P \gtrsim n^{-1}$, then $\tau_c \ll T_n$ for circular orbits ($l \simeq n$) and $\tau_c \ll T_n/n$ for highly eccentric orbits ($l \simeq 0$). Hence $\tau_c \ll n^{-2}$ covers electron ejection from all orbits. For nonionizing collisions, P [Eq. (203)] cannot become arbitrarily small, which could occur for quasi-elastic or l-changing collisions. At thermal energies, $v_1 \simeq n^{-1}$, which is greater than $v_3 \simeq 10^{-4}$ a.u. for most n of interest, and $\tau_c \simeq A_1 n$ such that (203) implies that $P \gg A_1/n^2(l+\frac{1}{2})$. The angular momentum change ΔL for n fixed due to e–A implusive encounters at distance R_{12} from B^+ must therefore satisfy

$$\Delta L \simeq P\langle R_{12}\rangle \simeq \tfrac{1}{2}P[3n^2 - l(l+1)] \gg \tfrac{1}{2}A_1[3n^2 - l(l+1)]/n^2(l+\tfrac{1}{2}) \tag{204}$$

which is, in general, fulfilled only at the highest initial l when the permitted $\Delta L \gg A_1/l$. Small initial l require from (204) large changes $\Delta L \gg A_1$ for impulse model validity (because then the momentum imparted by the core on the highly elliptical orbits becomes considerably strengthened over that for circular orbits). Detailed discussion on the implication of (203a) to angular momentum changes is given in Ref. 47 (second listing).

Condition C

Cross section (183) is an upper bound for any process based on e–A encounters and described by QIA, SQ, or any derivative of QIA. Thus,

$$\sigma_{\text{tot}}^{(13)}(v_3) = \begin{cases} \langle v_1 \sigma_{13}^{\text{T}}(v_1)\rangle / v_3, & v_1 \gg v_3 \quad (205a) \\ \sigma_{13}^{\text{T}}(v_3), & v_1 \ll v_3 \quad (205b) \end{cases}$$

Condition D

The second assumption listed at the beginning of this section implies that 1 and 2 behave as separate and as independent scatterers. This is valid provided that $R_{12} \simeq n^2 a_0 \ll A_{1,2}$ the scattering amplitudes for (1–3) and (2–3) collisions and that the reduced wavelength λ_{i3} for (i–3) relative motion is much less than R_{12} so that A_1 is not affected by A_2 and vice versa. In general, $\lambda_{13} \ll n^2$ at all energies even for $v_1 \ll v_3$, when $\lambda_{13} \simeq n$. Also, for (2–3) collisions at thermal energy, $\lambda_{23} \simeq k_3^{-1} \simeq 10^{-1}$. Hence, $R_{12}^2 \gg \lambda_{i3} A_i$ such that multiple scattering can be neglected. This condition is the one most easily and generally satisfied in Rydberg collisions, but it alone is not sufficient for QIA validity.

Condition E

Distortion of motion of 3 when interacting with 1 due to core 2 can, in general, be neglected except at thermal energies when the cross section for (2–3) encounters are large ($\sim 10^3$ Å^2). Impulse expression (32) must then be appropriately generalized. The result involves a nine-dimensional integral for the T matrix.[2]

Condition F

Quantal impulse approximation and its derivatives are based on (1–3) encounters whether or not they are distorted by V_{32} and contain no inelastic transitions that can arise in direct (2–3) encounters. Hahn[56] recently showed that the distortion correction to l-changing collisions owing to the 2–3 interaction is large for intermediate values of $n \lesssim 20$ and that the effect on the T matrix for 2–3 collisions through higher-order terms neglected in the customary impulse approximation can be very important. Lane and Preston[57] demonstrated via a direct numerical Monte Carlo classical simulation that the core is very important in the ionization of Rydberg atoms (Ar(ns), $n \geqslant 30$) by thermal dipole molecules (as HCl).

11.5c. Binary encounter methods for charged-particle–Rydberg collisions

When $|f_{13}|^2$ in (191) is replaced by $(4\hbar^4 Z^2/P^4 a_0^2)$, the differential cross section for on-the-energy-shell Coulomb scattering of the Rydberg electron 1 by an incident ion (or electron) of charge Ze, then the differential cross section for energy change ϵ reduces, for given initial speeds v_1 and v_3, to

$$\frac{d\sigma}{d\epsilon}(v_1, v_3) = \frac{4\pi Z^2 e^4}{3 v_1 v_3^2}\,[P_\ell^{-3} - P_{\rm u}^{-3}] \tag{206}$$

The momentum-change limits,

$$\begin{aligned} P_{\rm u} &= \min[p_1^+ = M_1(v_1 + v_1'), p_3^+ = M_{\rm AB}(v_3 + v_3')] \\ P_\ell &= \max[p_1^- = M_1|v_1' - v_1|, p_3^- = M_{\rm AB}|v_3 - v_3'|] \end{aligned} \tag{207}$$

are (p_1^-, p_1^+), (p_3^-, p_1^+), and (p_3^-, p_3^+), provided ϵ ($\geqslant 0$) is, respectively, within limits $[0, \epsilon^-]$, $[\epsilon^-, \epsilon^+]$, and $[\epsilon^+, \frac{1}{2}M_{\rm AB}v_3^2]$, where

$$\epsilon^{\pm} = \frac{4M_1M_3}{(M_1 + M_3)^2}\,[\tfrac{1}{2}M_{\rm AB}v_3^2 - \tfrac{1}{2}M_1 v_1^2 \pm \tfrac{1}{2}(M_3 - M_1)v_1 v_3] \tag{208}$$

are the kinetic energies transferred from 3 to 1, which is initially moving with speed v_1 directed, respectively, toward or away from 3. The energy range $\epsilon > \epsilon^+$ is entered provided $p_3^+ < p_1^+$, i.e., provided

$$2M_1v_1 - v_3(M_3 - M_1) > 0 \tag{209}$$

which ensures real final speeds v_1' and v_3'. Hence,

$$\frac{d\sigma}{d\epsilon}(v_1, v_3) = \frac{4\pi Z^2 e^4}{3v_1 v_3^2 \epsilon^3} I(\epsilon; v_1) \tag{210}$$

where

$$I(\epsilon; v_1) = v_1\left(v_1^2 + \frac{3\epsilon}{2M_1}\right), \qquad 0 \leqslant \epsilon \leqslant \epsilon^- \tag{211a}$$

$$= \left[v_3 + \left(v_3^2 - \frac{2\epsilon}{M_{AB}}\right)^{1/2}\right]^3 - \left[\left(v_1^2 + \frac{2\epsilon}{M_1}\right)^{1/2} - v_1\right]^3 \equiv (v_3 + v_3')^3 - (v_1' - v_1)^3, \qquad \epsilon^- \leqslant \epsilon \leqslant \epsilon^+ \tag{211b}$$

$$= \left(v_3^2 - \frac{2\epsilon}{M_{AB}}\right)^{1/2}\left(v_3^2 - \frac{\epsilon}{2M_{AB}}\right) \equiv v_3'\left(v_3'^2 + \frac{3\epsilon}{2M_{AB}}\right) \qquad \epsilon^+ \leqslant \epsilon \leqslant \tfrac{1}{2}M_{AB}v_3^2 \tag{211c}$$

which is zero for $\epsilon \geqslant \epsilon_{max} = \frac{1}{2}M_{AB}v_3^2$, the maximum energy that can be transferred.

For incident electrons, $\epsilon^+ = \epsilon^-$, and I is given by (211a) or (211c), where appropriate. For incident ions, only (211a) and (211b) are appropriate because (209) is never satisfied for $v_3 > v_1$. On integrating over the symmetrical velocity distribution $F_{nl}(v_1)$ appropriate to Rydberg state nl, the binary-encounter cross section for ionization by a charged particle is

$$\sigma_{nl}(v_3) = \frac{4\pi Z^2 e^4}{3v_3^2}\int_{I_n}^{\epsilon_{max}} \epsilon^{-3}\, d\epsilon \int_0^\infty v_1^{-1} F_{nl}(v_1) I(\epsilon; v_1)\, dv_1 \tag{212}$$

where I_n is the binding energy. For sufficiently high n, when $\frac{1}{2}M_1 v_1^2$ can be replaced via the Virial theorem by I_n, then, for incident electrons, $(M_3 = M_1)$,

$$\frac{d\sigma_e^{(e)}}{d\epsilon} = \frac{\pi e^4}{\frac{1}{2}M_1 v_3^2}\left(\frac{1}{\epsilon^2} + \frac{4}{3}\frac{I_n}{\epsilon^3}\right), \qquad 0 < \epsilon \leqslant \epsilon^{\pm} = \frac{1}{2}M_{AB}v_3^2 - I_n \tag{213}$$

which is the main contribution and

$$\frac{d\sigma_e^{(e)}}{d\epsilon} = \frac{\pi e^4}{\frac{1}{2}M_{AB}v_3^2}\left(\frac{v_3'}{v_1}\right)\left(\frac{1}{\epsilon^2} + \frac{2M_{AB}v_3'^2}{3\epsilon^3}\right), \qquad \epsilon^{\pm} < \epsilon \leqslant \frac{1}{2}M_{AB}v_3^2 \tag{214}$$

which is negligible, being operative only over the small end range I_n of energy change. Hence, for collisional ionization by electrons of energy E,

$$\sigma_{\rm I}^{(e)}(E)=\frac{\pi e^4}{E/I_n}\left[\frac{5}{3}\frac{1}{I_n^2}-\frac{(E/I_n)-\frac{1}{3}}{[(E/I_n)-1]^2}\right]\simeq\frac{An^4}{E/I_n}+\frac{B}{(E/I_n)^2} \tag{215}$$

which exhibits an n^4 dependence for asymptotic impact energies E *measured in units of the binding energy* I_n, where A and B are constants. Cross sections for ion impact can be similarly obtained from (211).

Although the binary-encounter formulae were derived here from the semiquantal treatment (Sect. 11.5a) for a general scattering amplitude $f_{13}(g,\psi)$ and (1-3) interaction, similar or identical expressions have been derived by more direct means[58, 59] for Coulomb interactions alone, or when f_{13} is a function only of momentum change P, and have received widespread application and review[58] in atomic collision physics. They can be generalized to include electron exchange,[58] which is not important, however, for Rydberg collisions. Whereas the effect of the core on the orbital speed v_1 is acknowledged via (212), its effect (acceleration and focusing) in the projectile can be acknowledged (but not rigorously justified) by replacing the outside factor E^{-1} in (206), (212), and (215) by $E+\alpha$, where α is 3.25,[58, 60] to agree with the classical three-body sections of Abrines et al.[10]

For e-metastable rare gas collisions, (212) yields[61] close agreement with experiment, with Born's approximation, and to within 10–20% of (215) with $\alpha=3.25$.[60] Apart from their simplicity, the main attraction of the binary-encounter formulae for Rydberg collisions is that they automatically include an infinite summation over all angular momenta l_ϵ of the ejected electron. In e-excited atom collisions., many l_ϵ are required for converged $d\sigma/d\epsilon$,[61] and the contributions from high-order multipoles are substantial in comparison to the dipole contribution (a feature exhibited also in direct excitation (Sect. 11.2d) and consistent with the quasi-classical nature of the Rydberg electron). Direct application of even Born's approximation of ionization is, therefore, time-consuming, and the binary-encounter expressions (206)–(215) for charged-particle impact or the more general semiquantal (SQ) formulae (189) or (191) for neutral impact represent very efficient methods for calculation of cross sections $\sigma_{\rm I}$ for ionization of Rydberg atoms. For neutral impact, SQ yields $\sigma_{\rm I}\simeq E^{-1}$ in accord with the correct Born–Bethe asymptotic limit when no projectile excitation occurs, and when f_{13} for e–A collisions is replaced by the corresponding Born value, SQ reproduces the Born cross sections for A–B(n) ionizing collisions.

The binary-encounter cross section (206) and the classical three-body cross section[10] for charged-particle collisions yield the incorrect E^{-1} asymptotic energy dependence in contrast to the correct Born–Bethe asymptote $E^{-1}(A\ln E+B)$. For excitation and ionization from excited states, the dipole contribution A is very small in comparison to B (even for e–He($2\,^{1,3}$S) collisions; see fifth listing in Ref. 8 and Refs. 19 and 61), so that $A\ln E$ becomes effective only at extremely high energies ($\gtrsim$1000 eV for e–He($2\,^3$S) collisions;

see Fig. 11.1) and the validity of the binary encounter result (206) is extended to much higher-impact energies.

11.5d. Extension of binary-encounter method to asymptotic energies

The binary-encounter (BE) method is applicable only when the collision time $\tau_c \simeq \rho/v \ll T_n$, the orbital period, a criteria that becomes invalid for distant encounters ρ, which are, therefore, not properly treated by BE so that, for asymptotic charged-particle collisions, the $\ln E$ dipole term, which originates from distant encounters, is absent in BE. This inadequacy of BE in its omission of adiabatic effects can be remedied either by using the Fourier components of the energy change ϵ in accord with proper correspondence arguments, as pointed out by Percival and Richards,[38] or, alternatively, as presented here, by invoking the Weizsäcker–Williams principle of virtual quanta.[37]

Here the electrostatic field of a passing charged particle at high energies is replaced by that arising from an equivalent flux of photons, which also involves a (negligible) magnetic interaction. For distant encounters, with $\rho \geqslant \rho_{\min}$,

$$d\sigma/d\epsilon = N(\hbar\omega)\sigma_I(\hbar\omega), \qquad \epsilon = \hbar\omega, \quad \rho \geqslant \rho_{\min} \tag{216}$$

where $N\,d\epsilon$ is the number of virtual photons in the equivalent photon field and σ_I the cross section for photoejection of the Rydberg electron of mass M_1 by a photon of energy ϵ. For a straight-line incident path, then[37]

$$N(\hbar\omega) = \frac{2}{\pi}\frac{Z^2e^2}{\hbar c}\left(\frac{c}{v_3}\right)^2\frac{1}{\epsilon}\left\{\chi K_0(\chi)K_1(\chi) - \frac{v_3^2}{2c^2}\chi^2[K_1^2(\chi) - K_0^2(\chi)]\right\} \tag{217}$$

where K_i are modified Bessel functions of argument

$$\chi = \frac{\omega\rho_{\min}}{v_3} = \frac{(\epsilon\beta)n^2a_0}{\hbar v_3} \equiv \left(\frac{I_1}{\frac{1}{2}M_1v_3^2}\right)^{1/2}\left(\frac{\beta\epsilon}{2I_n}\right) \tag{218}$$

in which $\rho_{\min}$ is taken as β times the Rydberg radius n^2a_0. Hence, for photon energies $\hbar\omega \gtrsim (1/n^2)(e^2/a_0)$, which contribute most to the integral cross section, $\chi \ll 1$, with the result that

$$N(\hbar\omega) = \frac{2}{\pi}\left(\frac{Z^2e^2}{\hbar c}\right)\left(\frac{c}{v_3}\right)^2\frac{1}{\hbar\omega}\left[\ln\left(\frac{1.123}{\chi}\right) - \frac{v^2}{2c^2}\right] \tag{219}$$

The photoejection cross section in terms of the differential oscillator strength (126) and binding energy I_n is

$$\sigma_I = 2\pi^2\alpha a_0^2\,(df/d\epsilon_{\text{au}}) \simeq 4\pi^2\alpha a_0^2 n(I_n/\epsilon)^3, \qquad \alpha = e^2/\hbar c \tag{220}$$

with the result that, for distant collisions,

$$\left(\frac{d\sigma}{d\epsilon}\right)_D = \frac{4\pi Z^2 a_0^2}{\frac{1}{2}M_1 v_3^2}\left(\frac{I_n}{\epsilon}\right)^4 n^3 \ln\left[\frac{1.123^2}{\beta^2}\left(\frac{2M_1 v_3^2}{I_1}\right)\left(\frac{I_n}{\epsilon}\right)^2\right] \tag{221}$$

which includes the ln term of quantal treatments. Because $\epsilon \gtrsim I_n$, the argument of this term is inappreciable unless $\frac{1}{2}M_1 v_3^2 \gtrsim I_1 \gg I_n$, i.e., at extremely high energies. The classical energy change ϵ in a Coulombic binary encounter with impact parameter ϵ is

$$\epsilon(\rho) = \frac{(Ze^2)^2}{\frac{1}{2}M_1 v_3^2}(\rho^2 + \rho_{\min}^2)^{-1} \tag{222}$$

so that for close collisions,

$$\left(\frac{d\sigma}{d\epsilon}\right)_C = 2\pi\rho\frac{d\rho}{d\epsilon} = \frac{\pi Z^2 e^4}{\frac{1}{2}M_1 v_3^2}\frac{1}{\epsilon^2} \equiv \frac{4\pi Z^2 a_0^2}{\frac{1}{2}M_1 v_3^2}\left(\frac{I_n}{\epsilon}\right)^2 n^4 \tag{223}$$

which agrees with the first term of (213) and which dominates (221) for high n and impact energy up to many I_n. The combination of (223) or of (206) in general, for close encounters with (221) for distant encounters yields the correct $A(1+B\ln E)/E$ dependence at high-impact energies.

A similar procedure for collisional $n \to n'$ transitions yields

$$\sigma_{nn'}(v_3) = N(\hbar\omega_{nn'})\sigma_A(\hbar\omega_{nn'}) \equiv (2\pi^2\alpha a_0 e^2) f_{nn'} N(\epsilon_{nn'}) \tag{224}$$

where σ_A is the photoabsorption cross section, such that the aid of (123) and (124) for the oscillator strength and of (219),

$$\sigma_{nn'}(v_3) = \frac{4}{3}\frac{n^4\pi a_0^2}{(\frac{1}{2}M_1 v_3^2)}\left(\frac{e^2}{a_0}\right)\frac{1+(s/n)}{s^3} \times J_s(s)J_s'(s)g(s)\ln\left[\frac{1.123^2}{\beta^2}\left(\frac{2M_1 v_3^2}{I_1}\right)\left(\frac{I_n}{\epsilon_{nn'}}\right)^2\right] \tag{225}$$

The ln term is therefore negligible when both n and s are large compared with unity and the binary-encounter result (215) appropriately modified by taking $d\epsilon = (I_1/n'^3)$ provides the dominant contribution. The result (225) essentially agrees with Percival and Richards,[38] who derived it from consideration based on detailed balance and on Fourier components of ϵ.

11.6. Monte Carlo procedures

Monte Carlo simulations of three-body collisions have been performed for rates of collisional excitation of Rydbergs by thermal electrons[11] and for the cross sections of collisonal ionization of H(n) by high-energy monochromatic electrons and protons.[10] For thermal electrons in a plasma, only a small fraction of collisions contribute to direct excitation and ionization, and the overall distribution of excited atoms among their discrete and continuous states is determined by the balance of upward and downward cascading. The energy transfers are small and the collisions are adiabatic in general. The procedure of Mansbach and Keck[11] focused on an equilbrium rate of collisional deexcitation $i \to f$, based on Monte Carlo simulation of the fraction of collisions causing deexcitation combined with a variational (or bottleneck) treatment of recombination[62] for the rate at which atoms cross some energy level S between i and f. In this sense, the method can be considered as complementary to that of Percival and associates[10] for direct excitation at much higher energies. These procedures were fully documented elsewhere.[10-12,62] It was noted by Vriens and Smeets[63] that the rates of Mansbach and Keck[11] are accurate only over a limited adiabatic range of electron temperatures; extrapolation to the sudden regime is without validity. (For more detail, Monte Carlo simulations of neutral–neutral collisions were described in Chap. 6).

11.7. Semiempirical electron-impact cross sections

In an effort to provide working formulae for cross sections over a wide range of impact energies without recourse to explicit application of the appropriate theoretical treatments in the previous sections, many authors have produced, to various degrees of compatibility, semiempirical fits based on available experimental and theoretical data and variations. The most recent semiempirical cross sections[60,63] for charged-particle impact should suffice until such time when implementation of the preceding detailed theories over relevant energy ranges becomes available, although the fits of Gee et al.[63] are based on the strong-coupling correspondence principle.

Based on the binary-encounter method of Sect. 11.5d, Vriens and Smeets[60] recommend

$$\sigma_n^{\mathrm{I}}(E) = \frac{\pi e^4}{(E + 3.25 I_n)} \left[\frac{5}{3 I_n} - \frac{1}{E} - \frac{2}{3} \frac{I_n}{E^2} \right] \tag{226}$$

as the cross section for collisional ionization of Rydberg atoms with ionization potential I_n by electrons of energy $E \gtrsim I_n$. This expression has the correct BE limit (215) at high E (the logarithmic term is negligible for excited states), reproduces the classical three-body Monte Carlo results (Sect. 11.6) to within statistical accuracy (10%) for all E, and agrees within 10 to 20% with the more elaborate binary encounter results of both Ton-That and Flannery[61] given by the full expression (212) for electron-impact ionization of metastable rare gases, which agree with available experiment, and of Roy and Rai[64] for ionization of ground-state alkali atoms.

The recommended cross section for $p \to q$ excitation by electron impact is[60]

$$\sigma_{pq}(E) = \frac{\pi a_0^2 (2R)}{E + (\alpha_{pq} R)} \left[A_{pq} \ln\left(\frac{E}{2R} + \beta_{pq} \right) + B_{pq} \right] \tag{227a}$$

where R is the Rydberg unit of energy (13.6 eV),

$$A_{pq} = (2R/E_{pq})f_{pq} \tag{227b}$$

in terms of the absorption oscillator strength f_{pq} for a transition of energy E_{pq}, and

$$B_{pq} = \frac{1}{p^3}\left(\frac{2R}{E_{pq}}\right)^2\left[1 + \frac{4}{3}\frac{I_p}{E_{pq}} + \frac{b_p}{p}\left(\frac{I_p}{E_{pq}}\right)^2\right] \tag{227c}$$

in terms of the ionization potential I_p and of b_p given as[60]

$$b_p = 1.41 \ln p - 0.7 - (0.51/p) + (1.16/p^2) - (0.55/p^3) \tag{227d}$$

In the absence of α_{pq} and β_{pq}, (227) provides the correct high-energy limit. A simple yet accurate expression for f_{pq} was given by Johnson[65] whose semiempirical cross sections are not as accurate as those just given. For $p-q \gtrsim p \gg 1$, then $B_{pq} \gg A_{pq}$, the last term in (227c) is negligible ($\ll 1$), and (227a) reduces to the correct binary-encounter limit, obtained for excitation by multiplying (213) by $2R/p^3$. Extrapolation to lower-impact energies E is provided by introducing in (227a) the parameters[63]

$$\beta_{pq} = \exp(-B_{pq}/A_{pq}) - (0.4E_{pq}/R) \tag{227e}$$

and

$$\alpha_{pq} = [8 + (23s^2/p^2)]/[8 + 1.1ps + (0.8/s^2) + 0.4(n^3/s)^{1/2}|s-1|] \tag{227f}$$

which depend only on properties of the Rydberg atom.

The preceding expressions yield excitation cross sections that are finite at threshold, a feature correct for excitation[66] because of dipole coupling between degenerate levels. Cross sections for $q \to p$ deexcitation are obtained from (227) by the use of the detailed balance relation.

Figure 11.4 illustrates the semiempirical excitation cross sections (227) of Vriens and Smeets[60] for various $p \to q$ transitions in atomic hydrogen together with those of Gee et al.[63] The overall agreement is good only at the higher-impact energies $E > 4I_p$, where Gee et al. claim accurate results.[63]

11.8. Summary

In this chapter we presented a comprehensive and unified account of the theory of Rydberg collisions with electrons, ions, and neutrals. The methods given ranged from quantal to classical descriptions, with necessary emphasis on the semiclassical analysis of the motion of either the (internal) Rydberg electron or (external) incident projectile, or both. The internal semiclassical wave functions not only provided both Bohr–Sommerfeld quantization and the Heisenberg correspondence principle, two valuable and essential assets for Rydberg atoms, but also lent themselves quite naturally in the action-angle variable representation to perturbation treatments based on the Hamilton–Jacobi equation. The fixed-center quantal method (Sect. 11.2b) led directly

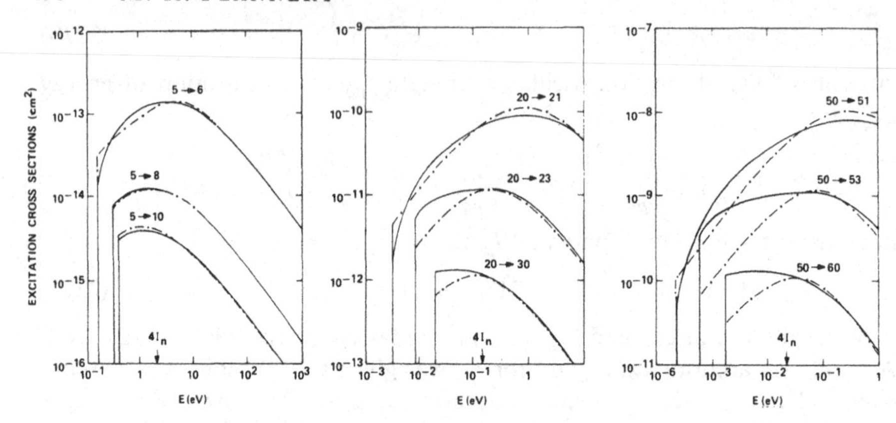

Fig. 11.4. Semiempirical cross sections[60] for $p \rightarrow n$ transitions in e–H(p) collisions for $p=5$, 20, and 50, where I_n is the ionization energy for H(p). Solid curves: Eq. (227a) from Ref. 60; broken curves: from Ref. 63.

to the full quantal impulse treatment and to its semiquantal and binary-encounter derivatives (Sect. 11.5). It also can be used for rearrangement and charge-transfer-ion–atom collisions.[9] The methods were presented here with the aim of providing a basic theoretical foundation for many future detailed investigations of various processes involving Rydberg atoms.

Thermal collisions that simply mix a small number of angular momentum l states within a given n can be accurately described by quantal or semiclassical close-coupling procedures (Sects. 11.2a and 11.3a), the only methods, apart from impulse treatments, that originate from studies of ground-state collisions. In A–B(n) thermal collisions, the speed of the Rydberg electron remains much greater than the A–B^+ relative speed so that, in order to examine the effects of the core B^+, some suitable molecular treatment that provides the response of the Rydberg electron to the charging field of the AB^+ ionic complex is required. For nl–nl' transitions, the close-coupling formalism of Arthurs and Dalgarno[17] is directly applicable, whereas for $n \rightarrow n'$ thermal collisions, the approach of Janev and Mikalov[67] is interesting.

Collisional transitions between neighboring or adjacent Rydberg levels n and n' are best handled in terms of the equivalent oscillator theorem (Sect. 11.4d) or its sudden and weak-coupling limits, when appropriate, for determination of the transition amplitudes. These amplitudes can then be inserted into a properly constructed semiclassical analysis (Sect. 11.3) of the collision cross section, e.g., the multichannel eikonal treatment[8] (Sect. 11.3a) for electron impact, the multistate orbital treatment[18] (Sect. 11.3a) for heavy-particle impact, or the familiar classical path and straight-line impact parameter approaches (Sect. 11.3b), which are the heavy-particle–high-energy limits of the previous semiclassical methods. The fixed-center quantal method

(Sect. 11.2b) is appropriate particularly for $nlm \rightarrow n'l'm'$ direct and rearrangement collisional transitions.

Collisional excitation involving large changes in n and ionization is best treated by the fixed-center analysis (Sect. 11.2b) and its quantal and semiquantal impulse derivatives (Sect. 11.5), which tend at high energy to the weak-coupling limit of perturbation-based procedures (such as close coupling). The semiquantal method essentially recasts the full quantal impulse treatment with a plane-wave description of the ejected electron in a form more suitable and efficient for the evaluation of A–B(n) ionization, including even ground-state targets B. For electron and ion impact, these impulse methods essentially reduce, with the use of the Rutherford cross section, to the standard binary-encounter approach[56-59] (Sect. 11.4d) with an appropriate momentum distribution for the Rydberg electron. For Rydberg collisions with ions or electrons, the $E^{-1} \ln E$ asymptotic term omitted by these impulse treatments is not important relative to the E^{-1} term but may be included by careful attention to its source (Sect. 11.5d).

Finally, classical procedures such as Monte Carlo computer simulations (Sect. 11.6) are effective for direct ionization and for charge-transfer collisions, which may also be described by the fixed-center treatment (Sect. 11.5d) and the impulse derivatives (Sect. 11.5).

Acknowledgment

This research was sponsored by the U.S. Air Force Office of Scientific Research under Grant No. AFOSR-80-0055.

References and notes

1. M. R. Flannery, *Ann. Phys.* (*N.Y.*) *61,* 465 (1970); *79,* 480 (1973).
2. M. R. Flannery, *Phys. Rev. A 22,* 2408 (1980).
3. A. K. Dupree and L. Goldberg, *Ann. Rev. Astron. Astrophys. 8,* 231 (1970).
4. I. L. Beigman, L. A. Vainshtein, and I. I. Sobel'man, *Zh. Eksp. Teor. Fix. 57,* 1703 (1969) [*Sov. Phys. JETP 30,* 920 (1970)].
5. I. C. Percival and D. Richards, *J. Phys. B 3,* 1035 (1970).
6. L. P. Presnyakov and A. M. Urnov, *J. Phys. B 3,* 1267 (1970).
7. D. Richards, *J. Phys. B 5,* L53 (1972).
8. M. R. Flannery and K. J. McCann, *J. Phys. B 7,* 2518 (1974); *7,* L522 (1974); *8,* 1716 (1975); *Phys. Rev. A 10,* 2264 (1974); *12,* 846 (1975).
9. M. R. Flannery, *J. Phys. B 8,* 2470 (1975).
10. R. Abrines and I. C. Percival, *Proc. Phys. Soc.* (*London*) *88,* 873 (1966); R. Abrines, I. C. Percival, and N. A. Valentine, *Proc. Phys. Soc.* (*London*) *89,* 515 (1966).
11. P. Mansbach and J. Keck, *Phys. Rev. 181,* 275 (1969).
12. I. C. Percival and D. Richards, in *Advances in Atomic and Molecular Physics,* eds. D. R. Bates and B. Bederson (New York: Academic, 1975), vol. 11, p. 1.

13. B. L. van der Waerden, *Sources of Quantum Mechanics* (Amsterdam: North Holland Publ., 1967).
14. H. A. Kramers, *Quantum Mechanics* (New York: Dover, 1964).
15. M. Born, *Mechanics of the Atom,* (London: Bell and Sons, 1927; reprinted New York: Ungar, 1960).
16. H. Goldstein, *Classical Mechanics* (Reading, Mass.: Addison-Wesley, 1980), chaps. 10 and 11.
17. A. M. Arthurs and A. Dalgarno, *Proc. R. Soc. London A256,* 540 (1960).
18. K. J. McCann and M. R. Flannery, *J. Chem. Phys. 63,* 4695 (1975); *69,* 5275 (1978).
19. M. R. Flannery, W. R. Morrison, and B. L. Richmond, *J. Appl. Phys. 46,* 1186 (1975).
20. M. R. Flannery and K. J. McCann, *J. Phys. B 12,* 427 (1979); *Astrophys. J 236,* 300 (1980).
21. L. Vainshtein, L. Presnyakov, and I. Sobel'man, *Zh. Eksp. Teor. Fiz. 45,* 2015 (1963) [*Sov. Phys. JETP 18,* 1383 (1964)].
22. M. R. Flannery, *J. Phys. B,* in press.
23. J. P. Coleman, in *Case Studies in Atomic Collision Physics,* eds. E. W. McDaniel and M. R. C. McDowell (Amsterdam, North Holland Publ., 1969), vol. 1, chap. 3, p. 101; R. G. Newton, *Scattering Theory of Waves and Particles* (New York: McGraw-Hill, 1966), p. 587.
24. H–H(n): F. R. Pomilla and S. N. Milford, *Astrophys. J. 144,* 1174 (1966); F. R. Pomilla, *Astrophys. J. 148,* 559 (1967).
25. e–H(nl): A. E. E. Rogers and A. H. Barrett, *Astrophys. J. 151,* 163 (1968); D. H. Sampson and L. B. Golden, *Astrophys. J. 161,* 321 (1970); *163,* 405 (1971); *170,* 169 (1971); *170,* 181 (1971).
26. H^+–H(nl) direct and charge-transfer collisions: N. Toshima, *J. Phys. Soc. Japan 42,* 633 (1977); *43,* 605 (1977); *43,* 610 (1977); *46,* 927 (1979); *46,* 1295 (1979); *47,* 257 (1979).
27. D. R. Bates in *Quantum Theory, I: Elements,* ed. D. R. Bates (New York: Academic Press, 1961), chap. 8, p. 252.
28. M. R. Flannery and K. J. McCann, *J. Phys. B 7,* 1158 (1974); and references therein.
29. I. C. Percival and D. Richards, *J. Phys. B 4,* 918 (1971); *4,* 932 (1971).
30. L. I. Schiff, *Quantum Mechanics* (New York: McGraw-Hill, 1968), chaps. 6 and 8.
31. F. J. Dyson, *Phys. Rev. 75,* 486 (1949).
32. W. Magnus, *Commun. Pure Appl. Math. 7,* 649 (1954); D. W. Robinson, *Helv. Phys. Acta. 36,* 140 (1963); P. Pechukas and J. C. Light, *J. Chem. Phys. 44,* 3897 (1966).
33. See, for example, L. D. Landau and E. M. Lifshitz, *Quantum Mechanics: Nonrelativistic Theory,* 2nd ed. (Reading, Mass.: Addison-Wesley, 1965), chap. VII.
34. See M. V. Berry and K. E. Mount, *Rep. Prog. Phys. 35,* 315 (1972).
35. L. D. Landau and E. M. Lifshitz, *The Classical Theory of Fields,* 4th rev. ed. (Elmsford, N.Y.: Pergamon Press, 1975), p. 181.
36. J. H. Van Vleck, *Phys. Rev. 24,* 330 (1924).
37. J. D. Jackson, *Classical Electrodynamics,* 2nd ed. (New York: Wiley, 1975).
38. I. C. Percival and D. Richards, *J. Phys. B 3,* 315 (1970).
39. D. H. Menzel, *Nature 218,* 756 (1968); *Astrophys. J. Suppl. Ser. 18,* 221 (1969).
40. I. C. Percival and D. Richards, *Astrophys. Lett. 4,* 235 (1970).

41. H. A. Bethe and E. E. Salpeter, *Quantum Mechanics of One- and Two-Electron Atoms* (New York: Springer-Verlag, 1957), p. 269.
42. M. R. Flannery and K. J. McCann, *Appl. Phys. Lett. 31,* 599 (1977).
43. P. F. Naccache, *J. Phys. B 5,* 1308 (1972).
44. E. de Prunelé, *J. Phys. B 13,* 3921 (1980).
45. I. C. Percival and D. Richards, *J. Phys. B 10,* 1497 (1977).
46. G. F. Chew, *Phys. Rev. 50,* 196 (1950); G. F. Chew and G. C. Wick, *Phys. Rev. 85,* 636 (1952); G. F. Chew and M. L. Goldberger, *Phys. Rev. 87,* 778 (1953).
47. M. R. Flannery, *J. Phys. B 13,* L657 (1980); *15,* 3249 (1982).
48. E. Fermi, *Nuovo Cimento 11,* 157 (1934).
49. V. A. Alekseev and I. I. Sobel'man, *Zh. Eksp. Teor. Fiz. 49,* 1274 (1965) [*Sov. Phys. JETP 22,* 882 (1966)].
50. K. Omidvar and H. L. Kyle, *Phys. Rev. A 2,* 408 (1970).
51. M. R. Flannery, *J. Phys. B 5,* 334 (1972); *Can. J. Phys. 50,* 61 (1972).
52. M. R. Flannery and K. J. McCann, *Phys. Rev. A 19,* 2206 (1979).
53. M. Matsuzawa, *J. Phys. B 10,* 1543 (1977); *12,* 3743 (1979).
54. A. P. Hickman, *Phys. Rev. A 19,* 994 (1979); *23,* 87 (1981).
55. E. de Prunelé and J. Pascale, *J. Phys. B 12,* 2511 (1979).
56. Y. Hahn, *J. Phys. B 15,* 613 (1982).
57. N. F. Lane and S. Preston, private communication.
58. L. H. Thomas, *Proc. Cambridge Philos. Soc. 23,* 713 (1927); M. Gryszinski, *Phys. Rev. 138,* A336 (1965); L. Vriens, in *Case Studies in Atomic Collision Physics I,* eds. E. W. McDaniel and M. R. C. McDowell (Amsterdam: North-Holland Publ., 1969), chap. 6, p. 337.
59. D. R. Bates and W. R. McDonough, *J. Phys. B 3,* L83 (1970); *5,* L107 (1972).
60. L. Vriens and A. H. M. Smeets, *Phys. Rev. 22,* 940 (1980).
61. D. Ton-That and M. R. Flannery, *Phys. Rev. A 15,* 517 (1977); D. Ton-That, S. T. Manson, and M. R. Flannery, *J. Phys. B 10,* 621 (1977).
62. J. C. Keck, *Adv. Chem. Phys. 13,* 85 (1967); in *Advances in Atomic and Molecular Physics,* eds. D. R. Bates and I. Esterman (New York: Academic Press, 1972), vol. 8, p. 39.
63. C. S. Gee, I. C. Percival, J. G. Lodge, and D. Richards, *Mon. Not. R. Astron. Soc. 175,* 209 (1976).
64. B. N. Roy and D. K. Rai, *Phys. Rev. A 8,* 849 (1973).
65. L. C. Johnson, *Astrophys. J. 174,* 227 (1972).
66. M. Galitis and R. Damburg, *Proc. Phys. Soc. (London) 82,* 192 (1963); C. Bottcher, *J. Phys. B 5,* 2199 (1972).
67. R. K. Janev and A. A. Mikalov, *Phys. Rev. A 20,* 1890 (1979).

12

Experimental studies of the interaction of Rydberg atoms with charged particles

J.-F. DELPECH

12.1. Introduction

Charged particles play an important role in most situations involving ionized gases and plasmas. Highly excited atoms are often also present in these situations; they may be produced either by excitation (direct or stepwise) or as the result of recombination processes that lead to the neutralization of electron-ion pairs.

The effect of highly excited atoms in plasmas may thus be extremely important: they may govern the macroscopic properties of the plasma, for example, through their photoionization processes;[1] furthermore, their collisional and radiative properties play a dominant role in the extremely complex sequence of processes called collisional-radiative recombination.[2]

In addition, they are of great practical interest because of their importance in plasma diagnostics (as, for example, in the radio recombination lines used in astrophysics or in the line- and continuum-intensity measurements used in laboratory plasmas). Collisional properties of high Rydbergs interacting with ions are also of importance in the physics of neutral beam injection and the heating of magnetically confined plasmas.

It is thus not surprising that most of our experimental information concerning these processes has come initially from studies in plasma physics, where experimental conclusions were drawn from detailed comparisons between computed and observed excited state populations either in an unperturbed plasma[3,4] or in a plasma perturbed by photoionization of highly excited levels.[1]

Recently, more direct experiments involving collisions of high Rydbergs with electrons and ions have been undertaken. These will be discussed in this chapter, with the theoretical aspects of Rydberg-ion and Rydberg-electron collisions discussed in Chap. 11. In most of what follows, the identification and/or detection of highly excited atoms relies on the technique of field ionization, which is also described in detail in Chap. 9 of this volume.

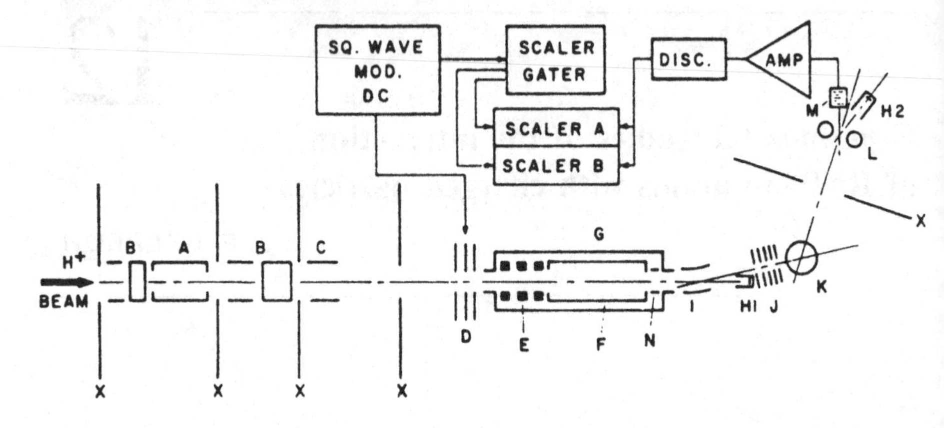

Fig. 12.1. The merged-beam apparatus of Koch and Bayfield: (A) electron-transfer gas cell; (B, I, L) electrostatic deflection; (C) ion-beam gating plates; (D) n-band-defining electric field; (E, N) electric-potential change rings; (F) voltage-labeled interaction region; (G) liquid-He cryopump; (H1, H2) Faraday cups; (J) electrostatic filter lens (Ref. 6); (K) deflection magnet; (M) particle multiplier; (X) differential pumping apertures. (From Koch and Bayfield[5])

12.2. Collisions of highly excited atoms with ions

In this section, we shall be concerned with the following three types of reactions that can take place in the collision of a Rydberg atom $A(n)$ with an ion of charge q, B^{+q}.

First, collisional excitation of $A(n)$ to a different level n':

$$A(n) + B^{+q} \rightarrow A(n') + B^{+q} \qquad (n' \neq n) \tag{1}$$

Second, collisional ionization of $A(n)$, which is similar to Eq. (1) except that the product now lies in the continuum of A^+:

$$A(n) + B^{+q} \rightarrow A^+ + e + B^{+q} \tag{2}$$

Third, charge exchange in which A gives an electron to B:

$$A(n) + B^{+q} \rightarrow A^+ + B^{+q-1}(n') \tag{3}$$

and the resultant B ion may itself be excited to level n'.

In 1975, Koch and Bayfield[5] reported the results of the first experimental study of H^+-H ($n \simeq 47$) collisions using a merged-beam apparatus and center-of-mass collision energies between 0.4 and 61 eV.

They used a single ion source (Fig. 12.1) producing a pure H^+ beam with energy near 11 keV and an energy spread of less than 20 eV.[6] A fraction of the ions in this beam are converted to neutral $H(n)$ atoms by electron-transfer collisions on xenon atoms. The resultant beam, containing both H^+ and $H(n)$, is passed through an axial electric field modulated at 1 kHz between 105 V cm^{-1} (ionizing field for $n \gtrsim 50$) and 171 V cm^{-1} (ionizing field for

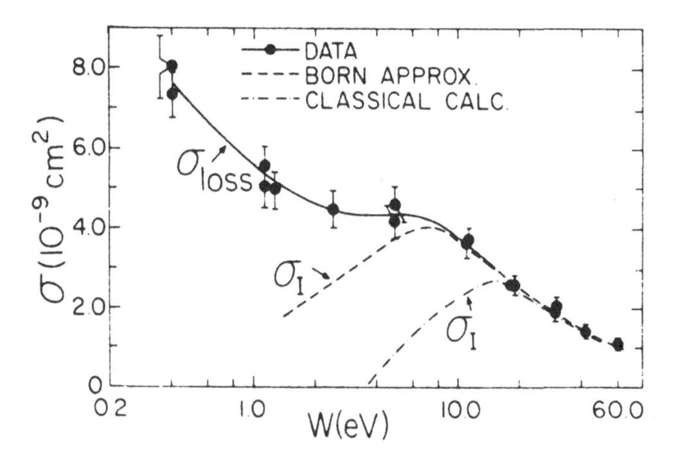

Fig. 12.2. Electron-loss cross section σ_{loss} for collisions between H^+ and an atom H (high n) in the nominal n band $44 \leqslant n \leqslant 50$ as measured by Koch and Bayfield. A smooth curve connects the data points. The cross section scale is uncertain by a factor of about 2. (From Koch and Bayfield[5])

$n \gtrsim 44$). This defines a narrow band of principal quantum numbers between 45 and 50, which can be characterized adequately by its central value $n = 47$. The fraction of the total neutral-atom flux at the beam energy and within this band is known to about 15%.

The H^+ and $H(n)$ beams, both at essentially the same energy, enter a very low pressure region (pressure measured to be $\lesssim 10^{-11}$ Torr), where the H^+ beam energy is decreased in a nearly uniform (to 25%) axial electric field that is too weak to ionize H ($n \simeq 47$). Merged-beam collisions then take place between $H(n)$ at the initial beam energy and H^+ at a lower energy, resulting in center-of-mass energies between 0.3 and 60 eV.

The H^+ and $H(n)$ beams interact for 50 cm in a volume kept at a constant electric potential with respect to ground. Any hydrogen atom that is ionized inside the interaction region is accelerated and the corresponding flux is analyzed in charge, in energy, and in momentum and is sent to two scalers, each of which is gated on opposite half-cycles of the modulating voltage. The difference between these two signals is then due to collisional electron loss from highly excited hydrogen atoms H ($44 \lesssim n \lesssim 50$), where electron loss is the sum of two processes, electron transfer between the two collision partners [reaction (3)] and ionization. Ionization itself is due either to the direct process of reaction (2) or to the indirect process of reaction (1) when n' is large enough for $H(n')$ to be subsequently ionized by the accelerating electric field.

The measured cross sections (shown in Fig. 12.2) are larger by a factor of about 3.5 than the theoretical predictions for the sum of reactions (2) and (3). This reflects, in part, the uncertainty in the cross section scale that arises primarily because of the uncertainty in the determination of the beam overlap

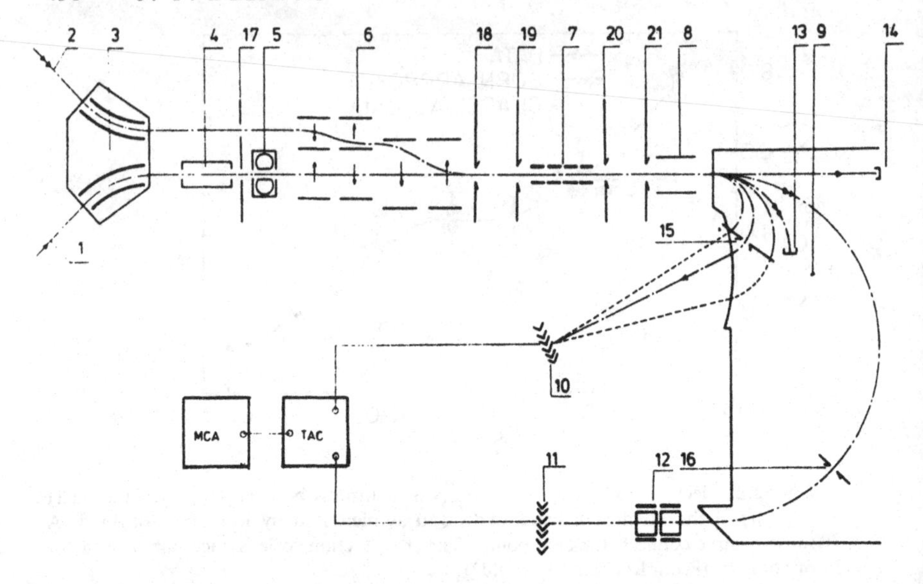

Fig. 12.3. Schematic diagram of the apparatus used by Burniaux et al. (not to scale): (1) hydrogen beam, (2) helium beam, (3) mass filter, (4) neutralizer, (5) ionizer, (6) beam mixer, (7) interaction region, (8) Einzel lens, (9) analyzer, (10) H^+ detector, (11) He^+ detector, (12) quadrupole doublet, (13) He^{2+} probe, (14) biased H probe, (15) movable H^+ slit, (16) He^+ slit, (17–21) diaphragms, (TAC) time-to-amplitude converter, (MCA) multichannel analyzer. (From Burniaux et al.[8])

and, in part, the contribution of indirect ionization through reaction (1) followed by field ionization. This contribution is, however, difficult to evaluate at energies below about 8 eV, where the electron velocity and the collision velocity are the same for $n=47$. Above this energy, we may expect a $1/E$ dependence for reaction (1), and the shape of the curve would thus not be substantially distorted. At a collision energy of 20 eV, using reasonable assumptions about the experimental conditions, the use of the analytic formula of Lodge et al.[7] shows that the cross section for reaction (1) would be about 10^{-9} cm^2 and should thus contribute about half of the total measured cross section, bringing agreement between theory and experiment well within error bars.

Burniaux and co-workers[8] studied the nonsymmetric charge-transfer reaction

$$^3He^{2+} + H(n) \rightarrow {}^3He^+ + H^+ \tag{4}$$

in the relative energy range 0.25–478 eV for excited hydrogen atoms with principal quantum numbers ranging from 8 to 24. The experimental approach was a merging-beam technique with coincident detection of the reaction products.

The beam of atomic hydrogen is produced by charge exchange of a H^+ beam in argon (Fig. 12.3). In an auxiliary experiment, the authors determined that the population of excited H(n) states does indeed have a $1/n^3$ dependence, as expected. The electrostatic mixer that merges the ion beam into the neutral beam has fields of ~ 1000 V cm^{-1} and acts on the atoms for about 3×10^{-7} s. Thus states with principal quantum numbers lower than $n=7$ or 8 decay radiatively and states with $n > 30$ are ionized in the merging field. A carefully defined ionizing field is applied to the H(n) beam before the interaction region in order to modify, in a well-controlled way, the excited state population of the beam.

Ions formed in the interaction region emerge with characteristic kinetic energies and momenta and can thus be separated from ions produced elsewhere by a magnetic analyzer. The primary He^{2+} beam is collected on a probe and the neutral hydrogen beam is detected by secondary emission. The H^+ and He^+ particles are detected by two multichannel plates connected to two time-to-amplitude converters and to a multichannel analyzer. Coincident ion pairs produced in beam–beam collisions thus appear as a peak above the flat spectrum generated by the uncorrelated ion yield from the beam–residual-gas interaction.

Experiments have been conducted at 12 values of the relative energies from 0.25 to 478 eV, corresponding to relative velocities ranging from 8×10^5 to 3.5×10^7 cm s^{-1}, and for 5 values of the ionizing field, from 1 to 60 kV cm^{-1}, thus effectively retaining states up to $N=30$ for the lowest field and $N=10$ for the highest field.

The quantity actually measured is

$$S(N) = \sum_{n=8}^{N} p(n)\sigma(n) \tag{5}$$

where $\sigma(n)$ is the charge-transfer cross section for excited state H(n) and $p(n)$ the population of this state in the beam. It is then easy to deduce from this quantity an average cross section for charge transfer to He^{2+} from excited hydrogen and principal quantum numbers ranging from n_1 to n_2 (Fig. 12.4):

$$\sigma(n_1, n_2) = \frac{S(n_2) - S(n_1)}{p(n_1) + \cdots + p(n_2)} \tag{6}$$

To check the validity of their results, the authors used $^4He^{2+}$ ions instead of $^3He^{2+}$ ions in some experimental runs. At the same relative velocity, the energy of the hydrogen beam was different for the two isotopes and thus also the population of excited states by a factor as large as three on the high-velocity side, through the dependence of the H^+–Ar charge-exchange process on velocity. The cross section was nevertheless found to be independent of the mass of the ion (Fig. 12.5). This constitutes a stringent test of the validity of the experimental method.

Over most of their range of relative energy, these results correspond to relatively low values of collision velocity ($v/v_e \lesssim 1$, where v is the collision velocity

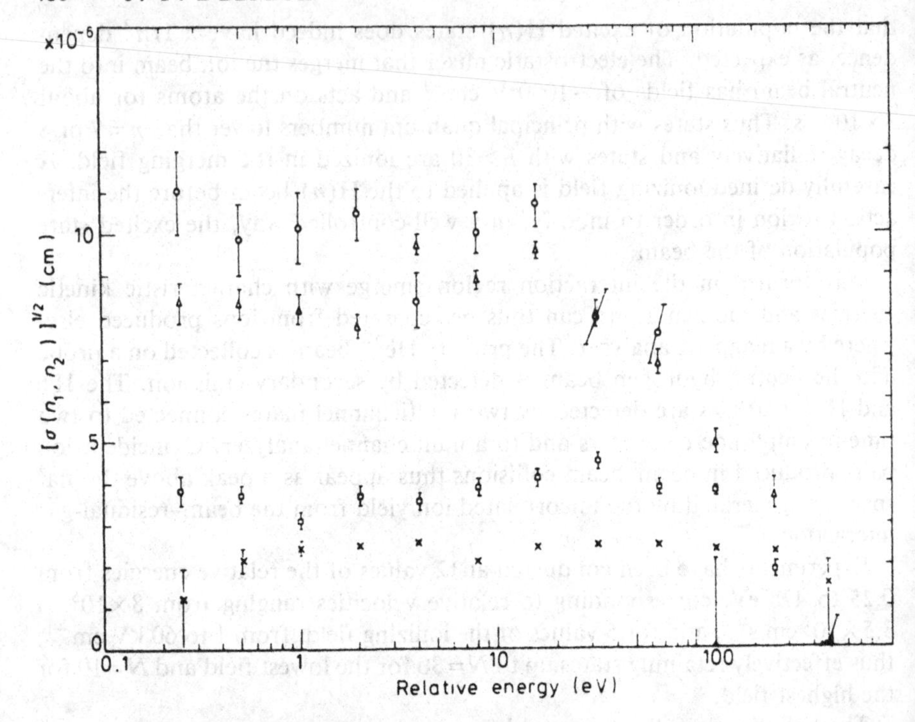

Fig. 12.4. Average exchange cross section $\sigma(n_1, n_2)$ in $H+{}^3He^{2+}$ deduced from the experimental data of Burniaux et al. by Eq. (6), using the independently measured energy-dependent excitation factor of H. The open circles denote $\sigma(20, 24)$; the open triangles, $\sigma(15, 19)$; the open squares, $\sigma(10, 14)$; the crosses, $\sigma(8, 9)$. (From Burniaux et al.[8])

and $v_e = 2.19\times 10^8/n$ cm s^{-1} the electron orbital velocity) and are in reasonable agreement with the low-velocity limit for the charge-exchange cross section of Olson[9]

$$\sigma(n) = 5.5\pi n^4 a_0^2 q \text{ cm}^2$$

where a_0 is Bohr's radius and q the ionic charge number. This is especially true for the $\sigma(15, 19)$ and $\sigma(10, 14)$ cross sections. The $\sigma(8, 9)$ and $\sigma(20, 24)$ cross sections are likely to be overestimated and underestimated, respectively, in this experiment, as discussed by the authors in their original publication.[8]

Kim and Meyer[10] studied the interaction of a beam of highly excited hydrogen atoms ($9 \leq n \leq 24$) with a N^{3+} beam at a relative collision velocity of 2.8×10^8 cm s^{-1}, which is substantially higher than the range of velocities covered in the experiments of Koch and Bayfield[5] and Burniaux et al.[8]

The fast hydrogen beam is produced in a H_2O neutralizer cell from 40-keV protons (Fig. 12.6), where the residual hydrogen ions are swept out by a field of about 300 V cm^{-1}. The beam then passes through a field ionizer[11] with a

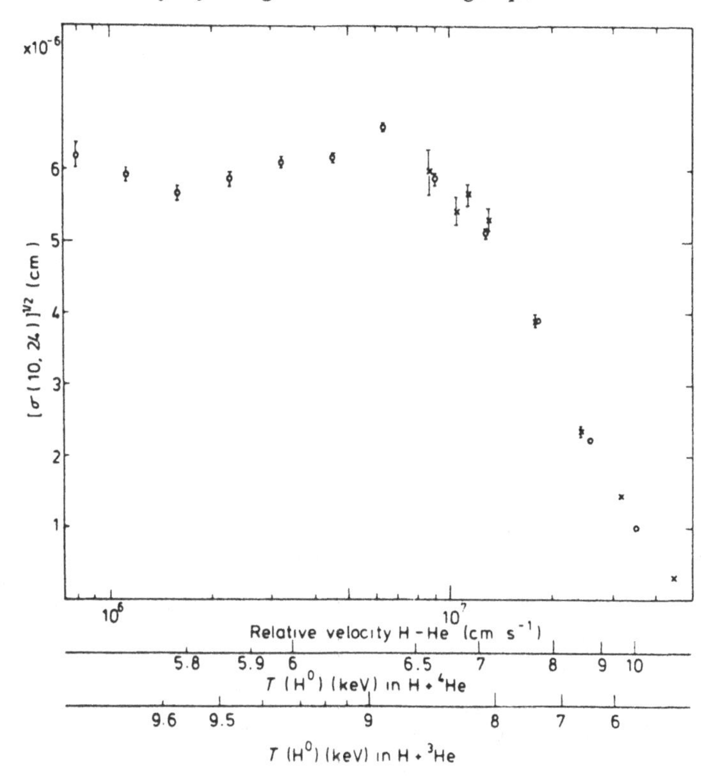

Fig. 12.5. Average exchange cross section in $H+{}^3He^{2+}$ (circles) and in $H+{}^4He^{2+}$ (crosses) at the same relative velocity. The two additional scales on the horizontal axis specify the kinetic energy T of the hydrogen atoms, which was different for ^3He and ^4He measurements and led to populations of the excited states differing by a factor of three at the high-velocity side. (From Burniaux et al.[8])

maximum field of 107 kV cm^{-1}, sufficient to ionize all states down to $n=9$. In an auxiliary experiment, the authors verified the expected $1/n^3$ dependence for the excited state population in the beam.

The N^{3+} beam crosses the hydrogen beam at 90° in a deflector-modulator cell where the collisionally produced H^+ ions are steered toward a detector by a field of 1 to 1.5 kV cm^{-1}.

Similarly to the experiment previously described, the apparatus measures a quantity [cf. Eq. (5)]

$$S(N) = \sum_{n=1}^{N} p(n)\sigma(n)$$

where $\sigma(n)$ is now the cross section for production of an H^+ ion in a $H(n)$-N^{3+} collision (Fig. 12.7). It should, however, be noted that the fields in the deflector-modulator cell are sufficient to ionize Rydberg atoms with $n>25$–28 and that two channels are thus available:

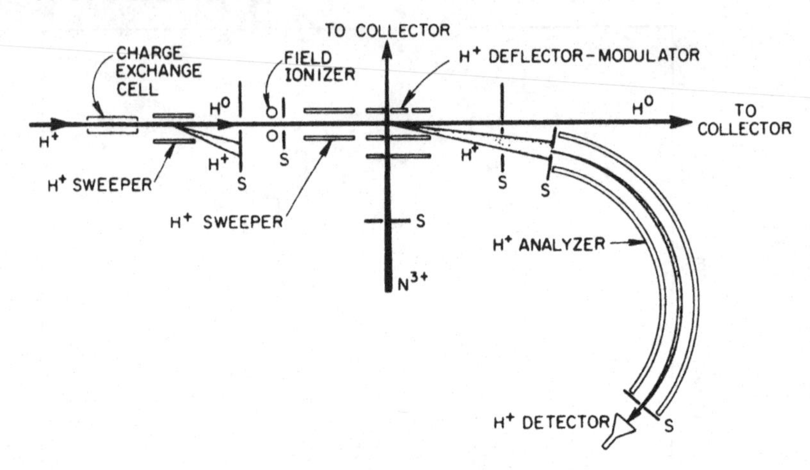

Fig. 12.6. Schematic layout of the experimental apparatus used by Kim and Meyer. The symbol S represents the location of beam-limiting apertures. The field applied by the H^+ sweepers is $\sim 300 \text{ V cm}^{-1}$. The deflector-modulator consists of three parallel-plate electrodes with a rectangular hole for the N^{3+} beam. The upper electrode is split, and the proton trajectory is determined by the combined dc fields provided by the deflection voltages applied on the downstream portion of the upper electrode and the central electrode. With this arrangement, a desired deflection can be obtained for a range of suitable voltage combinations. In addition to a dc deflection voltage, a sawtooth modulating voltage was applied to the deflector-modulator. Proton counts were recorded in a multichannel analyzer as a function of sawtooth voltage for a given dc deflector voltage. (From Kim and Meyer[10])

The direct channel

$$H(n) + N^{3+} \rightarrow H^+ + e + N^{3+} \tag{7a}$$

The indirect channel

$$H(n) + N^{3+} \rightarrow H(n' > 25\text{–}28) + N^{3+} \tag{7b}$$

where $H(n')$ is subsequently ionized by the electric field. (Electron-transfer collisions should be negligible in these fast collisions where the collision velocity exceeds the relevant orbital electron velocities by at least one order of magnitude; see Ref. 9.)

Assuming that $\sigma(n)$ has a power-law dependence, the measurements yield a cross section

$$\sigma(n) = 1.19 \times 10^{-15} n^{3.12} \text{ cm}^2 \tag{8}$$

for $9 \leq n \leq 25$.

This cross section does not scale as n^2. However, this should not be surprising because the contribution of reaction (7b) (which was neglected by the authors) should in fact be substantial for the highest quantum numbers. Use of the analytical formula of Lodge et al.[7] indicates that the cross section for

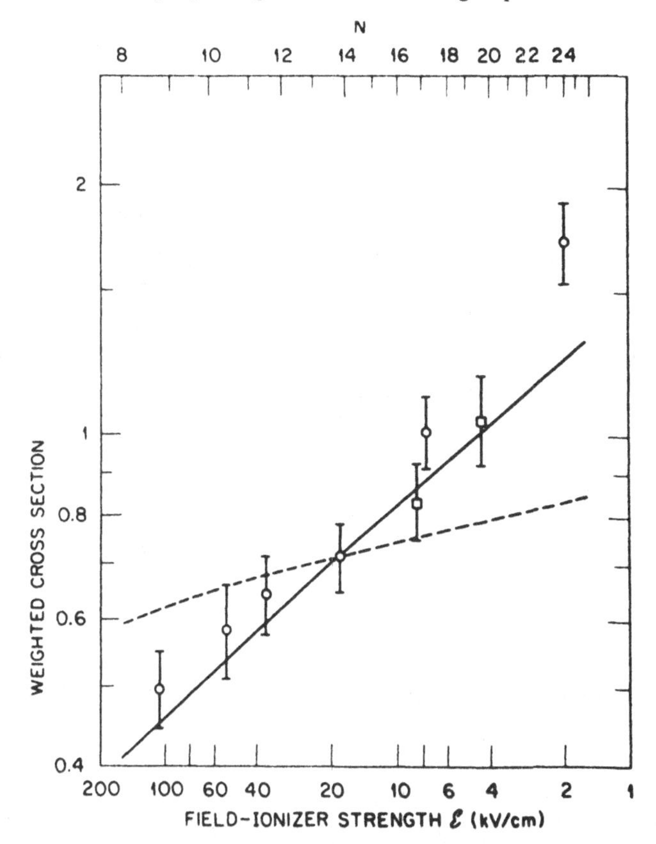

Fig. 12.7. The population-weighted cross section $S(N)=\Sigma_1^N p_n \sigma_n$ as a function of N, the upper cutoff of p_n, selected by field ionizer. Points depicted by circles were obtained by electrostatic field ionization; those indicated by squares by magnetic field ionization. The solid line shows least-squares results; the broken line the expected results, neglecting the indirect channel, for a cross section for the direct-channel scaling as n^2. Both lines are normalized to $N=14$. (From Kim and Meyer[10])

reaction (7b), starting from $n \leqslant 20$, is about 50% of their measured $S(N=20)$ (Fig. 12.7) and these experimental results may therefore not be inconsistent with the n^2 scaling rule.

12.3. Collisions of highly excited atoms with electrons

12.3a. Collisions with fast electrons

Schiavone and co-workers[12] reported in 1977 the results of an experiment primarily designed to measure the electron-impact excitation at energies between 30 and 300 eV of ground-state helium atoms to highly excited Rydberg levels. In the course of their experiments, they found evidence of l-changing in colli-

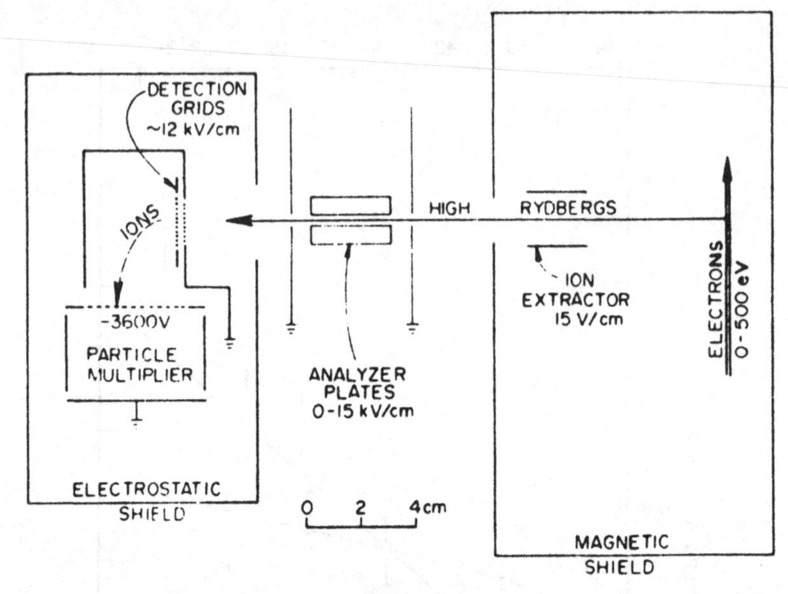

Fig. 12.8. Schematic diagram of the apparatus used at Bell Labs for the study of electron-impact excitation of helium. (From Schiavone et al.[12])

sions between fast electrons and Rydberg atoms and they were able to deduce cross sections from their data.

In their apparatus (Fig. 12.8), highly excited helium atoms (in the range $20<n<80$) were formed by collisions of electrons with He atoms. At these electron energies, low-l states (S, P, and D) are thought to dominate in the initial electron-impact excitation.

The atoms then traveled along the beam through an electric field analyzer that selectively ionized the atoms according to their principal quantum number. The remaining highly excited atoms reached the detector after traveling 20.1 cm from the electron beam with a corresponding mean time of flight of 108 μs.

Because of radiative decay, very few of the low-l atoms having $n<50$ were expected to reach the detector. However, the observed signals indicated that the number of surviving atoms with $n<50$ was large. This shows that the number of states with higher values of l, which have much larger radiative lifetimes than S, P or D states, is much larger than expected.

The authors interpret this result as being due to the high-l states populated via collisions of low-l Rydberg atoms with electrons. In this interpretation, a newly excited atom would undergo one or several (n, l) changing collisions before leaving the interaction region inside the magnetic shield.

The observed dependence of the signal on electron current supports this interpretation, and their analysis leads to the conclusion that, within a factor of two, the l-changing collisions have a cross section

$$\sigma_{\Delta l}(\text{cm}^2) \simeq 5 \times 10^{-15}(n^4/E)\ln(100En^2) \tag{9}$$

where the incident electron energy E is in electron volts.

Although initial and final l values cannot be specified in this experiment, this cross section seems to be in reasonable agreement with the theoretical predictions of Herrick[13] using the impact-parameter method.

Since the publication of this work, the effect of 300-K blackbody radiation on the population distribution of high-Rydberg atoms was found to be very important in some situations, and it was not clear at first how this would influence the interpretation of these experimental results.

The situation was recently clarified by Schiavone, Tarr, and Freund.[14] They pointed out that Δl transitions for $\Delta n=0$ would be very slow because of the almost complete degeneracy of higher-l states. Thus most transitions would involve $\Delta n=\pm 1$ and $\Delta l=\pm 1$, but these transfers themselves would have very little effect on the outcome of the experiment because there is hardly any population difference between neighboring n states. They concluded that blackbody radiation had no noticeable effect in their experimental situation.

The same group[14] conducted similar studies in neon and argon and found the l-changing cross sections for collisions with electrons, estimated to be $\sim 10^{-10}$ cm^2 for Ne and Ar, to be fairly consistent with those found in He in view of the large experimental uncertainties.

Recently, Stebbings and co-workers[15] undertook a preliminary series of experiments designed to study directly collisions between fast (25-eV) electrons and sodium Rydberg atoms in a single, well-defined, initial high-Rydberg state. A fraction of the atoms contained in a low-density sodium atom beam was excited, in zero electric field, to a selected Rydberg state by two-step laser photoexcitation via transitions of the type $3\,^2S_{1/2} \rightarrow 3\,^2P_{3/2} \rightarrow n\,^2D_{3/2,5/2}$. These atoms then interacted with electrons present in a weak crossed electron beam having a current density of ~ 1 μA cm^{-2}. Following an interaction time of ~ 8 μs, the electron beam was turned off and an increasing field applied across the interaction region in order to detect and identify the Rydberg states present via selective field ionization (SFI).[15] In order to identify effects from electron–Rydberg collisions, SFI data were recorded at the same delay after laser excitation, both with and without use of the crossed electron beam. Because both data sets contained the same contribution from states populated by collisions with background gas or by interactions with background thermal radiation, differences between them must have resulted from the ionization of products formed in electron–Rydberg collisions. Typical data for Na(36d) atoms are presented in Fig. 12.9, which encompasses the range of field strengths over which adiabatic ionization of sodium atoms with $n \simeq 36$ is expected. It is evident from Fig. 12.9 that collisions with electrons result in the appearance of a new ionization feature. Unique identification of the states of the atoms comprising this feature is not at present possible, although it is consistent with the ionization of 37p states.

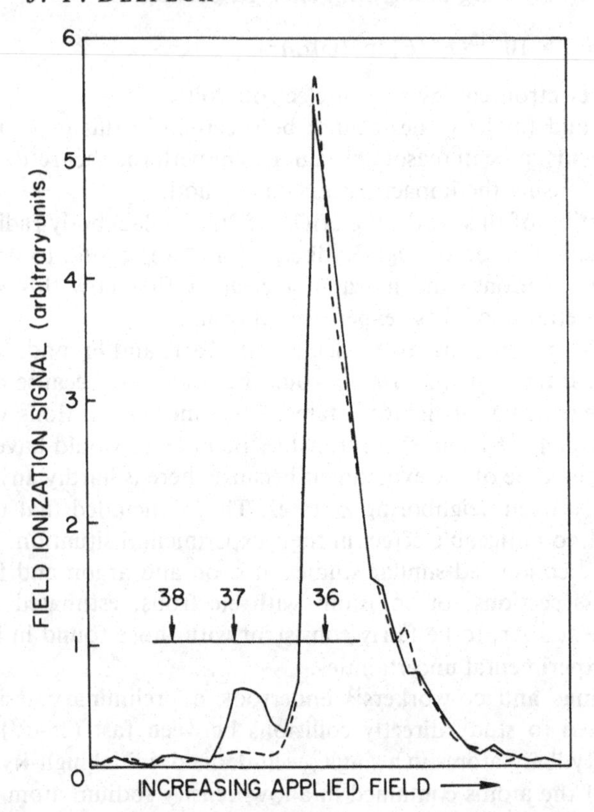

Fig. 12.9. Selective field ionization spectra obtained 8 μs after excitation of Na(36d) atoms in the presence (solid line) and absence (broken line) of a crossed 25-eV electron beam. The arrows indicate the threshold field strengths for adiabatic ionization of sodium atoms with different principal quantum numbers.

Measurement of the time development of this feature permits the estimation of a lower bound to the corresponding cross section, yielding the value

$$\sigma \simeq 6.6 \times 10^{-10}\ \mathrm{cm}^2$$

for 25-eV electrons. This cross section is remarkably close to the value of the Na(36d → 37p) cross section predicted theoretically at the same electron energy by Percival and Richards,[16] $\sigma = 9.2 \times 10^{-10}\ \mathrm{cm}^2$. However, this agreement may be fortuitous, and much further experimental work in this area is clearly required.

12.3b. Collisions with slow electrons

In the experiments reported by the Orsay group,[17,18] cell techniques were used to study the collision of highly excited helium atoms with thermal electrons

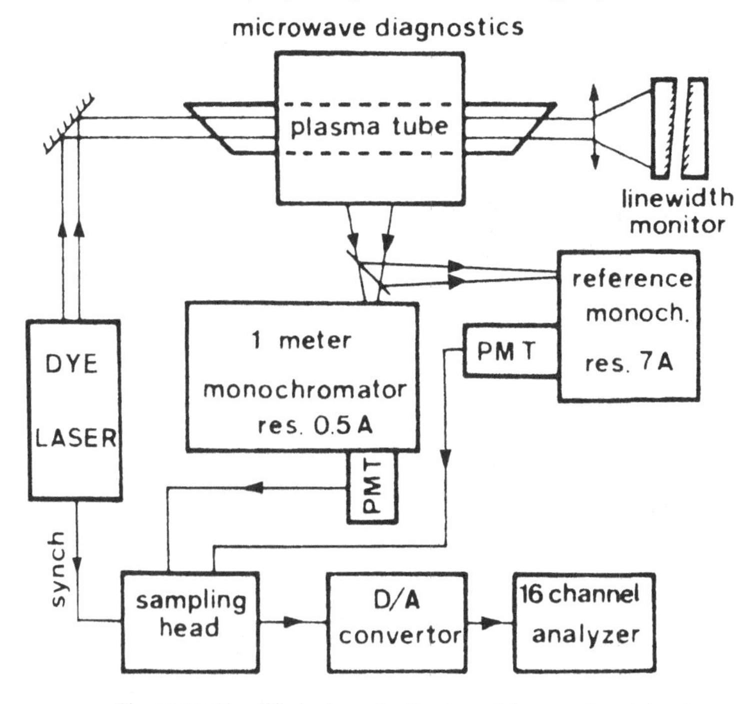

Fig. 12.10. Simplified schematic diagram of the experimental system used at Orsay; the sampling head and 16-channel analyzer have now been replaced by a fast photon-counting system. (From Gauthier et al.[19])

(Fig. 12.10). The He(n ^{3}P) states were produced by direct photoexcitation of He(2 ^{3}S) metastable atoms present in a high-purity, room-temperature, stationary helium afterglow. After cessation of the laser pulse, the populations of levels of principal quantum number n between 3 and 17 were followed by fluorescence spectroscopy with a 3.5-ns resolution.

Excited levels of helium are, of course, not exactly hydrogenic for the lower-l values. However, it was found that the collisional coupling between the singlet and triplet systems is very weak, whereas collisional coupling between the l sublevels of a given n level is extremely fast within a given multiplicity. The total population of level n thus remains always statistically distributed among its l sublevels of same multiplicity, and most of the population always remains in high-l states, which have the larger statistical weight and are essentially hydrogenic.

From an analysis of the temporal evolution of the population of all neighboring Rydberg levels (including the laser-pumped level) after laser excitation of a particular level (see Fig. 12.11), the Orsay group deduced total depopulation rate coefficients induced by electron collisions for principal quantum numbers $8 \leqslant n \leqslant 17$ (Fig. 12.12).

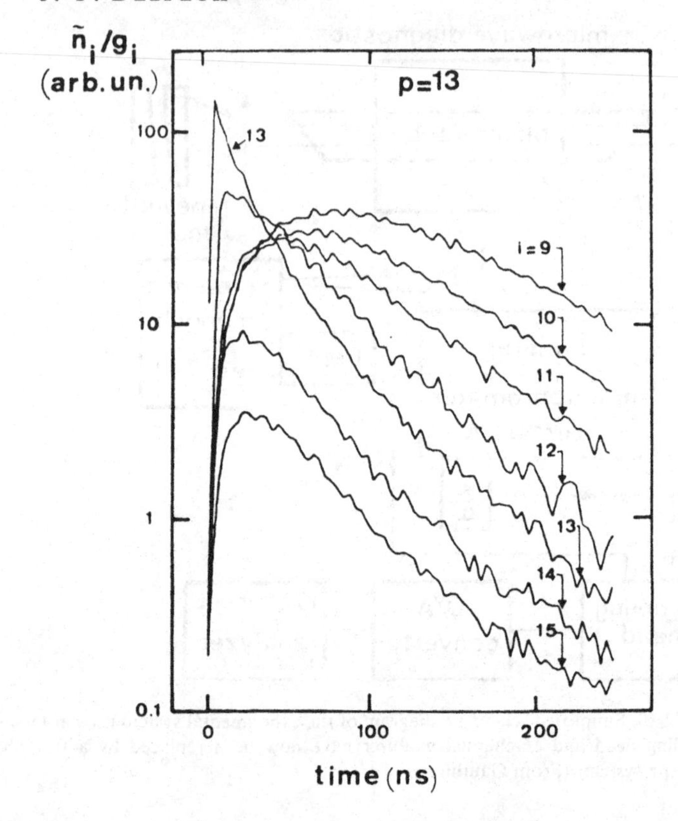

Fig. 12.11. Temporal evolution of the populations of levels 9–15 after selective excitation of level 13 by a laser pulse tuned to the $2\,^3S$–$13\,^3P$ transition at 2645 Å. The lines connect the experimental points for each level. Experimental conditions are $n_e = 2\times10^{11}$ cm^{-3}, $n_0 = 6.6\times10^{16}$ cm^{-3}, and $T_e = 400$ K. (From Devos, Boulmer, and Delpech[18])

These total rate coefficients lie slightly lower, by a factor of 0.64, than the predictions of Mansbach and Keck's theory[21] and of Johnson's semiempirical formulas.[22] Their variations with electron energy agree equally well with these two theories (see Ref. 23 and Fig. 12.13). However, the theories of Johnson and of Mansbach and Keck are quite different in physical content; the first remains within the dipole approximation and predicts that transitions with $\Delta n = \pm 1$ are strongly favored, whereas the second does not lead to such a propensity rule. These different predictions for the electron-induced individual transfer rate coefficients are shown in Fig. 12.14 together with experimental results for electron-induced transfer from $n = 14$ to $n' = 9$–12, 14, and 15. Similar results were obtained from level $n = 10$. They show conclusively that individual electron-induced collisional transfers are in good agreement with the

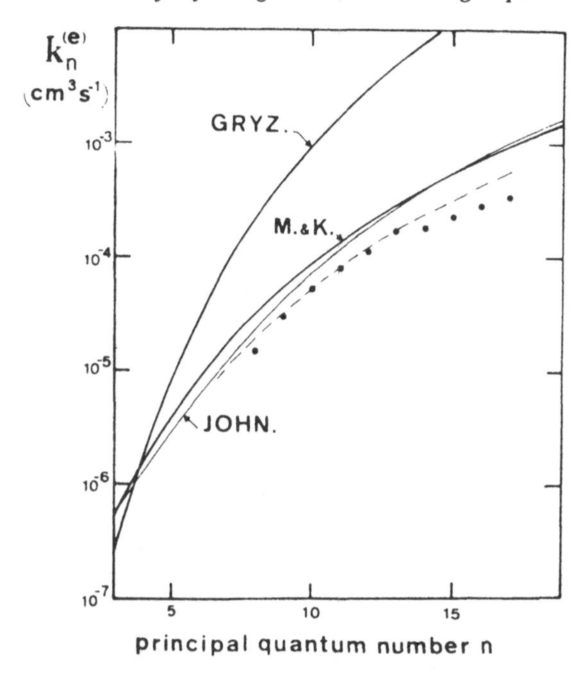

Fig. 12.12. Measured depopulation rate coefficients induced by electron collisions for principal quantum numbers n from 8 to 17 (dots) compared with the theoretical predictions of Gryzinski (GRYZ.),[20] Mansbach and Keck (M. & K.),[21] and Johnson (JOHN.).[22] The dashed line shows the empirical approximation of the experimental results. (From Devos, Boulmer, and Delpech[18])

classical predictions of Mansbach and Keck and that the dipole approximation is indeed not valid at such low electron energies, as expected.

Addendum

Since the writing of this chapter, extensive work has been published by MacAdam et al.[24] on l changing in sodium Rydberg atoms induced by ion collisions near the matching velocity, i.e., when the velocity of the charged particle nearly matches the mean orbital velocity of the bound atomic nl electron. The l-changing cross section is found to vary with n as n^{β}, where $\beta = 5.12 \pm 0.20$ at matching velocities. The cross section magnitude for

$$Ne^+ + Na(28d) \rightarrow Ne^+ + Na(28, l \geqslant 3)$$

at 1000 eV is $\sim 5 \times 10^{-8}$ cm^2, in agreement with existing theories. This exceeds the geometric cross section $n^4 \pi a_0^2$ by a factor of 10^3 and indicates that passage of an ion closer to the Rydberg atom than 30 orbital radii suffices to induce l changes in the velocity matching region.

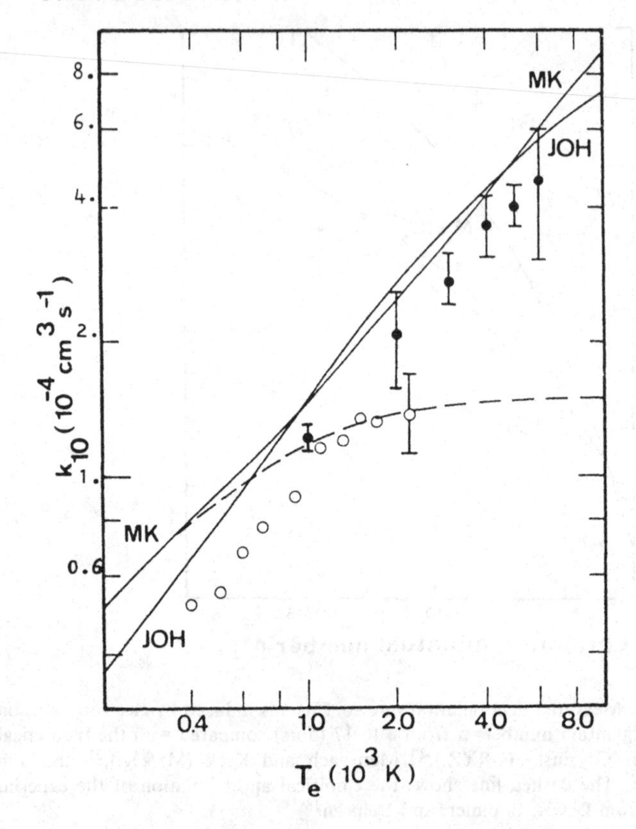

Fig. 12.13. Measured and calculated $10\,^3P$ depopulation rate as a function of electron temperature. The curves labeled MK and JOH represent the theoretical prediction of Mansbach and Keck[21] and of Johnson,[22] and the dashed line corresponds to the classical MK results neglecting ionization. Open circles are the results of Ref. 18; filled circles are obtained with an improved experimental system. (From Baran et al.[23])

These same authors[25] also observed charge transfer from laser-excited Na Rydberg atoms of principal quantum number $n = 20$–40 to Rydberg states of Na, He, Ne, and Ar projectiles. The final state distributions are relatively strongly peaked and their positions and widths change as the laser is tuned to various initial states or as the projectile energy is varied.

Acknowledgments

It is a pleasure to acknowledge the helpful discussions I had with Ron Olson as well as the generous hospitality of SRI International and the University of Maryland during the preparation of this manuscript.

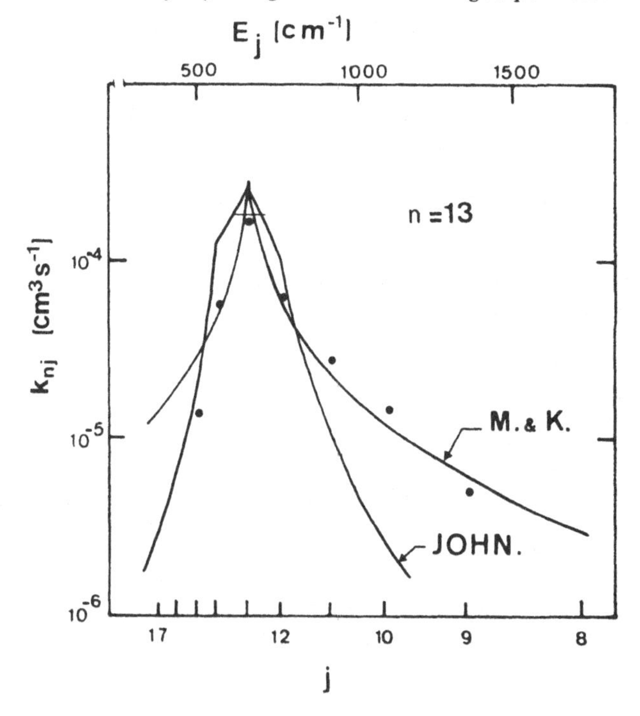

Fig. 12.14. Measured transfer rate coefficient induced by electron collisions from level p = 13 to levels of principal quantum number from 9 to 15; upper scale shows corresponding ionization wavenumber. Curves labeled M. & K. and JOHN. show theoretical predictions of Mansbach and Keck[21] and of Johnson.[22] For the laser-pumped level, the dot corresponds to the measured depopulation rate coefficient and the horizontal line to the sum of the independently measured transfer rate coefficients. (From Devos, Boulmer, and Delpech[18])

References and notes

1. J. Boulmer, F. Devos, J. Stevefelt, and J.-F. Delpech, *Phys. Rev. A 15,* 1502 (1977).
2. J. Stevefelt, J. Boulmer, and J.-F. Delpech, *Phys. Rev. A 12,* 1246 (1975).
3. L. C. Johnson and E. Hinnov, *Phys. Rev. 187,* 143 (1969).
4. J. Stevefelt and F. Robben, *Phys. Rev. A 5,* 1502 (1972).
5. P. M. Koch and J. E. Bayfield, *Phys. Rev. Lett. 34,* 448 (1975).
6. H. D. Zeman, K. Jost, and S. Gilad, *Rev. Sci. Instrum. 42,* 485 (1971).
7. J. G. Lodge, I. C. Percival, and D. Richards, *J. Phys. B 9,* 239 (1976).
8. M. Burniaux, F. Brouillard, A. Jognaux, T. R. Govers, and S. Szucs, *J. Phys. B 10,* 2421 (1977).
9. R. E. Olson, *J. Phys. B 13,* 483 (1980).
10. H. J. Kim and F. W. Meyer, *Phys. Rev. Lett. 44,* 1047 (1980).
11. F. W. Meyer and H. J. Kim, in *Electronic and Atomic Collisions,* eds. N. Takayanagi and H. Oda (Kyoto: Society for Atomic Collision Research, 1979), p. 552.

12. J. A. Schiavone, D. E. Donohue, D. R. Herrick, and R. S. Freund, *Phys. Rev. A 16,* 48 (1977).
13. D. R. Herrick, *Mol. Phys. 35,* 1211 (1978).
14. J. A. Schiavone, S. M. Tarr, and R. S. Freund, *Phys. Rev. A 20,* 71 (1979).
15. T. H. Jeys, G. W. Foltz, K. A. Smith, E. J. Beiting, F. G. Kellert, F. B. Dunning, and R. F. Stebbings, *Phys. Rev. Lett. 44,* 390 (1980); G. W. Foltz, E. J. Beiting, T. H. Jeys, K. A. Smith, F. B. Dunning, and R. F. Stebbings, *Phys. Rev. A 25,* 187 (1982).
16. I. C. Percival and D. Richards, *J. Phys. B 10,* 1497 (1977).
17. J.-F. Delpech, J. Boulmer, and F. Devos, *Phys. Rev. Lett. 39,* 1400 (1977).
18. F. Devos, J. Boulmer, and J.-F. Delpech, *J. Phys. (Paris) 40,* 215 (1979).
19. J.-C. Gauthier, J.-P. Geindre, J.-P. Moy, and J.-F. Delpech, *Phys. Rev. A 13,* 1781 (1976).
20. M. Gryzinski, *Phys. Rev. 115,* 374 (1959).
21. P. Mansbach and J. C. Keck, *Phys. Rev. 181,* 275 (1969).
22. L. C. Johnson, *Astrophys. J. 174,* 227 (1972).
23. G. Baran, J. Boulmer, F. Devos, and J.-C. Gauthier, in *Electronic and Atomic Collisions,* eds. K. Takayanagi and N. Oda (Kyoto: Society for Atomic Collision Research, 1979), p. 952.
24. K. B. MacAdam, R. Rolfes, and D. A. Crosby, *Phys. Rev. A 24,* 1286 (1981).
25. K. B. MacAdam, *Comments At. Mol. Phys. 11,* p. 53 (1981); K. B. MacAdam and R. Rolfes, Contributed papers, *XIII ICPEAC, Gatlinburg* (1981), p. 1115; K. B. MacAdam and R. Rolfes, *J. Phys. B 15,* L243 (1982).

13

Rydberg studies using fast beams

PETER M. KOCH

13.1. Introduction

The use of fast beams with laboratory kinetic energy on the order of 10^3 to 10^7 eV for the production and study of hydrogen atoms bound only by about 10^{-1}–10^{-2} eV has a long history. Already in the early part of this century, the canal ray light source - the low-vacuum predecessor of the modern high-vacuum fast-beam apparatus - was being used for emission spectroscopic studies of a wide variety of atomic species in low-lying Rydberg states, about $n \lesssim 8$ for hydrogen. Reviews by Ryde[1] and Andrä[2] describe and give references to much of this pioneering work.

In the 1960s and early 1970s, a resurgence of interest in the study of H(high-n) atoms in fast beams was motivated largely by the possibility of using "Lorentz-ionized" Rydberg states of its heavy isotopes to fuel and heat plasma devices being used for thermonuclear fusion experiments.[3,4] Detection methods based on electric field ionization of atoms (see Sect. 13.3a) permitted study of higher-lying n manifolds with unresolved substates and led to the measurement of cross sections for production of H(high-n) and He(high-n) atoms by various collision processes, some of which will be reviewed in Sect. 13.2. References 5–7 are important reviews covering this period of research. The thermonuclear plasmas, however, have now been made sufficiently dense to permit the use of collisional ionization of fast D(1s) atoms (or possibly D^- ions) for fueling and heating.[8-10] These can be produced much more efficiently than D(high-n) atoms.

This chapter will emphasize the most recent phase of the study of H(high-n) atoms in fast beams. Various combinations of collisions, field-ionization, and laser-excitation techniques have made it possible to produce useful beams of atoms in narrowed n bands, in individual n manifolds, or even in individual Stark substates. Sections 13.2–13.4 will review these techniques. Sections 13.5–13.7 will review the results of subsequent intense field and collision experiments that used these specially prepared beams. Because these experiments and the relevant theory cover a wide range of atomic physics, the restricted length of the present chapter does not allow each to be covered in great depth. Neither does it permit the results of fast-beam studies of multielectron Rydberg atoms (with the occasional exception of helium), ions, and molecules to be covered.

We mention that laboratory studies of thermal H(high-n) atoms will undoubtedly increase in number and sophistication as ultraviolet laser sources continue to improve. Quite recently, three- and two-quantum excitations of thermal H(1s) atoms to unresolved substates of individual $n \lesssim 35$ manifolds were reported.[11,12]

Hydrogen Rydberg atoms

The lowest-order model for a Rydberg atom is the nonrelativistic, spinless (Bohr) hydrogen atom. Described by a pure Coulomb potential, $V(r) = Zr^{-1}$ a.u., its energy levels are n^2 degenerate. Additional internal interactions in real atoms break some of this degeneracy. For a single Rydberg series that does not interact strongly with any others, the shifted energies $E(n;q)$ are easily parametrized in terms of a set of numbers $\delta(n;q)$, called (single-channel) quantum defects,[13] which vary weakly with principal quantum number n but more strongly with other quantum numbers q:

$$E(n;q) = -Z^2[n - \delta(n;q)]^{-2}\mathrm{Ry}_\mathrm{X} \tag{1}$$

where Ry_X is the Rydberg constant for the reduced mass appropriate for atom X. For two or more interacting Rydberg series, multichannel quantum-defect theory can often be used to parameterize the position of energy levels.[13-15]

In hydrogen, the largest additional internal interactions are those that produce relativistic (Dirac) fine structure. To order $(Z\alpha)^2$, the term energies are

$$E(n;j) = -\frac{Z^2}{n^2}\mathrm{Ry}_\mathrm{H}\left[1 + \frac{(Z\alpha)^2}{n}\left(\frac{1}{j+\frac{1}{2}} - \frac{3}{4n}\right)\right] \tag{2}$$

where j is the total (orbital plus spin) angular momentum quantum number.[16] One may easily combine Eqs. (1) and (2) to parametrize hydrogen energy levels with quantum defects as

$$\delta(n;j) = \frac{(Z\alpha)^2}{2}\left[\frac{1}{j+\frac{1}{2}} - \frac{3}{4n}\right] \tag{3}$$

For $Z=1$ and $n \gg 1$, the largest is $\delta(j=\frac{1}{2}) = 2.7\times10^{-5}$. That it is so small compared with one merely shows that fine-structure energy shifts in hydrogen are minuscule compared with the differences in energy between neighboring Bohr levels. Hyperfine and quantum-electrodynamic shifts are even smaller.[16] Some of these were measured in low-lying excited states of hydrogen(ic) systems with a variety of fast-beam methods.[17-20]

In an experimental apparatus, H(high-n) atoms are exposed to applied and stray electric fields that mix opposite parity levels and produce Stark shifts. Lüders[21] derived a useful formula

$$F_s \simeq 10^5 (Z/n)^5 \text{ V cm}^{-1} \tag{4}$$

for the field F_s needed to produce Stark shifts within a given n manifold that are on the order of the fine-structure splittings. For example, only $F_s \simeq 10^{-3}$ V cm^{-1} is required for H($n=40$). When $F \gg F_s$, Stark shifts are huge compared with fine-structure splittings, and the parabolic quantum numbers $\{n, n_1, n_2, |m|\}$ satisfying $n = n_1 + n_2 + |m| + 1$ provide the best labels for the levels.[16] In Stark substate-resolved studies (see Sect. 13.4d and 13.5), one must be careful to avoid loss of substate definition via nonadiabatic transitions in regions of low and nonconstant F, such as in the fringe field regions of parallel plates.[22] In principle, a magnetic field **B** could also be used for Zeeman substate-resolved studies, but usually one cannot avoid problems associated with the often nonnegligible motional electric field $\mathbf{F}_{\text{mot}} = \gamma \mathbf{v} \times \mathbf{B}$ that is seen by fast-beam particles whose trajectories are not perfectly collinear with **B**.

13.2. Production of hydrogen Rydberg atoms in a fast beam

13.2a. Electron-transfer collisions of H^+

During the 1960s and 1970s, a large number of experimental and theoretical studies were devoted to the electron-transfer collision

$$\text{H}^+ + \text{X} \rightarrow \text{H(high } n) + \text{X}^+ \tag{5}$$

over the proton energy range $E_k = 10^3$–10^6 eV, where X is taken to be an atomic or molecular target in its ground electronic state. Because neither energy considerations nor rigorous selection rules prohibit the population of any of the bound H(high-n) substates when $E_k \gg \Delta E$ (where the collision energy defect $\Delta E = |I_f - I_i| \sim 10$ eV is given by the difference in ionization potentials I of the exchanged electron in the initial (i) and final (f) states), both theory and experiment have to deal with the large number of substates present when $n \gg 1$. The experiments vary according to the attempt made to resolve the substates; detection has been based on radiative decay ($n \leqslant 6$) and on electric field ionization ($n \geqslant 6$) (see Sect. 13.3). Currently under development for measurement of substate-resolved cross sections are methods based on transitions driven by laser or microwave sources.[23] These will be described briefly in Sect. 13.4.

Most theoretical studies of Eq. (5) concentrated, for $n \gg 1$, on the fundamental case

$$\text{H}^+ + \text{H(1s)} \rightarrow \text{H}(nlm) + \text{H}^+ \tag{6}$$

and employed the Oppenheimer–Brinkman–Kramers (OBK), Born, and other related high-energy approximations, which were reviewed recently by Belkic et al.[24] These are justified only when the relative collision velocity $v \gg 1$ a.u., the H(1s) orbital velocity; this corresponds to $E_k \gg 25$ keV. In this limit, these theories predict an n^{-3} scaling[25] of the substate-integrated sum $\sigma_{10}(n;E_k) = \sum_{lm}\sigma_{10}(nlm;E_k)$ for the total cross section for electron transfer into an individual n-manifold:

$$\sigma_{10}(n;E_k) = n^{-3}\sigma_c(E_k) \tag{7}$$

where $\sigma_c(E_k)$ is an n-independent reduced cross section. The total cross section for transfer into all bound states is $\sigma_{10}(E_k) = \sum_{n=1}^{\infty}\sigma_{10}(n;E_k)$.

For the case of multielectron target atoms,[26-28] the same n-scaling prediction follows for $n \gg 1$ when the rather drastic assumption is made that the target electrons move in independent hydrogenic orbitals.[28] At lower values of n, there can be significant departures from this rule if the energy defect ΔE is small for electron transfer into a low-lying H(n) excited state. For example, in H^+-Cs collisions,[29] $\Delta E(n=2) = 0.47$ eV is the smallest, and the H($n=2$) states are populated with largest probability when $E_k \lesssim 3$ keV.

Il'in et al.[6,30] first showed how the n^{-3} scaling of Eq. (7) and the n^{-4} scaling of the ionization electric field F [Eq. (10)] could be used to measure $\sigma_c(E_k)$ for various targets X in Eq. (5). In the limit of n as a continuous variable, their theory implied that the detected field-ionization signal S would be linearly proportional to $F^{1/2}$, a dependence that has been verified by several groups performing experiments at $E_k > 4$ keV.[31-35] The $\sigma_c(E_k)$ values are obtained from the slopes of such curves. High-resolution scans recorded by a number of investigators for $6 \leqslant n \lesssim 28$ have displayed structure due to partially resolved n manifolds; individual substates were not resolved.[5-7,32,36,37] In unpublished work on $E_k = 10$ keV H^+-Xe collisions, the author has observed $S \propto F^{1/2}$ to the smallest value of $F \simeq 15$ V cm^{-1} that was studied. This corresponds to ionization of $n \simeq 80$.

Taken together, these results imply that Eq. (7) is valid for $10 \lesssim n \lesssim 80$ and for $E_k \gtrsim 4$ keV, although different experiments have probed different ranges of F (and, therefore, n) and of E_k. (For $2 \leqslant n \leqslant 5$, see Sect. 13.2c.) Notice that $E_k = 4$ keV is at least a factor of six below the lower limit of validity of the OBK theory.

The OBK theory generally gives absolute σ_{10} values that are much too large but may give reasonable cross section ratios. For H^+-Ne and He^+-Xe collisions, King and Latimer[35] found agreement above 30 keV between experiment and Hiskes's[27] "prior" and "post" OBK theoretical values for fractional H($n=11$) production.[27] Below 20 keV, both theoretical ratios were about a factor of two too high. Better theory is needed for $E_k \lesssim 25$ keV.

Bohr,[38] using statistical arguments, and Drukarev,[39] using arguments based on the Born approximation, derived a simple scaling rule for predicting the energy E_k^{peak} at which the cross section peaks for process (5):

$$E_k^{\mathrm{peak}} \propto \Delta E \qquad \text{when} \quad I_i \simeq \Delta E \gg I_f \tag{8}$$

Figure 13.1 shows that Eq. (8) has been verified for a number of targets

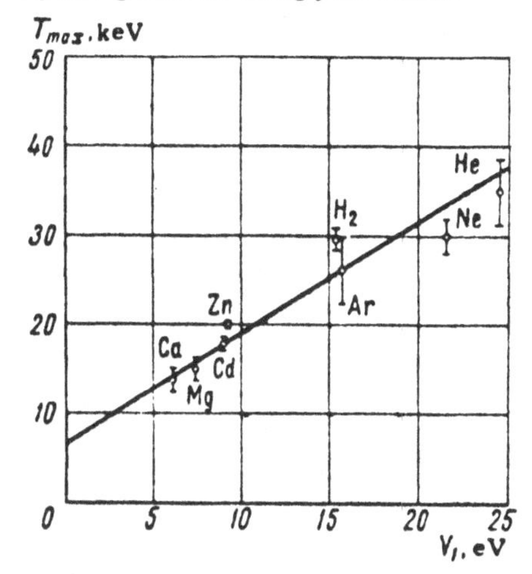

Fig. 13.1. Proton energy T_{max} corresponding to the maximum σ_c^{max} of the cross section for production of H(high-n) atoms by electron-transfer collisions [Eq. (5)] vs. the first ionization potential V_i of the target species. (From Oparin, Il'in, and Solov'ev[40])

with 6 eV $\lesssim I_i \simeq \Delta E \lesssim 25$ eV. More recent Kr and Xe data[35] and H(1s) data[32] are consistent with Fig. 13.1.

Equation (8) is a scaling law different from the more well known Massey adiabatic criterion,[41] which Drukarev[39] showed is valid only when $|I_i^{1/2} - I_f^{1/2}| \ll I_i^{1/2}$. It predicts $E_k^{peak} \propto (\Delta E)^2$.

Oparin et al.[40] also graphically showed the dependence of the measured maximum values $n^3\sigma_c^{max}$ of the cross sections for process (5) as a function of $I_i \simeq \Delta E$. The empirical formula expressing this dependence

$$n^3\sigma_c^{max} = 5.4 \times 10^{-13}\ \mathrm{cm}^2/[I_i(\mathrm{eV})]^{5/2} \tag{9}$$

was shown to be in reasonable agreement with cross sections estimated using simple reasoning due to Bohr.[38] Figure 13.1 and Eqs. (8) and (9) show that weakly bound targets (alkali metal vapors) have the largest $n^3\sigma_c^{max}$ values, which occur at the lowest E_k^{peak} values.

Table 13.1 is a compilation of references to most published measurements of $\sigma_c(E_k)$ for process (5).[30-37, 40, 42-44] As is explained in the footnote to the table, some measurements included "thick-target" yields for which collision processes leading both to production and to loss (see Sect. 13.7b) of H(high-n) atoms were important. The only detailed study of their angular distributions in thin or thick targets was reported by Kingdon et al.[43]

Very few measurements of the cross sections for He(high-n) production in He^+–X electron-transfer collisions have been reported. What data do exist show that at the same relative velocity in the range $v \simeq 1\text{–}3 \times 10^8$ cm s^{-1}, the H(high-n) and He(high-n) substate-integrated production cross sections are within a factor of two of each other.[6, 45]

Table 13.1. Guide to publications containing measured cross sections for the production of fast H(high-n) atoms by electron-transfer collisions of protons with atomic and molecular target gases or vapors, Eq. (5)

Reference	Measurements reported[a]
30	He (50, ot; 60), Ne (50, ot; 60), Ar (50, ot; 60), Kr (50, ot; 60) Xe (50, ot), H_2 (25–100, t; 120–200), N_2 (60, 120, 180), CO_2 (60, 120; 100t), H_2O (18, 42, ot), Li (18)
31	Li (10–180), Na (10–180, t), K (10–180, t), Cs (10–180, t), He (15–120), Ne (15–120, t), Ar (15–180), H_2 (20–180), CO_2 (30–180, t)
32	Ar (10–65), H (7–60)
33	Ba (6–30), K (6–30), Mg (6–30), Tl (6–30), H_2 (5–30, t), N_2 (5–30, t), C_8F_{16} (4–25, t), Freon 114 (6–30), H_2O (5, t), NH_3 (5, t)
34	He (28–140), Ar (10–130), H_2 (30–120)
35	He, Ne, Ar, Kr, Xe, H_2, N_2 (all 4–20)
36	H_2 (25–100, t)
37	He, Ne, Ar, Kr, Xe, H_2, O_2, N_2, NH_3 (all 20–150, t)
40	Mg (10–180, t), Ca (10–180), Zn (10–180), Cd (10–180, t)
42	Mg (5–70, t), Ne (15–60)
43	Mg, C_8F_{16}, H_2O, H_2 (all D^+, 10–140, ot)
44	He, Ne, Ar, Kr, Xe, H_2, NH_3, C_2H_2, C_6H_6 (all 20–150)

[a] Notation: the chemical symbol of the target species is followed by parentheses containing the energy in kiloelectron volts of the incident proton (or deuteron, if listed) and the letter "t" if thick-target measurements were also made and/or the letters "ot" if only thick-target yields were measured; no letter implies that only thin-target cross sections were reported. A number after a semicolon indicates that additional thin-target measurements were made at that energy.

13.2b. Experimental study of $H^+ + H(1s) \rightarrow H(13 \leqslant n \leqslant 28) + H^+$ collisions

Because of the fundamental importance of H^+–H(1s) collisions, we shall separately cover the measurement by Bayfield et al.[32] of the cross section for electron-transfer production of an unmeasured distribution of substates in the above manifolds over the energy range $E_k = 7$–60 keV. Figure 13.2 compares the data with the results of two theoretical calculations. The experimental curve peaks at $E_k = 25$–30 keV, in agreement with the scaling rule, Eq. (8), and Fig. 13.1. The OBK theoretical curve is at least 10 times too large and also fails to predict the correct position of the peak.[24,25] Because the data extended only up to $v = 1.5$ a.u., however, the theory was not really tested in its domain of validity $v \gg 1$ a.u.

The curve labeled CBA was obtained by Band.[46] His theory, based on the Born approximation, used the n^{-3} scaling rule, Eq. (7), and treated the non-orthogonality of initial and final (rearranged) states, a problem frequently

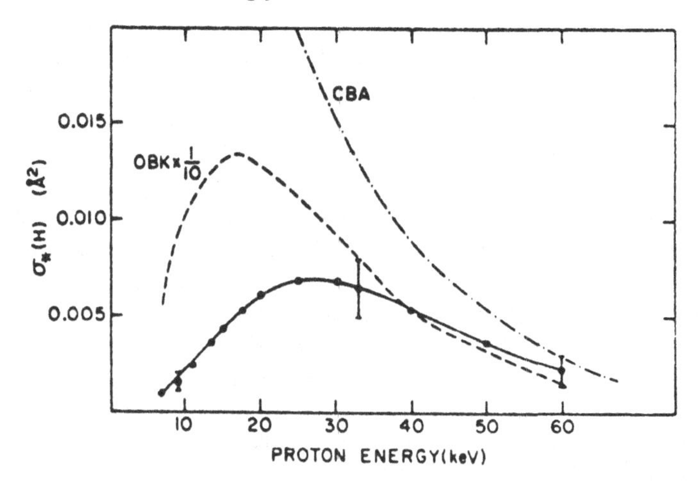

Fig. 13.2. Measured cross section for production of H(n) atoms within the n band $13 \leqslant n \leqslant 28$ in H^+-H(1s) electron-transfer collisions (filled circles). The total estimated uncertainty was ±30%. A solid line was drawn by eye through the data. Dashed curve is the theoretical result from an OBK approximation divided by 10. Dash-dot curve (CBA) is a theoretical curve calculated by Band.[46] (From Bayfield, Khayrallah, and Koch[32])

cited for this theory. Belkic et al. (Ref. 24, p. 303), however, criticized the usual orthogonality arguments and maintained that the only important property is the asymptotic orthogonality of wave functions. Within experimental error, there is agreement between Band's curve and the high-energy end of the experimental curve, but the accurate theoretical treatment of this and other inelastic H^+-H(1s) collision processes in the energy range near and below the cross section peak continues to be a challenge.

13.2c. Substate-resolved studies using fluorescence detection of H($2 \leqslant n \leqslant 5$)

Detection mechanisms based on radiative decay have been used to measure partial cross sections for production of some substates of H($n \leqslant 5$). Differences in radiative lifetimes have been used to separate different nearly degenerate contributions to Balmer transitions ($n' \geqslant 3$ to $n=2$).[47-51] Lyman transitions ($n' \geqslant 2$ to $n=1$) have also been used.[52] Stray electric-field-induced mixing of different fine-structure components has effectively prevented resolution of individual $n' \geqslant 4$ zero-field levels with $l' > 0$.[42,52,53]

Hydrogen(ns) production in H^+-He collisions has been found to obey n^{-3} scaling, Eq. (7), above E_k approximately equal to a few kilo electron volts for $n=2$-4 and above $E_k \simeq 100$ keV for $n=1$. The available data were summarized graphically by Bayfield.[54] References 52 and 53 contain more recent data.

Risley et al.[52] found that production of np ($n=2,3$) or Stark-induced np, nd ($n=4,5$) mixtures of hydrogen levels in 2 to 15 keV H^+-Ar collisions was not consistent with n^{-k} scaling with $k=3$ only. Measured k values ranged from 2.1 to 4.7 and depended on n and collision energy. They invoked a molecular picture of the collision process for a qualitative explanation of the data.

The Born–OBK predictions that $l=0,1$ partial cross sections should dominate those for $l \geqslant 2$ have been tested only with H($n=3$) experiments, none including H(1s) targets.[24,26-28] For He, Ne, Ar, H_2, N_2, and O_2 targets, measured partial cross sections decrease monotonically with increasing l for all $E_k \gtrsim 25$ keV.[47-49] For lower energies, the l ordering may be different. For example, at $E_k=10$ keV, Hughes et al.[47] found that H(3d) production dominated for Ar, N_2, H_2, and O_2 targets. Their 3p data[47] for He and Ar, however, are as much as a factor of four below more recent data.[48-50,52]

Knize et al.[55] recently used radio-frequency fields to modify the substate populations of H($n=3$) atoms produced by 49 keV H^+-N_2 electron-transfer collisions. A detailed theoretical analysis of line shapes permitted the extraction of ratios of $n=3$ substate populations, but the large number of substates and rf-induced resonances at higher n-values will make it difficult to extend this method.

This limitation also holds true when measurement of collision-induced coherence and polarization parameters are used to obtain (Doppler-free) information on the complete collision amplitude for nearly degenerate angular momentum, Stark, or Zeeman substates.[2,56]

13.2d. Other collision mechanisms for production of fast H(high-n) atoms

The collision processes that will be mentioned in this section have been observed to produce fast H(high-n) atoms,but they have not been developed for use as sources of such atoms for additional experiments.

Dissociation of molecular ions. Both H_2^+ and H_3^+ ions can be dissociated in collisions with gas targets or foils to produce H(high-n) atoms at nearly the same velocity.[7,43,57] Large velocity and angular spreads of the product H(high-n) atoms and the presence of excited molecules and molecular ions are problems.

Electron transfer from foils. Berkner et al.[58] studied production of D(high-n) atoms in collisions of 8- to 100-keV D^+ ions with C, Mg, Nb, and Au foils. Lucas et al.[59] obtained indirect information on H(high-n) production in 0.5- to 2.5-MeV H^+-carbon foil interations.

Electron detachment of H^-. Risley et al.[60] reported a recent study of electron detachment. For production of np or a mixture of parabolic states in the $n=2$-5 manifolds in 1- to 6-keV collisions of H^- with He and Ar, an n^{-3} scaling law was found not to fit all the data.

13.3. Detection of $n \gtrsim 6$ Rydberg atoms in a fast beam

13.3a. Static electric field ionization (EFI)

The first experimental observation of EFI of hydrogen by Traubenberg et al.[61] a half-century ago antedated by two years Oppenheimer's[62] theoretical prediction of this tunneling effect.[63,64] The detection sensitivity in this classic experiment, however, was not high because photons, not particles, were detected. The rapid decrease of signal with n limited this method of detection to $n \leqslant 8$. Furthermore, the poor vacuum in the canal ray tube meant that collision effects were important.[1,2,65]

Riviere and Sweetman pioneered the modern electric field detection method for weakly bound systems in which at least one fast ionized or dissociated particle is detected with high sensitivity in high vacuum. The method was first used to detect He^- ions and vibrationally excited molecular hydrogen ions,[66] later to detect H(high-n) atoms.[67] References 5–7 are early reviews of static or motional $\mathbf{v} \times \mathbf{B}$ (Lorentz) electric field detection methods. It was realized early on by both theorists[68] and experimentalists[5,36,37] that EFI alone could not be used to resolve one individual Stark substate in the midst of the large number of H(high-n) levels produced by fast collisions.

Each hydrogen Stark substate labeled by parabolic quantum numbers $\{n, n_1, n_2, |m|\}$, where $n = n_1 + n_2 + |m| + 1$, ionizes at a rate $\Gamma_\mathrm{I}(F)$ which increases approximately exponentially with F (see Sect. 13.5c).[16] Below a classical threshold F_cl, it must do so by quantal tunneling through a potential barrier.[64] Bailey et al.[68] calculated and graphically presented $\Gamma_\mathrm{I}(F)$ curves for a large number $n \leqslant 25$ substates for $F < F_\mathrm{cl}$. More recent calculations treat $F \geqslant F_\mathrm{cl}$ as well (see Chap. 2 and Ref. 69). Banks and Leopold[70-72] obtained accurate estimates for F_cl, which they displayed in the useful graphical form[72] shown in Fig. 13.3. The graph displays contours of the function

$$\tilde{\Phi}_\mathrm{cl}(u, v) \simeq n^4 F_\mathrm{cl}/Z^3 \text{ a.u.} \tag{10}$$

Quantization rules $u = (n_2 + \frac{1}{2})/n$ and $v = |m|/n$ relate the classical variables (u, v) to parabolic quantum numbers $(n_2, |m|)$. The overall n^{-4} scaling of F_cl with n is built into the scaled function $\tilde{\Phi}_\mathrm{cl}(u, v)$. The parabolic state $\{n, 0, n-1, 0\}$ at the bottom of each n manifold has the lowest $n^4 F_\mathrm{cl} \simeq 0.13$ a.u.; that at the top, $\{n, n-1, 0, 0\}$, has the highest $n^4 F_\mathrm{cl} \simeq 0.38$ a.u. Of course, quantal tunneling lowers observed thresholds below these values. This will be clearly shown in Fig. 13.10 for $\{40, 39, 0, 0\}$ and other substates.

Various experimental schemes have been used for static EFI detection of fast H(high-n) atoms. A practical upper limit for the field maintained without arcs by metallic electrodes separated by more than a few millimeters is 10^5 V cm^{-1}.[73] Because of reduced electron mobility, heated glass electrodes[74]

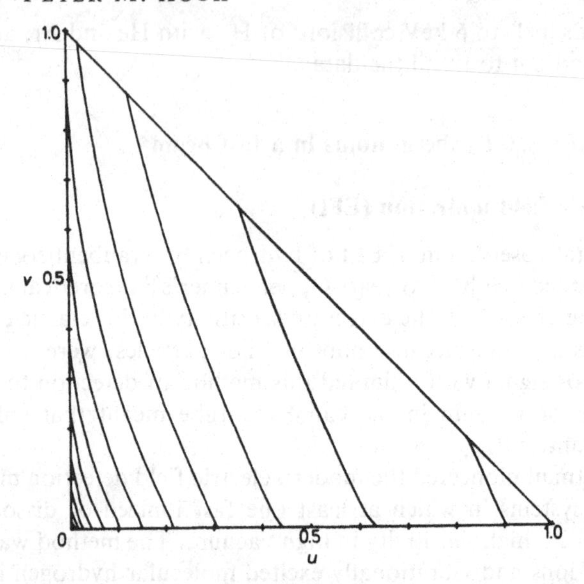

Fig. 13.3. Contours of the function $\tilde{\Phi}_{cl}(u, v)$ [Eq. (10)] that may be used to find the classical threshold electric field F_{cl} for ionization of substates of hydrogenic atoms. The first contour on the extreme right (which terminates at $u \simeq 0.9$, $v = 0$) corresponds to $\tilde{\Phi}_{cl} = 0.135$ and increases toward the origin in steps of 0.02. Quantization rules given in the text relate classical variables (u, v) and parabolic quantum numbers $(n, n_2, |m|)$. (From Banks and Leopold[72])

would allow this to be raised to $\sim 5 \times 10^5$ V cm^{-1}, which is sufficient for EFI of $H(n \geqslant 6)$.[68] Fields parallel and transverse to the beam axis have both been used for EFI. In the parallel case, the beam must travel through and may possibly strike edges of apertures in the electrodes that disturb the field uniformity;[5] use of meshes improves field uniformity but increases the number of struck "edges." Emerging ions or electrons are dispersed in energy but remain approximately parallel to the axis.

In the transverse case, plane-parallel electrodes may be used for superior field uniformity in the central region away from the fringe fields.[75] Because ions and electrons are deflected off the beam axis by an amount that depends on where in the field the ionization took place, special precautions must be taken to ensure a known collection efficiency. Kim and Meyer[76] used one such scheme in a recent collision experiment.

Bayfield et al.[32] described a method that used a longitudinal "ionizer" field and a transverse, more precise "preionizer" field to combine the advantages and counteract the disadvantages of each type.

13.3b. The energy-labeling technique

One important consequence of the n^{-3} scaling rule, Eq. (7), for production of H(high-n) atoms is the minuscule intensity $I(n^*)$ of atoms in a given $n^* \gg 1$

level compared with either the incident proton-beam intensity I_+ or the total neutral-beam intensity $I_0 = \sum_n I(n)$. Energy labeling can be used preferentially to detect experimental signals $S(n^*)$ while suppressing background signals S'.[77] Consider a mixed beam of protons and H(1s) and $H(n^* \gg 1)$ atoms all at the same beam energy E_k incident on an interaction region biased at a "label" potential V_L above ground potential. Only those protons produced by fast H atoms that are subsequently ionized by static or microwave EFI or by collisions[78] with background or other gases inside the voltage-labeled interaction region will be accelerated upon leaving the region to a labeled energy $E_L = E_k + eV_L$.[77] These consist of $S(n^*)$ and energy-labeled background S'_L; to minimize the latter, one usually wants the gas density inside the interaction region to be as low as possible. Protons produced elsewhere are background S'_k at the unlabeled, primary energy E_k. Charge-, "high-pass" energy-, and momentum analyses can be used to separate $S(n^*) + S'_L$ from S'_k + primary beams.[77] When $I(n^*)$ can be fully modulated through the use of a modulated static, microwave, or laser electric field, lockin detection may be used to isolate $S(n^*)$ from S'_L. Typical numbers in the Yale experiments were $I_+ \sim 6 \times 10^{12}$ s^{-1}, $I_0 \sim 10^{12}$ s^{-1}, $I(n^* \sim 50) \sim 10^6$ s^{-1}, $S(n^*) \sim 10^2$–10^6 s^{-1}, $S'_k \sim 10^8$–10^{12} s^{-1}, and $S'_L \sim 10^2$–10^6 s^{-1}.

13.3c. Microwave multiphoton ionization used for H(high-*n*) detection

If the finite energy or angular acceptance of an energy analyzer makes it desirable to ionize a H(high-*n*) beam without focusing or greatly increasing the fractional energy spread $\delta E_k / E_k$ of the resultant proton beam, an oscillating field $\mathbf{F}(t) = \mathbf{F}_0 \cos \omega t$ should be used. The maximum velocity dispersion produced by the ionization field is $\delta v = 2eF_0/m\omega$. For $E_k = 10$ keV, $\omega/2\pi = 10^{10}$ Hz, and $F_0 = 2$ kV cm^{-1} [sufficient to ionize $H(n \gtrsim 22)$], this corresponds to an energy dispersion $\delta E_k < 5$ eV. Use of high-Q copper cavities with small holes for beam entrance and exit minimizes the microwave power required to produce a given F_0. A "label" potential can easily be applied to each electrically floating cavity in order to permit energy-labeled ion detection.

13.4. Modification of the *n*-distribution produced by collisions

Because electron-transfer collisions produce an n^{-3} weighted beam of fast H(high-*n*) atoms with a generally unknown distribution of substates, additional interactions are needed to prepare atoms in an *n* band or in a single (Stark) substate.

13.4a. Spontaneous radiative decay and radiative cascade

Hiskes et al.[79] published graphs of radiative lifetimes of $H(n \leqslant 25)$ atoms in spherical and parabolic basis states. Additional results and discussions of selection rules, matrix elements, and *n* scaling are found in Refs. 16, 65, and

80. For the time scale $T \lesssim 10^{-6}$ s of typical fast-beam experiments, radiative decay of H($n \gtrsim 10$) states is not a significant problem.

To avoid the problem of radiative cascade filling of a lower level by a populated upper level, electric field ionization may be used, as will be shown in Fig. 13.5, to destroy all atoms with $n > n' \simeq (8F)^{-1/4}$ a.u. Then, one of the laser schemes that will be described in Sect. 13.4d is used to repopulate a level with $n'' > n'$. With no population in the levels with $n > n''$, cascade from above is absent. Figure 13.5 will show this for $n'' = 31$.

13.4b. Interactions with 300-K blackbody photons

Farley and Wing[81] calculated the n dependence of the total depopulation rate $R(n)$ of H(n) atoms induced by 300-K blackbody photons. The peak is at $R(n=11) \simeq 4 \times 10^4$ s^{-1}, and $R(n) < 10^4$ s^{-1} for $n \leqslant 6$ and $n \geqslant 36$. Because Stark substate purity was needed only during a time scale $T = 1\text{–}4 \times 10^{-7}$ s in the experiments performed in the laboratory of the author (see Sect. 13.5), it is understandable that no blackbody photon-induced effects were observed. A more complete discussion of these effects is presented in Chap. 5.

13.4c. The use of electric field ionization for level definition

Electric field ionization (EFI) provides a simple method for isolating in an n^{-3} weighted distribution of H(high-n) atoms only those in an "n band" ($n_a \lesssim n \lesssim n_b$) a few units wide. Atoms in such an n band are useful mostly for collision experiments that are relatively insensitive to small changes in n.[36,76,77,82,83] When the undefined H(n) beam is passed through a region of static or microwave electric field F modulated between F_a and F_b, atoms with $n_a \gtrsim (K/F_a)^{1/4}$ and $n_b \gtrsim (K/F_b)^{1/4}$, respectively, do not contribute to a subsequent experimental signal. Direct subtraction of the F_a and F_b signals or, equivalently, lockin detection at the frequency of the modulation between F_a and F_b yields the n-band-dependent part of the signal. For a static F, the constant $K \simeq \frac{1}{8}$ a.u. $= 6.4 \times 10^8$ V cm^{-1} is consistent with published measurements[5-7,32,36,37,84] and quantal theory.[68]

Koch[85] extended this technique to define a small number of Stark substates in a single H(n) manifold, $n = 10$. These were subsequently laser excited to $n \gtrsim 20$. A spectrum showing the substate definition that can be obtained with this method was shown in Fig. 6 of Ref. 22.

13.4d. Laser excitation of H(high-n) and He(high-n) levels

Because of high resolution and reasonable efficiency, continuous laser (or millimeter-wave) excitation of excited H or He atoms produced by collisions is the best method for production of fast H(high-n) or He(high-n) atoms in individual n manifolds or sublevels. Having fantastically low duty factors, $\sim 10^{-6}$, pulsed lasers are poorly matched to fast beams whose typical atom density is $\lesssim 10^5$ cm^{-3}.

Fixed-frequency lasers are useful for two reasons. First, a beam of photons with laboratory frequency ν_L is Doppler shifted in the particle rest frame to

$$\nu_D = \frac{1 - \beta \cos\theta}{(1 - \beta^2)^{1/2}} \nu_L \tag{11}$$

where θ is the crossing angle for the two beams and $\beta = v/c$. Thus ν_D may be tuned with variation of v or θ. Second, the energy levels themselves may be tuned with an externally applied electric or magnetic field.

When $\beta \ll 1$, $\nu_L \simeq \nu_D \equiv \nu$ and the frequency spread $\Delta\nu$ in the particle rest frame can be expressed as the quadrature sum of four terms:

$$\Delta\nu_1 = \nu\beta \frac{\Delta v}{v} |\cos\theta| \tag{12}$$

$$\Delta\nu_2 = \nu\beta \, \Delta\Phi \, |\sin\theta| \xrightarrow{\theta = 0, \pi} \sim \nu\beta(\Delta\Phi)^2 \tag{13}$$

$$\Delta\nu_3 = \Delta\nu_L \tag{14}$$

$$\Delta\nu_4 \simeq v/\pi L \tag{15}$$

where the particle beam has fractional velocity spread $\Delta v/v$, $\Delta\Phi = (\Delta\phi_L + \Delta\phi_P)/2$ is one-half the sum of the divergence angles of the laser and particle beams, $\Delta\nu_L$ the linewidth of the laser source, L the length of the photon–atom-beam interaction region, and $\Delta\nu_4$ a frequency spread introduced by a finite time of interaction $T = L/v$.

Resolution requirements alone do not fix the choice of θ. Low laser power or small transition matrix elements may require that T be maximized with a collinear-beam geometry, $\theta = 0$ or π, in order to maximize the laser-induced transition probability. For a fixed v, $\theta = 0, \pi$ also minimizes $\Delta\nu_2$. For a fixed Δv, the decrease in $\Delta\nu_1$ with increasing v was called "velocity bunching" by Kaufman.[86]

Table 13.2 summarizes the important parameters for two laser excitation schemes used to produce fast H(high-n) and He(high-n) atoms. In the first, Doppler velocity-tuned UV Ar III laser photons were used to excite an H(2s) atom beam to each of the individual levels H(n), $40 \leqslant n \leqslant 55$.[87,88] [A tuning spectrum is shown in Ref. 87. With a more energetic Ar III laser line, photo-ionization of H(2s) was also studied.[87]] Stark tuning of transitions to resolve upper substates was not feasible because of electric quenching of the lower H(2s) level (Ref. 16, pp. 286–8).

The second scheme uses the $^{12}C\,^{16}O_2$ laser, which has a rich spectrum of over 50 lines in the P and R branches of two bands near 10^3 cm^{-1} (10^{-1} eV).[89] Additional lines are available with isotopic substitution[89] and "sequence-band" transitions.[90] Bayfield[91] pointed out that the CO_2 laser frequencies would be useful for driving $n = 10$ to high-n transitions in both hydrogen and other Rydberg atoms and molecules.

Table 13.2. Summary of the demonstrated collinear laser, fast-beam methods for production of hydrogen and helium Rydberg states

Atom	H	He	H(Stark levels)
Reference	87, 88	93, 94	23, 85, 96
Laser type (cw)	Ar III	$^{12}C\,^{16}O_2$	$^{12}C\,^{16}O_2$
Laser line	$\lambda = 363.79$ nm(air)	>40 lines in 9-μm, 10-μm bands	>40 lines in 9-μm, 10-μm bands
Transitions	$H(2s) \rightarrow H(n \gtrsim 40)$ also $H(2s) \rightarrow H^+ + e^-$	$He(1s10l)^{2S+1}L_J \downarrow He(1sn'l')^{2S+1}L'_{J'}$ $(n' \gtrsim 20)$	$H(n=10) \rightarrow H(n \gtrsim 20)$; $H(n=9) \rightarrow H(16 \lesssim n \lesssim 20)$; $H(n=7) \rightarrow H(n=9)$; $H(n=7) \rightarrow H(n=10)$
Typical beam energy E_k (keV)	10	10	10
Rf ion source ΔE_k (eV)	~20	~25	~20
Particle-beam divergence $\Delta\phi_P$ (mrad)	3.5	3.5	3.5
Laser-beam divergence $\Delta\phi_L$ (mrad)	<1	3	3
Crossing angle θ	0	0	0
$\Delta\nu_1$ [Eq. (12)]	4 GHz (0.12 cm^{-1})	90 MHz (0.003 cm^{-1})	150 MHz (0.005 cm^{-1})
$\Delta\nu_2$ [Eq. (13)]	15 MHz	1 MHz	2 MHz
$\Delta\nu_3$ [Eq. (14)]	several GHz (multimode)	several MHz (unstabilized)	$\lesssim$1 MHz (stabilized)
$\Delta\nu_4$ [Eq. (15)]	<1 MHz ($L \simeq 1$ m)	<1 MHz ($L \simeq 0.5$ m)	<5 MHz ($L \simeq 10$ cm)
Doppler velocity tuning useful?	Yes (necessary)	Yes	Yes (not crucial)
Stark tuning of substates?	No [H(2s) quenching]	Yes	Yes, very useful (double resonance)
High-$\|m\|$ states accessible	No (selection rules)	Yes (singlets and triplets)	Yes

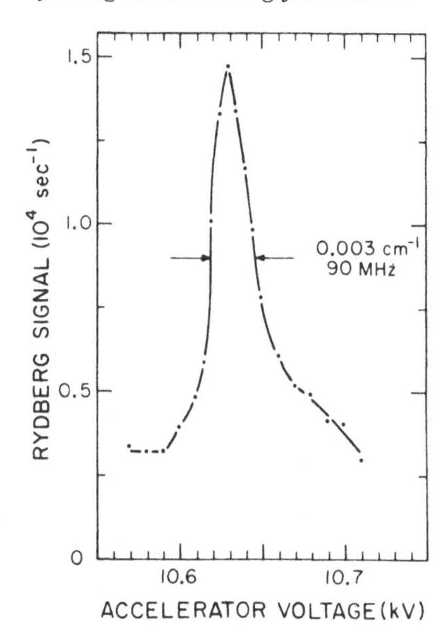

Fig. 13.4. A Doppler-tuning spectrum for the ^{4}He(1s10p) $^1P_1^o$–^{4}He(1s41s) $^1S_0$ transition induced by 9μm(P36) $^{12}C\,^{16}O_2$ laser photons. The voltage scale does not include a constant extraction potential in the He^+ ion source. The high-voltage tail is probably caused by a low-energy tail in the He(10p) kinetic energy distribution. See Table 13.2 for an analysis of the resolution.

Doppler-velocity-tuned CO_2 laser photons have been used to produce H($n \gtrsim 20$) atoms[83,88,92] in individual n manifolds having an unmeasured substate distribution as well as $n \gtrsim 30$ helium Rydberg atoms in individual singlet and triplet quantum substates.[93,94] Figure 13.4 shows an example of a singlet transition. The corresponding triplet transition is shown in Fig. 2 of Ref. 94.

Table 13.2 shows that the lower ν_L and $\Delta\nu_L$ of the CO_2 laser lead to much better resolution than with a UV laser. Use of a low-energy-spread ion source[95] ($\Delta E_k \lesssim 1$ eV) would lower $\Delta\nu_1$ to a few megahertz.

Koch[85] and Koch and Mariani[23,96] developed collinear CO_2 laser methods for the production and detection of individual Stark substates $\{n, n_1, n_2, |m|\}$ of H(high-n) atoms, including, for the first time, those with large values of the magnetic quantum number $|m|$.[23] Several transverse electric fields F_i, $i=1,2,\ldots$, shown in Figs. 13.5 and 13.6, were used to ionize or to Stark-tune energy levels at any convenient, fixed-beam energy. This was much simpler than Doppler velocity tuning. The spatial arrangement of the F_i were seen by the monoenergetic atoms as a well-defined sequence of time-varying fields in their rest frames.

Figure 13.5 shows the essence of the double-resonance method described in Ref. 23 that used two fields, $F_1 \simeq 45$ kV cm^{-1} and F_3 less than or approximately equal to several kilovolts per centimeter, to Stark-tune sequential,

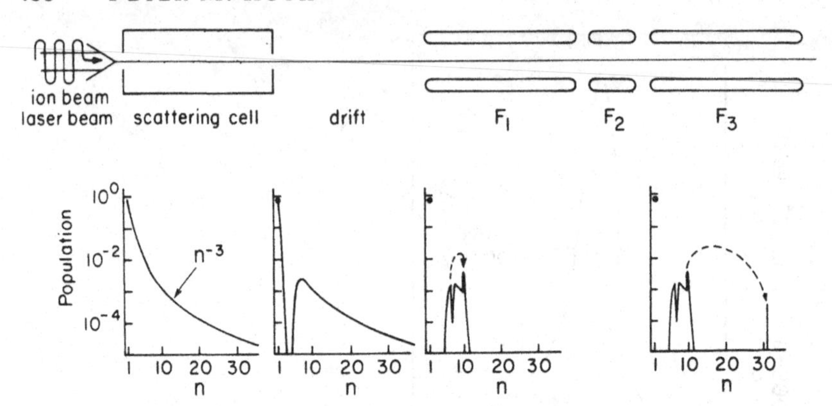

Fig. 13.5. *Top*. Principal features (not to scale) of the apparatus used to drive single- and double-resonance transitions among hydrogen Stark substates (see also Fig. 13.6). The $^{12}C\,^{16}O_2$ laser beam was linearly polarized parallel to each electric field and drove $\Delta|m|=0$ transitions between the $n_a=7$ and $n_b=10$ manifolds in F_1 and between the $n_b=10$ and $n_c \gtrsim 20$ manifolds in F_3. The H(n_c) atoms were detected by microwave multiphoton ionization (not shown). *Bottom*. Approximate fractional n populations at various points along the beam line for an $E_k \simeq 10$ keV neutral hydrogen beam produced by H^+-Xe electron-transfer collisions. The laser-produced notch at $n=7$ and bump at $n=10$ are exaggerated in the figure. (From Koch and Mariani[23])

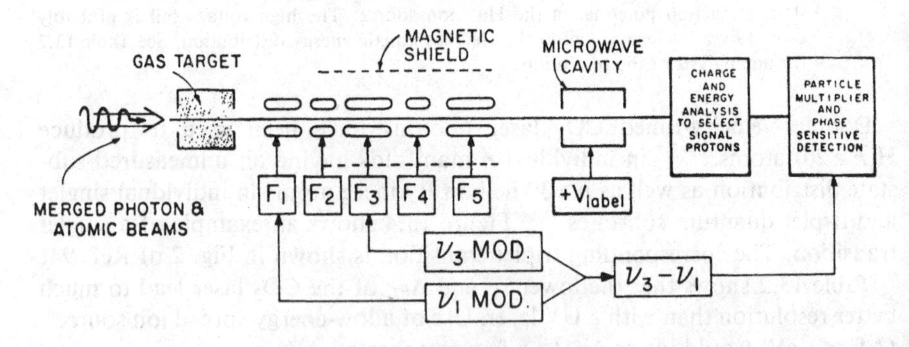

Fig. 13.6. A schematic diagram (not to scale) of the apparatus used to produce and to study CO_2-laser-excited hydrogen Stark substates. Because the fast H(n) atoms drifted through the precise spatial electric field regions F_1–F_5 at constant speed, each experienced the same, precisely reproducible, temporal sequence of fields in its rest frame. The cw laser beam could also be modulated at frequency ν_L (see also Fig. 13.5).

laser-induced ($n_a=7$ to $n_b=10$) and ($n_b=10$ to $n_c \gtrsim 20$) one-photon transitions, respectively. The H(n_c) atoms were subsequently detected (Sects. 13.3b and 13.3c). To facilitate lockin detection, the laser beam could be chopped at $\nu_L \simeq 300$ Hz and F_1 modulated by $\Delta F_1 \simeq 10$–30 V cm^{-1} at $\nu_1 = \nu_L/3 \simeq 100$ Hz. Depending on which field was kept constant while the other was swept slowly and on the demodulation frequency ν_{df} of the lockin, different kinds of spectra were obtained. Three examples are shown in Figs. 13.7 and 13.8.

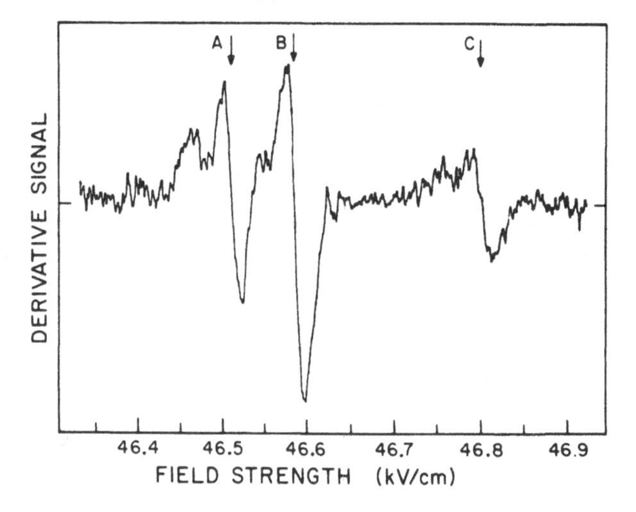

Fig. 13.7. A double-resonance Stark spectrum recorded as a function of F_1 (Fig. 13.5) for fixed $F_3 = 221$ V cm^{-1}. Because F_1 was modulated, lockin detection produced derivative signals. The transitions labeled by the letters are:

A: $\{7,3,3,0\} \xrightarrow{F_1} \{10,2,7,0\} \xrightarrow{F_3} \{31,4,26,0\}$

B: $\{7,2,2,2\} \xrightarrow{F_1} \{10,1,6,2\} \xrightarrow{F_3} \{31,3,25,2\}$

C: $\{7,1,1,4\} \xrightarrow{F_1} \{10,0,5,4\} \xrightarrow{F_3} \{31,2,24,4\}$

The arrows show the predicted resonant values of F_1 based on diagonal Padé approximants through [4/4]. (From Koch and Mariani[23])

Figure 13.7 was obtained at $\nu_{\rm df} = \nu_1$ with fixed F_3 and F_1 (modulated) swept slowly.[23] The three ($n_a = 7$ to $n_b = 10$) and the three ($n_b = 10$ to $n_c = 31$) Stark transitions shown in the figure legend were all degenerate to first order in F_1 and F_3, respectively. Higher-order Stark shifts were insufficient to resolve the latter in F_3 but were large enough to resolve the former $\Delta|m| = 0$ transitions between $|m| = 0, 2, 4$ states in F_1. The $\delta F_1 \simeq 100$ V cm^{-1} width of the peaks corresponds to an effective energy resolution $\Delta\nu/c \simeq 0.3$ cm^{-1}, which is about 60 times worse than $\Delta\nu_1$ in the fourth column of Table 13.2. To reach that limit, the field inhomogeneity would have to be reduced from $\Delta F_1/F_1 \simeq 0.2\%$ to ~ 30 ppm. Assuming perfect voltage stability and nominal plate separation $g = 0.64$ cm, this would require optical quality plates with $\Delta g \simeq 200$ nm.[97]

Figure 13.8 was obtained as F_3 (unmodulated) was swept slowly, and F_1 was modulated to Stark-switch the resolved transition $\{7,0,6,0\} \rightarrow \{10,0,9,0\}$ on and off resonance. The double-resonance spectrum in Fig. 13.8*b*, which records the output of a lockin tuned to the beat frequency $\nu_{\rm L} - \nu_1 \simeq 200$ Hz, displays peaks for only those transitions to $n_c = 33, 34$ Stark sublevels that originated on the single, tagged $n_b = 10$ sublevel. Figure 13.8*a* was obtained simultaneously with another lockin tuned to $\nu_{\rm L}$. Insensitive to the substate selectivity provided by the F_1 modulation, this single-resonance spectrum recorded transitions between many of the $n(n+1)/2 = 55$ Stark substates

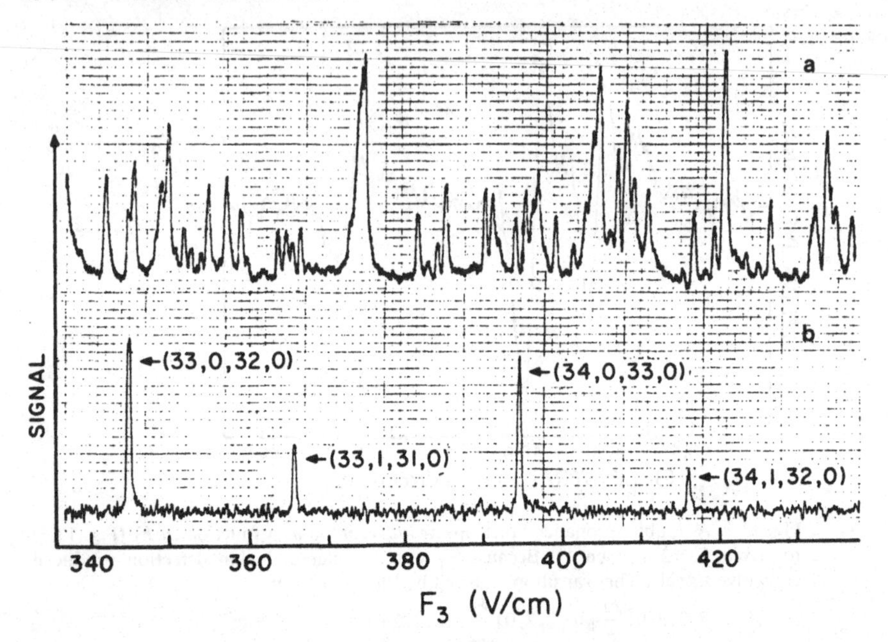

Fig. 13.8. Two Stark-tuning spectra recorded simultaneously with different lockin amplifiers for CO_2 laser excitation of H($n \simeq 34$) Stark substates. The method shown schematically in Fig. 13.5 was used. Trace (*a*): "single-resonance" spectrum, which corresponded to transitions in F_3 originating on a large number of H($n=10$) substates; trace (*b*): "double-resonance" spectrum, which corresponded to H($n=10$)–H($n \simeq 34$) transitions in F_3 originating on only the $n=10$ substate defined by the $\{7,0,6,0\} \rightarrow \{10,0,9,0\}$ transition driven in F_1.

(ignoring spin) in the $n_b=10$ manifold and the more than 500 substates in each of the $n_c \simeq 33$ manifolds.[16]

The $\Delta F_3 \simeq 0.4$ V cm^{-1} FWHM of the peaks in Fig. 13.8*b* corresponds to an energy width $\Delta\nu/c \simeq 0.03$ cm^{-1}, about six times worse than $\Delta\nu_1$ in the fourth column of Table 13.2. Improvement of $\Delta F_3/F_3$ to <200 ppm would meet this limit. To avoid the asymmetric lineshape (high-voltage tail) clearly evident in Fig. 2 of Ref. 85, it would be necessary to avoid pumping or detection of transitions pumped in the weaker fringe fields of the parallel plates.[75] This may be accomplished by crossing the laser and particle beams at a shallow angle θ such that the crossing zone lies completely inside the homogeneous field region. We have recently modified our apparatus in this manner. Beams from two independent cw CO_2 lasers cross the atomic beam in F_1 and F_3, respectively. The respective crossing angles, $\theta_1=0.188$ rad and $\theta_3=0.072$ rad, are too small to increase significantly the frequency spread $\Delta\nu_2$ [Eq. (13) and Table 13.2].

Carbon dioxide laser frequencies are also useful for driving transitions between $n_a=7$ to $n_b=9$ and $n_b=9$ to $16 \lesssim n \lesssim 20$ Stark states. Other useful sources would be a cw dye or diode laser tunable near $\lambda=0.9\ \mu m$ ($n_a=3$ to $n_b \gtrsim 7$), a cw F-center laser tunable near $\lambda=3\ \mu m$ ($n_a=5$ to high n), a cw CO laser step-tunable near $5\ \mu m$ ($n_a=5$ to $n_b=7$ and $n_b=7$ to high n), and a cw far-infrared laser or millimeter sources (high n to high n').

These double-resonance methods will be very useful for measuring partial cross sections for the production, rearrangement, and ionization of Stark substates in various collision processes *and* how they depend on F. Reference 23 discussed some possibilities. A very interesting application would be for measurement of the Stark substate distributions in the final $H(n \gtrsim 4)$ excited states observed recently by Clark et al.[98] in the UV laser photodetachment resonances in the negative ion H^-.[99]

13.5. Fast-beam laser spectroscopy of hydrogen Stark substates

13.5a. Some preliminary remarks

Because the nonrelativistic theory of the hydrogen Stark effect (the solution of the one-electron Schrödinger equation for the pure Coulomb–Stark potential $V_{CS}=-Zr^{-1}+Fz$) is covered in detail in Chap. 2, it will not be reviewed in depth here. Rather, a few chosen comments about this "... simplest quantum-mechanical problem for which there is no known [complete] analytic solution"[100] will suffice.

The Schrödinger equation for V_{CS} separates in parabolic coordinates (ξ, η, ϕ). When $F=0$, the negative energy motions along the "radial" coordinates ξ, η are finite and the $E<0$ spectrum is discrete. When $F>0$, a potential barrier appears in the η equation.[101] Because wave-mechanical motion through or over the barrier to $\eta=\infty$ permits eventual decay (field ionization) of states corresponding to an electron wave packet initially localized near the nucleus (small ξ, η), only resonances, or scattering states, in a continuous energy spectrum are left. Because the $F=0$ problem is completely understood, the $F>0$ Stark effect has always been a favorite testing ground for theories of resonances. Having properties that change continuously with $F>0$, the Stark substates are experimentally tunable resonances.

A resonance below the barrier ($n^4F<n^4F_{cl}$, Fig. 13.3) whose ionization rate $\Gamma_I(F)$ is not too large corresponds physically to an individual quasi-stationary substate localized near the Stark-shifted energy $E_R(F)$. Titchmarsh[102] explained this mathematically as "spectral concentration" near each complex eigenvalue $E(F)=E_R(F)-i\Gamma(F)/2$.[101] Above the barrier, individual resonances ionize too rapidly to be well localized in energy. In this case, Luc-Koenig and Bachelier[103] emphasized the importance of the distribution of oscillator strength for the interpretation and numerical calculation of intense-F Rydberg atom photoabsorption resonances recently observed near $E=0$.[104-106]

References 1, 16, and 65 give useful reviews of early experimental and theoretical Stark-effect studies. More modern low-F studies were reviewed in

Ref. 107. Koch[22] recently reviewed some of the work described in this section. The recent theoretical literature on the Stark effect is extensive. The interested reader should consult the sources just mentioned as well as Refs. 108–115.

13.5b. Precise laser, fast-beam measurements of $E_R(F)$

The carbon dioxide laser-excitation methods reviewed in Sect. 13.4d have allowed high-resolution absorption measurements[85] to be made in much higher-lying Stark manifolds than those studied long ago in emission ($n \leqslant 8$).[1,61]

Reference 85 describes a comparison of high-resolution measurements of the absolute energy of Stark states in the $n=25, 30$ manifolds with theoretical values obtained with high-order Rayleigh–Schrödinger perturbation theory derived by Silverstone[113] (to order $N=25$, later to $N \simeq 150$ by Silverstone et al.[114]). The magnitude of the parameter $n^4 F$, the same one that is so important for field ionization [Eq. (10) and Fig. 13.3], was shown to govern the divergence properties of the asymptotic energy series.

Table 13.3 revises slightly some of the wave number values presented in the insets of Figs. 3 and 4 of Ref. 85. These are the same revisions presented previously in Refs. 22 and 96. Although a small error was made by Koch in the original perturbation theory calculations, the revisions do not affect any of the conclusions of Ref. 85. Table 13.3 also includes the (unrevised) numerical theoretical results calculated by Luc-Koenig and Bachelier,[69] unfortunately using an erroneous atomic unit of F that was 71 ppm below the value au$(F) = 5.142\,250(17) \times 10^9$ V cm^{-1} that is obtained using the 1973 tabulation of the fundamental constants.[116] We estimate that the errors in their $E_U^{\mathrm{num}}(F_3)$ values are negligible compared with those produced by propagating the dominant source of experimental error, the 0.1% uncertainty in F_3. Thus, good agreement between experiment and theory holds.

It is useful to mention the reduced-mass corrections that must be made in Stark-effect calculations in hydrogenic systems.[22,69] (When the nuclear charge $Z>1$, see Ref. 22.) For $Z=1$, equations valid for an infinitely massive nucleus and electron mass m_e can be made valid for reduced mass μ if the following substitutions are made: m_e is replaced by $\chi\mu$, R by R/χ, F by $\chi^2 F$, and E by χE, where $\chi \equiv m_e/\mu$. For H, $\chi - 1 = 5.4462 \times 10^{-4}$; for D, $\chi - 1 = 2.7212 \times 10^{-4}$; for positronium, $\chi = 2$. For example, equations in atomic units for $E_R(F)$ and $\Gamma_I(F)$, respectively, become new equations for $\chi E_R(\chi^2 F)$ and $\chi \Gamma_I(\chi^2 F)$ in atomic units. For a given value F in volts per centimeter, we must use a corrected field in atomic units, $\chi^2 F(\mathrm{V\ cm^{-1}})/\mathrm{au}(F)$. One could refer to this as a physical atomic unit of electric field.

Silverstone and Koch[117] investigated numerically the use of Padé approximants for obtaining improved perturbative estimates of $E_R(F)$.[118] A Padé approximant $[L/M]$ to $E_R(F)$ is the quotient of two polynomials of degrees L and M in F whose power series agrees with that of $E_R(F)$ through order $L+M$. The Lth order truncated sum of perturbation theory is itself the Padé

Table 13.3. Comparison of experiment and theory for two laser-driven transitions between hydrogen Stark substates

Quantity	Transition I	Transition II
Lower state $\{n, n_1, n_2, \|m\|\}$	$\{10, 8, 0, 1\}$	$\{10, 0, 9, 0\}$
Upper State $\{n, n_1, n_2, \|m\|\}$	$\{25, 21, 2, 1\}$	$\{30, 0. 29, 0\}$
Beam energy E_k (eV)	7474(20)	7474(20)
$^{12}C\,^{16}O_2$ laser line	10 μm (R24)	10 μm (P24)
Atom frame photon energy k_R (cm^{-1})	974.576(5)	936.803(5)
Atom frame resonant F_3 (V cm^{-1})	2514(3)	689.3(7)
$n_U^4 F_3$ (a.u.)	0.191	0.109
$n_L^4 F_3$ (a.u.)	0.005	0.002
Calc RSPT[a] ($N>4$): $E_L(F_3)$ (cm^{-1})	$-1083.945(13)^{b,c}$	$-1100.753(4)^{b,c}$
Exp: $E_U(F_3)=k_R+E_L(F_3)$ (cm^{-1})	$-109.369(14)^{c,d}$	$-163.950(7)^{c,d}$
N-order RSPT[a] for $\{U\}$ (cm^{-1})	$S_U^{(6)}=-110.222(63)^{b,c}$ $S_U^{(7)}=-108.352(63)^{b,c}$	$S_U^{(24)}=-163.97(5)^{b,c}$
Exp. RSPT[a]	$E_U(F_3)-S_U^{(6)}=0.85(6)^c$ $E_U(F_3)-S_U^{(7)}=-1.02(6)^c$	$E_U(F_3)-S_U^{(24)}=0.02(5)^c$
Numerical calculation[e] $E_U^{num}(F_3)$ (cm^{-1})	$-109.385(63)^f$	$-163.924(50)^f$
Numerical calculation[e] $\Gamma_U^{num}(F_3)$ (cm^{-1})	1.21×10^{-5}	5.59×10^{-7}
$E_U(F_3)-E_U^{num}(F_3)$ (cm^{-1})	$0.02(6)^b$	$-0.03(5)^b$

[a] Rayleigh-Schrödinger perturbation theory (RSPT); energy calculated using results of Ref. 113.
[b] Quoted error is spread of theoretical values due to uncertainty in F_3.
[c] Numerical values revised from Ref. 85.
[d] Quoted error is quadrature sum of calculated errors in k_R and $E_L(F_3)$.
[e] Obtained by Luc-Koenig and Bachelier,[69] with an atomic unit of F that was 71 ppm too low.
[f] Quoted error is our estimate of the spread of theoretical values[e] due to experimental uncertainty in F_3.

approximant $[L/0]$. Figure 13.9 shows one case for $n_1-n_2\simeq n$ in which the diagonal sequence $[\frac{1}{2}N/\frac{1}{2}N]$ gave a four-significant-figure increase in accuracy over the oscillating perturbation series $[N/0]$ truncated at its smallest term $N=7$.[117] Furthermore, the diagonal sequence appeared to converge to the experimental result. We note that Graffi et al.[119] proved the convergence of the diagonal Padé sequence for the restricted one-dimensional Stark problem but that no proof has been reported for the full three-dimensional problem.

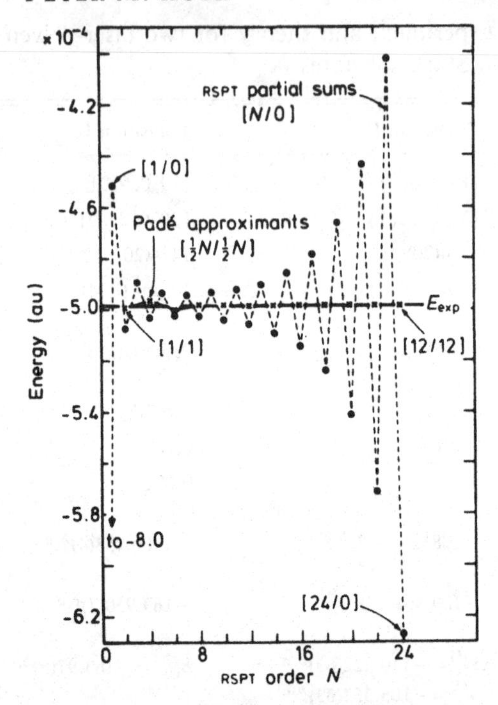

Fig. 13.9. Solid line: measured energy E_{exp} (converted to atomic units) for the Stark substate {25, 21, 2, 1} in $F = 2514(3)$ V cm^{-1}, which is a scaled field $n^4F = 0.191$ a.u. The measured wave numbers are given in Table 13.3. Filled circles: successive sums of the Rayleigh-Schrödinger perturbation theory, which are also the Padé approximants [N/0]; the asymptotic divergent nature of this sequence is obvious. Crosses: diagonal Padé approximants [½N/½N]; $E_{\text{exp}} - [12/12] = 0.4 \pm 2.9 \times 10^{-7}$ a.u., where the error is caused predominantly by the experimental uncertainty in F. (From Silverstone and Koch[117])

Damburg and Kolosov (Ref. 120 and also Chap. 2) made additional comparisons. They concluded that diagonal Padé approximants give higher accuracy in some cases and the perturbation series [N/0] truncated at its smallest term gives higher accuracy in other cases, particularly when $|n_1 - n_2| \ll n$.

13.5c. Precise laser, fast-beam measurements of $\Gamma_{\text{I}}(F)$

To measure $\Gamma_{\text{I}}(F)$ precisely for a number of $n = 30, 40$ Stark substates, Koch and Mariani[96] extended the techniques used to measure $E_{\text{R}}(F)$. Figure 13.6 shows schematically the five electric fields that were used for collinear laser excitation and subsequent tunneling field ionization in F_5 ($n^4F_5 < n^4F_{\text{cl}}$; Fig. 13.3) of the Stark substates. Details and results were presented in Ref. 96. Experimental and analytical asymptotic and numerical theoretical results for six $n = 40$ substates are shown in Fig. 13.10. The analytic results [Eq. (6) of

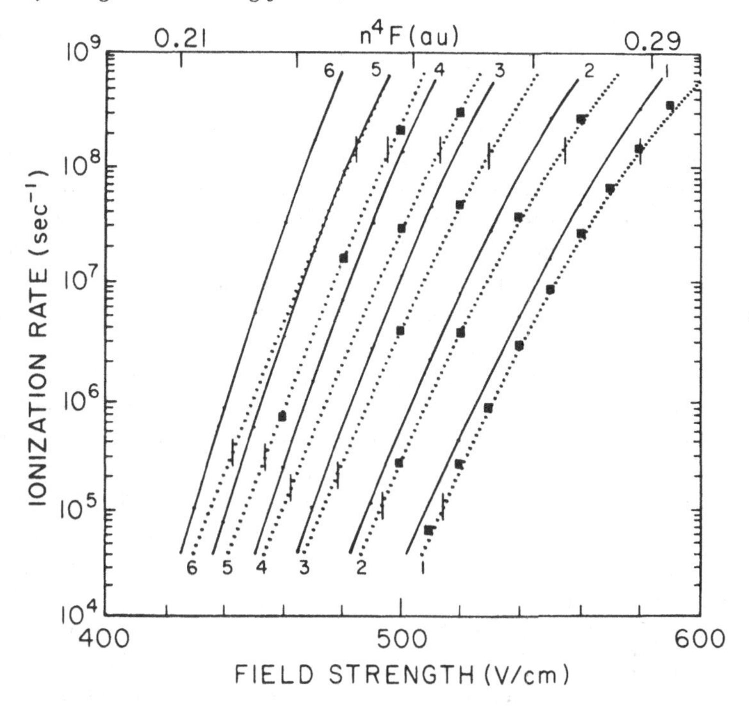

Fig. 13.10. Dotted lines: experimental $\Gamma_I(F)$ curves for the hydrogen Stark substates 1:{40, 39, 0, 0}; 2:{40, 38, 0, 1}; 3:{40, 38, 1, 0}; 4:{40, 37, 1, 1}; 5:{40, 37, 2, 0}; 6:{40, 36, 2, 1}. The vertical tick marks bound the experimentally significant portion of each curve. Above each upper tick mark, the experiment gave only a lower bound on $\Gamma_I(F)$. Below each lower mark, it gave only an upper bound. Solid lines: theoretical curves calculated with the use of a semiempirical, asymptotic formula [Eq. (6) of Ref. 121] with allowance for reduced-mass corrections. Squares: numerical theoretical results calculated for an infinitely massive nucleus.[69] (From Koch and Mariani[96])

Ref. 121] were not generally in good agreement; discrepancies as large as a factor of 10 were noted. To within the estimated $\pm 15\%$ uncertainty of the experiment (dominated by the $\pm 0.2\%$ uncertainty in F_5), the numerical theoretical results were in good agreement, but there is an open question about the magnitude of reduced-mass corrections to $\Gamma_I(F)$. The estimated corrections presented in Ref. 96 and herein revise the erroneous preliminary conclusions presented in Ref. 22, p. 192. The experiment gave a 3% upper limit for the difference in $\Gamma_I(F) \simeq 10^7$ s^{-1} at fixed F between hydrogen and deuterium atoms in the substate {40, 39, 0, 0}.[96] Estimates based on an analytic formula predicted that the H value would be ~3% above the D value and ~6–9% above the case for an infinite nuclear mass, respectively.

13.5d. Crossing of hydrogen Stark substates

Because of a conserved dynamical symmetry, nonrelativistic hydrogen Stark substates with the same $|m|$ value and from different n manifolds can cross.[69, 122, 123] In Rydberg states of multielectron atoms, the core electrons break this symmetry and induce avoid crossings and complicated ionization behavior in static[124] and time-slewed[125] fields. In hydrogen, the symmetry is broken predominantly by relativistic (fine-structure) interactions.[16, 21, 65] Because the associated quantum defect $\delta(j) \lesssim 10^{-5}$ is so small (Sect. 13.1b), the symmetry is broken very weakly, and it is understandable that no broken-symmetry-induced ionization effects were observed experimentally with $n = 30, 40$ substates.[96] It may be possible to use the perturbative method of Komarov et al.[126] to estimate the avoided crossing gaps.

Hatton[123] predicted that some *nonrelativistic,* "long-lived" Stark substates should avoid crossing. His analysis was based on the eigenvalues n_i, $i=1,2,3$ (which correspond to parabolic quantum numbers n_1, n_2, and $|m|$) of a complete set of commuting operators that count wave function nodes.[69] Two substates a and b were predicted to avoid crossing if $n_i^a \leqslant n_i^b$ for $i=1,2,3$ or if $n_i^a \geqslant n_i^b$ for $i=1,2,3$. However, if two states predicted to avoid crossing approach each other only when F is greater than the value of F_{cl} (Fig. 13.3) for one or both of them, his assumptions of classically bound motion (in the η direction) and of "long-lived" states are no longer valid. This, in fact, appears to be the general case. Consider, for example, $a=\{30,0,29,0\}$ and $b=\{31,0,30,0\}$ for which $n_i^b \geqslant n_i^a$, $i=1,2,3$. These states approach each other near $F=2.0$ kV cm^{-1}, or $n^4F=0.36$ a.u. for $n=31$. This is nearly a factor of three above the classical threshold $n^4F_{\mathrm{cl}} \simeq 0.13$ a.u. (Fig. 13.3). A proper avoided crossing analysis would have to include the very large width of the rapidly ionizing level(s);[124] this was ignored in Ref. 123.

13.6. Rydberg atoms in strong electromagnetic fields

13.6a. Microwave ionization and excitation

Bayfield and Koch[84] introduced a new method for studying intense-field multiphoton processes with laboratory fields that were actually quite weak.[127–131] By reducing the Coulomb atomic binding field proportional to n^{-4} to very small values with the use of H(high-n) atoms, microwave electric fields $F(t)=F_0 \cos \omega t$ with F_0 greater than or approximately equal to tens of volts per centimeter could be used to study the n, F_0, and ω dependences of multiphoton ionization and excitation.[84, 87, 88, 132–134]

Figure 13.11 gives one example of the experimental results for H($n=48$) atoms.[88] The ω dependences of the comparable rates for multiphoton ionization (b) and for multiphoton excitation (a) to higher-lying states (with $n \lesssim 60$) were measured with an axially directed TE_{10}-mode waveguide electric field. The atom beam passed through ~8-mm-diameter holes in opposing sides of the waveguide. These holes perturbed the field so that F_0 was within ~5% of its constant peak value $F_0 = 80$ V cm^{-1} only for about the central half of the 1-cm-wide waveguide. For the ~10-keV atoms, this meant a transit-time-

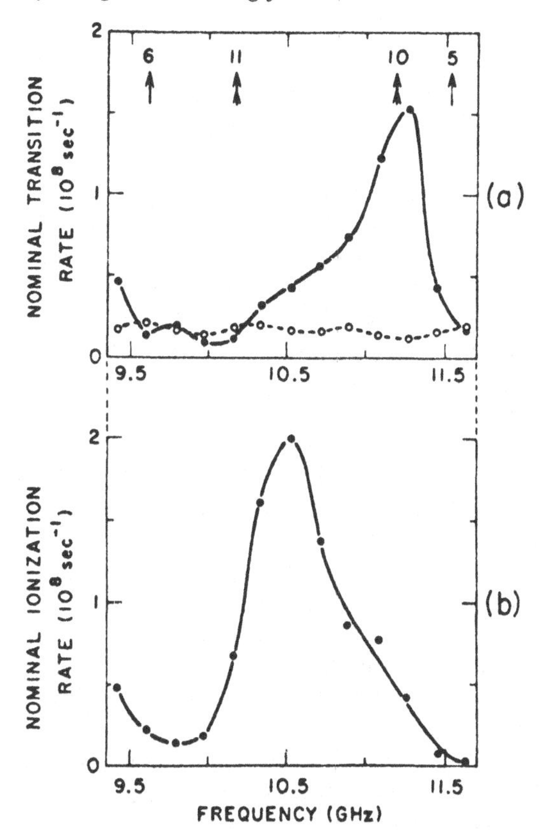

Fig. 13.11. The frequency dependence of the absolute rates for (*a*) excitation and (*b*) ionization of H($n=48$) atoms in a microwave field whose peak electric field strength was $F_0=80$ V cm^{-1}, or $n^4F_0=0.083$ a.u. The parameter ω/ω_{at} ranges from 0.16 to 0.19. The open circles in (*a*) are background signals. Single- and double-headed arrows show the position of multiphoton transition frequencies between the field-free $n=48$ manifold and field-free $n=49$ and $n=50$ manifolds, respectively, for the number of photons shown. (From Bayfield, Gardner, and Koch[88])

induced frequency spread $\Delta\nu_T \simeq 0.05$ GHz. The resonance widths 0.5–1 GHz observed for $n=48$ and other n values ($45 \leqslant n \leqslant 57$) were significantly larger than this $\Delta\nu_T$, so they are physically significant.[88] The observed resonance frequencies for ionization and excitation were not the same, and they shifted with n_0. Ionization was found to occur at peak fields $n^4F_0 \simeq 0.08$ a.u., significantly below the lowest classical threshold $n^4F_{cl} \simeq 0.13$ a.u. (Fig. 13.3).

These experiments stimulated subsequent theoretical activity. Delone et al.[135] characterized the microwave ionization of H(high-n) atoms as "diffusion of the electron along atomic states highly perturbed by the field" (p. 223). Their estimated diffusion time, however, was criticized by Meerson et al.,[136] who proposed an ionization model based on the classical effect of stochastic instability of a nonlinear oscillator. Percival and co-workers[137-139] used a classical, Monte Carlo trajectory model and obtained quantitative agreement with some of the experimental results, but they concluded that "much more work needs to be done...to obtain a clear picture of the physical process responsible for ionization" (Ref. 139, p. 36).

That it treats the interaction of the atom with the field to all orders, rather than perturbatively, is a distinct advantage of their method. Neglect of all quantal effects is the principle disadvantage. Resonant transitions between discrete quantum states cannot be treated, but the theory does allow for classical resonance phenomena. From considerations of the magnitude of classical actions, it was claimed that the effects of quantal barrier penetration were negligible for high n.[137] Clearly this is not true in the $\omega \rightarrow 0$ static limit. For example, Fig. 3 of Ref. 96 shows that appreciable tunneling ionization, $\Gamma_1(F) \simeq 10^7$ s^{-1}, of the Stark substate $\{30, 0, 29, 0\}$ was observed at $F \simeq$ 720 V cm^{-1} ($n^4F \simeq 0.113$ a.u.), which is 13% below the classical threshold $F_{cl} = 827.52$ V cm^{-1} ($n^4F_{cl} = 0.13035$ a.u.) calculated (Fig. 13.3 and Ref. 140) for an infinite nuclear mass. Similar statements apply for the Stark substates in Fig. 13.10. Semiclassical theories[141] based on a model introduced by Keldysh[131] have been developed to make the theoretical connection between ω-dependent tunneling and multiphoton ionization, but their relevance for the H(high-n) microwave experiments is questionable.

Just how rich and complicated the *classical* motions can be when the ratio of field and electron orbital frequency is $\omega/\omega_{at} \simeq 1$ (a range near that of the experimentally obtained[88] $0.13 \leqslant \omega/\omega_{at} \leqslant 0.26$) was shown by plots of planar spatial orbits and of the time dependence of the "compensated energy" E_c.[137-139] Four different classes of orbits were found: C1, orbits on invariant tori in phase space that may never ionize; C2, orbits that ionize very rapidly; C3, orbits that make transitions to extra highly excited (EHE) states and then ionize; and C4, orbits that end up in other EHE states and seem to remain much more stable against ionization.

Future experiments should attempt to test these predictions by measuring the time dependence of the ionization probability. Because the theory implies that the EHE states in the C4 orbits have a final $n_f \gtrsim 5n_0$, it will be important to avoid their ionization by stray fields or collisions.

The work of Percival and co-workers[137-139] and Meerson et al.[136] links the microwave ionization and excitation processes in H(high-n) atoms to the much more general subject of the onset of stochastic (also called chaotic, ergodic, or statistical) motions in nonlinear classical systems. How these onsets are affected when a system is quantized is currently being investigated in models of laser multiphoton dissociation of molecules.[142] Zewail concluded that an important underlying theoretical issue is "the question of what precise relationship, if any, exists between classical chaos and quantum mechanical motion. Clearly we still do not know the precise meaning of quantum ergodicity" (Ref. 143, p. 33). Because the classical and quantum-mechanical motions of the field-free hydrogen atom are known exactly, this system may be a useful testing ground for intense-field studies of these issues.

13.6b. Quantal structure of H(high-n) atoms in an intense oscillatory field

Blochinzew[144] first considered the quantal structure of a hydrogen atom in a sinusoidal electric field. More recently developed theories were discussed in Refs. 127, 145–148. One effect of a linearly polarized field is to produce sidebands displaced by integral units $\pm p\hbar\omega$ of the photon energy above and below the zero-field H(n) levels. For a given value of the permanent dipole moment d, each amplitude is proportional to

$$J_p(dF_0/\hbar\omega) \tag{16}$$

the value of the Bessel function of order p whose argument is the ratio of the peak dipole energy to the photon energy. (Later we will use p_i and p_f to refer to the order of the sideband for initial and final states with principal quantum numbers n_i and n_f, respectively.) The theoretical treatments concentrated on individual n manifolds and did not explicitly treat the influence of the continuum. The previous section implies that ionization would become important when $n^4F_0 \gtrsim 0.08$ a.u. for $\omega/\omega_{at} \simeq 1$. In plasma physics experiments, harmonic[146] and nonharmonic[149] hydrogen satellites were observed at lower values of n^4F_0.

Bayfield[150] proposed a laser fast-beam experiment for the simultaneous production and detection of H(high-n) sideband levels in a single microwave field. Koch[151] reported observation of the $p_f = +1, 2, 3, 4$ sidebands of H($n_f = 48$) in a related experiment, which used, instead, two separated microwave fields (a TM_{020} cavity (A) followed by TM_{010} cavity (B), both at $\omega/2\pi =$ 9.91 GHz) to decouple the sideband-production process in (A) from the ionization–detection process in (B). Because Ref. 151 is not widely available, it is useful to describe the results of that experiment here.

The $^{12}C\,^{16}O_2$ laser photons from the 9μmP(12) line were Doppler velocity tuned into resonance with the ($n_i = 10$ to $n_f = 48$) field-free transition (see Sect. 13.4d) at a beam energy $E_k = 9.60$ keV. The H($n_f = 48$) atoms were detected (Sects. 13.3b and 13.3c) by ionization in the downstream cavity (B). At certain other discrete energies E'_k, transitions were detected only when the microwave electric field strength $F_0(A)$ in the upstream cavity (A) was nonzero. Occurring at 8.23 keV ($p_f = +1$), 7.15 keV ($p_f = +2$), 6.04 keV ($p_f = +3$), and 5.10 keV ($p_f = +4$), they were interpreted as transitions from the $p_i = 0$ sideband of $n_i = 10$ to the p_f sideband of $n_f = 48$ on the basis of the Doppler-shifted CO_2 laser photon energies. Alternatively, they could be viewed as multiphoton transitions involving the absorption of one laser photon and the net absorption of p_f microwave (dressing) photons.

Figure 13.12 shows the dependence of the $p_f = 1, 2$ signals on $[F_0(A)/$

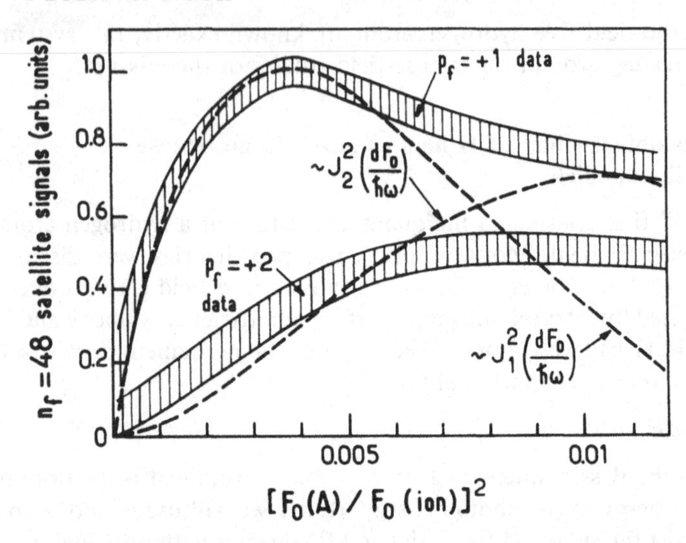

Fig. 13.12. Cross-hatched areas: experimental signals for CO_2-laser-induced transitions from the $p_i = 0$ sideband level of H($n_i = 10$) to the $p_f = 1, 2$ sideband levels, respectively, of H($n_f = 48$) in a 9.91-GHz electric field whose peak field strength was $F_0(A)$. $F_0(\text{ion}) \simeq 90$ V cm^{-1} was the field strength observed to ionize the H($n_f = 48$) atoms. Dashed curves: theoretical curves proportional to the square of the Bessel functions of order p_f. The normalization is discussed in the text. (Data from Koch[151])

$F_0(\text{ion})]^2$, a parameter proportional to the microwave power in cavity (A). The $F_0(\text{ion}) \simeq 90$ V cm^{-1} was the measured ionization field strength. The dashed lines in the figure show those theoretical curves based on Eq. (16) that were obtained by adjusting the magnitude and argument of the first maximum of the square of J_1 to agree with the position and height of the $p_f = +1$ peak. With the same normalization parameters, the square of J_2 was used to plot the $p_f = 2$ curve. The fitted dipole moment is $d_f = 2.6 \times 10^3$ a.u., or, in more physical terms, $d_f/n_f^2 \simeq 1$ for $n_f = 48$. This result is quite reasonable; the *permanent* dipole moment in the oscillatory field was about one scaled atomic unit (1 a.u./n^2). It follows that only the $p_i = 0$ sideband of $n_i = 10$ (i.e., the zero-field level) contributed to the experiment if we assume that it had a similar scaled dipole moment $d_i/n_i^2 \simeq 1$. Then the argument of the Bessel functions for $n_i = 10$ is $(d_i F_0/\hbar\omega) \ll 1$, and only $J_0^2 \simeq 1$ is nonzero.

The dashed curves reproduce the $p_f = 1, 2$ data well only for F_0 values up to the peak in the $p_f = 1$ curve.[151] This discrepancy at higher values of F_0 was also observed in more extensive, subsequent experiments by Bayfield et al.,[148] whose Fig. 2 shows similar behavior for the 7.829-GHz, $p_f = -1$ sideband of H($n_f = 44$). They speculated that it was caused by contributions made by atoms having more than one significant value of d_f. The use of atoms prepared in an individual (Stark) substate may allow cleaner experiments to be performed than those with mixed, unresolved substates.

Because the maximum $F_0(A)$ in Fig. 13.12 was only 11% of F_0(ion), it is obvious that many interweaving patterns of sidebands of a large number of n manifolds are populated at F_0 values required for ionization. Because this produces a nearly continuous quantal density of states, it is probably reasonable for classical mechanics to describe important aspects of the intense microwave-field excitation and ionization processes.

13.7. Collision studies with fast H(high-n) atoms

A few general remarks will help to introduce the large topic of collisions with fast H(high-n) atoms. Let a Rydberg atom whose Rydberg electron has a rms orbital velocity $v_0 = Zn^{-1}$ a.u. collide at relative velocity v with a target particle. When the velocity ratio $v/v_0 \gg 1$, the collision may reasonably be called "fast." Similarly, "intermediate" and "slow" collisions are characterized by $v/v_0 \simeq 1$ and $v/v_0 \ll 1$, respectively. A natural cross section scale is often the geometric size $n^4 \pi a_0^2$ of the Rydberg atom unless the Rydberg electron (or core ion) scatters "quasi-freely" with the core ion (or Rydberg electron) acting as a distant spectator. Then the relevant free-particle cross section may be a more suitable scale. Because of the additional long-range Coulomb interaction, collisions of Rydberg atoms with charged particles generally have much larger cross sections than those at the same v/v_0 with neutral targets. Chapters 6, 8, and 11 in this book treat the theory of some Rydberg atom collision processes in more detail.

13.7a. Collisions of H(high-n) atoms with ions of charge q

Two merged-beam experiments[77,82] and one crossed-beam experiment[76] were reported. Each used H(high-n) atoms in an n band, $n_a \lesssim n \lesssim n_b$ (Sect. 13.4c), conveniently characterized by its central n value $\bar{n} \simeq (n_a + n_b)/2$. The quantal substate distributions were never measured, but a calculation[152] implied that transitions induced by spatially varying fields $F \gtrsim 10^3$ V cm^{-1}, which were seen as time-varying fields in the atomic rest frames, could have made them statistical. Each experiment had such fields. That appreciable populations of high-$|m|$ Stark levels of fast H($n=7$) atoms were subsequently observed spectroscopically is consistent with the statistical distribution.[23] Its classical analog, the microcanonical distribution, has been used in most classical calculations of collision cross sections.[153-156]

Koch and Bayfield[77] used the merged-beam method for the first ion-H(high-n) collision experiment. Figure 13.13 shows their measured cross sections σ_{loss} for electron-loss (electron transfer σ_{et} plus ionization σ_{I}) in H$^+$-H($44 \lesssim n \lesssim 50$) collisions, which covered the slow, intermediate, and fast regimes $0.26 \leqslant v/v_0 \leqslant 3.3$. Classical scaling rules derived by Abrines and Percival,[154] which simultaneously scale the collision energy as n^{-2} and the cross section as n^4, were used to obtain the theoretical results shown for $\bar{n} = 47$ from

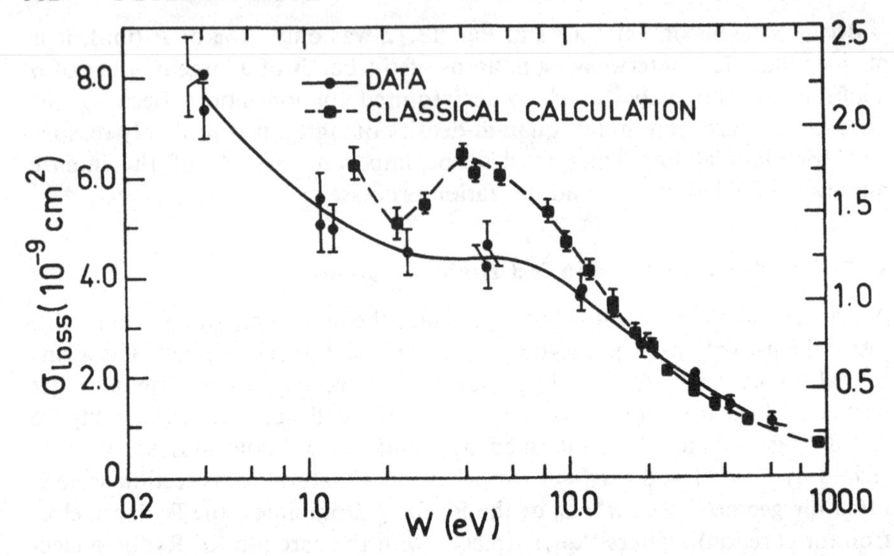

Fig. 13.13. Circles: the measured cross section σ_{loss} for electron loss in collisions of $H(44 \leq n \leq 50)$ atoms with protons vs. the center-of-mass collision energy W.[77] Squares: the classically scaled, classical Monte Carlo theoretical σ_{loss} values tabulated in Ref. 155. The experimental (left cross section scale) and theoretical (right cross section scale) curves have been normalized to each other with a factor of 3.48.

the $n=1$ classical, Monte Carlo σ_{loss} values tabulated by Banks et al.[155] (Olson's recent $q=1$ classical values agree with the latter over their common range $1 \leqslant v/v_0 \leqslant 4$.[156]) The curves have been normalized by an additional factor of 3.48, which is comparable to the estimated factor-of-two uncertainty in the experimental σ_{loss} scale (left vertical axis). We would like to have more experimental data, especially near 4 eV ($v/v_0=0.84$), and more theoretical data below ~1 eV ($v/v_0 \lesssim 0.42$), but the reasonable agreement confirms the validity of the classical scaling rules.[154]

Reasonable experimental bounds on σ_{et} at $W \lesssim 20$ eV ($v/v_0 \lesssim 1.9$) were shown in Ref. 77 to be consistent with classically scaled classical theory. Toshima[157] recently derived quantal results for σ_{et} that also obey classical scaling. Neglect of non-energy-resonant electron transfer channels was given as a possible reason for his results being somewhat below experiment and classical theory.[157]

Olson's classical Monte Carlo calculations support this conclusion.[156] For the case $v/v_0=1$, $q=1$, and $n=20$ closest to the experiment,[77] the cross section $\sigma_{et}(n')$ for each of the nonresonant channels between $n'=18$–25 was greater than one-half of σ_{et} $(n'=n)$, which was the largest.

Burniaux et al.[82] measured σ_{et} for the asymmetric process

$$He^{2+} + H(\bar{n}) \rightarrow He^{+}(n') + H^{+} \tag{17}$$

for four incident n bands centered on $\bar{n}=8.5$, 12, 17, 22 and for final $He^+(n')$ ions with $n' \lesssim 40$. For their center-of-mass collision energy range $0.25 \leqslant W \leqslant 478$ eV, v/v_0 covered the ranges from 0.03 to 1.36 for $\bar{n}=8.5$, and 0.08 to 3.51 for $\bar{n}=22$. In comparisons among their low-energy n-band data and previous He^{2+}–H(1s) data, they found semiquantitative agreement with classical scaling. Within errors due to comparisons of graphical data, Koch found reasonable agreement between their $\bar{n}=12$ and $\bar{n}=17$ data[82] and Olson's classical, Monte Carlo results (Ref. 156, Fig. 1, $q=+2$, $v/v_0 \geqslant 1$). More theoretical results are needed for $v/v_0 < 1$. Their experimental evidence[82] for efficient electron transfer into nonresonant channels n' about 5 to 10 units below $2n$ is also consistent with the later theoretical conclusions of Olson.[156] More experimental and theoretical comparisons for widely varying values of v/v_0, q, and n and n' are needed, however.

Kim and Meyer[76] observed electron loss by H(high-n) atoms colliding with N^{3+} ions at the fixed relative energy $W=40$ keV amu^{-1}. Electric field ionization was used to vary the maximal value N of principal quantum numbers n that contributed to the measured σ_{loss} over the range $9 \leqslant N \leqslant 24$. Their data corresponded to the fast velocity range $v/v_0 \simeq 11$–30, where ionization was expected to be the dominant electron-loss mechanism. From their data analysis, they inferred a relation

$$\sigma_{\text{loss}} = 1.19 \times 10^{-15} n^{3.12} \text{ cm}^2 \qquad (9 \leqslant n \leqslant 24) \tag{18}$$

that was inconsistent with the n^2 scaling of classical and Born theories of ionization. For example, Olson[156] found that his classical cross section approached the asymptotic form

$$\sigma_{\text{loss}} = (6n^2q^2/v^2)\pi a_0^2 \tag{19}$$

when $v/v_0 \gtrsim 3$ for $q=+1$, and when $v/v_0 \gtrsim 6$ for $q=+10$.

The collisions actually took place in an electric field $F=1$–1.5 kV cm^{-1} that was sufficient to ionize atoms H(n') with $n' \gtrsim 27$ ($n'^4F \gtrsim 0.13$ a.u.; Fig. 13.3). With explicit classical calculations, Olson[158] found that a two-step loss mechanism not considered by the authors of Ref. 76 could reproduce the approximate n^3 scaling of the experiment and give cross sections within a factor of two of those observed experimentally: N^{3+}–H(n) n-changing collisions with geometric-sized cross sections produced H($n' \gtrsim 27$) atoms, which were subsequently ionized by the apparatus electric field. In classical Monte Carlo calculations, Olson and MacKellar[159] found that explicit inclusion of the electric field raised some ionization cross section values by two orders of magnitude and reduced some electron-transfer values fourfold.

Recent experimental σ_{loss} data for fast ($v/v_0 \simeq 10$–36) collisions of H($10 \leqslant n \leqslant 24$) atoms with N^{q+} and O^{q+} ions ($1 \leqslant q \leqslant 4$) are consistent with the q^2 and v^{-2} dependences of Eq. (19).[160]

13.7b. Collisions of fast H(high-n) atoms with neutral targets

For H(high-n)-H(1s) collisions with $v/v_0 \gg 1$, Butler and May[161] first predicted that the cross section σ_I for ionization of the Rydberg electron should be approximately equal to the total scattering cross section σ_e for free electrons colliding with H(1s) at the same relative velocity $v_e = v$ and, therefore, should be n independent. More general theories were subsequently developed by Smirnov,[162] de Prunelé and Pascale,[163] and Matsuzawa.[164]

Indirect measurements of electron loss by fast H(high-n) atoms were first made by Riviere and Sweetman[36] (H_2 target; $v/v_0 \simeq 13$–28) and later by Oparin et al.[40] (Na, Mg, Cd, Ne targets; $v/v_0 \simeq 7$–25). The inferred σ_{loss} values were found to be comparable to σ_e values for gas targets but to be somewhat below those for the metal-vapor targets.

Koch[83] carried out the first detailed test of the theoretical predictions by measuring the total cross sections for ionization σ_I of D($35 \lesssim n \lesssim 50$) atoms and for destruction σ_d (sum of ionization, excitation to $n \gtrsim 61$, and deexcitation to $n \lesssim 28$) of laser-excited D($n^* = 46$) atoms in 6- to 13-keV collisions ($v/v_0 = 16$–23 for $n^* = 46$) with 300-K N_2 molecules. N_2 was chosen because the $N_2^-(^2\Pi_g)$ compound-state-induced structure in σ_e provided an unmistakable ruler for comparison.[165]

Figure 13.14 shows a new comparison of the σ_I and σ_d data of Ref. 83 with the recent σ_e data published by Kennerly,[165] whose quoted uncertainties are lower than those of Ref. 166. To within their estimated ±20% uncertainties, σ_I and σ_d, respectively, agree with σ_e and $\sigma_e + \sigma_{10}$, which confirms the independent, quasi-free scattering of both particles in the Rydberg atom. The estimated[83] "quasi-free" electron scattering energy resolution

$$\Delta E/\bar{E} \simeq 4/\sqrt{3}nv \,(\text{a.u.}) \tag{20}$$

gave a $\Delta E \simeq 0.3$ eV, about 20–30 times worse than that of high-resolution free-electron scattering experiments, $\Delta E \simeq 15$ meV.[165,167] In principle, ΔE could be lowered by greatly increasing n or by preparing electron orbits having small values of $\mathbf{v} \cdot \mathbf{v}_0$ (large values of $|m|$ relative to a $\hat{\mathbf{z}}$ axis parallel to $\mathbf{v}$). Neither scheme appears to be very practical. Rather, these Rydberg–neutral target collisions are interesting in their own right.

Future experiments should investigate the effect of Ramsauer minima in free e^--Ar, -Kr, -Xe collisions near $v_e = 0.2$ a.u. in the quasi-free electron scattering process.[168] High orbital momentum components will contribute to electron partial waves with $l > 0$, whose phase shifts $\delta(l) \neq \pi$ when $\delta(l=0) = \pi$. This, as well as small angle scattering[162] of the core ion by its polarization interaction with the rare gas targets, will tend to raise σ_I in the region of the minimum of σ_e. The strongly coupled regime $v/v_0 \simeq 1$ lying between thermal- and fast-beam experiments is unexplored and needs to be investigated with these and other targets.

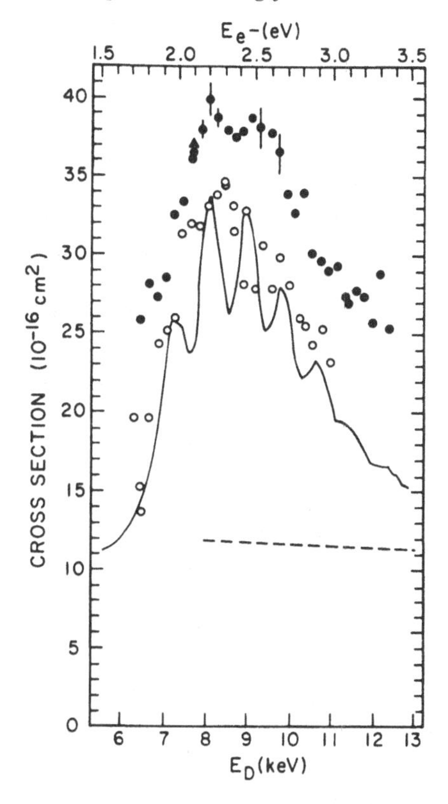

Fig. 13.14. Open circles: measured cross section for ionization of D($35 \lesssim n \lesssim 50$) atoms in collisions with N_2 vs. the kinetic energy E_D of the deuterium atom. Filled circles (triangle): measured cross section σ_d for destruction of $n^* = 46$ ($n^* = 71$) D atoms in collisions with N_2. Solid line: Kennerly's[165] measured total cross section for free electron-N_2 scattering vs. E_e– (eV). The E_D and E_e– scales have been arranged to correspond to the same collision velocity. Dashed line: measured cross section σ_{10} for electron transfer in H^+-N_2 collisions[78] plotted vs. $E_{D^+} = 2E_{H^+}$. (After Koch[83])

13.8. Future prospects

With a long past and an active present, the future looks exciting for Rydberg studies using fast beams. The marriage of laser spectroscopic techniques to fast-beam collision techniques has allowed the preparation of beams of H(high-n) atoms in individual Stark substates, including those with large values of $|m|$. The resolved substates may be used for additional spectroscopic or collision studies. It will be very important to extend the present spectroscopic measurements from the tunneling regime below the classical ionization threshold to near the zero-field ionization limit where resonant field-dependent structure was observed in photoabsorption studies with alkali[104, 106] and rare gas[105] atoms.

The spectroscopic experiments are naturally leading to new collision experiments. It would be interesting to measure the partial cross sections for the production of individual substates as a function of collision process, energy, target species, *and* electric field strength.[169] Equally interesting would be the study of the properties of individual substates in subsequent collisions. Examples would include substate changing, *n* changing, ionization, or electron-transfer collisions. The same electric field ionization, laser, and microwave spectroscopic techniques used to prepare the initial substates would also be useful for probing the distribution of final substates.

Because of motional electric fields, the intense magnetic field problem (see Chap. 3) is not well suited to fast-beam study. Rather, this method would be ideally suited to spectroscopic studies of hydrogen atoms in intense crossed electric and magnetic fields. A magnetic field oriented perpendicular to the particle beam velocity vector would produce a transverse electric field F_{mot} in the particle rest frame. Additional electric field plates would provide a variable field to be added to or subtracted from F_{mot}. Theoretical work on the intense crossed-field problem is in its infancy.[170–176]

Experiments, followed by theory, have shown how little we understand the details of the interaction of Rydberg atoms with an intense oscillatory field. Classical methods have given us a glimpse of the complexity of the microwave excitation and ionization process in hydrogen atoms, which is not unlike the multiphoton excitation and dissociation process in polyatomic molecules. Theoretical questions associated with the concept of classical vs. quantal ergodicity[143,177] may be resolved by careful experimental and theoretical studies of a real system whose zero-field structure is well understood, both classically and quantum mechanically. The hydrogen atom is one such system.

Though omitted from this chapter, fast-beam studies of Rydberg states of multielectron atoms or ions are extensive.[2] Microwave spectroscopy of helium Rydberg atoms in a fast beam[178] will complement and extend the extensive previous studies that used thermal atoms excited by electron impact.[179–182] In our laboratory we have recently begun a series of Stark spectroscopy, field ionization, and collision experiments using helium Rydberg atoms. They are excited to individual singlet or triplet substates with the same kind of CO_2 laser double-resonance techniques (with Stark tuning of substrates) that were described earlier in this chapter for experiments with hydrogen atoms.

A final speculative notion for the future involves the use of fast beams to study planetary atoms, those with two or more, possibly weakly interacting, Rydberg electrons that may have long radiative and autoionization decay lifetimes relative to lower-lying levels.[177,183] The helium planetary atom would be the simplest such system and, therefore, the one of greatest theoretical interest. With nearly 80 eV of internal energy and orbits whose overlaps with the $He(1s^2)\,^1S_0$ ground state are small, such states would not be easily excited by direct photoexcitation involving a small number of photons. One could, in principle, envision producing such states by the combination of one or more energetic collisions (to pour in the excitation energy and angular momentum) and subsequent laser excitations (to provide the resolution needed to observe individual quantum states). If long-lived states of the helium planetary atom could be produced, one could look forward to an exciting series of spectroscopic and collision experiments.

Acknowledgments

The major part of this research was supported by the National Science Foundation. The author gratefully acknowledges receipt of A. P. Sloan Foundation (1978–82) and Yale Junior Faculty (1980–1) Fellowships and the hospitality during 1980–1 of S. Haroche and the other members of Laboratoire de Spectroscopie Hertzienne de l'Ecole Normale Supérieure, where this chapter was written. Ms. C. Marthouret and Ms. L. Liptak cheerfully typed the first and final drafts of the manuscript, respectively. F. Wuilleumier (LURE-Orsay) facilitated the preparation of two figures and was an enthusiastic and capable guide for the author's introduction in 1980–1 to the world of experiments involving synchrotron radiation. Yale University undergraduates E. Rock, C. Hanson, and J. Bowlin contributed to several of the experiments reviewed herein. D. Mariani contributed skill and enthusiasm to much of the recent work at Yale and proofread the manuscript. The author especially thanks his wife, Nancy, for her moral support during the preparation of this chapter.

References and notes

1. N. Ryde, *Atoms and Molecules in Electric Fields* (Stockholm: Almqvist and Wiksell, 1976).
2. H. J. Andrä, in *Progress in Atomic Spectroscopy,* eds. W. Hanle and H. Kleinpoppen (New York: Plenum, 1979), part B.
3. J. R. Hiskes, *Nucl. Fusion 2,* 38 (1962).
4. D. R. Sweetman, *Nucl. Fusion, Suppl. 1,* 279 (1962).
5. A. C. Riviere, in *Methods of Experimental Physics,* ed. L. Marton (New York: Academic Press, 1968), vol. 7A.
6. R. N. Il'in, in *Atomic Physics,* eds. S. J. Smith and G. K. Walters (New York: Plenum, 1973), vol. 3, p. 309.
7. N. V. Federenko, V. A. Ankudinov, and R. N. Il'in, *Zh. Tekh. Fiz. 35,* 585 (1965) [*Sov. Phys. Tech. Phys. 10,* 461 (1965)].
8. F. L. Ribe, *Rev. Mod. Phys. 47,* 7 (1975).
9. W. S. Cooper, *Bull. Am. Phys. Soc. 24,* 1193 (1979).
10. C. F. Burrell, W. S. Cooper, R. R. Smith, and W. F. Steele, *Rev. Sci. Instrum. 51,* 1451 (1980).
11. G. C. Bjorklund and R. R. Freeman, in *Laser Spectroscopy,* eds. H. Walther and K. W. Rothe (New York: Springer-Verlag, 1979), vol. IV.
12. H. Zacharias, H. Rottke, J. Danon, and K. H. Welge, *Opt. Commun. 37,* 15 (1981).
13. M. J. Seaton, *Comments At. Mol. Phys. D 2,* 37 (1970).
14. K. T. Lu and U. Fano, *Phys. Rev. A 2,* 81 (1970).
15. U. Fano, *J. Opt. Soc. Am. 65,* 979 (1975).
16. H. A. Bethe and E. E. Salpeter, *Quantum Mechanics of One- and Two-Electron Atoms* (New York: Springer-Verlag, 1957).

17. S. R. Lundeen and F. M. Pipkin, *Phys. Rev. Lett. 46,* 232 (1981).
18. G. Newton, D. A. Andrews, and P. J. Unsworth, *Phil. Trans. R. Soc. London 290,* 373 (1979).
19. K. A. Safinya, K. K. Chan, S. R. Lundeen, and F. M. Pipkin, *Phys. Rev. Lett. 45,* 1934 (1980).
20. J. Bollinger, S. R. Lundeen, and F. M. Pipkin, Abstracts of Contributed Papers, VII ICAP, MIT (4–8 August 1980), p. 161.
21. G. Lüders, *Ann. Phys. (Leipzig), Ser. 6, 8,* 301 (1951).
22. P. M. Koch, in *Atomic Physics,* eds. D. Kleppner and F. M. Pipkin (New York: Plenum, 1981), vol. 7, p. 181.
23. P. M. Koch and D. R. Mariani, *J. Phys. B 13,* L645 (1980).
24. Dz. Belkic, R. Gayet, and A. Salin, *Phys. Rep. 56,* 279 (1979).
25. H. Schiff, *Can. J. Phys. 32,* 393 (1954); J. D. Jackson and H. Schiff, *Phys. Rev. 89,* 359 (1953); R. M. May, *Nucl. Fusion 4,* 207 (1964), and references therein.
26. J. R. Hiskes, *Phys. Rev. 137,* A361 (1965).
27. J. R. Hiskes, *Phys. Rev. 180,* 146 (1969).
28. A. V. Vinogradov, A. M. Urnov, and V. P. Shevel'ko, *Zh. Eksp. Teor. Fiz. 60,* 2060 (1971) [*Sov. Phys. JETP 33,* 1110 (1971)].
29. P. Pradel, F. Roussel, A. S. Schlachter, G. Spiess, and A. Valence, *Phys. Rev. A 10,* 797 (1974).
30. R. N. Il'in, B. I. Kikiani, V. A. Oparin, E. S. Solov'ev, and N. V. Federenko, *Zh. Eksp. Teor. Fiz. 47,* 1234 (1964) [*Sov. Phys. JETP 20,* 835 (1965)].
31. R. N. Il'in, V. A. Oparin, E. S. Solov'ev, and N. V. Federenko, *Zh. Tekh. Fiz. 36,* 1241 (1966) [*Sov. Phys. Tech. Phys. 11,* 921 (1967)].
32. J. E. Bayfield, G. A. Khayrallah, and P. M. Koch, *Phys. Rev. A 9,* 209 (1974).
33. R. H. McFarland and A. H. Futch, *Phys. Rev. A 2,* 1795 (1970).
34. R. LeDoucen and J. Guidini, *Rev. Phys. Appl. 4,* 405 (1969).
35. R. F. King and C. J. Latimer, *J. Phys. B 12,* 1477 (1979).
36. A. C. Riviere and D. R. Sweetman, in *Atomic Collision Processes,* ed. M. R. C. McDowell (Amsterdam: North-Holland Publ., 1964), pp. 734–42.
37. R. LeDoucen, J. H. Lenormand, and J. Guidini, *J. Phys. (Paris) 31,* 965 (1970).
38. N. Bohr, *K. Dans. Vidensk. Selsk. Mat. Fys. Medd. 18* (8) (1948).
39. G. F. Drukarev, *Zh. Eksp. Teor. Fiz. 52,* 498 (1967) [*Sov. Phys. JETP 25,* 326 (1967)].
40. V. A. Oparin, R. N. Il'in, and E. S. Solov'ev, *Zh. Eksp. Teor. Fiz. 52,* 369 (1967) [*Sov. Phys. JETP 25,* 240 (1967)].
41. J. B. Hasted, *Physics of Atomic Collisions* (London: Butterworth, 1964), Sect. 12.2.
42. K. H. Berkner, W. S. Cooper III, S. N. Kaplan, and R. V. Pyle, *Phys. Rev. 182,* 103 (1969).
43. J. Kingdon, M. F. Payne, and A. C. Riviere, *J. Phys. B 3,* 552 (1970).
44. R. LeDoucen, J. H. Lenormand, and J. Guidini, *C.R. Acad. Sci., Ser. B 271,* 437 (1970).
45. R. N. Il'in, V. A. Oparin, I. T. Serenkov, E. S. Solov'ev, and N. V. Federenko, *Zh. Eksp. Teor. Fiz. 59,* 103 (1970) [*Sov. Phys. JETP 32,* 59 (1971)].
46. Y. B. Band, *Phys. Rev. A 8,* 2866 (1973).
47. R. H. Hughes, C. A. Stigers, B. M. Doughty, and E. D. Stokes, *Phys. Rev. A 1,* 1424 (1970).
48. J. C. Ford and E. W. Thomas, *Phys. Rev. A 5,* 1694 (1972).
49. J. C. Ford and E. W. Thomas, *Phys. Rev. A 5,* 1701 (1972).

50. H. R. Dawson and D. H. Loyd, *Phys. Rev. A 15,* 43 (1977).
51. J. L. Edwards and E. W. Thomas, *Phys. Rev. A 2,* 2346 (1970).
52. J. S. Risley, F. J. deHeer, and C. B. Kerkdijk, *J. Phys. B 11,* 1759 (1978); and references therein.
53. D. H. Loyd and H. R. Dawson, *Phys. Rev. A 19,* 948 (1979).
54. J. E. Bayfield, in *Atomic Physics,* eds. G. zu Putlitz, E. W. Weber, and A. Winnacker (New York: Plenum, 1975), vol. 4, p. 397.
55. R. K. Knize, F. M. Pipkin, and S. R. Lundeen, *Phys. Rev. Lett. 49,* 315 (1982).
56. J. Burgdörfer, *J. Phys. B 14,* 1019 (1981).
57. R. H. Hughes, D. B. Kay, C. A. Stigers, and E. D. Stokes, *Phys. Rev. 167,* 26 (1968).
58. K. H. Berkner, I. Bornstein, R. V. Pyle, and J. W. Stearns, *Phys. Rev. A 6,* 278 (1972).
59. M. W. Lucas, W. Steckelmacher, J. Macek, and J. E. Potter, *J. Phys. B 13,* 4833 (1980).
60. J. S. Risley, F. J. deHeer, and C. B. Kerkdijk, *J. Phys. B 11,* 1783 (1978); and references therein.
61. H. Traubenberg, R. Gebauer, and G. Lewin, *Naturwissenschaften 18,* 417 (1930).
62. J. R. Oppenheimer, *Phys. Rev. 31,* 66 (1928).
63. C. Lanczos, *Z. Phys. 68,* 204 (1931).
64. E. U. Condon, *Am. J. Phys. 46,* 319 (1978).
65. E. U. Condon and G. H. Shortley, *The Theory of Atomic Spectra* (Cambridge University Press, 1951).
66. A. C. Riviere and D. R. Sweetman, *Phys. Rev. Lett. 5,* 560 (1960).
67. A. C. Riviere and D. R. Sweetman, Proceedings, VI ICIPG, Paris (1963), p. 105.
68. D. S. Bailey, J. R. Hiskes, and A. C. Riviere, *Nucl. Fusion 5,* 41 (1965).
69. E. Luc-Koenig and A. Bachelier, *J. Phys. B 13,* 1743 (1980).
70. D. Banks and J. G. Leopold, *J. Phys. B 11,* L5 (1978).
71. D. Banks and J. G. Leopold, *J. Phys. B 11,* 37 (1978).
72. D. Banks and J. G. Leopold, *J. Phys. B 11,* 2833 (1978).
73. L. L. Alston, ed., *High-Voltage Technology* (Oxford University Press, 1968).
74. H. Gould, *Phys. Rev. A 14,* 922 (1976).
75. R. G. E. Hutter, *J. Appl. Phys. 18,* 797 (1947).
76. H. J. Kim and F. W. Meyer, *Phys. Rev. Lett. 44,* 1047 (1980).
77. P. M. Koch and J. E. Bayfield, *Phys. Rev. Lett. 34,* 448 (1975).
78. H. Tawara and A. Russek, *Rev. Mod. Phys. 45,* 178 (1973).
79. J. R. Hiskes, C. B. Tarter, and D. A. Moody, *Phys. Rev. 133A,* 424 (1964).
80. K. Omidvar, NASA Technical Memorandum No. 82015, Goddard Space Flight Center, September 1980.
81. J. W. Farley and W. H. Wing, *Phys. Rev. A 23,* 2397 (1981).
82. M. Burniaux, F. Brouillard, A. Jognaux, T. R. Govers, and S. Szucs, *J. Phys. B 10,* 2421 (1977).
83. P. M. Koch, *Phys. Rev. Lett. 43,* 432 (1979).
84. J. E. Bayfield and P. M. Koch, *Phys. Rev. Lett. 33,* 258 (1974).
85. P. M. Koch, *Phys. Rev. Lett. 41,* 99 (1978).
86. S. L. Kaufman, *Opt. Commun. 17,* 309 (1976).
87. P. M. Koch, L. D. Gardner, and J. E. Bayfield, in *Beam Foil Spectroscopy,* eds. D. J. Pegg and I. A. Sellin (New York: Plenum, 1976), vol. 2.
88. J. E. Bayfield, L. D. Gardner, and P. M. Koch, *Phys. Rev. Lett. 39,* 76 (1977).

89. C. Freed, L. C. Bradley, and R. G. O'Donnell, *IEEE J. Quant. Elec. QE-16,* 1195 (1980).
90. J. Reid and K. Siemsen, *J. Appl. Phys. 48,* 2712 (1977).
91. J. E. Bayfield, in *The Physics of Electronic and Atomic Collisions,* eds. J. S. Risley and R. Geballe (Seattle: University of Washington Press, 1975).
92. J. E. Bayfield, *Rev. Sci. Instrum. 47,* 1450 (1976).
93. P. M. Koch, Abstracts of Contributed Papers, V ICAP, Berkeley (26–30 July, 1976), pp. 232–3.
94. P. M. Koch, *Opt. Commun. 20,* 115 (1977).
95. J. K. Layton and G. D. Magnuson, *Rev. Sci. Instrum. 43,* 1546 (1972).
96. P. M. Koch and D. R. Mariani, *Phys. Rev. Lett. 46,* 1275 (1981).
97. J. C. Zorn and T. C. English, *Adv. At. Mol. Phys. 9,* 244 (1973), sect. V.C.2.
98. D. A. Clark, H. C. Bryant, K. B. Butterfield, C. A. Frost, J. B. Donahue, P. A. M. Gram, M. E. Hamm, R. W. Hamm, and W. W. Smith, *Bull. Am. Phys. Soc. 25,* 1137 (1980).
99. M. E. Hamm. R. W. Hamm, J. B. Donahue, P. A. M. Gram, J. C. Pratt, M. A. Yates, R. D. Bolton, D. A. Clark, H. C. Bryant, C. A. Frost, and W. W. Smith, *Phys. Rev. Lett. 43,* 1715 (1979).
100. D. F. Blossey, *Phys. Rev. B 2,* 3976 (1970).
101. L. D. Landau and E. M. Lifshitz, *Quantum Mechanics* (*Non-Relativistic Theory*) (Elmsford, N.Y.: Pergamon Press, 1977).
102. E. C. Titchmarsh, *Eigenfunction Expansions Associated with Second-Order Differential Equations* (Oxford University Press, 1958), Part II.
103. E. Luc-Koenig and A. Bachelier, *J. Phys. B 13,* 1769 (1980).
104. R. R. Freeman and N. P. Economou, *Phys. Rev. A 20,* 2356 (1979).
105. B. E. Cole, J. W. Cooper, and E. B. Saloman, *Phys. Rev. Lett. 45,* 887 (1980).
106. T. S. Luk, L. DiMauro, T. Bergeman, and H. Metcalf, *Phys. Rev. Lett. 47,* 83 (1981); and references therein.
107. K. J. Kollath and M. C. Standage, in *Progress in Atomic Spectroscopy,* eds. W. Hanle and H. Kleinpoppen (New York: Plenum, 1979), part B.
108. M. Hehenberger, H. V. McIntosh, and E. Brändas, *Phys. Rev. A 10,* 1494 (1974); N. Hoe, B. D'Etat, J. Grumburg, M. Caby, E. Leboucher, and G. Couland, *Phys. Rev. A 25,* 891 (1972); M. B. Kadomtsev and B. M. Smirnov, *Zh. Eksp. Teor. Fiz. 80,* 1715 (1981) [*Sov. Phys. JETP 53,* 885 (1981)].
109. T. Yamabe, A. Tachibana, and H. J. Silverstone, *Phys. Rev. A 16,* 877 (1977).
110. W. P. Reinhardt, in *Electronic and Atomic Collisions,* eds. N. Oda and K. Takayanagi (Amsterdam: North-Holland Publ., 1980).
111. E. Harrell and B. Simon, *Duke Math. J. 47,* 845 (1980).
112. J. Killingbeck, *Rep. Prog. Phys. 40,* 963 (1977).
113. H. J. Silverstone, *Phys. Rev. A 18,* 1853 (1978).
114. H. J. Silverstone, B. G. Adams, J. Cizek, and P. Otto, *Phys. Rev. Lett. 43,* 1498 (1979).
115. M. T. Grisaru and H. N. Pendleton, *Ann. Phys.* (*N.Y.*) *79,* 518 (1973).
116. E. R. Cohen and B. N. Taylor, *J. Phys. Chem. Ref. Data 2,* 663 (1973).
117. H. J. Silverstone and P. M. Koch, *J. Phys. B 12,* L537 (1979).
118. G. A. Baker, *Essentials of Padé Approximants* (New York: Academic Press, 1975).
119. S. Graffi, V. Grecchi, S. Levoni, and M. Maioli, *J. Math. Phys. 20,* 685 (1979).
120. R. J. Damburg and V. V. Kolosov, *J. Phys. B 14,* 829 (1981).
121. R. J. Damburg and V. V. Kolosov, *J. Phys. B 12,* 2637 (1979).

122. K. Helfrich, *Theor. Chim. Acta 24,* 271 (1972).
123. G. J. Hatton, *Phys. Rev. A 16,* 1347 (1977).
124. M. G. Littman, M. L. Zimmerman, and D. Kleppner, *Phys. Rev. Lett. 37,* 486 (1976).
125. T. H. Jeys, G. W. Foltz, K. A. Smith, E. J. Beiting, F. G. Kellert, F. B. Dunning, and R. F. Stebbings, *Phys. Rev. Lett. 44,* 390 (1980).
126. I. V. Komarov, T. P. Grozdanov, and R. K. Janev, *J. Phys. B 13,* L573 (1980).
127. J. E. Bayfield, *Phys. Rep. 51,* 317 (1979); *Prog. Quant. Electr. 6,* 219 (1980).
128. P. Bräunlich, in *Progress in Atomic Spectroscopy,* eds. W. Hanle and H. Kleinpoppen (New York: Plenum, 1979), part B.
129. N. B. Delone, *Usp. Fiz. Nauk. 115,* 361 (1975) [*Sov. Phys. Usp. 18,* 169 (1975)].
130. P. Lambropoulos, *Adv. At. Mol. Phys. 12,* 87 (1976).
131. L. V. Keldysh. *Zh. Eksp. Teor. Fiz. 47,* 1945 (1964) [*Sov. Phys. JETP 20,* 1307 (1965)]; see also H. R. Reiss, *Phys. Rev. A 22,* 1786 (1980).
132. P. M. Koch, L. D. Gardner, and J. E. Bayfield, Abstracts of Contributed Papers, IX ICPEAC (1975), pp. 473–4.
133. P. M. Koch, Abstracts of Contributed Papers, International Conference on Multiphoton Processes, University of Rochester (1977), pp. 72–3.
134. J. E. Bayfield, in *Multiphoton Processes,* eds. J. H. Eberly and P. Lambropoulos (New York: Wiley, 1978), pp. 191–8.
135. N. B. Delone, B. A. Zon, and V. P. Krainov, *Zh. Eksp. Teor. Fiz. 75,* 445 (1978) [*Sov. Phys. JETP 48,* 223 (1978)].
136. B. I. Meerson, E. A. Oks, and P. V. Sasorov, *Pis'ma Zh. Eksp. Teor. Fiz. 29,* 79 (1979) [*Sov. Phys. JETP Lett. 29,* 72 (1979)].
137. J. G. Leopold and I. C. Percival, *Phys. Rev. Lett. 41,* 944 (1978).
138. J. G. Leopold and I. C. Percival, *J. Phys. B 12,* 709 (1979).
139. D. A. Jones, J. G. Leopold, and I. C. Percival, *J. Phys. B 13,* 31 (1980).
140. J. G. Leopold, private communication.
141. A. M. Perelomov, V. S. Popov, and V. P. Kuznetsov, *Zh. Eksp. Teor. Fiz. 54,* 841 (1968) [*Sov. Phys. JETP 27,* 451 (1968)]; and references therein.
142. R. B. Walker and R. K. Preston, *J. Chem. Phys. 67,* 2017 (1977).
143. A. Zewail, *Phys. Today 33,* 27 (1980) and references therein.
144. D. Blochinzew, *Phys. Z. Sowjetunion 4,* 501 (1933).
145. P. A. Cohn, P. Bakshi, and G. Kalman, *Phys. Rev. Lett. 29,* 324 (1972); P. Bakshi, G. Kalman, and A. Cohn, *Phys. Rev. Lett. 31,* 1576 (1973); V. P. Gavrilenko and E. A. Oks, *Zh. Eksp. Teor. Fiz 80,* 2150 (1981) [*Sov. Phys. JETP 53,* 1122 (1981)].
146. W. W. Hicks, Lawrence Radiation Laboratory Report No. LBL–2470, 1974.
147. N. B. Delone, B. A. Zon, V. P. Krainov, and V. A. Khodovoi, *Usp. Fiz. Nauk 120,* 3 (1976) [*Sov. Phys. Usp. 19,* 711 (1976)].
148. J. E. Bayfield, L. D. Gardner, Y. Z. Gulkok, and S. D. Sharma, *Phys. Rev. A 24,* 138 (1981).
149. T-J. A. Nee and H. R. Griem, *Phys. Rev. A 14,* 1853 (1976).
150. J. E. Bayfield, Abstracts of Invited Papers, European Study Conference on Multiphoton Processes, Seillac, France, April (1975), unpublished.
151. P. M. Koch, Postdeadline paper presented at the International Conference on Multiphoton Processes, University of Rochester (6–9 June, 1977), unpublished.
152. P. M. Koch, unpublished thesis, Yale University, 1974.
153. I. C. Percival and D. Richards, *Adv. At. Mol. Phys. 11,* 1 (1975).
154. R. Abrines and I. C. Percival, *Proc. Phys. Soc. 88,* 40 (1966).

155. D. Banks, K. S. Barnes, and J. McB. Wilson, *J. Phys. B 9,* L141 (1976).
156. R. E. Olson, *J. Phys. B 13,* 483 (1980).
157. N. Toshima, *J. Phys. Soc. Japan 47,* 257 (1979).
158. R. E. Olson, *Phys. Rev. A 23,* 3338 (1981).
159. R. E. Olson and A. D. MacKellar, *Phys. Rev. Lett. 46,* 1451 (1981).
160. F. W. Meyer and H. J. Kim, Abstracts, European Conference on Atomic Physics, Heidelberg (6–10 April, 1981), vol 5A, part II, p. 924.
161. S. T. Butler and R. M. May, *Phys. Rev. 137,* A10 (1965).
162. B. M. Smirnov, in *The Physics of Electronic and Atomic Collisions,* eds. J. S. Risley and R. Geballe (Seattle: University of Washington Press, 1976), pp. 701–11.
163. E. de Prunelé and J. Pascale, *J. Phys. B 12,* 2511 (1979).
164. M. Matsuzawa, *J. Phys. B 13,* 3201 (1980).
165. R. E. Kennerly, *Phys. Rev. A 21,* 1876 (1980); and references therein.
166. D. Mathur and J. B. Hasted, *J. Phys. B 10,* L265 (1977).
167. R. J. Van Brunt and A. Gallagher, in *Electronic and Atomic Collisions,* ed. G. Watel (Amsterdam: North-Holland Publ., 1978), pp. 129–42; K. Rohr and F. Linder, *J. Phys. B 9,* 2521 (1976).
168. E. W. McDaniel, *Collision Phenomena in Ionized Gases* (New York: Wiley, 1964), p. 118.
169. K. Omidvar, *Phys. Rev. 153,* 121 (1967).
170. L. P. Kotova, A. M. Perelomov, and V. S. Popov, *Zh. Eksp. Teor. Fiz. 54,* 1151 (1968) [*Sov. Phys. JETP 27,* 616 (1968)].
171. Yu. N. Demkov, B. S. Monozon, and V. N. Ostrovskii, *Zh. Eksp. Teor. Fiz. 57,* 1431 (1970) [*Sov. Phys. JETP 30,* 775 (1970)].
172. G. F. Drukarev and B. S. Monozon, *Zh. Eksp. Teor. Fiz. 61,* 956 (1971) [*Sov. Phys. JETP 34,* 509 (1972)].
173. L. A. Burkova, I. E. Dzyaloshinskii, G. F. Drukarev, and B. S. Monozon, *Zh. Eksp. Teor. Fiz. 71,* 526 (1976) [*Sov. Phys. JETP 44,* 276 (1976)].
174. H. Crosswhite, U. Fano, K. T. Lu, and A. R. P. Rau, *Phys. Rev. Lett. 43,* 963 (1979).
175. A. R. P. Rau, *J. Phys. B 12,* L193 (1979).
176. J. C. Gay, *Comments At. Mol. Phys. 9,* 97 (1980).
177. D. R. Herrick and F. H. Stillinger, *Comments At. Mol. Phys. 2,* 57 (1976), see pp. 60–1; also *Phys. Rev. A 11,* 446 (1975).
178. D. R. Cok and S. R. Lundeen, *Phys. Rev. A 23,* 2488 (1981).
179. W. H. Wing and K. B. MacAdam, in *Progress in Atomic Spectroscopy,* eds. W. Hanle and H. Kleinpoppen (New York: Plenum, 1978), part A.
180. T. A. Miller and R. S. Freund, in *Advances in Magnetic Resonance,* ed. J. S. Waugh (New York: Academic Press, 1977), vol. 9.
181. J. W. Farley, K. B. MacAdam, and W. H. Wing, *Phys. Rev. A 20,* 1754 (1979); *Erratum 21,* 2185 (1980).
182. M. Rosenbluth, R. Panock, B. Lax, and T. A. Miller, *Phys. Rev. A 18,* 1103 (1978).
183. I. C. Percival, *Proc. R. Soc. (London) A353,* 289 (1977).

Index